A Handbook of Globalisation and Environmental Policy

A Handbook of Globalisation and Environmental Policy

National Government Interventions in a Global Arena

Edited by

Frank Wijen

Assistant Professor of Strategic Management, Erasmus University, Rotterdam, The Netherlands
Research Fellow, Globus, Tilburg University, The Netherlands

Kees Zoeteman

Professor of Sustainable Development and Globalisation, Tilburg University, The Netherlands

Jan Pieters

Senior Economics Advisor, Ministry of Housing, Spatial Planning and the Environment, The Hague, The Netherlands

Edward Elgar
Cheltenham, UK • Northampton, MA, USA

Published by
Edward Elgar Publishing Limited
Glensanda House
Montpellier Parade
Cheltenham
Glos GL50 1UA
UK

Edward Elgar Publishing, Inc.
136 West Street
Suite 202
Northampton
Massachusetts 01060
USA

A catalogue record for this book
is available from the British Library

ISBN 1 84376 913 1 (cased)

Printed and bound in Great Britain by MPG Books Ltd, Bodmin, Cornwall

Contents

PART IV: THE ROLE OF GOVERNMENT IN INTERNATIONAL AND SUPRANATIONAL FORUMS

PART V: NEW DIRECTIONS

List of Figures

List of Tables

List of Boxes

About the Authors

Maarten Arentsen is Managing Director of the Center for Clean Technology and Environmental Policy at the University of Twente. He is involved in research projects on energy policy, energy market reform, and (green) energy innovation. Dr Arentsen has published in national and international books and journals and lectures in undergraduate and postgraduate programmes.

Pedro Conceição is Deputy Director of the Office of Development Studies of the United Nations Development Programme (UNDP) in New York. Prior to joining the UNDP, he taught at the Technical University of Lisbon (UTL). Dr Conceição holds an MSc degree from UTL and a PhD from the University of Texas at Austin.

Theo de Bruijn is a Senior Research Associate at the Center for Clean Technology and Environmental Policy at the University of Twente and an Associate Professor at Saxion University of Professional Education. He is interested in processes of industrial transformation and regional sustainability strategies. Dr de Bruijn is also the European Coordinator of the Greening of Industry Network.

Chris Dutilh has been responsible for Unilever's environmental affairs in the Netherlands since 1989, providing him with experience in environmental management and life cycle assessment. He has worked at Unilever since 1976. Dr Dutilh holds a degree in Chemistry and a PhD in Biochemistry and has published various articles on sustainable development.

Daniel Esty is a Professor at the Yale Law School and the Yale School of Forestry and Environmental Studies. He held senior positions at the US Environment Protection Agency. Dr Esty is the author or editor of eight books, including *Greening the GATT,* as well as numerous articles on environment and trade, global governance, competitiveness, security, and development issues.

Nicole Gerard is a Writer and Programme Coordinator at Miami Dade College and a Law and Policy Consultant to FERN, the EU forest campaign.

She used to work as a Consultant to Greenpeace International. Dr Gerard, a French national, holds a PhD in Environmental Law from Edinburgh University and an MSc in International Affairs from Columbia University.

Eric Harkink is a PhD Candidate at Twente University, studying R&D performance management in networked organisations. He worked as a Junior Researcher at Tilburg University. Mr Harkink obtained an MSc degree in International Business from Tilburg University in 2002. His thesis dealt with sustainable development in the oil and gas industry.

Niek Hazendonk is a Policy Advisor at the Dutch Reference Centre for Agriculture and Nature and has been involved in all national policy plans on landscape and physical planning. He worked as a Landscape Researcher and as a Landscape Architect, in Spain and the Netherlands. Mr Hazendonk studied Landscape Architecture at Wageningen University.

Jan Hol is Executive Vice-President for Communication with the Nuon energy company, a Lecturer at communication courses, and author of several books. He was Vice-President of Public Relations with Royal Ahold, Director of Internal Communications with KLM, and Director of New Media for the NCRV broadcasting company. Mr Hol studied Political Sciences, Mass Communication, and Educational Sciences.

John Hontelez has been Secretary General of the European Environmental Bureau since 1996. He used to be Chairman of Friends of the Earth International. Mr Hontelez has been an active environmentalist since 1970, first nationally and, from 1986 on, at the European and global levels, supporting the environmental movement in Central and Eastern Europe.

Maria Ivanova is Director of the Global Environmental Governance Project at the Yale Center for Environmental Law and Policy. She worked at the OECD in Paris and the Swedish Environmental Protection Agency in Stockholm. Ms Ivanova, a Bulgarian national, holds degrees from Mount Holyoke College and Yale University, where she is completing her doctorate.

Marion Jansen is a Counsellor in the WTO's Economic Research and Statistics Division. She was an Advisor to the British Government and the European Commission. Dr Jansen studied Economics in Konstanz and Toulouse and holds a PhD in Economics from the Universitat Pompeu Fabra. She published on development, government regulation, international finance, and labour markets.

Tom Jones is Head of Global and Structural Policies at the Environment Directorate of the OECD in Paris, focusing on interrelations of economic and environmental issues. Previous OECD responsibilities were related to globalisation and the environment, natural resource management, and transport issues. Prior to joining the OECD in 1989, he was a Senior Water Economist with Environment Canada.

Inge Kaul is Director of the Office of Development Studies of the United Nations Development Programme (UNDP) in New York. She was Director of the Human Development Report Office and held other senior positions at the UNDP. Dr Kaul was also a UN Policy Analyst and an Assistant Professor at Konstanz University, where she obtained her PhD.

Alexander Keck is an Economist in the WTO's Economic Research and Statistics Division. He worked at the German Federal Ministry for the Environment, the Office of the Director-General of the United Nations Industrial Development Organization, and the European Parliament. He holds advanced degrees in Economics, Environment & Development, and Finance from the Universities of Heidelberg, Cambridge, and New York.

René Kemp is a Senior Research Fellow at the Maastricht Economic Research Institute on Innovation and Technology and Senior Advisor for TNO-STB. Together with Jan Rotmans, he developed the approach to transition management that has been adopted by the Dutch national government. He has published widely on the topic of innovation for the environment and environmental policy.

Alkuin Kölliker is a Research Associate at the University of Bielefeld, focusing on multi-speed integration in Europe and the theory of collective goods. He was a Research Fellow at the Max Planck Institute. Dr Kölliker holds a degree in Economics from the University of Bern and a PhD in Political Science from the European University Institute in Florence.

Ludwig Krämer is Head of the Unit for Legal Affairs and Governance of the Directorate-General Environment of the European Commission. He is an Honorary Professor at Bremen University, a Visiting Professor at University College London, and a Lecturer at the College of Europe. Dr Krämer wrote some 15 books and 150 articles on EC environmental law.

Duncan Liefferink is a Lecturer in Political Science of the Environment at the University of Nijmegen. His main research interests are in European and comparative environmental politics. Dr Liefferink holds a PhD from

Wageningen University and is the author of *Environment and the Nation State* (1997) and *The Europeanisation of National Environmental Policy* (2004).

Susan Martens is a Researcher at the Environmental Policy Group at Wageningen University. Her main research interests are environmental sociology and policy, sustainable consumption, sustainable building and dwelling, public participation, civil-society development, non-governmental organisations, and China. Ms Martens studied Environmental Policy at the University of Nijmegen.

Claude Martin has been Director General of WWF International since 1993 and an Advisor to the governments of China, Ghana, and Switzerland. He has worked for WWF since the early 1970s: as a researcher in India, a Director in Ghana, and Chief Executive of WWF Switzerland. Dr Martin holds an MSc in Biology and a PhD in Wildlife Biology.

Henk Massink has been a Senior Policy Advisor at the Department of International Affairs of the Ministry of Agriculture, Nature and Food Quality since 2000. He deals with strategic matters in national and international agricultural policy, such as the relation between sustainability and trade development. Mr Massink studied Theology in Leiden and Philosophy in Amsterdam.

Hans Opschoor is Rector of the Institute of Social Studies in The Hague and Professor of Environmental Economics at the Free University in Amsterdam. He was Review Editor of the IPCC and a member of international scientific committees and national delegations. Dr Opschoor studied economics in Rotterdam and holds a PhD from the Free University in Amsterdam.

Saskia Ozinga is Director of FERN, the EU forest campaign that she created in 1995. She is also Coordinator of the Northern Support Office of the World Rainforest Movement and Coordinator of Forest Movement Europe. She worked for Friends of the Earth Netherlands. Ms Ozinga holds MSc degrees in Biology and Healthcare from Utrecht University.

Jan Pieters is a Senior Economic Advisor at the Directorate of Strategy and Administration of the Directorate-General for Environment of the Dutch Ministry of Housing, Spatial Planning and the Environment. Former positions with the Ministry include Representative at the OECD in Paris (1995-1999) and Head of the Department of Economic Affairs of the Directorate Administrative Affairs (1989-1995).

Diahanna Post is a Doctoral Candidate in Political Science at the University of California at Berkeley. Her thesis deals with the influence of international food safety standards on domestic and regulatory policies. Ms Post has published on environmental standards in Eastern Europe and genetically modified foods in Europe and the US.

Britta Rendlen is Manager for Member Relations at the World Business Council for Sustainable Development. Her background is in arts management and corporate and private philanthropy. Ms Rendlen holds an MBA from the Kellogg School of Management at Northwestern University, a BA in Performing Arts Management, and an associate degree in Social Sciences.

Luc Soete is Professor of International Economics at Maastricht University and Director of the Maastricht Economic Research Institute on Innovation and Technology. Dr Soete has degrees in Economics (1972, summa cum laude) and Development Economics (1973, summa cum laude) from the University of Ghent and a DPhil from the University of Sussex (1978).

Gert Spaargaren is a Senior Researcher and Professor of Environmental Policy for Sustainable Lifestyles and Consumption at Wageningen University. His main research interests and publications are in the field of environmental sociology, sustainable consumption and behaviour, and the globalisation of environmental reform. Dr Spaargaren studied Sociology and holds a PhD from Wageningen University.

Björn Stigson has been President of the World Business Council for Sustainable Development since 1995. He worked as a Financial Analyst with the Swedish Kockums Group, was President and CEO of the Swedish Fläkt Group, and Vice-President and a Member of ABB Asea Brown Boveri's Executive Management Group. Mr Stigson has served on several boards and advisory councils.

Charlotte Streck is a Counsel for International and Environmental Law at the World Bank in Washington DC and an Adjunct Lecturer at the University of Potsdam. She worked for the Global Public Policy Project and (co-) authored several books and articles. Dr Streck was educated in Law and Biology at the Universities of Berlin, Regensburg, Freiburg, and Córdoba.

Michael Toffel is a Doctoral Candidate at the Haas School of Business at the University of California at Berkeley, studying corporate environmental strategy and evaluating environmental policy. He worked as an Industrial Manager in Environment, Health, and Safety. Mr Toffel's articles have been published in *California Management Review, Journal of Industrial Ecology,* and *Corporate Environmental Strategy.*

Gerard van Dijk is affiliated with the Reference Centre of the Dutch Ministry of Agriculture, Nature and Food Quality. He worked at a regional office of the Ministry (1975-1991), at the Ministry's International Nature Conservation Affairs Division, and was seconded to UNEP (2000-2002). Mr Van Dijk studied at Wageningen University.

Joost van Kasteren has been a freelance Science Writer since 1984. He worked as a Science Writer at Delft Technical University and was Editor of a Dutch technology magazine. Mr Van Kasteren holds a degree in Molecular Science and has written books and articles on food, agriculture, and sustainable development.

Jan van Vliet is a Senior Advisor to and Programme Manager at the Reference Centre of the Dutch Ministry of Agriculture, Nature and Food Quality. He worked in several (executive) functions at the Ministry (1979-1995) and at a regional department of the Ministry (1995-1998). Mr Van Vliet studied at Wageningen University.

Sietske Veenman is a Researcher at the Nijmegen School of Management, University of Nijmegen. Her research interests focus on European environmental politics; she is currently engaged in a major comparative research project on the impact of international institutions and trade on the convergence of national environmental policies. Ms Veenman studied environmental policy at the University of Nijmegen.

Rob Visser has been Head of the Environment, Health, and Safety Division of the OECD since 1998. He was Principal Administrator in the OECD Chemicals Division, a Toxicologist in hospitals, with TNO, and at the Dutch Ministry of Health and the Ministry of Spatial Planning, Housing and the Environment. Dr Visser holds a PhD from Leiden University.

David Vogel is a Professor at the Haas School of Business and at the Department of Political Science at the University of California at Berkeley and a Visiting Professor at INSEAD. He has published extensively on government regulation, trade and environment, and comparative business-government relations. Dr Vogel's books include *Trading Up* and *National Styles of Regulation.*

Konrad von Moltke is Adjunct Professor of Environmental Studies at Dartmouth College and a Visiting Professor at the Free University in Amsterdam. He taught at the State University of New York and published extensively on medieval history, comparative education, and international environmental policy. Dr von Moltke studied Mathematics at Dartmouth and Medieval History in Munich and Göttingen.

Rifka Weehuizen is a Researcher at the Maastricht Economic Research Institute on Innovation and Technology. She worked at the Dutch Ministry of Education, Culture and Science at the Foresight Institute (STT) in The Hague. Ms Weehuizen obtained a degree in History and Business Administration from the University of Groningen in 1996.

Frank Wijen is an Assistant Professor of Strategic Management at Erasmus University, Rotterdam, with research interests in organisation theory and the practice of private and public environmental management. He worked as a Senior Researcher, a Marketing Manager, and an Entrepreneur. Dr Wijen holds a PhD in Management from Tilburg University and an MSc in International Management from Maastricht University.

Pieter Winsemius is a Member of the Netherlands Scientific Council for Government Policy and a Professor at Tilburg University. He worked at McKinsey & Company, was Dutch Minister for Spatial Planning, Housing, and the Environment, and wrote several books. Dr Winsemius studied Physics at Leiden University and obtained PhD and MBA degrees (the latter from Stanford University).

Kees Zoeteman is Professor of Sustainable Development and Globalisation at Tilburg University. He served as Deputy Director General at the Dutch Ministry of the Environment, chaired the Board of the European Environment Agency, and was a Member of the China Council on Environment and Development. Dr Zoeteman has had a lifelong interest in reconciling economy and nature.

Preface

The genesis of this book lies in a protracted discussion on the policy challenges that the national government encounters when it is being exposed to more global spheres of influence. We felt that the institutional and social settings in which governments develop national policies have changed dramatically, especially over the last decade. Whereas the 'visible hand' used to have a firm grip on its subjects, the government is now itself subject to a host of international and supranational regulations, the economic power of large business organisations, and public opinion, which is fed by increasingly assertive citizens.

We suspected that these important changes might call for a reconsideration of existing policies: instruments that used to be effective in the past may no longer be tailored to the prevailing circumstances. While the policy challenge, therefore, is huge, we felt that insights into this topic were highly scattered and that adequate policy responses were not readily available. The lack of sufficiently detailed and comprehensive information that is needed to develop adequate national policies in a globalised arena spurred us to explore the topic in greater depth. We decided to focus on environmental policy for practical reasons, since covering all areas would weaken our effort and we were best acquainted with this field, and because many environmental issues have a regional or global dimension.

Early in 2002, we embarked on a joint venture between Globus, the Institute for Globalisation and Sustainable Development at Tilburg University, and the Dutch Ministry of Housing, Spatial Planning and the Environment. We invited renowned experts from academia, the government, supranational organisations, business, and non-governmental organisations to write papers on recent developments and their underlying drives in specific areas related to the interface of globalisation and environmental policy. These papers were presented at the International Conference on Globalization and National Environmental Policy that was held in Veldhoven, the Netherlands, from 22 to 24 September 2003 (for more information, see www.uvt.nl/globalizationconference).

After the Conference, several rounds of revision and editing were undertaken to enhance the richness of the contributions and to make them as coherent and consistent as possible, though a certain degree of diversity has remained owing to the heterogeneity of subtopics and perspectives.

The aim of the book is to provide the reader with a wide-ranging and in-depth understanding of key aspects of national environmental policy-making in a globalised setting. The book seeks to cross disciplinary boundaries and to bridge theory and practice. The targeted audience is diverse, ranging from newcomers to the field seeking to obtain a comprehensive overview to seasoned experts wishing to acquire specific insights or to broaden their views. The book aims to be relevant to diverging constituencies: academics with an interest in internationalisation, public administration, or environmental issues; governmental policy-makers involved in environmental and/or social issues; policy-makers in supranational organisations; corporate managers, corporate environmental support staff, and liaison officers of business associations; and policy-makers in non-governmental organisations.

The entire venture, the tangible outcome of which is the present book, could not have been realised without the invaluable support from a considerable number of individuals and organisations. The Dutch Ministry of Housing, Spatial Planning and the Environment (in particular, the Strategy and Administration Department of the Directorate-General for the Environment as co-organiser), together with other public and private sponsors, provided generous financial support in organising the Conference and realising this book.

Several of our Globus colleagues showed an extraordinary commitment. Petra van der Ham displayed enormous dedication and precision in managing key aspects of the Conference and in realising this book, especially the incorporation of textual changes and the implementation of the layout. Ludo van Dun and Hans van Poppel eschewed no effort to conscientiously adapt illustrations and implement editorial changes most accurately, especially in the delicate final stages. Eric Harkink willingly created and tailored numerous laborious illustrations. Several linguistic experts from the translation department at Tilburg University (in particular, Patricia Gouldner and Rikkert Stuve) greatly improved the readability and linguistic aspects of the different chapters. Ilse van Eck, Wiebe Vos, Paulien van der Straaten, and Suzanne Verheij fulfilled important supporting roles in organising the Conference. Paul van Seters' moral support throughout the process was also of great importance.

Madelon de Ruiter and Maries Dijk of the Dutch Ministry of Spatial Planning, Housing and the Environment raised support and publicity at their Ministry. Last but not least, Dymphna Evans and Matthew Pitman of Edward

Elgar Publishing showed great openness to our initiative as well as professional and kind support.

We sincerely thank all of these individuals and organisations for their invaluable help in shaping and improving the present volume. We wish the reader a pleasant and instructive journey throughout this book, and welcome any constructive feedback (please contact us at globus@uvt.nl).

The editors
Rotterdam, Tilburg, and The Hague, October 2004

1. Globalisation and National Environmental Policy: An Overview

Kees Zoeteman, Frank Wijen, and Jan Pieters[1]

SUMMARY

After outlining the scope, target audience, and structure of the book, we review the literature on globalisation and environmental policy, especially the impact of globalisation on the environment and changes in environmental governance in relation to increasingly global spheres of influence. This is followed by a succinct representation of the essential points of all contributions to this volume. While each chapter has its own distinct focus and perspective, common themes have been identified in major outcomes and future directions: the delicate and multifaceted relation between economic globalisation and environmental protection, changes in the prioritisation of environmental issues, shifts in governance mechanisms, dealing with reduced sovereignty, prospects for existing and new policy instruments, and finding a balance between globalisation and national environmental policy. These findings lead to conclusions with respect to the commensurability of different governance levels and the compatibility of different policy areas.

INTRODUCTION

In contemporary history, national governments have fulfilled a central role in governance. They are the highest authorities in relatively sovereign states, developing national policies and implementing them through lower governmental bodies such as provinces and municipalities. Yet, the authority of national governments has come under pressure. Foreign spheres of

[1] We are indebted to Paul Dekker and Paul van Seters for their constructive comments on an earlier version.

influence have constrained governments' external sovereignty: supranational institutions such as the World Trade Organization (WTO) and the European Union (EU) prescribe a significant and increasing number of national regulations. At the same time, governmental authority has been eroded 'from below'. Companies that are 'footloose' because of their presence in a multitude of countries – and whose turnover may exceed the gross product of national economies – may pressurise governments to obtain favourable treatment, while domestic firms may lobby their governments to avoid regulations with adverse international competitive effects. Through telecommunication and imports, for example, citizens and consumers have easy access to foreign cultures and their artefacts and connect more easily with citizens abroad, especially through the internet. This widening scope for citizens and the intensifying interference of supranational institutions with civic issues have often widened the gap between citizens and ruling authorities.

This phenomenon of globalisation is particularly apparent in the domain of environmental policy. For example, a large majority of national environmental regulations in the EU consists of implementing directives from Brussels, where the European Commission (EC) resides, while the WTO has bounded national environmental policies. This can be understood against the backdrop of cross-boundary environmental problems, calling for higher-level interventions, and the possible impact of environmentally inspired government interventions on other policy areas, in particular, a common European market or global market. Multinational companies, both driving and thriving on more global economic exchanges, may play different national governments off against one another to obtain lenient environmental regulations or, conversely, may apply stringent environmental standards required for one market around the globe. The environmental conditions under which imported products have been produced are generally not transparent to consumers, hampering their well-informed decision-making. And citizens may boycott – or otherwise campaign against – foreign or multinational companies they perceive as 'dirty'.

The arena in which national governments operate, therefore, has undergone important changes, which may have reduced the effectiveness of formerly well-functioning environmental policies. The increasingly global or regional spheres of influence may call for a reconsideration of policy instrument in order to be in a better position to meet the novel opportunities and constraints that national governments presently face. The central aim of this book is to improve our understanding of the impact of global spheres of influence on the scope of national environmental policies, and to explore effective policy responses to these new opportunities and threats. A better understanding of these interactions may lead to more enlightened visions and better management of power relations.

While many salient publications on the relations between national environmental policy and economic globalisation have appeared, either their scope tends to be limited to specific aspects or their perspective tends to be biased. Our purpose was to provide a comprehensive book on the interface of globalisation and national environmental policy, addressing the impact of the actors and factors shaping globalisation on the scope of action for national environmental policy-makers. The book means to be broad in scope by covering all major aspects, multiple in level of analysis by incorporating macro-, meso-, and micro-levels, heterogeneous in perspectives by representing diverging social viewpoints, and complementary as to disciplines by drawing on insights from economics, law, sociology, political science, public administration, and environmentology. We also aimed to obtain in-depth insights. We were interested both in conceptual contributions pushing back the theoretical frontier and in new case studies providing detailed insights into specific issues. Leading experts from academia, supranational organisations, government, business, and non-governmental organisations (NGOs) were invited to write original contributions on relevant aspects.

Besides providing a comprehensive account of our subject, we were also interested in identifying the degree of commensurability between different levels of influence (global, regional, national, and local) as well as the compatibility of different areas (economic development, jurisdictional competence, political discretion, and environmental protection). Clarifying the tensions between regional or global spheres of influence, on the one hand, and national or local forces, on the other, helps us to address the question whether existing national environmental policies and instruments are commensurable with the opportunities and threats presented by globalisation. Teasing out the compatibility and prioritisation of economic and environmental imperatives against the backdrop of existing legal and political structures is also crucial to understanding what lies within and what lies beyond the bounds of the possible in environmental policies. National governments that wish to design effective environmental policies should consider the options and constraints of the different areas. The present book aims to shed some light on actual and potential governance mechanisms and policy instruments in the environmental field that reflect the changes in the distribution of powers brought about by economic, political, and social globalisation.

While the book aims to provide the reader with both breadth and depth, it has been designed predominantly from the perspective of a medium-sized 'developed' country. As a result, all 43 authors originate from Europe and North America, and the setting of many case studies is Europe, with an emphasis on the Netherlands. This is partly related to the relatively advanced status of environmental policies in these regions, but it has obviously led to a Western bias.

The book's targeted audience is primarily academics and policy-makers. It may be useful for students and scholars, both newcomers who wish to obtain a comprehensive overview and those with advanced knowledge who wish to obtain a deeper understanding of specific issues. Policy-makers in international and supranational organisations, national and local governments, companies and trade associations, and NGOs may use it as a source of inspiration for future policies. The book obviously targets those interested in environmental problems, though there may be close parallels with social issues. While its scope is international, this book may also be relevant to readers concerned with domestic areas of interest, as these are increasingly exposed to foreign influences.

Turning to the theoretical embedding of the book, we now review literature dealing with the key issues of globalisation, national environmental policies, and governance, including their interrelations. As these issues are addressed extensively throughout the book, the literature review is succinct. Then we outline the structure of the book and the main points of each individual chapter. Finally, we pull the diversity of insights together by identifying and discussing commonalities in the variety of contributions.

THEORETICAL ISSUES

For centuries, the fates of many nation states have been intertwined. Foreign military interventions, migration, and international trade have long histories.[2] The degree of interconnectedness has fluctuated over time, with eras of relative isolation and autarky altering with periods of more intensive international interactions.[3] Over the past few decades, political developments – including the collapse of communist regimes and the elimination of trade barriers – and technological innovations – especially in the fields of transport and communication – have paved the way for increased internationalisation of economic, political, and cultural activities, leading to the (perceived) compression of space and time and the emergence of relational networks at the global level.[4] Globalisation has many facets,[5] though the economic

2 Hobsbawm, 1975; Landes, 1998; Lechner and Boli, 2000; McNeill and McNeill, 2003; Schaeffer, 2003.

3 Baker et al., 1998; Frankel, 2000; McNeill and McNeill, 2003; Reinicke, 1998; Scholte, 2000; Waters, 2001.

4 Anheier et al., 2001; Castells, 1996; Held, 2004; Inglehart, 1997; Lechner and Boli, 2000; Waters, 2001.

5 Schaeffer, 2003; Scholte, 2000; Waters, 2001.

dimension has been discussed most often. During the 1990s, international economic transactions (in particular, international trade and foreign investment) rose exponentially,[6] though internationalisation is spread unevenly: many activities take place within or between regional clusters such as parts of Europe, North America, and Japan.[7] In the academic literature, the economic and political interdependence of states has long been recognised,[8] though literature on globalisation witnessed an upsurge in the 1990s.[9] Some authors highlight the amenities of economic globalisation, in terms of enhanced choice and lower prices of products as well as higher national incomes.[10] Others argue that globally unleashed market forces have caused or enhanced many social evils, such as increased income inequality and infringement of sovereignty and democracy.[11]

The human impact on the natural environment rose dramatically with the expansion of economic activities,[12] though there are substantial differences between countries as to their claims on natural resources and pollution of the environment.[13] Sensitised by alarming publications about the depletion of natural resources and environmental degradation[14] and by popular concerns for the visible environmental consequences of industrialisation (such as air, water, and soil pollution), national governments in Western countries started developing environmental policies (i.e., strategic courses of action to solve or contain problems related to ecological resources and systems) from the 1970s onwards. Environmental policy issues appeared prominently on the academic agenda in the 1980s, and the field flourished from the 1990s onwards.[15] In the economics literature, environmental issues are classic examples of externalities because costs and benefits accrue to different parties.[16] Likewise, collective action is hampered because individual actors lack adequate information or do not have the incentives to protect common environmental goods.[17] The protection of natural resources also faces legal problems, especially in transboundary and international settings.[18]

---

6 Esty and Gentry, 1997; Frankel and Rose, 2002.
7 Dunning, 2000; Ohmae, 1995; Rugman, 2001.
8 Hobsbawm, 1975; Keohane and Nye, 1977; Wallerstein, 1974, 1980.
9 For overviews, see: Lechner and Boli, 2000; Levy-Livermore, 1998; Michie, 2003.
10 Bhagwati, 2002; Moore, 2003; Norberg, 2003.
11 Hertz, 2001; Klein, 2000; Landes, 1998; Mullard, 2004; Stiglitz, 2002; Van Seters et al., 2003.
12 EEA, 2003; UNEP, 2003; Wackernagel and Rees, 1996.
13 World Economic Forum, 2002.
14 Carson, 1962; Meadows et al., 1972.
15 For overviews, see: Bromley, 1995; Lesser et al., 1997; Mäler and Vincent, 2003; Sutherland, 2000; Tietenberg et al., 1999; Van den Bergh, 1999, 2002.
16 Lesser et al., 1997; Tietenberg, 2003.
17 Hardin, 1968; Kaul et al., 1999; Kölliker, 2002; Tietenberg, 2003.
18 Birnie and Boyle, 2002.

Interactions between globalisation, on the one hand, and environmental issues and national policies, on the other, have also been widely documented. The impact of economic globalisation on the natural environment has been studied extensively.[19] Positive effects have been framed in terms of the use of environmentally less harmful products and processes, often associated with income rises as a consequence of international specialisation and trade. Negative effects are primarily defined in terms of scale: the use of natural resources and pollution increase when economic activities expand and when inputs and products are transported more frequently and over longer distances. The overall environmental impact may be positive or negative, depending on the prevailing circumstances.[20]

Likewise, there is a considerable body of research on the effect of the stringency of national policies on international competitiveness and investment decisions.[21] Those embracing the 'race to the bottom' hypothesis argue that international policy competition drives polluting economic activities to 'pollution havens' (i.e., countries with lax environmental regimes), leading to a policy competition in which countries structurally decrease their environmental standards to attract or keep business within their national borders. According to the 'regulatory chill' hypothesis, states refrain from adopting more stringent regulation in order not to deter actual and potential investors. By contrast, the 'race to the top' hypothesis holds that stringent product standards in certain countries are increasingly being exported, especially by multinational companies diffusing the company-wide enforcement of stringent standards in order to reap economies of scale or preserve a favourable corporate or brand image.

Another strand of research has focused on governance issues. Some argue that globalisation has eroded the formal power of national governments, not only in industrialised countries but also and especially in developing countries.[22] In their view, power has shifted from states, which used to operate at the regulatory apex, to multinational companies, which now control a large share of global trade and which may be subject to privileged tax and regulation arrangements. Some have argued that private – profit or not-for-profit – organisations may assume, or have taken over, formerly public environmental roles. Under the banner of 'sustainable business' or 'corporate social responsibility', business organisations may take environmentally

---

[19] Antweiler et al., 2001; Copeland and Taylor, 2003; Esty and Gentry, 1997; Frankel and Rose, 2002; IISD and UNEP, 2000; Nordström and Vaughan, 1999; Zarsky, 1999.

[20] Jenkins et al., 2003.

[21] Boyce, 2004; Copeland and Taylor, 2003; Esty and Geradin, 1998; Jaffe et al., 1995; Mabey and McNally, 1999; Mani and Wheeler, 1998; OECD, 2002; Vogel, 1995, 1997.

[22] Held et al., 1999; Hertz, 2001; Nayyar, 2002; Ohmae, 1995; Opschoor et al., 1999; Sassen, 1996; Strange, 1996.

benign actions or engage in global self-regulation, for example, through environmental management standards such as ISO 14001.[23] Alternatively, civil society, often taking concerted action through NGOs and increasingly organised on a regional or global scale,[24] may take on a prominent role in 'civilising' globalisation, in order to make economic developments more compatible with social and environmental imperatives or even have them reinforce one another.[25] Hybrid forms here include governance modes such as 'voluntary agreements' between industry and government[26] and 'public-private partnerships' involving governments, companies, and NGOs.[27] Others advocate a global environmental government,[28] arguing that, since many environmental issues are transboundary or global in nature, only a global institute can coordinate national actions and internalise national externalities. Yet others recognise the need for global or regional action but argue that national governments will remain the governing institutions.[29] International environmental regimes of cooperating states may then be effective in solving global or regional collective action problems.

STRUCTURE AND SYNOPSIS

Each of the five parts of this book deals with specific elements of the central topic. Part I has a strong conceptual focus, elucidating theories and terminology pertaining to major aspects of globalisation and environmental policy. Besides, it sketches recent empirical developments to provide the reader with the latest state of affairs. Part II explores different societal perspectives and presents contributions by representatives from the business and civic sectors providing viewpoints of major societal constituencies. Where the previous part describes and analyses the behaviour of companies and NGOs, often from an academic perspective, we also wished to let these societal actors, who shape or affect processes of globalisation and environmental policy, speak for themselves. Parts III and IV consist of case studies, the former dealing with global or regional influences on domestic environmental policies, and the latter focusing on national contributions to

---

[23] Brunsson and Jacobsson, 2000; Cooperrider and Dutton, 1999; Cutler et al., 1999; Holliday et al., 2002; Powell and Clemens, 1998; Winsemius and Guntram, 2002.

[24] Anheier et al., 2001; Arts, 1998; Florini, 2000.

[25] Dunning, 2003; Etzioni, 1990; Florini, 2003; Giddens, 2001; Sklair, 2002; Spaargaren et al., 2000; Van Seters et al., 2003.

[26] Carraro and Lévêque, 1999; Mol et al., 2000; OECD, 2003.

[27] Holliday et al., 2002; Osborne, 2000; Reinicke, 1998; Rischard, 2002; Warner and Sullivan, 2004.

[28] Esty and Ivanova, 2002; Group of Lisbon, 1993; Sandel, 1996; UNDP, 1999.

[29] Barrett, 2003; European Commission, 2004; Gray, 1999; Haas et al., 1993; Kaul et al., 2003; Young, 1994, 1999.

global or regional forums. The final part (V) provides challenging views of new possible avenues for more effective national, international, or supranational governance of environmental issues.

The different parts and chapters are self-contained. They can be understood independently without relying on preceding parts and chapters, though the reader may assimilate their contents more easily by adhering to the order of the book. It may be especially helpful to take in the conceptual frames traced in Part I before proceeding to other parts. For reasons of convenience, a short summary has been provided at the start of each chapter, allowing the reader to catch the chapter's essence at a glance. While all authors received the same guidelines pertaining to the structure and contents of their contributions,[30] the wide range of topics, levels, and perspectives has unavoidably led to a measure of heterogeneity in the form and substance of the different chapters.

Tom Jones kicks off Part I with an overview of major conceptual relations between globalisation and national environmental policy (Chapter 2). He also sketches recent trends in economic globalisation, discusses empirical implications for national regimes, and concludes that economic and environmental imperatives can be (made) compatible.

The important issue of national incentives for action is central to Alkuin Kölliker's contribution (Chapter 3). He identifies different types of environmental goods and different trade effects, and argues that their combined effect predicts governmental responses.

Pedro Conceição and Inge Kaul then focus on global public goods, whose collective availability and non-rivalry extend to the global level (Chapter 4). In their view, the nature of many public goods is not technically given but shaped by human perspectives and actions. They also discuss and contextualise different possibilities for financing global environmental actions.

Marion Jansen and Alexander Keck discuss at length the crucial relationship between multinational trade regulations and the discretionary possibilities and constraints of national environmental policies (Chapter 5). They argue that environmental actions that distort free international trade are not necessarily illicit, though WTO jurisprudence leaves considerable uncertainty as to the acceptability of specific national measures.

René Kemp, Luc Soete, and Rifka Weehuizen discuss policies geared towards environmentally benign innovations (Chapter 6). They describe national innovation policy instruments and their context specificity and argue

[30] Guidelines included: delimitation of topic, definition of key terms, description and analysis of the past (10-20 most recent years) and present situation, approach taken to cope with recent changes, and future developments.

that globalisation has engendered a governance void, which should be filled by international NGOs.

The contribution by Kees Zoeteman and Eric Harkink focuses on national and corporate sustainability attitudes (Chapter 7). Drawing on a study of two sectors, they point out salient parallels between the attitudes of multinational companies and their home countries.

Gert Spaargaren and Susan Martens highlight the role of citizen-consumers in a globalised setting (Chapter 8). They review literature on ecological citizenship, sustainable consumption, and global civil society. Referring to lifestyle studies, they argue that national environmental policies can be effective if different aspects of citizen-consumer behaviour are considered simultaneously.

The focus then shifts to the interaction between local and central levels of environmental decision-making (Chapter 9). David Vogel, Michael Toffel, and Diahanna Post compare regulative interactions in the EU and the United States (US) in three environmental areas. They conclude that national or state governments can play an important role when diffusing their stringent regimes through federal regulations.

The final contribution of this part (Chapter 10) deals with globalisation and environmental governance in developing countries. Hans Opschoor discusses sustainable development, globalisation, institutions, and types of agents. Focusing on Africa, he argues that the lack of effective governance in developing countries can be compensated to some extent by an active role of civil society.

Part II consists of three contributions by business representatives and two by NGO leaders. Björn Stigson and Britta Rendlen make a case for sustainable business (Chapter 11). They describe the role of the World Business Council for Sustainable Development (WBCSD) in this process and indicate how government and business can contribute to the realisation of sustainable development.

Jan Hol from Nuon, a major Dutch energy supplier, then discusses the success factors in renewable energy in the Netherlands (Chapter 12). He argues that the prospects are dim in a liberalised European energy market and highlights Nuon's strategic response to liberalisation.

Chris Dutilh describes how and why the multinational Unilever company is involved in sustainable business operations (Chapter 13). He identifies different types of actors, highlights the difference between consumers and citizens, and underscores that government policy should be geared towards bridging the gap between the two.

According to Claude Martin, globalisation has threatened the environment (Chapter 14). He indicates how a globally operating NGO like WWF has addressed this challenge and pleads for cooperative platforms such as partnerships.

In order to defend environmental interests at the EU level, local NGOs have joined forces through the European Environmental Bureau (EEB, Chapter 15). John Hontelez explains the modus operandi of the EEB and argues that the best way of warranting environmental interests is to anchor them into European legislation.

Part III covers our first set of case studies, dealing with the impact of globalisation and regionalisation on domestic environmental policies. Maarten Arentsen and Theo de Bruijn discuss the interactions of EU regulations and Dutch national policy for four energy dossiers (Chapter 16). They conclude that a country's relative power position and international similarity of viewpoints are important determinants of national discretion.

Saskia Ozinga and Nicole Gérard argue that globalisation has aggravated the problem of illegal logging (Chapter 17). They explore possible remedies and conclude that more targeted regulation and independent monitoring would help to reduce this important problem.

Joost van Kasteren points to the scale-up effects of globalisation, which has rendered the development and marketing of environmentally benign pesticides for a small market such as the Netherlands no longer attractive (Chapter 18). While the Dutch government is facing not only these economic but also regulative restrictions at the supranational level (EU, WTO), several national policy options may render crop protection more environment-friendly.

Henk Massink, Gerard van Dijk, Niek Hazendonk, and Jan van Vliet provide a detailed account of the qualitative and quantitative changes in Dutch agriculture and the local environment in connection with free global trade (Chapter 19). They conclude that the intensification and scaling-up of agricultural production has led to reduced biodiversity and to landscape changes, and that the national government has several options to offset the negative environmental consequences.

Part IV turns to national interventions in the international arena. Sietske Veenman and Duncan Liefferink analyse the influence of three proactive countries on EU climate policies (Chapter 20). They conclude that these countries proceed differently in terms of the directness of their approach and the purposefulness of their influence and that they may combine different strategies.

Ludwig Krämer discusses the functioning, achievements, and shortcomings of the European Community's environmental policy (Chapter 21). He argues that a more centralised system would not have led to improved environmental performance because – against the backdrop of national implementation differences – environmental results depend essentially on the political will to act.

The Organisation for Economic Co-operation and Development (OECD) has elaborated a joint system for testing and registering new chemicals (Chapter 22). Rob Visser argues that, when data obtained in one country are also recognised in other (OECD) countries, as occurs in the chemicals system, important advantages accrue to different societal strata.

Frank Wijen and Kees Zoeteman analyse the strengths and weaknesses of the Kyoto climate regime (Chapter 23). After exploring four scenarios for future political constellations and indicating possibilities for creating leverage, they conclude that more appropriate incentives should be built into future climate policies.

The final part (V) bridges the present and the future. Daniel Esty and Maria Ivanova address environmental protection from a global governance perspective (Chapter 24). After discussing interactions between economic globalisation and environmental protection, they make a case for a global environment mechanism: a light institutional superstructure drawing on public policy networks and applying modern information technology.

Charlotte Streck's contribution is devoted to the different types of international policy networks in which national governments are involved (Chapter 25). She discusses the opportunities and threats of global public policy networks and argues that these may constitute an effective complementary governance mechanism.

According to Konrad von Moltke, economic globalisation and environmental globalisation, though both governed by different rules, are intertwined (Chapter 26). The challenge is to design international institutions that consider environmental and economic imperatives in a more integrated fashion.

Finally, Pieter Winsemius reflects on the impact of different schools of public administration on past and present environmental policies in the Netherlands (Chapter 27). He then argues that a small country can have a significant impact on the international arena by creating a breeding ground for environmental NGOs.

OUTCOMES AND FUTURE DIRECTIONS

While the richness and specificity of the different chapters does not allow for shorthand, facile conclusions, some connecting threads can be woven.

Findings highlighted by several authors and complementarities between various contributions may be summarised under the following headings.[31]

The Delicate, Multifaceted Relation between Ecology and the Economy

Generally speaking, there is no statistical evidence of relations between the globalisation of economic activities and positive or negative environmental changes. While hard-and-fast statements should be avoided because causalities are very hard to establish, the negative effects of international trade and foreign direct investment (FDI) on the environment have sometimes been overstressed, as has been the presumed negative impact of environmental measures on economic development.[32] Global economic activities may adversely affect the environment,[33] but this may be outweighed by the regional or global diffusion of stringent local environmental regimes.[34] In any case, we need better assessment and communication of the impact of trade and investment on the environment, on the one hand, and of the competitive effects of national environmental regulations, on the other hand.[35] We may conclude, therefore, that the pursuit of economic and environmental objectives does not necessarily involve trade-offs.

Economic globalisation and environmental protection can be made compatible if a number of conditions are met:[36]

- Trade, FDI, and environmental policies pull in the same direction and create synergies, for example, by abolishing those subsidies that lead to trade distortion and environmental disruption, and by promoting the international trade of environment-friendly products.
- Environmental externalities are internalised, so that costs and benefits are allocated correctly, in particular by applying full-cost accounting.
- Environmental policies are in place timely, implying that polluting now while generating growth and cleaning up later is not a sustainable strategy.

Yet, conflicts may arise between economic and environmental policies. Environmental regulations – particularly those developed by virtue of the subsidiarity principle (i.e., national discretion to regulate issues that do not necessarily call for supranational measures) – are often tailored to local circumstances and may clash with free-trade imperatives, which call for a

[31] The references in this section concern arguments made by the contributors to this volume.
[32] Jones.
[33] Ozinga and Gerard; Van Kasteren.
[34] Vogel et al.
[35] Jones; Kölliker.
[36] Jones; Martin; Stigson and Rendlen; Von Moltke.

uniform regulative framework.[37] As economic objectives are often prioritised, it is important to design environmental policies that have minimal trade effects.[38] However, the fact that national environmental regulations distort international trade does not necessarily involve rejection of the environmental rules: if they serve environmental objectives that are important both ecologically and in relation to the economic costs incurred, and if there are no alternatives with less unfavourable economic effects, the WTO may authorise them.[39]

Local or national environmental regimes may spill over to the federal level or to other countries. Evidence of both positive and negative effects is provided. Several authors show instances of a 'race to the bottom'.[40] In their views and studies, regionalisation or globalisation leads to unleashed market forces that do not allow for environmental considerations and render scaling-up inevitable. As a result, the environment is harmed: more fossil fuels are being exploited, biodiversity is increasingly reduced, the illegal logging of tropical forests is aggravated, and environment-friendly pesticides for small markets fail to be developed.

By contrast, there is also evidence of a 'race to the top' in the US and the EU.[41] Stringent local standards in proactive states spread to other states via the central level, especially in the EU. However, a further tightening of EU environmental regulations relies on the optimistic assumption that there is the political will to do so.[42] One way of raising environmental performance is to stimulate the diffusion of environment-friendly technologies, in which innovation policy instruments should be tailored to the prevailing circumstances.[43] Stringent environmental standards may also spread through private actors: proactive multinational companies (such as Unilever) may set corporate standards that apply worldwide, for all their subsidiaries.[44]

These empirical findings and a literature review[45] lead us to conclude that national environmental standards tighten up if:

- There is the political will to protect the environment.
- Civil society, company head offices, and politicians expect or demand higher standards.

---

37 Arentsen and De Bruijn; Krämer; Von Moltke.
38 Kölliker.
39 Jansen and Keck.
40 Hol; Martin; Ozinga and Gerard; Van Kasteren.
41 Krämer; Vogel et al.
42 Krämer.
43 Kemp et al.
44 Dutilh; Stigson and Rendlen.
45 Jones.

- Non-environmental aspects (such as access to inputs or markets and ease of governance) dominate over (the costs of) environmental standards in business location decisions.
- Market demands or government regulations force global business to engage in technological innovation, which is subsequently applied worldwide to exploit economies of scale.
- NGOs pressurise private companies – for example, by boycotting brands or companies – in order to compensate for the relative absence of the state.

A systematic solution for the tension between trade liberalisation and environmental protection is proposed in two different directions. One possibility is to reform the WTO in such a way that environmental concerns are more fully integrated into economic decisions.[46] The other approach is to negotiate and facilitate environmental issues in a specialised World Environment Mechanism.[47] The former approach would seem to hold out better perspectives in the short term, although environmental problems are often more complex than trade issues.

Changes in Domestic Environmental Issues

Once industrialised countries have solved or contained their local problems such as soil pollution, the more persistent international environmental problems remain an item on the political agenda. While domestic problems with an international character such as international waste transports, transboundary water pollution, and climate change have been around for a long time, national governments are now facing new opportunities and constraints in solving them. For example, national governments can impact the EU agenda; at the same time, they see their discretion reduced through imposed EU legislation.[48]

Collective action is hazardous, especially at the international level. Yet, transboundary problems of a public or semi-public nature must be addressed at the regional or global level,[49] in which it is important to note that the nature of environmental problems is amenable to human interpretation.[50]

Though some transboundary environmental problems can and should be solved differently in a globalised world, some domestic problems have been created or aggravated by globalisation. Free trade may lead to higher production and consumption levels, scaling-up, and the geographic

[46] Jansen and Keck; Martin; Von Moltke.
[47] Esty and Ivanova.
[48] Arentsen and De Bruijn; Veenman and Liefferink.
[49] Kölliker.
[50] Conceição and Kaul.

reallocation of activities, which, in their turn, may affect the national environment.[51] Moreover, globalisation has accelerated the exploitation of precious natural resources such as tropical forests and has hampered their environmentally sound management.[52] Furthermore, global suppliers may be discouraged from supplying environment-friendly products to small markets.[53] Problems that have been engendered or aggravated by globalisation are much more difficult to solve, especially when important economic forces are at work, as in the case of agriculture, natural resources, and chemicals.

Changes in Governance Mechanisms

Globalisation has led to important changes in the distribution of power and governance modes. Some important developments are discussed below.

New position of societal actors

Multinational companies increasingly operate in regional clusters and global alliances. The WBCSD is a prominent example of a cooperative platform of environmentally proactive companies.[54] Local NGOs have also joined forces in international networks, such as the EEB, in response to the shift of regulative power from local and national levels to the supranational level, such as the EU.[55] Other NGOs (such as WWF) are global players. They focus on global environmental problems, even if these often involve local action.[56] The behaviour of individual citizens is susceptible to global influences.[57] Citizens also organise themselves increasingly on a global scale, sometimes only for one-off events.[58] These – ephemeral or more structural – civil networks are constructed with the aid of the internet and attract huge media attention. The negative publicity and boycotts engendered by (perceived) local environmental missteps have the potential to negatively affect the global corporate image and market performance of multinational companies.[59]

New supranational forums

The process in which many societal actors increasingly move to international arenas for information exchange, negotiation, policy-making, and

---

[51] Massink et al.
[52] Ozinga and Gerard.
[53] Van Kasteren.
[54] Stigson and Rendlen.
[55] Hontelez.
[56] Martin.
[57] Spaargaren and Martens.
[58] Winsemius.
[59] Dutilh; Ozinga and Gerard; Zoeteman and Harkink.

enforcement – which are beyond direct national control – has put pressure on national governments to add a networking role to their traditional roles as regulators and supervisors. There is a multitude of formal and informal networks in which governments seek alliances for their specific interests and exert influence in the power centres of supranational organisations such as the EU and the United Nations (UN).[60] The effectiveness of relational networks may benefit from a central organism that facilitates and coordinates actions outside extant power structures.[61]

Apart from political interactions in the more classic 'public-public' policy networks,[62] governments increasingly participate in 'public-private' networks to exchange information, share facilities, and jointly manage products, processes, or regions. Such platforms for governments, companies, and NGOs can be effective if they meet certain conditions for cooperation, including expected benefits for all parties involved, willingness to realise common goals, and readiness to make individual resources collectively available.[63]

Growing complexity and reduced impact of nation states

Increasing numbers of international and supranational directives, conventions, and related regulations impose more and more obligations on individual states and limit their sovereignty.[64] Furthermore, because ever more state and non-state actors are involved in consultation and decision-making, the impact of small and medium-sized individual states on the development of international policy is continuously dwindling.[65]

Hence, new strategies are needed for proactive governments to enhance the impact of their initiatives. Strategies to achieve good results include prioritisation and linkage of issues, coalition formation with other delegations, domestic preparation of package deals with other ministries, and investment in scientific knowledge.[66] Economy-environment compatibilities and trade-offs should be analysed in advance, and a sufficiently independent position of environmental interests should be aimed at in economy-driven international or supranational negotiations.[67]

Governments can also develop joint initiatives with like-minded nations and global companies.[68] Such initiatives may include agreements on more

---

60 Streck.
61 Esty and Ivanova.
62 Arentsen and De Bruijn; Veenman and Liefferink; Vogel et al.
63 Conceição and Kaul; Martin; Opschoor; Streck; Visser; Zoeteman and Harkink.
64 Arentsen and De Bruijn; Jansen and Keck; Krämer; Van Kasteren.
65 Hol; Hontelez; Winsemius.
66 Veenman and Liefferink; Wijen and Zoeteman; Winsemius.
67 Esty and Ivanova; Jones; Kölliker; Von Moltke.
68 Zoeteman and Harkink.

sustainable performance in combination with export-credit guarantees, co-financing, eco-labelling, and public-performance rating.[69] Grants to certified companies for managing global commons could also be part of new arrangements.

Coping with Reduced Sovereignty

The infringement of globalisation on national discretion has led to the following changes with respect to governments' environmental policies.

Transparency and accountability

The development of more international and supranational policies has increasingly turned nation states into the implementers of international mandates.[70] The primacy has shifted away from national instruments, such as command-and-control measures and taxes, to regional or global regulations, such as EU directives, and 'soft law', like OECD guidelines, aiming – among other things – at enhancing transparency and polluter accountability. National authorities, therefore, are transiting from a directing to a reporting role. This requires:

- Increasing transparency by streamlining the reporting obligations of local governmental bodies and business organisations, as supranational authorities will monitor the incidence of inconsistencies, and reporting results may generate novel or altered obligations.
- Enhancing accountability by adopting and enforcing standards for environment-friendly behaviour and by developing novel policy instruments that are flexible and cost-effective.
- Improving the operational capacity of national and local governmental bodies to implement national commitments.
- Generating national mechanisms to prevent non-compliance sanctions to which countries may be subjected, for example, by imposing carbon dioxide emission caps on domestic industries.

When nation states have recurrently experienced supervision by higher authorities and the threat of impending sanctions, they may anticipate the collective supervision of the implementation of international mandates by setting their ambition levels below those they would otherwise strive for. National environmental authorities may counteract pressures for a more conservative approach by being sensitive to knowledge provided by

[69] Conceição and Kaul; Ozinga and Gerard.

[70] Arentsen and De Bruijn; Hol; Hontelez; Jansen and Keck; Krämer; Van Kasteren.

progressive colleagues in other countries and by exchanging information on the availability of cost-effective solutions that show the feasibility of ambitious objectives, such as the EU's 'open method of coordination'.

Promotion of self-enforcing mechanisms

Many international agreements are not respected because the actors involved or targets aimed at do not face the appropriate incentives. This goes particularly for global public and common pool environmental resources.[71] In the absence of an effective world government, governments may then look for self-enforcing mechanisms. Market forces can sometimes be structured in ways that stimulate environment-friendly solutions, for example, through eco-labelling.[72] If this is not the case, global public agreements should be conceived in such a way that it is in the nation states' own interest to participate, for example, because participation provides access to restricted markets. If public goods can be turned into club goods that are non-rival in nature and available only to participants, collective action problems can be reduced or even solved.[73] A third course of action is to develop instruments that are applicable both domestically and internationally, and that allow for cost-effective solutions while achieving ambitious environmental objectives, for example, through internationally tradable permits or regional covenants.

Need for improved communication with citizens

When the physical or psychic distance between supranational policy-makers and those experiencing their impacts widens, states may lose their legitimacy towards their own citizens. Realising international agreements may involve making compromises and having to give up some national or local political priorities, which may give rise to general feelings of alienation. The low degree of trust many citizens have in their governments and in supranational administrative bodies is an indication of this phenomenon.[74] Indeed, many citizens feel that their government's environmental policies do not match their lifestyles.[75] While there may be good reason to challenge the sovereignty of citizen-consumers, national governments should also communicate supranational decisions to their citizens more actively and voice idiosyncratic preferences of local or national constituencies at the international level.[76] To enhance communication, efficient technologies like the internet could be used

---

[71] Conceição and Kaul; Kölliker.
[72] Jones; Martin; Stigson and Rendlen.
[73] Conceição and Kaul; Wijen and Zoeteman.
[74] Hontelez; Martin.
[75] Spaargaren and Martens.
[76] Dutilh; Spaargaren and Martens.

to inform citizens about the substance and form of supranational or international policies, national ambitions, and local repercussions.

Prospects for Environmental Policy Instruments

Conventional public policy instruments are licences, standards, levies, subsidies, taxes, and tradable permits. To avoid competitive distortions for domestic industries and to avoid supranational sanctioning of unachieved targets, nations may choose, or be forced, to downgrade their ambition levels and use the conventional toolkit sparingly.[77] Yet, we have also witnessed the collective upgrading of many environmental standards in the EU and the US.[78] It is unclear whether this ratcheting-up process will continue. The US presently lacks the willingness at the federal level to endorse nation-wide dissemination of stringent local standards, while no EU member state is currently pushing hard to get stringent national norms spilled over to other European countries.

In these circumstances, proactive national governments can still take a number of actions:

- Avoid national participation in a race to the bottom, for example, by adhering to internationally agreed process standards.[79]
- Stimulate race-to-the-top effects by supporting regional or global product standards such as chemicals classifications and international eco-labelling systems.[80]
- Foster the development of NGOs to expose environmental evils,[81] act as countervailing powers,[82] create international leverage,[83] and/or fill a global governance void.[84] NGOs themselves should also be critically monitored to make sure that their actions are in the best interest of the environment.
- Create and foster informal networks and partnerships around 'coalitions of the willing', involving public and private actors at the local, national, regional, and global levels.[85]
- Promote voluntary codes of conduct by favouring or rewarding companies that have an environmental code certification and stimulate multinational

[77] Hol; Martin; Ozinga and Gerard; Van Kasteren.
[78] Krämer; Vogel et al.
[79] Massink et al.
[80] Martin; Ozinga and Gerard; Visser; Vogel et al.
[81] Ozinga and Gerard.
[82] Opschoor.
[83] Winsemius.
[84] Kemp et al.
[85] Esty and Ivanova; Martin; Streck; Zoeteman and Harkink.

companies whose headquarters are within the national territory to adopt universal environmental standards for all their subsidiaries.[86]
- Tailor 'green' innovation policy instruments to the prevailing circumstances, anticipate economies of scale in technological innovations, and bridge the financial gap between present investments and future pay-offs.[87]
- Link environmental policies to other, high-priority policy areas requiring the same types of measures, for example, focusing on public health to reduce air pollution.[88]

Finally, there are many ways to develop the entrepreneurial role of small and medium-sized nations, including: (1) the development of leadership as a go-between for larger states and the use of convening and secretariat functions; (2) the promotion of corporate social responsibility via multinational companies and NGOs operating within the national borders by providing environmental venture capital, think tanks, and centres of excellence.[89]

Balancing Globalisation and National Environmental Policy

Globalisation has enhanced, if not created, an intertwining of issues, actors, and states. At the same time, the centres of gravity have shifted beyond national borders: environmental decisions are increasingly being taken by supranational organisations such as the EU and the WTO and by large multinational companies. In order to develop and implement effective environmental policies and to maintain or restore legitimacy towards their citizens, governments need to take appropriate and concerted action, both domestically and internationally. In choosing the appropriate administrative level, governments should not think primarily in terms of the nature of a problem itself, for example, a global issue such as climate change, but rather in terms of the forum where solutions to this problem could be most effectively addressed, which might initially involve a regional regime for the climate issue.

National governments are losing classic policy options: long-term targets are controversial in the international arena; subsidies may have to be restricted to avoid international trade distortions; levies and taxes are more difficult to effectuate; and national covenants may fit poorly with international accountability imperatives. There is also less room for new initiatives in areas that have already been regulated in an international

86 Martin; Stigson and Rendlen; Zoeteman and Harkink.
87 Kemp et al.
88 Wijen and Zoeteman.
89 Winsemius.

framework. Yet, new partnerships may emerge, and innovative policy instruments can be developed. This requires an entrepreneurial role for public authorities to use the opportunities that globalisation offers.

Not all policy areas are equally vulnerable to loss of ambition. Special concern is needed for:

- Environmental issues lacking owners: global commons such as oceans, ice caps, climate, and the ozone layer.[90]
- Natural resources representing large economic interests, for example, the trade in rare species or tropical timber, which are reflected in international economic agreements that lack accompanying environmental measures.[91]
- Environmental problems causing local harm but requiring large-scale technological solutions, such as low-emission vehicles and crop protection.[92]
- Environmental issues requiring international enforcement of measures, for example, mitigating climate change.[93]
- Problems lacking environmental champions who implement local measures and push towards acceptance of progressive policies at the central level, such as Germany in the EU and California in the US.[94]
- Environmental issues sensitive to corruption, such as the exploitation of natural resources, especially in developing countries.[95]

Given these changes and the limited possibilities for national governments to resist the effects of globalisation, we need shifts towards instruments that can be used nationally. Such shifts may include:

- The reduced use of those levies and other instruments that distort international markets.[96]
- Increased cross-border cooperation with foreign governments, multinational companies, and global NGOs.[97]
- Exploration of and engagement in partnerships.[98]
- Increased use of higher-level (i.e., federal, regional, or global) standards instead of lower-level (i.e., national or local) standards.[99]

---

[90] Conceição and Kaul; Kölliker; Ozinga and Gerard; Wijen and Zoeteman.
[91] Martin; Ozinga and Gerard.
[92] Conceição and Kaul; Van Kasteren; Vogel et al.
[93] Kölliker; Wijen and Zoeteman.
[94] Krämer; Veenman and Liefferink; Vogel et al.
[95] Opschoor; Ozinga and Gerard.
[96] Jansen and Keck; Jones; Stigson and Rendlen.
[97] Conceição and Kaul; Martin; Visser; Zoeteman and Harkink.
[98] Dutilh; Martin; Opschoor; Streck.
[99] Visser; Vogel et al.; Wijen and Zoeteman.

- Greater emphasis on the education of and communication with civil society.[100]
- Promotion of self-regulation and private initiatives with environmental robustness.[101]
- Stimulating the application of self-enforcing mechanisms in international agreements.[102]

As outlined above, the commensurability of governance levels and the compatibility of different areas are of central importance. Figures 1.1 and 1.2 recapitulate – in a strongly simplified way, for the sake of clarity – interactions between globalisation and national environmental policies with respect to these issues. Figure 1.1 represents the situation in which nation states are relatively autonomous. The national government operates at the top of the regulative hierarchy. It has a high degree of discretion to develop national environmental policy, which is often enforced through legislation and other types of regulation, such as voluntary agreements. These policies are then implemented by local authorities, who issue and control environmental licences to local companies. Civil society, often organised through local NGOs, exerts moral pressure on local and national governments as voting citizens and economic influence on local companies as employees and consumers. Business lobbies the national government in order to avoid too stringent or costly regulations. While supranational organisations, such as the UN, exist, their influence is relatively limited. In this situation, the dominant influence – represented by thick arrows – is of a formal nature and goes from the national government via local government to business: public authorities largely impose command-and-control measures on business. The focus is predominantly on local environmental problems on the basis of a local cost-benefit analysis, as these can be addressed by the competent authorities; transboundary environmental issues are largely taken for granted. The sovereignty situation is characterised by a relatively high degree of convergence between jurisdictional competence, the level of political influence, and the scale of economic activities, which are predominantly located at the national and sub-national levels.

Globalisation dramatically alters the arena in which environmental policy is developed and implemented (Figure 1.2). National borders have become much more permeable. Business has internationalised its activities and footloose multinational companies are subject to different national regimes. Local business lobbies national government and – via supranational trade associations – supranational organisations to stress the importance of a 'level

[100] Dutilh; Spaargaren and Martens.
[101] Jones; Stigson and Rendlen; Zoeteman and Harkink.
[102] Wijen and Zoeteman.

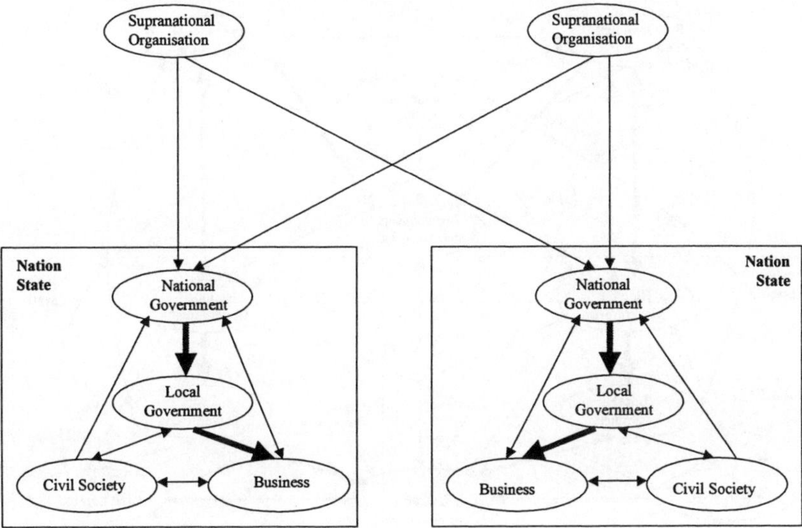

Figure 1.1 Influences in the national sovereignty system

playing field', i.e., the absence of different competitive positions owing to divergent environmental regulations. Civil society still has contacts with local companies as well as local and national governments, but consists increasingly of global citizens who raise their voice against unacceptable behaviour of foreign governments, citizens, and companies, often orchestrated by supranational NGOs. Consumers increasingly purchase products from around the world, whose marketing is often supported by global brands. The national government has lost a considerable degree of control 'from below', as the behaviour of its subjects is increasingly driven by foreign influences. At the same time, the national government's formal power is also reduced 'from above': supranational organisations (the WTO and the EU, in particular) constrain the discretion of the national government through multilateral regulations. While the national government can still initiate its own policies, it more often implements multilateral agreements that reflect the divergent interests of a considerable number of countries. Consequently, the national government often becomes a relatively small player in a large arena, while the increased distance between regulators and citizens as well as the increased divergence of interests may cause feelings of alienation. On the basis of 'imported' regulations and its 'own' preferences, the national government then develops and implements environmental policy, its choice of instruments being constrained not only by supranational rules but also by the necessity to consider the international competitiveness of domestic industries. Local

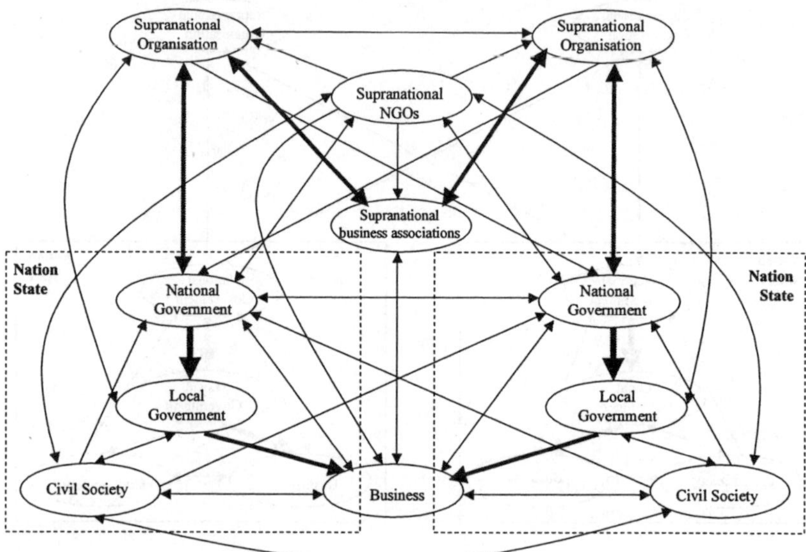

Figure 1.2 Influences in the globalisation system

governments still implement national policy, though they may also have direct contacts with supranational bodies. While national and local governments remain important for companies, the business community increasingly interacts directly with supranational organisations, for example, by lobbying supranational bodies and entering into partnerships with supranational NGOs. The national government has not only lost part of its sovereignty but has also gained new opportunities: it consults, negotiates, and coalesces – through formal and informal contacts – with foreign governments, supranational organisations, multinational companies, and multinational NGOs. These new or intensified contacts enable the government to address transboundary, regional, or global environmental problems. Whereas hierarchical relations dominate in the case of national sovereignty, networking governs the globalised arena. Globalisation has brought about an intertwining of foreign and domestic influences. At the same time, with national jurisdiction being constrained by territorial boundaries, and economic, political, and social influences transgressing those boundaries, different spheres of influence are growing more disjunctive. This may hamper environmental protection, especially if states can be played off against one another by powerful economic actors. However, the environment may also benefit from spill-overs of proactive environmental practices and multilateral cooperation. The balance for the environment is context-dependent, in which the conception of governance modes and incentives for different societal actors are of crucial importance.

In conclusion, the studies in this volume show that the impact of globalisation on national environmental policy is indeed huge. While some changes have occurred in a short period, many have taken place more progressively and may have remained unnoticed to many. This book shows the multiple factors and their interactions that have led to actual and desired changes in the ways in which national governments can intervene effectively in the global arena in order to protect the environment.

REFERENCES

Anheier, H., M. Glasius, and M. Kaldor (eds) (2001), *Global Civil Society 2001*, Oxford: Oxford University Press.

Antweiler, W., B. Copeland, and S. Taylor (2001), 'Is free trade good for the environment?', *American Economic Review*, **91**(4), 877-908.

Arts, B. (1998), *The Political Influence of Global NGOs: Case Studies on the Climate and Biodiversity Conventions*, Utrecht: International Books.

Baker, D., G. Epstein, and R. Pollin (eds) (1998), *Globalization and Progressive Economic Policy*, Cambridge: Cambridge University Press.

Barrett, S. (2003), *Environment & Statecraft: The Strategy of Environmental Treaty-Making*, Oxford: Oxford University Press.

Bhagwati, J. (2002), *Free Trade Today*, Princeton: Princeton University Press.

Birnie, P. and A. Boyle (2002), *International Law and the Environment*, 2nd ed., Oxford: Oxford University Press.

Boyce, J. (2004), 'Green and brown? Globalization and the environment', *Oxford Review of Economic Policy*, **20**(1), 105-128.

Bromley, D. (ed.) (1995), *The Handbook of Environmental Economics*, Oxford: Blackwell.

Brunsson, N. and B. Jacobsson (2000), *A World of Standards*, Oxford: Oxford University Press.

Carraro, C. and F. Lévêque (1999), *Voluntary Approaches in Environmental Policy*, Dordrecht: Kluwer.

Carson, R. (1962), *Silent Spring*, Boston: Houghton Mifflin.

Castells, M. (1996), *The Information Age: Economy, Society and Culture: Volume I: The Rise of the Network Society*, Malden: Blackwell.

Copeland, B. and S. Taylor (2003), *Trade and the Environment: Theory and Evidence*, Princeton: Princeton University Press.

Cooperrider, D. and J. Dutton (eds) (1999), *Organizational Dimensions of Global Change: No Limits to Cooperation*, Thousand Oaks: Sage.

Cutler, C., V. Haufler, and T. Porter (eds) (1999), *Private Authority and International Affairs*, Albany: State University of New York Press.

Dunning, J. (ed.) (2000), *Regions, Globalization, and the Knowledge-Based Economy*, Oxford: Oxford University Press.

Dunning, J. (ed.) (2003), *Making Globalization Good: The Moral Challenges of Global Capitalism*, Oxford: Oxford University Press.

EEA (2003), 'Europe's environment: The third assessment: Environmental assessment report no. 10', *http://reports.eea.eu.int/environmental_assessment_report_2003_10/en*, Copenhagen: European Environment Agency.

Esty, D. and B. Gentry (1997), 'Foreign investment, globalisation, and environment', in OECD, *Globalization and Environment: Preliminary Perspectives,* Paris: Organisation for Economic Co-operation and Development.

Esty, D. and D. Geradin (1998), 'Environmental protection and international competitiveness', *Journal of World Trade,* **32**(5), 5-46.

Esty, D. and M. Ivanova (eds) (2002), *Global Environmental Governance: Options & Opportunities,* New Haven: Yale Center for Environmental Law and Policy.

Etzioni, A. (1990), *The Moral Dimension: Toward a New Economics,* New York: Free Press.

European Commission (2004), *Stimulating Technologies for Sustainable Development: An Environmental Technologies Action Plan for the European Union,* COM(2004) 38 final, Brussels: European Commission.

Florini, A. (ed.) (2000), *The Third Force: The Rise of Transnational Civil Society,* Tokyo: Japan Center for International Exchange.

Florini, A. (2003), *The Coming Democracy: New Rules for a New World,* Washington DC: Island Press.

Frankel, J. (2000), 'Globalization of the economy', NBER working paper 7858, *http:www.nber.org/papers/w7858,* Cambridge: National Bureau of Economic Research.

Frankel, J. and A. Rose (2002), 'Is trade good or bad for the environment? Sorting out the causality', NBER working paper 9201, *http://www.nber.org/papers/w9201,* Cambridge: National Bureau of Economic Research.

Giddens, A. (ed.) (2001), *The Global Third Way Debate,* Cambridge: Polity Press.

Gray, B. (1999), 'The development of global environmental regimes: Organizing in the absence of authority', in D. Cooperrider and J. Dutton (eds), *Organizational Dimensions of Global Change: No Limits to Cooperation,* Thousand Oaks: Sage.

Group of Lisbon (1993), *Limits to Growth,* Lisbon: Gulbenkian Foundation.

Haas, P., R. Keohane, and M. Levy (eds) (1993), *Institutions for the Earth: Sources of International Environmental Protection,* Cambridge: MIT Press.

Hardin, G. (1968), 'The tragedy of the commons', *Science,* **162**, 1243-1248.

Held, D. (ed.) (2004), *A Globalizing World? Culture, Economics, Politics,* 2nd ed., London: Routledge.

Held, D. et al. (1999), *Global Transformations: Politics, Economics and Culture,* Cambridge: Polity Press.

Hertz, N. (2001), *The Silent Takeover: Global Capitalism and the Death of Democracy,* London: Heinemann.

Hobsbawm, E. (1975), *The Age of Capital, 1848-1875,* London: Weidenfeld and Nicolson.

Holliday, C., S. Schmidheiny, and P. Watts (2002), *Walking the Talk: The Business Case for Sustainable Development,* Sheffield/San Francisco: Greenleaf/Berrett-Koehler.

IISD and UNEP (2000), *Environment and Trade: A Handbook,* Winnipeg/Stevenage: International Institute for Sustainable Development/United Nations Environment Programme.

Inglehart, R. (1997), *Modernization and Postmodernization: Cultural, Economic, and Political Change in 43 Societies,* Princeton: Princeton University Press.

Jaffe, A., S. Peterson, P. Portney, and R. Stavins (1995), 'Environmental regulation and the competitiveness of U.S. manufacturing: What does the evidence tell us?', *Journal of Economic Literature,* **33**, 132-163.

Jenkins, R., J. Barton, A. Bartzokas, J. Hesselberg, and H. Mereke Knutsen (2003), *Environmental Regulation in the New Global Economy: The Impact on Industry and Competitiveness,* Cheltenham: Edward Elgar.

Kaul, I., I. Grunberg, and M. Stern (eds) (1999), *Global Public Goods: International Cooperation in the 21st Century,* New York: Oxford University Press.

Kaul, I., P. Conceição, K. Le Goulven, and R. Mendoza (eds) (2003), *Providing Global Public Goods: Managing Globalization,* New York: Oxford University Press.

Keohane, R. and J. Nye (1977), *Power and Interdependence: World Politics in Transition,* Boston: Little and Brown.

Klein, N. (2000), *No Logo: No Space, No Choice, No Jobs,* London: Flamingo.

Kölliker, A. (2002), *The Impact of Flexibility on the Dynamics of European Unification,* Dissertation, Florence: European University Institute.

Landes, D. (1998), *The Wealth and Poverty of Nations: Why Some Countries are so Rich and Some so Poor,* New York: Norton & Company.

Lechner, F. and J. Boli (eds) (2000), *The Globalization Reader,* Malden: Blackwell.

Lesser, J., D. Dodds, and R. Zerbe (eds) (1997), *Environmental Economics and Policy,* Reading: Addison-Wesley.

Levy-Livermore, A. (ed.) (1998), *Handbook on the Globalization of the World Economy,* Cheltenham: Edward Elgar.

Mabey, N. and R. McNalley (1999), *Foreign Direct Investment and the Environment: From Pollution Haven to Sustainable Development,* Surrey: WWF UK.

Mäler, K. and J. Vincent (eds) (2003), *Handbook of Environmental Economics, Volume I: Environmental Degradation and Institutional Responses,* Amsterdam: Elsevier.

Mani, M. and D. Wheeler (1998), 'In search of pollution havens? Dirty industry in the world economy (1960-1995)', *Journal of Environment and Development,* **7**(3), 215-247.

McNeill, J. and W. McNeill (2003), *The Human Web: A Bird's-Eye View of World History,* New York: Norton & Company.

Meadows, D., D. Meadows, J. Randers, and W. Behrens (1972), *The Limits to Growth: A Report for the Club of Rome's Project on the Predicament of Mankind,* London: Earth Island.

Michie, J. (ed.) (2003), *The Handbook of Globalisation,* Cheltenham: Edward Elgar.

Mol, A., V. Lauber, and D. Liefferink (eds) (2000), *The Voluntary Approach to Environmental Policy: Joint Environmental Approach to Environmental Decision-Making in Europe,* Oxford: Oxford University Press.

Moore, M. (2003), *A World without Walls: Freedom, Development, Free Trade, and Global Governance,* Cambridge: Cambridge University Press.

Mullard, M. (2004), *The Politics of Globalisation and Polarisation*, Cheltenham: Edward Elgar.

Nayyar, D. (2002), *Governing Globalization: Issues and Institutions,* Oxford: Oxford University Press.

Norberg, J. (2003), *In Defense of Global Capitalism,* Washington, DC: Cato Institute.

Nordström, H. and S. Vaughan (1999), *Trade and Environment: Special Studies 4,* Geneva: World Trade Organization.

OECD (2002), *Environmental Issues in Policy-Based Competition for Investment: A Literature Review,* Paris: Organisation for Economic Co-operation and Development.

OECD (2003), *Voluntary Approaches for Environmental Policy: Effectiveness, Efficiency and Usage in Policy Mixes,* Paris: Organisation for Economic Co-operation and Development.

Ohmae, K. (1995), *The End of the Nation-State: The Rise of Regional Economies,* London: HarperCollins.

Opschoor, J., K. Button, and P. Nijkamp (eds) (1999), *Environmental Economics and Development,* Cheltenham: Edward Elgar.

Osborne, S. (ed.) (2000), *Public-Private Partnerships: Theory and Practice in International Perspective*, London: Routledge.

Powell, W. and E. Clemens (eds) (1998), *Private Action and the Public Good,* New Haven: Yale University Press.

Reinicke, W. (1998), *Global Public Policy Networks: Governing Without Government?,* Washington, DC: Brookings Institution Press.

Rischard, J. (2002), *High Noon: Twenty Global Problems, Twenty Years to Solve Them,* New York: Basic Books.

Rugman, A. (2001), *The End of Globalization: A New and Radical Analysis of Globalization and What it Means for Business,* London: Random House.

Sandel, M. (1996), *Democracy's Discontent: America in Search of a Public Philosophy,* Boston: Harvard University Press.

Sassen, S. (1996), *Losing Control? Sovereignty in an Age of Globalisation,* New York: Columbia University Press.

Schaeffer, R. (2003), *Understanding Globalization: The Social Consequences of Political, Economic, and Environmental Change,* Lanham: Rowman & Littlefield.

Scholte, J. (2000), *Globalization: A Critical Introduction,* Hampshire: Palgrave.

Sklair, L. (2002), *Globalization: Capitalism & its Alternatives*, 3rd ed., Oxford: Oxford University Press.

Spaargaren, G., A. Mol, and F. Buttel (eds) (2000), *Environment and Global Modernity,* London: Sage.

Stiglitz, J. (2002), *Globalization and its Discontents,* New York: Norton & Company.

Strange, S. (1996), *The Retreat of the State: The Diffusion of Power in the World Economy,* Cambridge: Cambridge University Press.

Sutherland, W. (2000), *The Conservation Handbook: Research, Management and Policy,* Oxford: Blackwell.

Tietenberg, T. (2003), *Environmental and Natural Resource Economics,* Boston: Addison-Wesley.

Tietenberg, T., K. Button, and P. Nijkamp (eds) (1999), *Environmental Instruments and Institutions: Environmental Analysis and Economic Policy, Volume 6,* Cheltenham: Edward Elgar.

UNDP (1999), *Human Development Report 1999,* Oxford: Oxford University Press.

UNEP (2003), 'GEO year book 2003', *http://www.unep.org/geo/yearbook/,* Nairobi: United Nations Environment Programme.

Van den Bergh, J. (ed.) (1999), *Handbook of Environmental and Resource Economics,* Cheltenham: Edward Elgar.

Van den Bergh, J. (ed.) (2002), *Handbook of Environmental and Resource Economics,* Cheltenham: Edward Elgar.

Van Seters, P., B. de Gaay Fortman, and A. de Ruijter (eds) (2003), *Globalization and its New Divides: Malcontents, Recipes, and Reform,* Amsterdam: Dutch University Press.

Vogel, D. (1995), *Trading Up: Consumer and Environmental Regulation in a Global Economy,* Cambridge: Harvard University Press.

Vogel, D. (1997), 'Trading up and governing across: Transnational governance and environmental protection', *Journal of European Public Policy*, **4**(4), 556-571.

Wackernagel, M. and W. Rees (1996), *Our Ecological Footprint: Reducing Human Impact on the Earth,* Gabriola Island: New Society.

Wallerstein, I. (1974), *The Modern World-System: Volume I: Capitalist Agriculture and the Origins of the European World-Economy in the Sixteenth Century,* New York: Academic Press.

Wallerstein, I. (1980), *The Modern World-System: Volume II: Mercantilism and the Consolidation of the European World-Economy, 1600-1750,* New York: Academic Press.

Warner, M. and Sullivan, R. (eds) (2004), *Putting Partnerships to Work: Strategic Alliances for Development between Government, the Private Sector and Civil Society,* Sheffield: Greenleaf.

Waters, M. (2001), *Globalization,* 2nd ed., London: Routledge.

Winsemius, P. and U. Guntram (2002), *A Thousand Shades of Green: Sustainable Strategies for Competitive Advantage,* London: Earthscan.

World Economic Forum (2002), '2002 environmental sustainability index', *http://www.ciesin.columbia.edu/indicators/ESI,* Geneva: World Economic Forum.

Young, O. (1994), *International Governance: Protecting the Environment in a Stateless Society,* Ithaca: Cornell University Press.

Young, O. (ed.) (1999), *The Effectiveness of International Environmental Regimes: Causal Connections and Behavioral Mechanisms,* Cambridge: MIT Press.

Zarsky, L. (1999), 'Havens, halos, and spaghetti: Untangling the evidence about foreign direct investment and the environment', in OECD, *Globalization and Environment,* Paris: Organisation for Economic Co-operation and Development.

PART I

Concepts and State of Affairs

2. Trade and Investment: Selected Links to Domestic Environmental Policy

Tom Jones[1]

SUMMARY

I provide an overview of key relations between economic globalisation and national environmental policies. The 1990s witnessed an upsurge in global economic flows (international trade, foreign direct investment (FDI), and portfolio investment). The environmental impacts of those economic changes can be broadly decomposed into their scale, structural, and technology components. Applied to the case of FDI in particular, there is little evidence that the net environmental effects have been negative, although there are some specific circumstances (particular sectors, industries, or locations) in which this may indeed be the case. National governments can work to ensure a positive result result by designing domestic environmental policies that support economic globalisation goals, and vice versa.

INTRODUCTION

Political interest in the relationship between economic globalisation and environmental policy grew rapidly during the 1990s. On the economic side, this interest was fuelled by an increasingly liberalised trade and investment agenda (Uruguay Round, NAFTA, the proposed OECD Multilateral Agreement on Investments, etc.). On the environment side, the impetus was provided by increased awareness that resolving global environmental problems (such as those covered by the Kyoto Protocol, the Convention on Biological Diversity, etc.) depended in many respects on structural changes occurring in the global economy. The 1992 World Summit on Sustainable

[1] Comments received on an earlier draft from Ken Ruffing and Frank Wijen are gratefully acknowledged. However, the views expressed here are the author's, and do not necessarily reflect the opinions of the OECD, the author's affiliation, or its member countries.

Development also played an important role in keeping the environmental aspects of globalisation on the policy agenda.

As an organisation committed to trade and investment liberalisation, and also one with a long-standing interest in environmental issues, the Organisation for Economic Co-operation and Development (OECD) has been interested in this debate from the beginning, focussing mainly on the potential for *synergy* within this relationship.

This chapter provides a brief overview of recent developments in the process of economic globalisation, with particular emphasis on the trade and investment parts of that process. Some of the ways in which globalisation may affect the environment in OECD countries are also summarised.[2] I conclude with a few thoughts about how synergies between economic globalisation and domestic environmental policy goals might be further developed in the future.

RECENT DEVELOPMENTS IN GLOBALISATION[3]

Economic globalisation is a process leading to increased economic interdependence (more global trade and capital flows, wider presence of international economic agreements, expanding opportunities for technology transfer, etc.) as well as changes in underlying economic structures (including reformed business organisation and altered consumption patterns). Globalisation has both endogenous and exogenous components. To the extent that globalisation results from technological change (such as development of the internet), it may be beyond the power of governments to control. However, globalisation is also fuelled by policy decisions of governments, such as decisions to liberalise trade and investment. To this extent, governments can influence both the pace and the direction of the globalisation process.

International trade and investment promote economic growth and competition. They also stimulate technology development and diffusion, and promote the structural changes necessary to make more efficient use of available resources. This benefits citizens through lower prices, greater product diversity, and increased purchasing power.

In a globalising economy, distances and national boundaries become smaller obstacles to econômic activity. Markets become increasingly

[2] The reverse question of how environmental quality may affect the process of economic globalisation is not addressed in depth, mainly because these effects are assumed to be relatively small.

[3] Much of the material for this section is based on OECD, forthcoming 2004. Some of the section is also based on OECD, 2001a.

interdependent through structural changes induced by the dynamics of increased trade, capital, transportation, and technology flows – changes for which the primary vehicles are often multinational enterprises. Overall, the scale of economic activity picks up, and the nature of that activity changes – both technologically and in structural terms.

The lowering of tariff and non-tariff barriers contributed – but was not the only contributor – to a steady rise in *international trade* in both goods and services throughout the 1990s (see Figure 2.1). The share of trade in total transactions remained consistently high, averaging 15 per cent of OECD gross domestic product (GDP) over this period. The share of trade in goods was four times the share of trade in services. The international division of production processes is also becoming more common, replacing older forms of trade that were once based on the importation of raw materials and exports of final goods. This increased specialisation is occurring through international subcontracting agreements with independent firms and within multinational firms (intra-firm trade). Intermediate goods now have substantial weight in total international trade, representing, for example, about half of the total trade between the European Union (EU) and other OECD member countries.

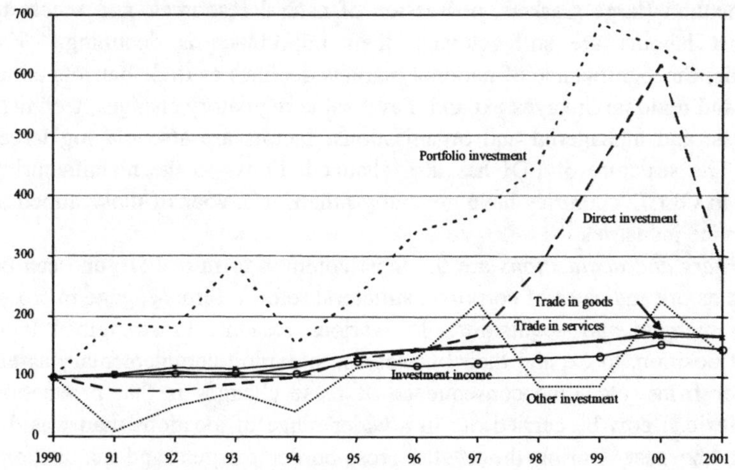

Figure 2.1 Trends in international trade and investment components[4]

[4] See OECD, forthcoming 2004. Represents (average imports + exports) or (average assets + liabilities). OECD data excludes the Czech Republic, Hungary, and the Slovak Republic for 1990-1992, Greece for 1998, Iceland and the Slovak Republic for 2001. Portfolio investments exclude financial derivatives.

The upsurge in *foreign direct investment* (FDI) and *portfolio investment* (see also Figure 2.1) was especially marked in the second half of the 1990s. Some forms of investment flows – especially portfolio flows, where investments consist of relatively shares in corporate equity – have also proven to be highly volatile, with periods of decline being followed by periods of rapid increase, and vice versa. Developing countries have increasingly been able to attract significant FDI flows. Beginning at an annual average of 46 billion US dollars (USD) in the 1988-1993 period, this had grown to more than 240 billion USD by 2000.[5] In relative terms, Official Development Assistance (ODA) flows are now much lower (53.1 billion USD in 2000),[6] even though ODA remains a significant factor in many individual countries. Most FDI is still occurring within the OECD area. During the period 1998-2000, Japan, the EU, and the United States (US) accounted for about three-quarters of global FDI flows.[7] Although the developing world is receiving an increasing share of these investments, the distribution of FDI in the developing countries remains unbalanced: 75 per cent of the inflows go to only eleven economies.[8] Of the total inflow to the developing world, Asia and Latin America receive about one-half and one-third, respectively, while Africa receives less than 5 per cent.[9] Although the traditional factors driving FDI location (large markets, possession of natural resources, and access to low-cost labour) are still relevant, their importance is declining.[10] For example, the significance of national markets declines as trade barriers come down and regional linkages expand. Favourable regulatory changes, technical progress, and managerial and organisational factors are also playing larger roles. The structure of FDI has also changed. Flows to the manufacturing sector in OECD countries have generally fallen, in favour of those aimed at the service industries.[11]

Mergers and acquisitions are the most common form of FDI, and can be viewed as one indicator of corporate structural reform. Firms engage in cross-border mergers and acquisitions for various reasons: to strengthen their market position, to expand their businesses, to exploit complementary assets in other firms, etc. One consequence of these changes is that production activities can now be carried out in a wider range of locations than was the case in the past. During the 1990s, cross-border mergers and acquisitions increased more than five-fold worldwide on a 'deal value' basis (see Figure

---

5 UNCTAD, 2001.
6 OECD, 2001a.
7 UNCTAD, 2001.
8 Argentina, Brazil, Mexico, Saudi Arabia, China, Indonesia, Malaysia, Singapore, Thailand, Hong Kong, and China.
9 UNCTAD, 2000.
10 UNCTAD, 2001.
11 UNCTAD, 2000.

2.2). The upsurge in both deal value and number of deals was especially significant in the late 1990s. As with investment activity more generally, some declines in mergers and acquisitions have been experienced in recent years. Mergers and acquisitions are taking place in manufacturing as well as in services, changing the shape of global industry. During the 1990s, the most active sectors at the global level were oil, automotive equipment, banking, finance, and telecommunications.

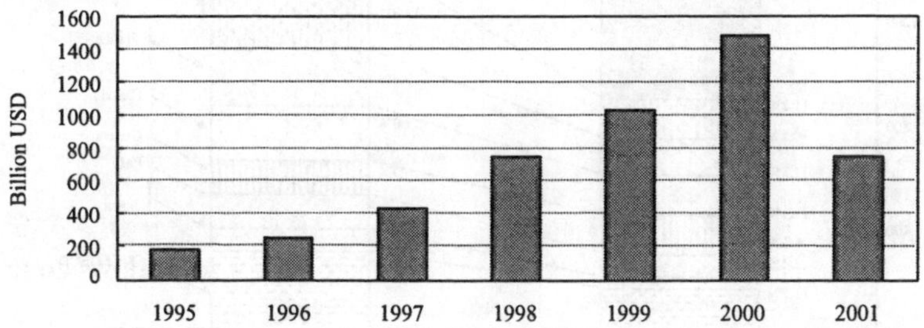

Figure 2.2 Trends in global cross-border mergers and acquisitions[12]

Between 1993 and 1999, the absolute value of production in firms under foreign control also increased in many OECD countries (see Figure 2.3). In some European countries, gross output under foreign control increased dramatically, particularly in Ireland (plus 180 per cent), Sweden (plus 170 per cent), and Finland (plus 140 per cent). The level of production of firms under foreign control also increased in both value and percentage terms in the US and the United Kingdom (UK).

In the mid-1990s, an average of 14 per cent of all *inventions* in any OECD country involved some element of foreign ownership. Seven per cent of patents were also the result of international cooperative research. The technology balance of payments is an indication of international technology transfers: licences, patents, know-how, research, and technical assistance. Unlike R&D expenditure, these are payments for production-ready technologies.

In most OECD countries, technology receipts and payments both rose sharply during the 1990s (see Figures 2.4 and 2.5). Overall, the OECD area maintained its position as net technology exporter vis-à-vis the rest of the world, although some parts of the OECD region (for instance, the EU) continued to run a deficit.

[12] OECD, forthcoming 2004.

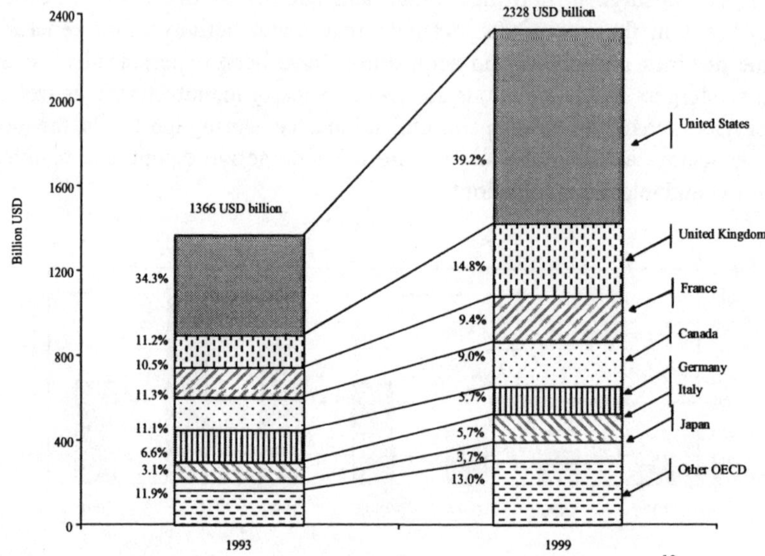

Figure 2.3 Trends in the share of affiliates under foreign control[13]

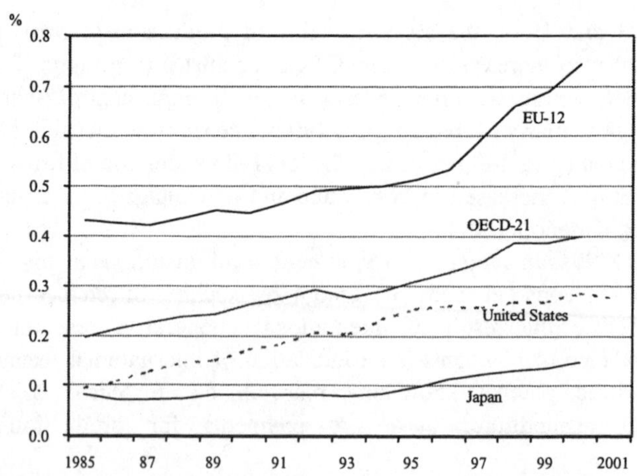

Figure 2.4 Trends in technology flows by geographical area[14, 15]

---

[13] OECD, 2002a.

[14] Technology flows consist of the average of technological payments and receipts. In Figure 2.4, they are expressed as a percentage of GDP in selected countries and regions.

[15] The EU-12 and OECD-21 flows in Figures 2.4 and 2.5 include intra-area flows and exclude Denmark and Greece; data are partly estimated. The OECD-21 flows in Figures 2.4 and 2.5 exclude the Czech Republic, Hungary, Iceland, Poland, the Slovak Republic, and Turkey. Figures 2.4 and 2.5 are based on OECD, 2002b.

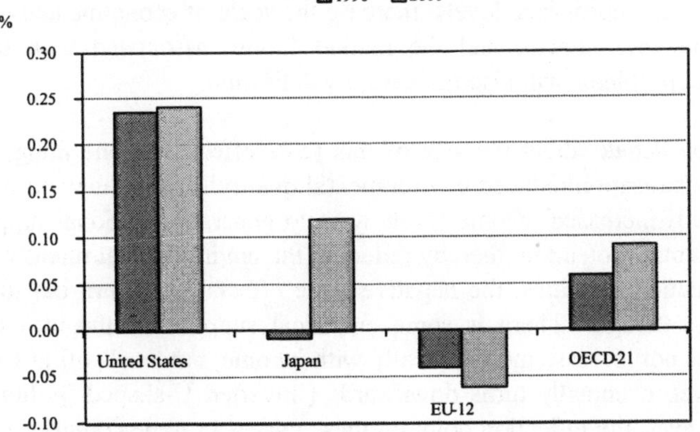

Figure 2.5 Changes in the technology balance of payments[16]

ENVIRONMENTAL EFFECTS OF ECONOMIC GLOBALISATION[17]

The environmental effects of trade and investment activity can be decomposed into scale effects (expansion of economic output), structural effects (reallocation of production and consumption), and technology effects (technological development and diffusion). In general, the scale effects are expected to be negative – a higher level of economic activity exacerbates any deficiencies in environmental policies which already exist, while the technological and structural effects are expected to be positive – the 'environment-friendliness' of new technologies or new economic structures is likely to be higher than the existing average. From an analytical perspective, what is most relevant is the net outcome of the three effects, not the individual parts.

The standard view in the literature is that trade and investment are not responsible in themselves for environmental problems – these problems stem from inadequate environmental controls on production and consumption activities. Trade and investment are merely the vehicles through which these inadequacies are (indirectly) expressed. On the other hand, these globalisation-related activities *will* exacerbate environmental problems if the environmental policies in place when globalisation occurs are not set – and

[16] Changes in the technology balance of payments are expressed as a percentage of GDP in 1990 and 2000.

[17] Most of the material for this section is based on OECD, 1997a.

implemented – at appropriate levels. Increase the scale of economic activities that harm the environment, and the market failure associated with their environmental problems will also increase, by definition.

There is some debate about the size of this scale effect. For one thing, the 'increased globalisation/reduced environmental quality' linkage may not be a constant one. If increased income levels were to generate additional support for environmental protection, thereby reducing the environmental intensity of current production activities, the negative scale effects could turn out to be less than first thought. There is some empirical support for the idea that pollution does not increase monotonically with income, but levels off at some point, and even eventually turns downwards ('inverted U-shaped' pollution curves). However, the inflection point on these curves in some countries may be much higher than current income levels, so the environment-intensity of production may need to continue to rise for an unacceptably long time in these countries. Even if the inverted-U relationship does hold for particular pollutants, and for broad groups of countries over particular periods of time, it may not be true indefinitely, nor for the world as a whole.

One structural shift to which economic liberalisation has already contributed is the increased participation of the newly industrialised economies in the international economy, and their general move out of primary-commodity production, into resource-processing, light manufacturing, or service activities. There is some evidence that, once countries begin to industrialise, economic liberalisation helps to render their economic structure less pollution-intensive than that of those countries whose economies remain closed.[18] On the other hand, even if these shifts are producing local environmental improvements, these improvements may be neutralised at the global level if other countries are now producing the more pollution-intensive products. In effect, the environmental impact may be *shifted*, rather than *reduced*.

Trade in environmentally preferred products (particularly eco-efficient capital equipment and its accompanying 'clean' production technologies) is another mechanism through which globalisation may ultimately benefit the environment. For example, trade liberalisation can expand the potential market for low-emission vehicles. It can also improve access to environmentally preferable raw material inputs, such as low-sulphur coal. Conversely, trade can also make some environmentally harmful products or technologies more accessible (for instance, hazardous substances), with negative consequences for the environment. Pollution-intensive trade flows in some countries *have* changed significantly in recent decades. However, these

[18] OECD, 1998a; OECD, 2001b.

shifts seem to have been due more to structural changes in the economy than to differences in environmental policies from country to country. For example, Finland and Austria maintain relatively high environmental standards *and* a high share of environmentally sensitive goods in their exports. It may, therefore, be that economic policies are changing the pollution-intensity structure of trade flows, but it does not seem that the stringency of national environmental policy regimes is necessarily at the root of these changes. At a more disaggregated level, there does appear to be some connection between higher levels of environmental stringency and reduced comparative advantages in the exportation of pollution-intensive goods. But here again, the data is not conclusive. For example, Germany (a relatively high-standard country) has maintained its economic advantage in the chemicals and metals industries, both of which have the highest environmental compliance costs within the manufacturing sector.

Another important issue is the effect of globalisation-induced transportation activities on the environment. Trade liberalisation and globalisation-related structural reforms in the transport sector are both likely to lead to lower freight costs across most transport modes in most countries. Even if transport costs happen to rise in the short term (for example, in Central and Eastern Europe), the longer-term pressure on costs is still likely to be downwards. Reduced prices, combined with the increased incomes that should result from a more efficient transport system, are, in turn, likely to result in new demands for transport services, which will increase pressure on the environment (noise, air pollution, congestion). These scale effects will be felt more strongly in some countries than in others (such as Alpine transit countries). A recent study estimated that, even after full implementation of all Uruguay-Round commitments, these effects would lead to an increase in the global volume of internationally traded goods of only some 3-4 per cent, with the volume of international freight transport increasing only slightly more (4-5 per cent).[19] This increase was then compared to macro-economic projections of total increases in transport to 2004 – the year that the Uruguay Round commitments should be fully implemented, which suggested a 71 per cent increase over 1992 levels. This is approximately 15 times the growth expected to result from trade liberalisation itself. In other words, the scale effect of trade liberalisation itself – a surrogate for globalisation – may be relatively small compared with the additional transport required to service 'normal' economic growth.

[19] OECD, 1998a.

The basic globalisation-environment linkage can, therefore, be summarised along the following key axes:[20]

- The linkage works in both directions. The economic changes brought about by globalisation can have an impact on the environment, but changes in environmental conditions can also have an impact on the economy. Both environmental *and* economic policies may, therefore, play a role in modifying the relationship.
- The linkage may lead to either positive or negative environmental results. It is the net environmental effects which are ultimately of most interest.
- General equilibrium effects are important. The net environmental effects of globalisation will depend on the aggregate interaction of any scale, technology, composition, product, and regulatory changes that are induced. The regional and/or sectoral aspects of these effects may be quite different from effects on countries or the impact at the level of the global economy.
- The effects of economic globalisation on the global environment are only part of the story. Local and regional environmental impacts are also important; perhaps even more so, given their typically larger role in forming political reactions to environmental problems.
- The time dimension is crucial. Some elements of a more globalised economy may initially seem benign for the environment, but the opposite may be true when these elements are viewed in a more dynamic context. Given the long-term nature of environmental problems, this is an important distinction. For example, as physical environmental constraints begin to 'bite' (as pollution expands or natural resources become more scarce), private economic costs may increase, making it more profitable for firms to reduce emissions on their own, even in the absence of government policies.
- Domestic environmental policies are not necessarily the same as domestic policies that affect the environment. Environmentally relevant policies do not originate only in Environment Ministries; they also include policies that originate in other departments which affect the environment (for instance, tax policy).

In conclusion, although not enough is known yet about the precise ways in which globalisation affects the environment (or vice versa), at least the news is not all bad. There is enough research suggesting that the effects of globalisation could be positive or neutral for the environment that it would seem unwise to assume the contrary from the outset. Furthermore, where negative effects are expected, the reason probably has more to do with failings in environmental policies than in economic ones. Therefore,

---

[20] OECD, 1998b.

globalisation is more likely to increase the need for good environmental policies than to reduce it.

ENVIRONMENTAL EFFECTS OF POLICIES RELATED TO FDI[21]

This section presents a summary of some recent OECD work aimed at understanding the way in which national policies affect one part of the globalisation process – the desire to attract new FDI. The focus is on two key issues: (i) whether national policies to enhance environmental regulatory regimes drive investment towards 'pollution havens' in a 'race to the bottom' dynamic; and (ii) whether the desire to attract new FDI inhibits the development of more proactive, stringent environmental regulations ('regulatory chill' effect), or encourages improvements in environmental regulations ('race to the top' effect). Implicitly, therefore, the question of how domestic environmental policies may be affected by some of the pressures related to economic globalisation is also addressed.

Race to the Bottom

No formal econometric models support the argument that *governments* will act contrary to their interests and compete for investment by adopting lower domestic environmental standards in order to attract investment.[22] For one thing, it is difficult to ascertain why host countries set low environmental standards in the first place.[23] Nevertheless, some anecdotal evidence exists to suggest that this practice does occur in some contexts. For example, Mabey and McNally found that a British company pressured regional authorities in India to de-notify one of India's three designated eco-fragile areas, so they could proceed with a port development.[24] Esty and Gentry reported that foreign energy companies in China have felt pressure to reduce environmental standards in order to satisfy local authorities that desired the lowest possible price for power generation.[25]

Viewed from the perspective of *investors* (i.e., not governments), environmental costs usually represent only a small percentage of firm costs.

[21] Much of the material for this section is based on OECD, 2001c. See also OECD, 1997b. Space considerations prohibit discussion of the effects of globalisation on domestic ('greenfield') investment – the focus here is, therefore, on international investment activities (e.g., FDI).
[22] Revesz, 1992.
[23] Neumayer, 2001.
[24] Mabey and McNally, 1999.
[25] Esty and Gentry, 1997.

Other factors are, therefore, usually more important in firms' investment-location decisions, including the availability of cheap labour,[26] significant natural-resource endowments,[27] the presence of an adequate infrastructure, the presence of a suitable industrial base,[28] market size,[29] and concerns about liability or consumer pressures in home countries.[30] Some authors even contend that investors will refrain from investing in countries that have lax environmental regulations.[31] It may also be more efficient for multinational investors to apply home-country standards to all their operations, including their foreign ones.[32] The efficiency and stability of regulatory frameworks, rather than of their environmental dimensions in particular, may be the more important determinants of investment-location decisions.[33]

On the other hand, there are some reports of 'pollution haven' behaviour in specific sectors or industries. Industries that face additional costs – due to public relations problems, the establishment of environment divisions in firms, liability and insurance claims, legal fees, occupational health and safety costs, and delays/costs resulting from EIA procedures – are the ones most likely to be affected by this issue.[34] These criteria are especially applicable to industries involved in activities related to hazardous substances. For example, Xing and Kolstad found some US chemical companies migrating to developing countries because of the lower stringency of environmental regulations governing sulfur-dioxide emissions in those countries.[35] Other examples of the industrial flight of heavily polluting sectors have been found in the case of asbestos tile and benzidine dye manufacturing facilities relocating to Mexico and Romania,[36] wood finishing firms moving to Mexico from California,[37] and wet processing in the tanning industry moving from Europe to Brazil.[38] When examining the pollution-haven hypothesis across the whole economy, mixed results also sometimes emerge. For example, Jha et al. found that there was industrial flight to some sectors in Poland that were heavily regulated in the home countries, but that a race to the top (see below) was also developing in other sectors of the Polish economy.[39]

---

26 Esty and Geradin, 1998; Low and Yeats, 1992; Mani and Wheeler, 1998.
27 Low, 1993.
28 Mani and Wheeler, 1998.
29 Wheeler and Mody, 1992.
30 Esty and Geradin, 1998; UNCTAD, 1999.
31 Revesz, 1994.
32 Esty and Geradin, 1998; Nordström and Vaughan, 1999; UNCTAD, 1993.
33 Markandya, 1999.
34 Clapp, 1998.
35 Xing and Kolstad, 1998.
36 Cairncross, 1990.
37 GAO, 1990.
38 Mabey and McNally, 1999.
39 Jha et al.,1999.

Race to the Top

The notion of regulatory chill asserts that countries will refrain from adopting more ambitious environmental standards at home than they otherwise might have, because they fear losing international competitiveness.[40] As in any case involving a counterfactual argument, it is difficult to provide analytical evidence that regulatory chill effects actually occur, but anecdotal data can be used to help gauge its potential significance (see also Box 2.1). For example, Esty and Geradin cited various examples of OECD governments' failing to enact energy tax regimes or other greenhouse gas-reduction measures, at least in part because of competitiveness concerns.[41] Mabey and McNally asserted that Brazilian tanneries specialising in low-quality products are trapped between being unable to compete with better-quality competitors in Europe, and lower-cost producers in Asia.[42]

Box 2.1 Export processing zones

An export processing zone (EPZ) is a specified area in which domestic policy instruments apply that are not generally applicable to other parts of the country. Recent estimates suggest that there are more than 850 EPZs worldwide.[43]

In studies examining the environmental impacts of FDI in EPZs, it has generally been concluded that free zones have caused few pollution problems, because the new facilities are often cleaner in environmental terms than their domestic counterparts. Sivalingam even found that there had been favourable effects on the Malaysian natural environment because the EPZs were well planned and built in accordance with the existing local building regulations.[44] Enterprises locating in these zones were also usually required to provide a green zone, unlike companies in the rest of the country, where backyard industries were often set up in residential areas, and which can be highly polluting.

On the other hand, there are some cautionary views here as well. For example, there have also been reports of pollution associated with the leather tanning industry in an EPZ in Mauritius.[45]

---

40 Nordström and Vaughan, 1999.
41 Esty and Geradin, 1998.
42 Mabey and McNally, 1999.
43 Ge, 1999.
44 Sivalingam, 1994.
45 UNIDO, 1999.

On the other hand, there are also reports of situations in which investment has actually improved the standard of environmental regulation. This is consistent with the 'Porter hypothesis', which asserts that stronger environmental policies can improve countries' competitiveness by fostering innovation and efficiency.[46] For example, China actively prefers investments in coal-fired power plants that apply clean combustion technologies.[47] Israel has also apparently adopted EU pesticide standards in order to gain increased access to EU markets.[48] Joint ventures have emerged in Indonesia to exploit opportunities for the infusion of chlorine-free processing in the pulp and paper industry.[49]

Owing to their central role in key investment decisions, multinational enterprises are likely to be key vectors for any efforts aimed at promoting a race to the top in environmental standards. Studies tend to find that multinational companies upgrade the environmental standards used at their plants, regardless of location.[50] The main areas of improvement occur in technology and environmental management systems. What then tends to happen is that a new-source bias emerges, whereby environmental procedures used in modern plants are more rigorous than those applied to older plants.[51] Dasgupta et al.[52] found that companies improve their environmental performance in response to more effective national regulations. Local community pressures seem to play a role as well.[53] However, some authors have also emphasised that stronger environmental performance may be more attributable to the transfer of new technology and facilities than to stricter environmental procedures per se.[54] Others caution that the infusion of cleaner technology may lead to more intensive production methods, which may be unsustainable over the long term.[55]

In summary:

- Governments do not generally lower environmental standards in order to attract new investments. On the contrary, countries which operate transparent and efficient environmental programmes are often quite successful in attracting new investments.

---

[46] Porter, 1999.
[47] Goldenman, 1999.
[48] Vogel, 2000.
[49] UNCTAD, 1999.
[50] Vogel, 1997; Vogel, 2000; Weiss, 1993.
[51] Jaffe et al., 1995.
[52] Dasgupta et al., 1998.
[53] Pargal and Wheeler, 1995; Wheeler et al., 1997; Zarsky, 1999.
[54] Dasgupta et al., 1998.
[55] Zarsky, 1999.

- Nor is there much evidence of 'dirty' firms moving to countries with lower environmental standards for reasons related to the ambition level of environmental policy. Some firms do move from high- to low-standard jurisdictions, but the reason they do so typically has little to do with the level of environmental stringency existing in either source or host countries.
- Most pollution-intensive FDI originating in industrialised countries flows towards other developed countries, rather than to developing ones. Even in the developing countries, the amount of inward investment in pollution-intensive industries is often a small proportion of total FDI receipts.
- Foreign investors sometimes bring modern technologies with them that improve local environments. Multinational firms frequently build state-of-the-art facilities with the latest (low-pollution) technologies. They also tend to employ advanced environmental management systems. Partly for these reasons, foreign investors are more likely to meet the environmental standards of the countries in which they operate than are domestic companies.
- On the other hand, there is evidence that some types of firm – especially 'resource-seeking' firms facing above-average pollution control costs – *will* invest abroad in order to take advantage of lower environmental costs. For these firms, outputs are relatively undifferentiated, so small price differences can translate into large changes in market share or profitability. Investment flows towards these industries may, therefore, be particularly susceptible to differences in environmental standards.
- Pollution havens seem to be associated more with protectionist economies than with environmentally tolerant ones. If anything, the available evidence suggests that the imposition of higher environmental standards tends to generate a technological response, rather than leading to capital flights.

GLOBALISATION AND NATIONAL ENVIRONMENTAL POLICIES

International trade and investment contribute to long-term economic growth and development, and thereby provide a solid foundation for the ability to achieve environmental goals. However, achieving these goals implies that environmental policies need to be designed and implemented so as to (i) encourage appropriate internalisation of environmental externalities, and (ii) promote coherence between environmental policies, on the one hand, and trade/investment liberalisation policies, on the other. Domestic environmental policy can help in both of these areas,[56] as the following suggestions show:

1. Governments should reform existing domestic policies that are both trade/investment-distorting and environmentally damaging. For example, certain forms of economic subsidy both intensify pressures on environmental resources and reduce the competitiveness of foreign products produced using the environmental resources that are being subsidised. Reform of these subsidies would usually represent a 'win-win' opportunity for both the environment and economic growth.
2. Governments should work to better understand the potential environmental impacts of trade and investment liberalisation. Although prior assessment of the environmental impacts of specific development projects is already required in most OECD countries, the application of Environmental Impact Assessment (EIA) procedures to trade and investment liberalisation initiatives is relatively new. Some OECD countries are already committed to undertaking environmental (sustainability) reviews of WTO[57] and regional/bilateral trade agreements, in order to identify those trade measures most likely to exacerbate environmental pressures, or areas where further liberalisation could yield additional environmental benefits. The OECD has also developed methodologies for carrying out these assessments. However, more needs to be done, both to help explain the environmental impacts associated with globalisation and to help environmental policy focus on the most environmentally sensitive sectors/activities.
3. Governments should use their own policy levers to encourage multinational producers based in their countries to internalise their environmental externalities. This could involve either regulatory approaches or market-based approaches (environmental taxes, tradable permits), but the ultimate goal should be to achieve the right level of environmental protection at the lowest cost.
4. Governments should also support private-sector markets in encouraging global producers/investors to internalise their environmental externalities, wherever they may be operating. There are already more than 60,000 multinational corporations, with more than 800,000 affiliates abroad, driving the global expansion of investment flows.[58] An even larger number of domestic firms has strong interests in opportunities for foreign trade. Formal regulation by government is, therefore, not the only tool capable of affecting corporate environmental conduct. More informal approaches (such as

---

[56] See also Objective 5 in OECD, 2001d.
[57] The World Trade Organization promotes free trade at the global level through multilateral agreements.
[58] UNCTAD, 2001.

voluntary codes of conduct, shareholder activism, and environmentally conscious funds) are examples. Each of these approaches seeks to influence environmental behaviour largely through market-driven forces. However, government policy can also encourage these activities, even if they do not depend at root on government action. For example, the OECD Guidelines for Multinational Enterprises represent a government effort to encourage (voluntary) private-sector practices that are compatible with environmental objectives.[59]

5. Governments should actively design domestic environmental policies so as to be consistent with accepted disciplines on trade and investment liberalisation at the international level. For example, a wide range of national product policies are used to support environmental goals, including eco-labelling, extended producer responsibility, and 'green' public purchasing. Each of these policies may generate adverse impacts on international trade or investment activities. For example, trade concerns may arise when eco-labels are applied in a manner that protects – even inadvertently – domestic producers to the detriment of foreign ones. In order for these policies to avoid becoming disguised market barriers, they need to be non-discriminatory, transparent, involve widespread consultation, and be non-protectionist in intent.

6. Governments need to do a better job of explaining the linkages between environmental protection and economic globalisation to their constituents. Three misconceptions in particular may be relevant in this context:

- There is a tendency to attribute more environmental ills to the globalisation process than is probably warranted. Environmental problems arise largely from deficiencies in environmental policies, rather than from globalisation per se.
- On the other hand, policies that support economic globalisation also need to recognise that improving environmental quality is often welfare-enhancing, implying that policies which have the effect of reducing the competitiveness of individual firms or sectors are not always inappropriate.
- The negative effects of environmental policies on competitiveness or economic growth may be exaggerated in any event. Sometimes, these effects are not negative in the first place; sometimes, they are not significant, even where they are negative. There may even be situations in which environmental policies are beneficial for competitiveness, via the dynamics of technology development.

[59] OECD, 2000.

REFERENCES

Cairncross, F. (1990), 'Cleaning up: A survey of industry and the environment', *The Economist*, **24**, 1-26.

Clapp, J. (1998), 'Foreign direct investment in hazardous industries in developing countries: Rethinking the debate', *Environmental Politics*, **7**(4), 92-113.

Dasgupta, S., H. Hettige, and D. Wheeler (1998), *What Improves Environmental Performance? Evidence from Mexican Industry*, Policy Research working paper 1877, Washington, DC: World Bank Development Research Group.

Esty, D. and B. Gentry (1997), 'Foreign investment, globalisation and environment', in OECD, *Globalization and Environment; Preliminary Perspectives*, Paris: OECD.

Esty, D. and D. Geradin (1998), 'Environmental protection and international competitiveness', *Journal of World Trade*, **32**(5), 5-46.

GAO (1990), *Report on the Furniture Finishing Industry*, Washington, DC.

Ge, W. (1999), *The Dynamics of Export Processing Zones*, UNCTAD Discussion Papers, Geneva: UNCTAD.

Goldenman, G. (1999), 'The environmental implication of foreign direct investment: Policy and institutional issues', in OECD, *Foreign Direct Investment and the Environment*, Paris: OECD.

Jaffe, A., S. Peterson, P. Portney, and R. Stavins (1995), 'Environmental regulation and the competitiveness of U.S. manufacturing: What does the evidence tell us?', *Journal of Economic Literature*, **33**(1), 132-163.

Jha, V., A. Markandya, and R. Vossenaar (eds) (1999), *Reconciling Trade and the Environment: Lessons from Case Studies in Developing Countries*, Cheltenham: Edward Elgar.

Low, P. (1993), 'The international location of polluting industries and the harmonization of environmental standards', in H. Munoz and R. Rosenberg (eds), *Difficult Liaison – Trade and the Environment in the Americas*, New Brunswick: Transaction Publishers.

Low, P. and A. Yeats (1992), 'Do dirty industries migrate?', in P. Low (ed.), *International Trade and the Environment*, Washington, DC: World Bank.

Mabey, N. and R. McNally (1999), *Foreign Direct Investment and the Environment: From Pollution Haven to Sustainable Development*, Surrey: WWF UK.

Mani, M. and D. Wheeler (1998), 'In search of pollution havens? Dirty industry in the world economy (1960-1995)', *Journal of Environment and Development*, **7**(3), 215-247.

Markandya, A. (1999), 'Overview and lessons learnt', in V. Jha, A. Markandya, and R. Vossenaar (eds), *Reconciling Trade and the Environment: Lessons from Case Studies in Developing Countries*, Cheltenham: Edward Elgar.

Neumayer, E. (2001), 'Pollution havens: An analysis of policy options for dealing with an elusive phenomenon', *Journal of Environment & Development*, **10**(2), 147-177.

Nordström H. and S. Vaughan (1999), *Trade and Environment: Special Studies 4*, Geneva: World Trade Organization.

OECD (1997a), *Economic Globalisation and the Environment*, Paris: OECD.

OECD (1997b), *Foreign Direct Investment and the Environment; An Overview of the Literature*, Paris: OECD.

OECD (1998a), *Open Markets Matter: The Benefits of Trade and Investment Liberalisation*, Paris: OECD.

OECD (1998b), *Globalisation and the Environment: Perspectives from OECD and Dynamic Non-Member Countries*, Paris: OECD.

OECD (2000), *The OECD Guidelines for Multinational Enterprises*, Paris: OECD.

OECD (2001a), *The Environmental Benefits of Foreign and Direct Investment: A Literature Review*, Paris: OECD.

OECD (2001b), *Sustainable Development: Critical Issues*, Paris: OECD.

OECD (2001c), *Environmental Issues in Policy-Based Competition for Investment: A Literature Review*, Paris: OECD.

OECD (2001d), *OECD Environmental Strategy for the First Decade of the 21st Century*, Paris: OECD.

OECD (2002a), *AFA Database*, Paris: OECD.

OECD (2002b), *TBP Database*, Paris: OECD.

OECD (forthcoming 2004), *Indicators of Economic Globalisation*, Paris: OECD.

Pargal, S. and D. Wheeler (1995), *Informal Regulation of Industrial Pollution in Developing Countries: Evidence from Indonesia*, Washington, DC: World Bank.

Porter, G. (1999), 'Trade competition and pollution standards: 'Race to the bottom' or 'stuck at the bottom?', *Journal of Environment and Development*, **8**(2), 133-151.

Revesz, R. (1992), 'Rehabilitating interstate competition: Rethinking the 'race to the bottom' rational for federal environmental regulation', *New York University Law Review*, **67**, 1210-1254.

Revesz, R. (1994), *Trade and Sustainable Development*, UNEP Environment and Trade Series, 1, Geneva: UNEP.

Sivalingam, G. (1994), *The Economic and Social Impact of Export Processing Zones: The Case of Malaysia*, ILO working paper, (66), Geneva: ILO.

UNCTAD (1993), *Environmental Management in Transnational Corporations*, New York: UNCTAD.

UNCTAD (1999), *Foreign Portfolio Investment and Foreign Direct Investment: Characteristics, Similarities and Differences – Policy Implications and Development Impact*, Geneva: UNCTAD.

UNCTAD (2000), *World Investment Report*, Geneva: UNCTAD.

UNCTAD (2001), *World Investment Report*, Geneva: UNCTAD.

UNIDO (1999), *Export Processing Zones: Principles and Practice*, Vienna: UNIDO.

Vogel, D. (1997), 'Trading up and governing across: Transnational governance and environmental protection', *Journal of European Public Policy*, **4**(4), 556-571.

Vogel, D. (2000), 'Environmental regulation and economic integration', *Journal of International Economic Law*, **3**(2), 265-280.

Weiss, E. (1993), 'Environmentally sustainable competitiveness: A comment', *Yale Law Journal*, **102**(2123).

Wheeler, D. and A. Mody (1992), 'International investment location decisions: The case of US firms', *Journal of International Economics*, **33**, 57-76.

Wheeler, D., R. Hartman, and M. Huq (eds) (1997), *Why Paper Mills Clean Up: Determinants of Pollution Abatement in Four Asian Countries*, Washington, DC: World Bank.

Xing, Y. and C. Kolstad (1998), *Do Lax Environmental Regulations Attract Foreign Investment?*, Economics working paper no. 6-95R, Santa Barbara: University of California.

Zarsky, L. (1999), 'Havens, halos, and spaghetti: Untangling the evidence about foreign direct investment and the environment', in OECD, *Foreign Direct Investment and the Environment*, Paris: OECD.

3. Globalisation and National Incentives for Protecting Environmental Goods: Types of Goods, Trade Effects, and International Collective Action Problems[1]

Alkuin Kölliker

SUMMARY

I address national incentives for protecting environmental goods in conjunction with the impact of economic globalisation. The central argument is that national incentives for environmental protection can be explained by combining the type of environmental goods to be protected (private, public, club, or common pool goods) with the effect of protection measures on international competitiveness (positive, neutral, or negative impact). The combination of the centripetal effects of different environmental goods (which are highest for club goods and lowest for common pool resources) and the competitive impact from environmental protection results in twelve possible outcomes of national incentives for environmental action, which are illustrated with empirical examples. The influence of globalisation and some of its driving forces on national incentives and legal possibilities for environmental protection are also discussed. I conclude with a brief discussion of four options for (re-)expanding the action space for national environmental policies in a global economic arena.

[1] I am indebted to Frank Wijen for his valid and constructive comments on an earlier version of this contribution. A number of other participants in the conference 'Globalisation and National Environmental Policy' (Veldhoven, 22-24 September 2003) also provided helpful suggestions. Finally, I would like to thank Christoph Engel and Frank Maier-Rigaud from the Max Planck Institute for Research on Collective Goods for a number of important additional remarks which helped me to revise the original draft.

INTRODUCTION

National Incentives for Environmental Protection

While nations may have similar preferences with regard to various environmental goods, their incentives to actually protect those goods differ from case to case. The result may be individual action (in the framework of national environmental policies), collective action (in the framework of international environmental agreements), or inactivity. If international environmental agreements already exist, participant countries face incentives to either stay or leave, while non-participants may have incentives to either remain outside or join. New entries and sustained participation give evidence of the centripetal effects of international environmental agreements, while exit threats and continued abstention can be interpreted as signs of centrifugal effects. Centripetal effects refer to the utility of participation in collective action, as opposed to abstention.[2]

A prominent example illustrates how the incentive structures for nations (or groups of nations) may vary across issue-areas and individual cases. The European Union (EU) has taken environmental action both by participating in the Kyoto Protocol on climate change, and by banning meat and meat products treated with growth-promoting hormones. The United States (US), by contrast, has withdrawn from the Kyoto Protocol and allows the use of growth-promoting hormones. Incentives for environmental action, however, appear to be entirely different in the two cases: in the case of the Kyoto Protocol, the EU has actively and almost desperately tried to bring in the US. In the case of the hormone ban, the EU did not push the US to adopt an equivalent policy, but faced strong US resistance to its own policy.

Which Factors Influence National Incentives?

The comparison between these two cases leads to a puzzle. Why is the EU unhappy about US abstention in the case of the Kyoto Protocol, but indifferent to its policy in the hormone case? And why is the US upset about EU policies in the hormone case but not in the Kyoto case? Two general questions which arise in this context are the following: which factors influence the initial choices (autonomous action, collective action, no action) of countries with regard to environmental protection in a given area? And what explains the centripetal or centrifugal effects of international environmental agreements once they are established? Theories on the provision of collective goods suggest that the incentives for actors to protect

[2] A more precise definition of centripetal (and centrifugal) effects follows in the next section.

environmental goods are influenced by the character of the goods in terms of public goods theory. According to international political economic theory, moreover, the trade effects of environmental protection may further influence these incentives. In addition to further developing and integrating these two strands of literature, the primary objective of this study was to give a theoretical answer (and some empirical illustrations) to the following, more specific questions: how do the nature of environmental problems and the trade effects of the solutions to these problems affect the incentives for countries to act individually, collectively, or not at all? How do these two explanatory factors influence the centripetal and centrifugal effects of international environmental agreements? And what is their combined impact on the difficulties of autonomous national action as well as on the severity of international collective action problems?

The answers to the previous questions help in finding a solution to a concluding question, relating to globalisation and national environmental policies. If globalisation is characterised by rapidly expanding trade and driven by, among other factors, international free-trade agreements, how do these factors affect the incentives and the legal possibilities of nations to protect environmental goods alone or in cooperation with other countries?

Types of Goods and Trade Effects of Environmental Protection

The focus of this study was thus on the possibilities and limits of national environmental policies, on the perspectives for the creation of international environmental agreements, and on the centripetal or centrifugal effects of such agreements on potential new participants. In this context, it is important to note that the research concentrated on collective action among countries and their governments, not among individuals, firms, or other non-state actors. Therefore, I investigated the character of environmental goods, as well as the competitive effects of environmental protection, exclusively from the viewpoint of countries as a whole. The latter were, hence, treated as the basic units of analysis.[3]

The theoretical analysis presented in this contribution is in three steps. The first two steps (section 2), focus on the impact of the character of environmental goods while neglecting trade effects. In the first step, reflecting insights from the existing literature on collective goods, it is argued that the incentives for national environmental protection are strongest in the case of environmental goods which represent predominantly private goods from the perspective of the countries involved (e.g., noise reduction, protection of the

---

[3] An example may underline the importance of this distinction: while national defence is a (non-excludable) public good for individual citizens, it is an (excludable) private good from the perspective of countries as a whole.

soil, protection of lakes). All other categories of environmental goods provide certain incentives for countries to engage in collective action, which explains why non-private goods are also referred to as collective goods. The severity of collective action problems, however, depends on the precise category of the good. In the second step, the attractiveness (i.e., the centripetal effects) of environmental agreements for potential participants is ranked, depending on the character of the environmental good to be protected. The ranking of four types of goods according to their centripetal effects was derived from a 'theory of differentiated integration' developed elsewhere and applied to various EU policies.[4] According to this new theory, centripetal effects are strongest in the case of club goods, followed by private goods and public goods. Common pool resources tend to lead to centrifugal rather than centripetal effects and, therefore, rank at the bottom.[5] In the third step (section 3), these type-of-good effects on national incentives are combined with the trade effects of environmental protection measures. Incentives for countries to engage in environmental protection depend not only on the character of environmental goods, but also on the impact of environmental protection measures on the international competitiveness of the domestic industry. In the analysis, I distinguish positive, negative, and neutral effects on competitiveness. Trade effects may, therefore, combine with type-of-good effects to either alleviate or exacerbate the collective action problem individual countries face when trying to protect environmental goods. As a result, centripetal effects are either weakened (in the case of competitive disadvantages) or reinforced (in the case of competitive advantages).

Significance, Empirical Illustrations, and Implications of Globalisation

The significance of the outlined theoretical analysis lies first of all, and at a general level, in the combination of the type-of-good effects with the trade effects of environmental protection. Both effects are important for understanding national incentives for environmental protection. Neglecting one of them may lead to the wrong conclusions. But together, the two factors may explain a substantial part of the variance in national incentives with regard to various environmental protection measures. A second, more specific, innovation presented in this contribution consists in the ranking of environmental goods according to the centripetal effects of international arrangements for their protection. This ranking provides a new measure for the severity of collective action problems. All in all, this contribution may

4 Kölliker, 2001a, 2002.

5 The ranking provided by the differentiated integration theory includes two further categories of goods, namely, excludable and non-excludable network goods. The reasons for not discussing them in the present contribution are explained below.

serve policy-makers as a rough guide to the incentive structures nations face in different fields and with regard to different instruments of environmental protection. This, in turn, may help nations to make the right institutional and policy decisions to improve environmental quality effectively and efficiently through either autonomous national action or collective international action. To the environmental research community, this chapter may provide a theoretical framework of analysis that permits broad comparisons, and hence learning, across a wide range of sub-fields of environmental protection.

The theoretical considerations on types of environmental goods and trade effects could solve the puzzle of diverging incentives for the EU in the cases of the Kyoto Protocol and the hormone ban. Those and other cases of environmental problems are used as empirical examples for various combinations of different types of goods and trade effects (section 4). It is argued that the Kyoto Protocol involves an international public good, while the hormone ban represents a private good for the countries involved. This explains why the Kyoto Protocol involves weaker centripetal effects than does the hormone ban. Trade effects reinforce these tendencies, because participation in the Kyoto Protocol may in some areas bring competitive disadvantages, while the hormone ban – acting as a barrier to imports – results in competitive advantages. In short, while the Kyoto Protocol combines a public good with competitive disadvantages, the hormone ban combines a private good with competitive advantages.

In the final section, I discuss some implications of the outlined theoretical mechanisms with regard to globalisation, national environmental policies, and the instruments of international environmental cooperation. I assess (1) how globalisation and its driving forces affect national incentives and the legal possibilities of environmental protection, (2) how countries can ensure the effectiveness of national environmental policies under the conditions of globalisation, and (3) under which conditions flexible, voluntary, and non-binding instruments may be useful in this context.

TYPES OF GOODS AND INCENTIVES FOR ENVIRONMENTAL PROTECTION

The value attached to environmental goods is a necessary but not sufficient condition for countries and their governments to take measures to protect them. National governments also need incentives to help them to transform preferences into action. A key objective of this study was to contribute to a better understanding of national incentives for environmental protection.

Explaining National Incentives for Environmental Protection

Why should better knowledge of the incentives of national governments be important for improving environmental quality? Knowing about incentives may help in identifying environmental measures likely to be adopted and implemented effectively within a given institutional framework. It may also help in shaping institutions, for example, to create the necessary incentives for specific environmental measures. I focus on two factors exerting a particularly powerful impact on the incentives for protecting environmental goods. The first factor, discussed in this section, is the nature of environmental goods in terms of public goods theory. The second factor, addressed in the next section, is the side-effects of environmental protection measures on competitiveness and trade. As pointed out in the introduction, both the nature of environmental goods and the trade effects are considered from the viewpoint of countries as a whole and the governments representing them, not from the perspective of individuals, firms, or other non-state actors.

This section proceeds as follows. I first present a classification of goods such as provided by public goods theory. Then follow some empirical examples of different types of international environmental goods. Next, I clarify which types of environmental goods countries can best protect autonomously, and for which goods there are incentives to act collectively. This can be established by comparing the gains for countries when acting autonomously (option 1) and the gains when taking part in collective action (option 2). However, because it fails to take into account the possibility of free-riding, this comparison does not allow proper assessment of how attractive it is for countries to actually participate in collective action. Therefore, the gains for countries when participating in collective action (option 2) are compared with the gains for them when abstaining from collective action established by other countries (option 3). Afterwards, I present a ranking of the centripetal effects of different types of goods. Finally, the findings are summarised with regard to the severity of international collective action problems.

Public Goods Theory and Types of Environmental Goods

Public goods theory uses two criteria to classify goods: excludability and rivalry in consumption.[6] It argues that the incentives for actors to provide goods depends on the types of goods that result from specific combinations of excludability and rivalry in consumption. Applied to the field of the

[6] The current economic literature on public goods originates in a seminal piece by Samuelson, 1954. For an authoritative overview of public goods theory, see Cornes and Sandler, 1996.

environment and to the level of nations, this means that the incentives of countries to protect environmental goods are influenced by the character of the goods in terms of public goods theory. In fact, international environmental agreements are often justified by arguing that environmental protection represents an international public good which requires collective international action, since individual countries lack the incentives to act autonomously.

In this study, I identify in detail the incentives for countries and their governments to *initiate* new environmental protection measures – either autonomously or collectively – or to remain inactive. Equally important, I identify the incentives to *join* already existing environmental protection measures, either by emulating unilaterally and autonomously the policies of other countries, or by signing up to international environmental agreements. To make the argument more easily understandable, as well as to underline its practical relevance, the illustrative examples are closely integrated with the presentation of the theoretical argument.

In order to classify environmental goods according to public goods theory, two questions must be answered. First, can countries (or groups of countries) protecting an environmental good exclude other countries from using that good? Second, is the utility that countries can derive from environmental protection measures affected if other countries can profit equally from the measures? The two questions are, in other words, whether an environmental good is excludable, and whether it involves rivalry in consumption. Environmental goods can be either excludable or non-excludable, and consumption can be either rival or non-rival. Combining the possible levels of excludability and rivalry in consumption, therefore, leads to four ideal types of goods (see also Figure 3.1). Private goods are characterised by excludability and rivalry in consumption, while public goods are non-excludable goods with non-rival consumption. Club goods combine excludability with non-rival consumption, and common pool resources are defined by non-excludability and rival consumption.[7]

Public policies can aim either at providing goods or at avoiding 'bads'. Providing a good increases the utility for actors compared with the status quo, while preventing the provision of a bad avoids reducing their utility. In analytical terms, these two cases are treated as symmetrical and equivalent, and no distinction is made between environmental policies improving environmental quality and environmental policies preventing a further

---

[7] Strictly speaking, club theory assumes rivalry in consumption to start as soon as the number of participants crosses a certain threshold. Since I am interested here in the 'pure' types of the four goods, it is assumed that consumption is non-rival, independently of the number of participants. Alternatively, it may be assumed that we look only at club goods with a number of participants which does not cross the rivalry threshold. On clubs and congestion, see Cornes and Sandler, 1996.

deterioration thereof.[8] I generally refer to the former case as 'providing an environmental good', while the latter is referred to as 'protecting an environmental good'.

		Excludability	
		Excludable	**Non-excludable**
Rivalry	**Rival**	**Private goods** Timber	**Common pool resources** Sea fish
	Non-rival	**Club goods** Environmental technology	**Public goods** Global climate

Figure 3.1 Types of goods

Examples of International Environmental Goods and their Protection

Environmental goods, as well as measures to protect them, do not always neatly correspond to one of the four categories of goods. Many have a mixed character. But the overall character of different environmental goods varies, often considerably. Therefore, the predominant character of environmental goods can often be plausibly established, at least in relation to other environmental goods. The following examples, which are also used later, illustrate this.

Private goods. Timber represents a natural resource with the character of a private good from the perspective of countries. Rivalry in consumption exists, but countries can exclude each other from using this resource. Domestic waste collection equally represents a private good from the perspective of countries. Other countries are clearly excluded from the benefits of local waste collection. Moreover, those benefits are rival: the costs of waste collection increase with the area covered. Prohibition of the use of growth-promoting hormones in meat production can also be considered a private

[8] This does not mean, however, that actors necessarily have the same intensity of preferences for providing environmental goods and preventing the occurrence of environmental bads. Experiments conducted by psychologists and economists suggest that actors have a status quo bias and a loss aversion. Kahneman et al. (1991: 205) suggested that '[a] revised version of preference theory would assign a special role to the status quo, giving up some standard assumptions of stability, symmetry and reversibility which the data have shown to be false'.

good for the countries concerned. The EU justifies the ban of hormones with the health risks of these hormones for the European population, which creates a perceived 'private' benefit for EU countries.

Common pool resources. High-sea fisheries and the prevention of overfishing involves a resource which is characterised by rivalry in consumption. Moreover, countries cannot exclude each other from using the resource of sea fish. This good has, therefore, the character of an international common pool resource.

Public goods. Climate protection represents an international public good for countries, since the benefits of climate protection are non-rival and non-excludable at the same time. The benefits of preventing global warming do not depend on the number of countries profiting from it. At the same time, no country can be excluded from those benefits. Contrary to the prevention of overfishing in the more narrow sense, the reduction of by-catches can be considered an international public goods problem. While no country can be excluded from the benefits of reduced by-catches, such by-catches do not involve rivalry in consumption. This is because fishermen from different countries compete for marketable fish, not for unprofitable by-catches.

Club goods. Environmental research represents a club good for countries, provided that full access to research results is restricted to the participating countries. (Otherwise, the research results would represent an international public good.) Exclusionary mechanisms may include both (international) intellectual property law and secrecy. Consumption is non-rival, because environmental technology can be used to improve environmental quality in one country without diminishing its utility for doing the same thing in other countries.

Autonomous Action versus Participation in Collective Action

The incentives for countries and their governments to protect environmental goods autonomously are fully intact in the case of (national) private goods, and autonomous action (option 1) leads to a (Pareto-) efficient outcome. For all other categories of goods (i.e., non-private or collective goods), collective action (option 2) is potentially more efficient than autonomous action. The reason for this can be summarised as follows. For non-excludable goods (public goods and common pool resources), collective action is superior to autonomous action, mainly because positive externalities can be internalised, which leads to greater efficiency. For example, the incentives for climate protection (an international public good) are much stronger if a country acts collectively with other countries rather than autonomously. But collective action is also superior to autonomous action when there are no positive externalities, provided that there is no rivalry in consumption. This is the case for club goods. The reason for the superiority of collective action lies in the possibility of sharing the costs of a good for which the costs of an additional unit of consumption are nil (which corresponds to the definition of non-rival consumption).

Participation versus Abstention with Regard to Collective Action

While collective action potentially leads to more efficient outcomes with regard to non-private goods than does autonomous action, the incentives for individual countries to participate in collective action – as opposed to abstaining from it – are not necessarily given and may vary sharply for different types of goods. In earlier work on legal differentiation in the context of the EU, I established a theory of differentiated integration, which includes a ranking of incentives for individual countries to participate in collective action, depending on the type of good involved. While limited space does not allow the full theoretical argument to be presented here, I provide a summary of the most important points.[9]

The theory suggests that the centripetal effects of international collective action on initially non-participating countries are strongest in the case of club goods. Private goods rank second, while public goods follow in the third position. Centripetal effects are weakest – and centrifugal effects strongest – in the case of common pool resources. The starting point of the theoretical argument leading to this conclusion is a more precise definition of centripetal effects. According to this definition, centripetal effects correspond to the likelihood of the utility of participation in collective action for initial non-participants turning positive as the number of participants increases. In his analyses of collective action problems, Thomas Schelling used diagrams representing separately the utility of cooperation and the utility of non-cooperation (or defection) for individual actors, depending on the number of participants.[10] Centripetal effects can be depicted in modified 'Schelling diagrams', which combine the utility curves for participation with the utility curves for non-participation (see Figure 3.2). This allows direct comparison of options 2 and 3 introduced above.

---

[9] For a more complete representation of the theory of differentiated integration, see Kölliker, 2001a, 2002. The theory as originally presented takes into account six rather than four types of goods. This is because it subdivides the category 'non-rival consumption' into two separate categories ('neutral' and 'complementary' consumption). The two categories of goods characterised by complementary consumption (excludable and non-excludable network goods) were not included in this study. The underlying reason is that empirical examples of environmental goods with a network character are rare. In the narrow sense, they may even be non-existent.

[10] Schelling, 1978.

Using such modified Schelling diagrams, it can be shown that the net utility of participation – and hence the centripetal effects – depends on the type of good. In order to substantiate this claim, the utility curve has to be split again into several components. On the production side are the costs of providing the good incurred by the individual participants. On the consumption side are the various effects of the good on both participants and non-participants. In line with the terminology used by economists, the effects on non-participants are referred to as 'external effects' (or 'externalities'). Analogously, I call the effects on the participants themselves 'internal effects'. The effects on non-participants can be divided into positive and negative external effects (or externalities); a similar distinction can be made between positive and negative internal effects.[11]

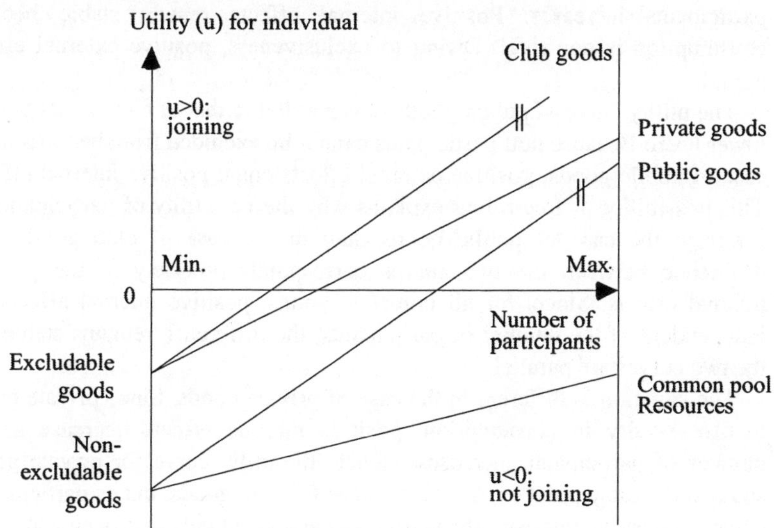

Figure 3.2 Types of goods and utility of participation

Ranking Types of Goods According to their Centripetal Effects

The development of cost curves, as well as of positive internal effects and positive external effects, depends on the type of good. This is because all of

[11] The following argument is based on the assumption that, independently of the number of participants, the *total* quantity of the good provided – rather than the quantity of the good or the cost of participation for each *individual* participant – is held constant. However, this is a presentational choice, which does not influence the conclusions with regard to the ranking of the different types of goods.

these elements are influenced by exclusiveness and rivalry in consumption. As a result, the utility curves combining production costs, and internal and external effects vary for different types of goods, which hence can be ranked according to their centripetal effects. Below, I summarise why and how the positions of the utility curves differ – under otherwise equal circumstances – for the four types of goods. I start with a discussion of club goods, continue with public goods and private goods, and conclude with common pool resources. In this rough summary, only the most important underlying assumptions are made explicit.

The utility curve of club goods slopes upwards, chiefly because an increasing number of participants allows the individual shares in the production costs to decrease. The other relevant components of the utility curve, positive internal and external effects, do not change as the number of participants increases. Positive internal effects remain stable because consumption is non-rival. Owing to exclusiveness, positive external effects are nil.

The utility curve of public goods runs parallel to that of club goods but at a lower level. Because non-participants cannot be excluded from benefits in the case of public goods, positive external effects equal positive internal effects. This possibility of free-riding explains why the net utility of participation is lower in the case of public goods than in the case of club goods. The difference between the two curves corresponds precisely to the positive internal effects. Since, for all non-rival goods, positive internal effects are independent of the number of participants, the difference remains stable and the two curves are parallel.

The situation is different in the case of private goods. Since private goods involve rivalry in consumption, positive internal effects decrease as the number of participants increases. Hence, the utiliy curve for private goods starts at the same point as the utility curve for club goods; but as participation expands, positive internal effects decrease and eventually approach zero. This is why, at a high number of participants, the utility curve for private goods approaches the utility curve of public goods.

The net utility of participation in the case of common pool resources is, in principle, similar to that in the case of public goods.[12] This is because, without introducing further modifications, the elements in the equations

[12] Both common pool resources and public goods are non-excludable goods. Therefore, positive external effects and positive internal effects are identical in the case of both common pool resources and public goods. The net utility of participation for each category of goods results from comparing the utility of participation with the utility of non-participation. In the equation defining the net utility of participation, positive internal and positive external effects cancel each other out (because they are, in principle, identical for all non-excludable goods). The result is that the equations for the utility curves of common pool resources and public goods are identical.

describing the net utility of participation for common pool resources and public goods are identical. But the special nature of common pool resources, characterised by a combination of non-exclusiveness and rivalry in consumption, leads to an effect which is often referred to as 'leakage'. This means that the benefits of cooperation may leak from the participants to the non-participants. Because they combine non-exclusiveness with rivalry in consumption, common pool resources involve such adverse incentive structures that collective action is mostly limited to the protection and common use of already existing resources, and does not usually involve the active production of resources.

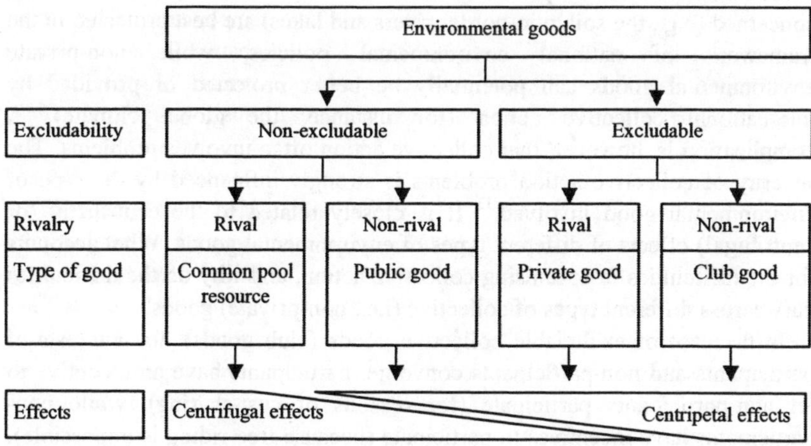

Figure 3.3 Types of goods and centripetal effects

This explains why the application of the concept of common pool resources to date has mostly been limited to natural resources, as opposed to man-made resources. Leakage tends to lead to situations in which, contrary to the situation with public goods, positive external effects exceed positive internal effects. This gap is small if only a few participants provide (or protect) the good, while many non-participant free-riders have to share the leaking benefits. But if increasing participation leaves few non-participants to free-ride and appropriate ever-growing shares of the leaked benefits, the net utility of participation in the case of common pool resources steadily decreases in comparison with that in the case of public goods. Hence, the utility curve for common pool resources runs below the utility curve for public goods, and the gap between the two increases with the number of participants. Figure 3.3 summarises the theoretical conclusions by representing the four types of (environmental) goods and their diverging centripetal effects in a different manner.

The Severity of Collective Action Problems

So far, it can be concluded that the character of environmental goods in terms of public goods theory helps us to understand under which circumstances countries can efficiently act alone, and under which circumstances international collective action is potentially more efficient. From the perspective of public goods theory, environmental resources that predominantly represent private goods from the perspective of the countries concerned (e.g., the soil, non-border rivers and lakes) are best protected in the framework of national environmental policies, while non-private environmental goods can potentially be better protected or provided by international collective action (for instance, the global climate). A complication is, however, that collective action often involves problems. The severity of collective action problems is strongly influenced by the type of environmental good involved.[13] It is closely related to the centripetal (or centrifugal) effects of different types of environmental goods. What accounts for the difficulties in organising collective action, and why do the difficulties vary across different types of collective (i.e., non-private) goods?

In the case of excludable collective goods (club goods), the interests of participants and non-participants converge: participants have an incentive to let non-participants participate (for reasons of cost-sharing), while non-participants have incentives to participate (because free-riding is impossible). The collective action problem here is, thus, not one of diverging interests, but one of transaction costs for the organisation of collective action, and one of finding acceptable cost-sharing arrangements.

In the case of non-excludable collective goods (common pool resources and public goods), collective action problems arise from diverging interests: non-participants are able to free-ride without sharing the costs of provision, while participants would like non-participants to pay an appropriate share of the provision costs. While the collective action problems with regard to production are similar for the two types of goods, they vary in respect of consumption. The reason for this is that different levels of rivalry in consumption can make free-riding either harmful or harmless with regard to the benefits of the good (while the free-riding problem with regard to cost-sharing remains in any case!). Free-riding in the case of rival consumption (common pool resources) is harmful, because goods consumed by free-riding

---

[13] While the literature on collective goods identifies the type of good as a key determinant of collective action problems, it also points to a number of other influential factors, such as the number and heterogeneity of actors (Keohane and Ostrom, 1994); see also Holzinger, 2002.

non-participants cannot be consumed by the participants. Countries enhancing their catches of sea fish while other countries limit their catches to prevent overfishing, provides an example of harmful free-riding in the case of a common pool resource. In the case of non-rival consumption (public goods), free-riding is harmless, because non-participants do not deprive the participants of the benefits of the good. Countries abstaining from the Kyoto Protocol clearly free-ride on the efforts made by the participant countries. But their free-riding is not harmful in the sense of depriving the participant countries from the expected benefits of reduced greenhouse gas emissions.

TRADE EFFECTS OF AND INCENTIVES FOR ENVIRONMENTAL PROTECTION

In the previous section, it was shown how the incentives of individual countries to protect environmental goods depend on the nature of these goods in terms of public goods theory. Type-of-good effects can theoretically be expected to have a strong impact on the capacity of countries to solve environmental problems either autonomously or collectively. But other influential factors exist. This section focuses on the effects of environmental protection on trade and competitiveness, which are then combined with the findings described in the previous section.

The Side-Effects of Environmental Protection on Trade

Like most other policies, environmental policies tend to have side-effects on other issue areas. Environmental policies may regulate products and their production, and thereby also affect the prices of production factors and products as well as the conditions for selling products. To the extent that national economies open up, environmental policies also influence trade and investment flows. Environmental policies thus can have effects on the competitiveness of domestic industries. As a result, governments may reconsider the costs and benefits of environmental policies, and hence their attitudes towards environmental protection measures. This is why it is important for environmental regulators to understand the side-effects of environmental policies on competitiveness. Environmental protection measures harming competitiveness discourage countries from acting autonomously and exacerbate collective action problems. Some scholars and practitioners point to this mechanism to justify strong international and/or European institutions for environmental action.[14] By contrast, protection

[14] See, for example, Moussis, 1998.

measures strengthening competitiveness may encourage autonomous action and facilitate collective action. But why do some environmental policies affect competitiveness negatively, while others have a positive impact or no effect at all? To answer this question, in the remainder of this section, the competitive advantages and disadvantages of environmental protection are briefly discussed, and the distinction between product and process standards, which plays an important role in this context, is explained. The potential combinations of type-of-good effects and trade effects are identified, and their ranking according to the severity of collective action problems is discussed. Empirical illustrations follow in the next section.

Competitive (Dis-)Advantages and Types of Standards

Environmental protection results in competitive disadvantages for the domestic industry if it leads to higher costs of production factors and products. Environmental regulation may concern either production processes or products. Both types of regulation impose additional costs on the products. This may negatively affect inward investment and/or demand for domestically produced goods and services. Competitive disadvantages provide individual countries with incentives to lower regulatory standards. Since there is a risk of different national regulations converging towards the lowest common denominator, a 'race to the bottom' may ensue. In the literature, this phenomenon is also referred to as the 'Delaware effect'. But the literature points out that the opposite phenomenon is also possible. Environmental regulation may result in competitive advantages for the domestic industry. This is the case if environmental regulation favours domestic producers by imposing higher costs of compliance on imported goods, thus constituting a technical barrier to trade. The underlying assumption here is that it is sometimes easier for domestic producers to comply with specific environmental rules than it is for non-domestic producers. If environmental regulation leads to competitive advantages for the domestic economy of one country, other countries are likely to follow. The resulting 'race to the top' with regard to environmental standards has also been referred to as the 'California effect'.[15]

Competitive advantages and disadvantages can be explained at a deeper level. The literature has identified as a crucial variable the distinction between product standards and process standards. While the literature

[15] For a summary of the background of the California effect, see Vogel, 1997. In this context, it should be noted that international convergence at a high level of environmental protection may also have roots other than barriers to trade for 'dirty' products. It is often argued, for example, that multinational firms apply uniform standards which are higher than the locally required level, for reasons which are related to economies of scale rather than market access.

predicts a race to the bottom where standards regulate production processes, it predicts a race to the top in the case of product standards. The enforcement of product standards is relatively easy in the case of both domestic and imported products. But as argued before, compliance may impose disproportionate costs on non-domestic producers and, therefore, lead to competitive advantages for the domestic industry. Process standards, by contrast, can be directly enforced only in the case of domestic producers. Non-domestic producers escape direct enforcement of process standards. However, they may be indirectly enforced by banning imports of goods produced without regard to process standards.[16] Yet, monitoring compliance is more difficult in the case of non-domestic producers than it is in the case of domestic producers. This may explain why process standards are more likely than product standards to lead to competitive disadvantages for the domestic industry. Nevertheless, some examples of process standards involving competitive advantages exist. The tuna-dolphin case, discussed below, is a case in point.

Combined Effects and Ranking of Collective Action

I have argued that both the type of environmental good to be protected and the side-effects of environmental protection measures on trade are important determinants of countries' incentives to engage in environmental protection by acting either alone or collectively. Combining the two factors should lead to a more comprehensive understanding of the problems of both autonomous national action and collective international action. Figure 3.4 shows all possible combinations of type-of-good effects and trade effects. It also indicates empirical examples for each combination, which are discussed in the following section. In the figure, the four types of goods are ranked according to their centripetal effects, which are very strong in the case of club goods (+), less strong in the case of private goods (=), weak in the case of public goods (–), and very weak in the case of common pool resources (– –). Trade effects are ranked according to the incentives for protecting the environment either unilaterally or in a group (as opposed to remaining inactive). Competitive advantages provide positive incentives (+) for

[16] Frank Wijen has pointed out that the extraterritorial enforcement of environmental standards can also be achieved by softer means, such as certification and labelling. The effectiveness of those instruments relies on a number of conditions. Assuming non-altruistic consumers, green labels are only effective if they signal a private benefit for individual consumers (for example, with regard to their personal health). This is generally not the case for process-standard labels. Exceptions include labels that simultaneously provide information on both production processes and products (for example, in the case of genetically modified food). Problems regarding information and trust add to the incentive problems: consumers must know the relevant green labels and trust the information they convey.

environmental action; competitive disadvantages provide negative incentives (–); and competitive neutrality leaves incentives unchanged (=). The combination of four different types of goods with three different kinds of trade effects leads to twelve possible cases. The most difficult case results in case 1: the protection of an international common pool resource, subject to very weak centripetal effects (– –), leads to competitive disadvantages (–). Collective action problems become less severe as we move downwards and to the right. The easiest situation for collective action (case 12) combines club goods, subject to strong centripetal effects (+), with competitive advantages (+).

		Environmental goods			
	Excludability	Non-excludable		Excludable	
	Rivalry in consumption	Rival	Non-rival	Rival	Non-rival
	Type of good	Common pool resource	Public good	Private good	Club good
	Centripetal effects	(– –)	(–)	(=)	(+)
Trade effects	Competitive disadvantages (–)	**1: – \ – – Sea fish (I)**	**4: – \ – Kyoto Protocol (I)**	**7: – \ = Sustainable timber harvesting**	**10: – \ + Environmental research (I)**
	Competitive neutrality (=)	**2: = \ – – Sea fish (II)**	**5: = \ – Kyoto Protocol (II)**	**8: = \ = Domestic waste collection**	**11: = \ + Environmental research (II)**
	Competitive advantages (+)	**3: + \ – – Sea fish (III)**	**6: + \ – Tuna-Dolphin case**	**9: + \ = Hormone ban**	**12: + \ + Environmental research (III)**

Figure 3.4 The severity of collective action problems

While every move to the right and/or downwards means that the collective action problem becomes less severe, the exact ranking of the twelve possible combinations depends on the relative strength of type-of-good effects and trade effects. If trade effects cannot alter the order of ranking resulting from the type-of-good effects, the latter can be considered dominant. For example, if common pool resources lead to a more severe collective action problem than do public goods, regardless of whether the competitive side-effects in the concrete case are positive, negative, or neutral, the type-of-good effects can be regarded dominant. The opposite case, characterised by the dominance of trade effects, is also possible. In this case, type-of-good effects cannot alter the basic ranking resulting from the trade effects: irrespective of which type of good is involved, it always matters more whether the competitive side-effects of environmental protection are positive, negative, or neutral. In between these two extreme cases, which are characterised by the dominance of either trade effects or type-of-good effects, are intermediary cases in which neither of the two effects is dominant in the sense of the above definition. Overall, the assumption of dominant type-of-good effects is empirically somewhat more plausible than the assumption of dominant trade effects. The numbering used in Figure 3.4, therefore, can, but does not have to be, interpreted as a ranking of the severity of collective action problems.[17]

ILLUSTRATIONS OF COMBINED EFFECTS

Below, I build upon the examples used above to illustrate different types of environmental goods in order to demonstrate how the four types of goods combine with positive, negative, and neutral trade effects. In some of these empirical examples, the competitive effects depend on the sub-field or the specific aspect under investigation. For this reason, the issues of high-sea fisheries, the Kyoto Protocol, and environmental research are split into two or three cases, representing different aspects with different competitive effects. The empirical illustrations include a number of cases which have received a high level of public attention, and, therefore, represent convenient points of reference for a discussion of type-of-good and trade effects on national incentives for environmental protection. Since the empirical cases are used to illustrate rather than test the theoretical claims regarding the severity of collective action problems, more rigorous selection criteria are not required.

Common Pool Resources

Common pool resource plus competitive disadvantages – High-sea fisheries (I). The most severe collective action problems arise when countries need to protect international common pool resources and when the measures to protect these environmental goods lead to competitive disadvantages for the

[17] If dominant type-of-good effects are assumed, the ranking of the severity of collective action problems (from most severe to least severe) follows the numbering of the twelve cases indicated in Figure 3.4: 1>2>3>4>5>6>7>8>9>10>11>12. If dominant trade effects are assumed, the ranking of the severity of collective action problems of the twelve cases is 1>4>7>10>2>5>8>11>3>6>9>12.

domestic industry (case 1: –\– –). This may happen if countries try to prevent overfishing with regard to sea fish that is traded internationally. The problem of high-sea fisheries is a common pool resource problem, which already in itself leads to grave collective action problems. Free trade in fish exacerbates these collective action problems and further reduces the incentives for acting autonomously. This is because measures against overfishing may increase the production costs for participating countries in comparison to those for non-participant countries. Under the above-mentioned conditions of open access (non-excludability) and free trade, countries trying to prevent overfishing are thus doubly penalised: not only do they catch less fish than other countries, but higher production costs may also make their products less competitive. A number of international fisheries commissions play an important role in the difficult task of managing international fisheries.[18] Some international common pool resource problems related to fisheries were solved by the adoption of a 200-nautical-mile exclusive economic zone in the framework of the international law of the sea in 1977. This step made 'national' private goods out of what before were international common pool resources.

Common pool resource plus competitive neutrality – High-sea fisheries (II). Collective action problems are somewhat less severe, but still serious, if the protection of common pool resources does not entail any trade effects (case 2: =\– –). This is the case if countries act to prevent overfishing in the case of sea fish that are not traded internationally, for a lack of free-trade arrangements or for other reasons. The argument also applies to whaling. The prohibition of trade in whales makes it less painful for countries to unilaterally restrict or stop whaling.

Common pool resource plus competitive advantages – High-sea fisheries (III). If the protection of a common pool resource goes hand in hand with competitive advantages, the incentives for autonomous action again increase (case 3: +\– –). With regard to high-sea fisheries, this may occur if countries combine rules to prevent overfishing with mechanisms to shield their domestic fishing industry from foreign competitors who do not respect such rules. The result is a competitive advantage for domestic fishermen. This said, it must not be forgotten that the common pool resource problems continue to exist and to make autonomous action relatively unattractive.

Public Goods

Public good plus competitive disadvantages – Climate protection (I). Difficult though they might be, the collective action problems in the context

---

[18] Peterson (1993) provided an overview of the management of fisheries through international commissions.

of the Kyoto Protocol are less severe than those of high-sea fisheries in at least one respect.[19] As argued above, the potential benefits of the Kyoto Protocol in terms of curbing global warming are non-rival rather than rival. Since those benefits are also non-excludable, the Kyoto Protocol represents an international public good. Measures to save carbon energy, including carbon dioxide taxes and participation in emissions trading schemes, are often expensive. As a result, energy-intensive industries exposed to international competition fear negative effects.[20] In this regard, the Kyoto Protocol combines a public good with competitive disadvantages (case 4: –\–). It hence poses collective action problems which are severe, although less so than the previously discussed combination of common pool resources and competitive disadvantages.

Public good plus competitive neutrality – Climate protection (II). The Kyoto Protocol may also be effective without jeopardising national competitiveness. This is the case when it affects households and industries protected from international competition. In such cases, the Kyoto Protocol represents a public good combined with competitive neutrality (case 5: =\–). Accordingly, the incentives for autonomous action are slightly greater than in the preceding case. The exemptions that national climate-protection legislation occasionally provides for energy-intensive sectors exposed to international competition underline the practical relevance of this distinction.

Public good plus competitive advantages – The US dolphin-protection standards. 'In eastern tropical areas of the Pacific Ocean, schools of yellowfin tuna often swim beneath schools of dolphins. When tuna is harvested with purse seine nets, dolphins are trapped in the nets. They often die unless they are released.' This is how the World Trade Organization (WTO) describes a problem of by-catches that was at the centre of a controversial GATT panel report involving a trade dispute between the US (as the defendant) and Mexico.[21] I come back to this issue below. Like the prevention of global warming, the reduction of by-catches can be characterised as an international public good. But while the implementation of the Kyoto Protocol is likely to go hand in hand with competitive disadvantages in some cases, and competitive neutrality in others, dolphin protection in the framework of the US Marine Mammal Protection Act is likely to produce competitive advantages for the domestic industry (case 6: +\–). Since the Marine Mammal Protection Act sets process rather

---

[19] On the negotiation and substance of the Kyoto Protocol, see Oberthür and Ott, 1999.

[20] Brack et al. (2000) comprehensively analysed the potential trade effects of climate protection policies. The European Commission (2001) estimated that the overall compliance costs of the Kyoto Protocol could range between 0.06 and 0.30 per cent of the EU gross domestic product.

[21] World Trade Organization, 2003a. See also Goode, 1998. For an overview of and related documents on the tuna-dolphin case, see Oceanlaw, 2003.

than product standards, its existence is in conflict with the race-to-the-bottom hypothesis for process standards.[22] According to the Marine Mammal Protection Act, the domestic fishing fleet must comply with certain dolphin-protection standards. This measure alone would expose the US fishing industry to competitive disadvantages with regard to their foreign competitors. As a result, the US would be subject to race-to-the-bottom pressures. In order to prevent this from happening, the Marine Mammal Protection Act shields the domestic fishing industry from such competition by allowing tuna imports exclusively from countries complying with US dolphin-protection standards. Consequently, the potential competitive disadvantages for the domestic industry are turned into competitive advantages. While the public-good problem is not eliminated, the incentives for autonomous environmental protection are better than in any of the previously discussed cases of public goods and common pool resources.

Private Goods

Private good plus competitive disadvantages – Sustainable timber harvesting. I have argued that timber has the character of a private good for countries. As they do in the case of sea fish, countries generally have an interest in harvesting timber in a sustainable fashion. The difference is that timber has the character of a 'national' private good. Harvesting timber does not, therefore, lead to an international collective action problem with regard to the type of good it represents. Yet, to the extent that sustainable harvesting methods lead to higher production costs for the domestic industry, such methods are likely to be subject to competitive pressure from timber-producing countries with lower standards. Such race-to-the-bottom pressure reduces national incentives to autonomously enforce sustainable harvesting methods. They can be countered by international standards for sustainable timber harvesting. The timber case illustrates a widespread fallacy, according to which collective action is considered unnecessary as long as the environmental good concerned represents a purely 'national' private good. If the protection of private goods leads to competitive disadvantages, the incentives for effective autonomous action may be insufficient (case 7: –\=).[23]

Private good plus competitive neutrality – Domestic waste collection. Although in different ways than timber, domestic waste collection also represents a private good for countries. Many countries impose standards with regard to waste collection on households, such as requirements for separating different categories of waste. Such standards may be a burden to

---

[22] I owe this point to Frank Wijen.

[23] Bourke, 1995. See also Barbier, 1995.

households, but they do not affect trade. Consequently, they result in neither competitive advantages nor disadvantages for the domestic industry (case 8: =\=).

Private good plus competitive advantages – The hormone ban case. The EU ban on the use of growth-promoting hormones in meat production constitutes a private good from the perspective of the EU.[24] The import ban on meat containing such hormones turns a potential competitive disadvantage for European farmers into a competitive advantage (case 9: +\=). Since the hormone ban combines the protection of a private good with competitive advantages, the incentives for the EU to maintain this measure remain intact.[25]

Club Goods

Club good plus competitive disadvantages – Environmental research (I). To the extent that environmental research represents a club good for the countries involved, the incentives for countries to participate in international environmental research programmes are intact, and no serious collective action problem exists. However, there are cases in which environmental research, representing a club good, may have negative trade effects on the domestic industry (case 10: –\+). This is the case if environmental research leads to the substitution of products in which the domestic industry is specialised, hence removing a comparative advantage.[26] An oil-producing and oil-exporting country probably has fewer incentives to promote research into alternative energy sources than has a country depending on oil imports.

Club good plus competitive neutrality – Environmental research (II). Environmental research does not always involve trade effects (case 11: =\+). There are no trade effects if the cleaner products resulting from environmental research are not traded internationally, or if environmental research is not strictly product-related in the first place. The latter may be the case, for example, when research is directed at safeguarding plants and animals that have no immediate value as tradable goods.

---

[24] Some background information on the EU and growth-promoting hormones has been published by the European Commission, 2003.

[25] As Konrad von Moltke has correctly observed, the EU hormone ban represents primarily a health issue rather than an environmental issue. However, health issues are often interconnected with environmental issues. While pollution is an environmental issue often closely related to health issues, food safety is a health issue often closely related to environmental issues. In the context of this study, I therefore considered the hormone ban case to be an issue with a strong component of environmental policy concerns in the wider sense.

[26] Frank Maier-Rigaud has drawn my attention to the substitution effects of environmental protection measures, including environmental research.

Club good plus competitive advantages – Environmental research (III). The incentives to produce environmental research results arc, of course, reinforced by positive competitive effects. Countries promoting new environmental technologies may provide their domestic industries with competitive advantages (case 12: +\+). This occurs when countries remove 'dirty' products from the domestic market, while helping the domestic industry to establish a monopoly in clean products. By contrast to other cases that have been discussed, these competitive advantages do not result from barriers to imports alone. Rather, they result from combining environmental standards with the promotion of new technologies. Since club goods encourage collective action without involving major collective action problems, it can be expected that international cooperation with regard to environmental research may generally develop more easily than cooperation involving other types of goods. There may indeed be some evidence of environmental-research cooperation emerging at an early stage and without major difficulties. One example is the Convention on Long-range Transboundary Air Pollution (LRTAP), which was signed in 1979 by 34 European governments and the European Community. An important early objective of the convention was the coordination of national research programmes.[27]

IMPLICATIONS FOR GLOBALISATION AND ENVIRONMENTAL POLICIES

The analysis so far has shown how the incentives of countries to protect environmental goods – either autonomously or collectively – are influenced both by the nature of these goods in terms of public goods theory and by the side-effects of environmental protection measures on trade. I now turn to some related questions and discuss to what extent they might be answered on the basis of the findings presented above. Three overlapping questions are addressed. First, how does globalisation affect environmental protection? Second, how can the effectiveness of domestic environmental policies be increased? And third, what are the scope and limits of flexible, voluntary, and non-binding international environmental agreements?

The Impact of Globalisation on Environmental Protection

In order to evaluate the effect of globalisation on environmental protection, a distinction must be drawn between economic globalisation, on the one hand,

[27] Haas et al., 1993.

and its driving forces, on the other.[28] According to an often-used and simple definition, globalisation is characterised by an accelerated increase in the mobility of products and production factors. Technology and international economic law are identified as two main driving forces of globalisation. The focus here is on international economic law, which promotes globalisation through 'negative integration'. Negative integration, often in the form of free-trade agreements, removes barriers to trade resulting from national legislation such as tariffs, quotas, technical barriers to trade, and discriminatory taxes.

Globalisation in the form of growing trade flows reinforces the competitive effects of environmental regulation. These effects of globalisation may go in either direction: competitive advantages as a result of environmental protection measures may make autonomous national policies more attractive, while competitive disadvantages may make autonomous environmental protection measures less attractive. Hence, while globalisation may make national environmental policies more attractive in some cases and less attractive in other cases, it does not necessarily affect the average level of environmental protection in any given country.

What has been neglected so far is the effect of international economic law, one of the driving forces of globalisation. International free-trade agreements prohibit or limit protectionist measures. They thereby may also restrict the use of environmental protection measures which create competitive advantages for the domestic industry and competitive disadvantages for foreign competitors.[29] Whereas trade flows themselves do not necessarily affect the average level of environmental protection afforded by individual countries, trade agreements introduce a bias. The combined impact of international trade and international free-trade agreements may hence negatively affect the overall incentives and possibilities of countries to protect the environment autonomously: while trade may limit the incentives for countries to act (in the case of environmental measures generating competitive disadvantages), trade agreements may prohibit action where countries have incentives to act (i.e., when environmental measures create competitive advantages). The GATT panel report on the tuna-dolphin dispute illustrates this. The panel found that the US dolphin-protection measures violated GATT rules.[30] The bias against national environmental policies that results from free-trade agreements may be reduced by the inclusion of

---

[28] Below, the term 'globalisation' essentially means 'economic globalisation'. It does not cover political, cultural, or social globalisation, whatever the precise definitions of those concepts may be.

[29] Oberthür and Ott, 1999.

[30] Goode, 1998. An overview of and related documents on the tuna-dolphin case can be found at Oceanlaw, 2003.

environmental safeguard clauses in free trade agreements, a possibility which is explored below.

Increasing the Effectiveness of Domestic Environmental Policies

Four options for environmental policies are now briefly discussed below. The first two options concern the international legal framework, the third concerns policy choices at the national level, and the fourth concerns the information available to actors. In game-theoretical terms, the outcomes with regard to the environment can be improved by changing the (international) rules of the game, by choosing intelligent (national) strategies, and by providing specific information to the players. It might at first appear paradoxical that increasing the scope for effective national environmental policy, as outlined in the first two options, should start with the international legal framework and hence with international collective action. In both cases, however, amending the international legal framework – or the application thereof – aims at restoring national sovereignty concerning environmental policies in areas in which international trade law may have limited it excessively.

The first option is the better enforcement and reinforcement of environmental safeguard clauses provided by international economic law. Such safeguard clauses allow countries to take environmental protection measures despite their potential effect as barriers to trade that might result in competitive advantages for the domestic industry.[31] At the same time, environmental safeguard clauses also require national measures that might constitute a barrier to trade to be motivated by environmental concerns, not by protectionism. In reality, it is often difficult to prove either of these. To the extent that environmental safeguard clauses are subject to judicial or quasi-judicial review (as in the EU and the WTO, respectively), international courts have a significant role to play in the application of such safeguard clauses.

The second option is to make sure that international trade law does not prevent individual countries from adopting measures implementing international environmental agreements. Critics assert that the current international system establishes a de facto superiority of international economic law over international environmental law. An important reason for this is that international economic law can be more easily enforced, thanks to traditional unilateral instruments (such as retaliation in trade policy) as well as new multilateral instruments (such as international courts). While countries can often break international environmental agreements without facing major consequences, the breaking of international trade agreements tends to be

[31] Examples of such clauses include Articles 30(4), 30(5), 95(4), 95(5), and 176 of the EC Treaty, as well as Article XX of the GATT.

punished more harshly. If international environmental law conflicts with international trade law, it is only logical for countries to disregard the former rather than the latter. It is not yet clear what is the best way to prevent or solve conflicts between international trade law and international environmental law. The international system is predominantly organised according to functional principles, and coordination between the different functional international organisations is weak. The single institutional framework provided by the United Nations is far too weak, not only with respect to the member states, but also with regard to its functional components. The situation is different in the EU, which has a much stronger single institutional framework.[32] The competences of the European Commission, the European Parliament, and the European Court of Justice cover nearly all aspects of the European Community pillar of the EU and an increasing number of parts of the other two pillars (foreign policy and criminal matters). At the international level, the only realistic, though imperfect, solution seems to be to integrate environmental concerns more strongly into the jurisdiction of the WTO.

The third option is for national environmental policies to circumvent the problems posed by international trade law. Individual countries could simply accept the modified action space given and refocus their attention on environmental problems and environmental protection measures with few or no competitive effects. Many specific environmental goods can be protected using different measures. If globalisation renders the application of particular instruments difficult – either because competitive disadvantages make them unattractive or because competitive advantages make them illegal in terms of international trade law, it may be possible to switch to the use of measures which are either unaffected or even encouraged by globalisation.[33] For example, governments may focus their efforts on measures minimising competitive disadvantages when implementing the Kyoto Protocol, for instance, by imposing limits on carbon dioxide emissions by households (heating and private cars) and industries that are not exposed to international competition. In line with the 'polluter pays' principle, which is also enshrined in EU law,[34] the costs of compliance would be borne by the households and industries concerned. In sectors exposed to international competition, limiting emissions might lead to competitive disadvantages if polluters pay. But to the extent that polluters escape the costs of reducing emissions, they will not be subject to competitive disadvantages. This is why subsidies may be the only viable way to cut emissions unilaterally in energy-intensive, export-oriented

---

[32] Article 3 of the EU Treaty provides 'The Union shall be served by a single institutional framework which shall ensure the consistency and the continuity of the activities carried out'.

[33] Verweij (2001) suggested such a strategy for the problem of global warming.

[34] See Article 174(2) of the EC Treaty.

industries. However, direct subsidies (e.g., for using emission-reducing equipment) may run into the barriers set up by EU and international competition law.[35] Indirect subsidies (e.g., through publicly funded research providing clean technology) are somewhat less problematic in this respect.

A fourth option for individual countries is to ignore the problems globalisation apparently causes for certain environmental policies, arguing that such problems are relatively minor. Although this proposal sounds trivial in theory, it might be of some significance in practice. Competitive disadvantages are an often-used argument against the implementation of national environmental policies. It is not unreasonable to think that such arguments, which assume strong regulatory competition, sometimes prevent environmental measures from being taken. However, some of the literature points out that the competitive effects of environmental regulation may often not be significant after all.[36] To the extent that this is true, governments may increase their environmental action space by presenting research and disseminating information which reduces the gap between public perceptions and reality with regard to regulatory competition in the environmental field.

Flexible and Non-Binding Environmental Agreements

Within the EU, and not only with regard to environmental policies, there is a discussion on 'softer' forms of regulation which show more respect for national sovereignty and autonomy. Such softer forms of regulation are characterised by their flexible, voluntary, and non-binding nature. In this context, 'flexible' means that not all member states need to participate; 'voluntary' stands for the fact that no participating country can be forced against its will to comply and decision-making must, therefore, be unanimous; and 'non-binding' means that countries are not legally required to implement decisions, while Community institutions are neither mandated nor entitled to enforce them.[37]

The effectiveness of flexible, voluntary, and non-binding EU measures is influenced by their centripetal effects. Such measures seem unfit for environmental policies involving centrifugal rather than centripetal effects.

---

[35] Article 8.2(c) of the WTO Agreement on Subsidies and Countervailing Measures allows 'assistance to promote adaptation of existing facilities to new environmental requirements imposed by law and/or regulations which result in greater constraints and financial burden on firms'. However, national subsidies can be challenged if they do not conform to a number of fairly restrictive conditions. For instance, the subsidy must be 'limited to 20 per cent of the cost of adaptation' (WTO, 2003b). See also Kölliker, 2001b.

[36] See, for instance, Vogel, 1997, and Faure, 2001. At a more general level, Krugman (1994) warned of what he portrayed as 'a dangerous obsession' with international competitiveness.

[37] 'Flexibility' may also refer to the differentiated implementation of EU directives. With regard to such flexibility in the field of environmental policies, see Scott, 2000.

They should, therefore, not be relied upon if environmental problems involve (non-excludable) international public goods or common pool resources, or if they go hand in hand with competitive disadvantages.[38] Flexible, voluntary, and non-binding measures are likely to be more appropriate in the case of (excludable) club goods, and if positive trade effects can be expected.[39]

CONCLUSIONS

The aim of this chapter was to explain how globalisation affects the incentives of countries and their governments to protect environmental goods autonomously or in cooperation with other countries. I thereby aimed to identify the possibilities and limits of national environmental policy as well as the necessity and difficulties of international environmental agreements, under the condition of globalisation. The study was largely theoretical in character, but empirical examples were used to illustrate the outlined mechanisms.

In the first step, the focus was on the protection of environmental goods, neglecting any side-effects. In the analysis, public goods theory was applied to collective action among countries at the international level. According to the theory of public goods, environmental goods that have the character of private goods from the perspective of countries, are best protected through national environmental policies. For all other categories of goods, international collective action may lead to better results. However, different types of non-private (or collective) goods involve different kinds of collective action problems. The success of collective action often depends on the attractiveness of international environmental agreements for initially non-participating countries. As I have shown elsewhere, different types of goods can be ranked according to their centripetal effects.[40] Club goods involve the strongest centripetal effects, followed by private goods and public goods. Common pool resources rank at the bottom of the list. In this step, the

---

[38] In her article on flexible integration and environmental policy in the EU, Müller-Brandeck-Bocquet (1997) considered competitive advantages for the participants in flexible arrangements among willing EU member states as a conditio sine qua non. In this context, it is interesting to note that Article 43 of the EU Treaty sets the condition that closer cooperation among member states 'does not constitute a barrier to or a discrimination in trade between the Member States and does not distort competition between them'.

[39] This refers to collective action problems among countries rather than among non-state actors. It is well-known that voluntary environmental agreements among non-state actors, or between non-state actors and the state, often work well *under the shadow of hierarchy*. The latter means that states demonstrate their readiness to impose effective environmental protection should voluntary agreements fail to produce the desired results. Grepperud and Pedersen (2003) presented a formal model of voluntary environmental agreements.

[40] Kölliker, 2001a, 2002.

incentives for autonomous and collective action were given by the nature of environmental problems. The boundaries of collective action problems were related to the boundaries of the environmental goods. I argued that 'national' private goods require autonomous domestic action, while 'international' non-private goods require collective international action.

In the second step, the side-effects of environmental protection on trade and the feedback effects of such trade effects on environmental protection were analysed. If trade effects are taken into account, the incentives of countries and their governments to protect environmental goods may change significantly. Other things being equal, individual countries are more willing to adopt environmental protection measures if those measures lead to competitive advantages, and less willing to do so if they lead to competitive disadvantages. Competitive advantages thus encourage autonomous action and facilitate collective action, while competitive disadvantages discourage autonomous action and exacerbate collective action problems.

It is important to note that the geographical scope of the environmental goods to be protected and the geographical scope of the trade effects resulting from environmental protection do not necessarily coincide. The two are basically independent from each other and hence can normally be expected to diverge. It can be concluded that the need for international collective action is not only determined by the geographical boundaries of the environmental goods concerned, but also by the geographical boundaries of the markets for the products affected by environmental protection measures. Taking into account the trade effects of environmental protection, thus, prevents a naïve application of the subsidiarity principle with an exclusive focus on the nature of the environmental goods. If environmental protection affects trade, the geographical scope of environmental problems is twofold. On one hand, it coincides with the geographical scope of the natural resources concerned. On the other hand, the scope of the markets for the affected products is equally relevant. If no traded goods are affected, the scope of the environmental problems is confined to the geographical scope of the environmental goods.

Globalisation is characterised by the accelerated increase in the mobility of products and production factors. Globalisation is driven not only by new technologies but also by international free-trade agreements. Increasing trade flows and stronger international free-trade commitments influence environmental policies in two separate ways. Increasing trade flows reinforce both the positive and the negative competitive effects of environmental protection measures. This leads to increased incentives for environmental protection in some cases and to lower incentives in other cases. While globalisation thus affects many environmental protection measures either positively or negatively, the average level of environmental protection need not be negatively affected. This conclusion must be modified when international free-trade commitments are taken into account. Environmental policies creating competitive advantages face the barrier of international economic law, which restricts policies with protectionist effects. Free-trade agreements restrict the ability of countries to adopt environmental protection measures leading to competitive advantages for the domestic industry. This tendency exists despite safeguard clauses in some agreements, such as in the EC Treaty. It leads to a bias, whereby the overall effect of globalisation on environmental protection may become negative.

Against this backdrop, four options for enlarging the action space of individual countries and their governments were discussed. The four options concern the relationship between international economic law and national environmental law, the relationship between international economic law and international environmental law, the national strategies for environmental protection, and the knowledge on the basis of which countries make decisions. The four options aim at (1) strengthening safeguard clauses allowing national environmental law exceptions from international economic law, (2) ending the de facto hierarchy which gives international economic law priority over international environmental law, (3) circumventing problems created by globalisation by focusing on environmental protection measures without negative competitive effects, and (4) generating and disseminating information on the actual extent of the competitive disadvantages of environmental protection measures, which tend to be overestimated, and proceeding to act where such effects are small or negligible.

REFERENCES

Barbier, E. (1995), 'Trade in timber-based forest products and the implications of the Uruguay Round', *Unasylva – An International Journal of Forestry and Forest Industries*, **46**(4), 3-10.

Bourke, J. (1995), 'International trade in forest products and the environment', *Unasylva – An International Journal of Forestry and Forest Industries*, **46**(4), 11-17, also published at *http://www.fao.org/docrep/v7850e/v7850e03.htm*.

Brack, D., M. Grubb, and C. Windram (2000), *International Trade and Climate Change Policies*, London: Earthscan.

Cornes, R. and T. Sandler (1996), *The Theory of Externalities, Public Goods and Club Goods*, Cambridge: Cambridge University Press.

European Commission (2001), *Communication from the Commission on the Implementation of the First Phase of the European Climate Change Programme*, COM(2001) 580 final, Brussels: European Commision.

European Commision (2003), 'Food safety: From the farm to the fork', *http://europa.eu.int/comm/food/fs/him/him_index_en.html*, Brussels: European Commision.

Faure, M. (2001), 'Regulatory competition vs. harmonization in EU environmental law', in D. Esty and D. Gerardin (eds) (2001), *Regulatory Competition and Economic Integration: Comparative Perspectives*, Oxford: Oxford University Press.

Goode, W. (1998), *Dictionary of Trade Policy Terms*, Adelaide: University of Adelaide/Centre for International Economic Studies.

Grepperud, S. and P. Andreas Pedersen (2003), 'Voluntary environmental agreements: Taking up positions and meeting pressure', *Economics and Politics*, **15**(3), 303-321.

Haas, P., R. Keohane, and M. Levy (eds) (1993), *Institutions for the Earth*, Cambridge: Massachusetts Institute of Technology.

Holzinger, K. (2002), *Transnational Common Goods: Strategic Constellations, Collective Action Problems, and Multi-Level Provision*, Habilitation, Otto-Friedrich-Universität Bamberg.

Kahneman, D., J. Knetsch, and R. Thaler (1991), 'The endowment effect, loss aversion, and status quo bias', *Journal of Economic Perspectives*, **5**(1), 193-206.

Keohane, R. and E. Ostrom (eds), 'Local commons and global interdepencence: Heterogeneity and cooperation in two domains', *Journal of Theoretical Politics*, **6**(4).

Kölliker, A. (2001a), 'Bringing together or driving apart the Union? Towards a theory of differentiated integration', *West European Politics*, **24**(4), 125-151.

Kölliker, A. (2001b), 'Public aid to R&D in business enterprises: The case of the United States from an EU perspective', *Revue d'Economie Industrielle*, **94**, 21-48.

Kölliker, A. (2002), *The Impact of Flexibility on the Dynamics of European Unification*, dissertation, Florence: European University Institute.

Krugman, P. (1994), 'Competitiveness – A dangerous obsession', *Foreign Affairs*, **73**(2), 28-44.

Moussis, N. (1998), *Handbook of European Union*, Rixensart: European Study Service.

Müller-Brandeck-Bocquet, G. (1997), 'Flexibele Integration – Eine Chance für die europäische Umweltpolitik?', *Integration*, **20**(4), 292-304.

Oberthür, S. and H. Ott (1999), *The Kyoto Protocol: International Climate Policy for the 21st Century*, Berlin: Springer.

Oceanlaw (2003), 'Internet guide to international fisheries – Bycatch: tuna/dolphin', *http://www.oceanlaw.net/netpath/page4-bycl.htm#bytuna*.

Peterson, M. (1993), 'International fisheries management', in P. Haas, R. Keohane, and M. Levy (eds), *Institutions for the Earth*, Cambridge: Massachusetts Institute of Technology, pp. 249-305.

Samuelson, P. (1954), 'The pure theory of public expenditure', *Review of Economics and Statistics*, **36**(4), 387-389.

Scott, J. (2000), 'Flexibility, "proceduralization", and environmental governance in the EU', in G. de Búrca and J. Scott (eds), *Constitutional Change in the EU: From Uniformity to Flexibility?*, Oxford: Hart.

Schelling, T. (1978), *Micromotives and Macrobehavior*, New York: Norton.

Verweij, M. (2001), 'A snowball against global warming: An alternative to the Kyoto Protocol', *Preprint of the Max Planck Project Group on Common Goods,* no.11.

Vogel, D. (1997), 'Trading up and governing across: Transnational governance and environmental protection', *Journal of European Public Policy*, **4**(4), 556-571.

World Trade Organization (2003a), 'Beyond the agreements: The tuna-dolphin dispute', *http://www.wto.org/english/thewto_e/whatis_e/tif_e/bey5_e.htm*.

World Trade Organization (2003b), 'Agreement on subsidies and countervailing measures', *http://www.wto.org/english/docs_e/legal_e/24-scm.pdf.*

4. Financing Global Public Goods: Responding to Global Environmental Challenges

Pedro Conceição and Inge Kaul[1]

SUMMARY

We address the question of how global public goods should be financed. Different types of goods are identified. We argue that the nature of goods is often not technically given but shaped by policy choices. Public goods include collective action components (CaCs); when issues shift from the national to the regional or global level, international CaCs add to their national counterparts. Four different production paths are identified in international CaCs, calling for different measures. We discuss six financial instruments: national-level arrangements to influence cross-border spillovers, cost-sharing of common international-level facilities, resourcing international-level intermediaries, managing global scarcities, levying user fees for services of international organisations, and levying user fees for nationally provided global services. It is argued that financial aid is fundamentally different from financing global public goods. We conclude that global public goods call for tailor-made financial arrangements, both within nation states and at the international level.

INTRODUCTION

Globalisation has brought the world closer together. A widening range of policy challenges stretches across borders. For example, the emergence of a new communicable disease – SARS – quickly led to threats beyond where the

[1] The views expressed are the authors' and do not necessarily reflect those of the Office of Development Studies of the United Nations Development Programme, with which they are affiliated. Please direct all comments and inquiries to inge.kaul@undp.org.

initial outbreak occurred. Financial crisis can rapidly spread from one country to the next, as the Asian financial crisis episodes in 1997 and 1998 so clearly illustrated. As a consequence, the world's agenda of international cooperation has changed in recent decades. Conventional concerns of foreign affairs, international relations, and foreign aid are increasingly sharing the political arena with a new set of issues, global public goods. These are issues that concern all and can often only be addressed adequately through cross-border collective action.

Many global environmental challenges fall into the category of global public goods. Mention can be made of global climate stability and the preservation of biodiversity. Global public goods are often seriously underprovided. Instead of enjoying global climate stability and biodiversity preservation, we are facing risks of global warming and a rapid decline in a number of the world's species. Put differently, many things or conditions that one would like to see in the global public domain as contributing positive utility to people's and countries' well-being present themselves as the opposite: as conditions that many perceive as 'bads', as detrimental to sustainable development.

The growing importance of global public goods and bads is an often-overlooked dimension of the debates on globalisation. While the concept of public goods is key to national public finance/economics debates and often informs discussions on the division of responsibilities between markets and states, the notion of global public goods is still a relatively fresh and unfamiliar one within the context of international cooperation. Therefore, in the next section, we introduce the concept of global public goods.

Although the notion of global public goods is of recent origin, we can already see that in the actual practice of international public policy-making and financing global challenges, use is made of a number of tools and instruments of public finance that have conventionally been used only within the national context. In the second section, we take stock of some of these new financing arrangements that have emerged at the international level in support of the provision of global public goods and explore how some of these innovations could be further complemented. We also show that domestic public finance choices are increasingly made 'with outside challenges in mind'.

Recent years have witnessed a rapid proliferation and diversification of international financing mechanisms: the creation of global-issue funds, such as the Global Environment Facility (GEF) and the Global Fund to Fight Aids, Tuberculosis, and Malaria (GFATM); the emergence of new markets, such as those for pollution or fishing permits; and new user fees, such as the surcharges levied by airlines, airport authorities, and other agents to meet the costs of enhanced world-wide civil aviation safety. Even today, however, international cooperation finance is, in the minds of many, still equated with 'foreign aid' (i.e., official development assistance, ODA). This signals that, so far, global-public-goods financing has, for the most part, happened in an ad hoc fashion, often in response to urgent challenges, such as the outbreak of a potentially global health epidemic or an international financial crisis. Yet, global-public-goods financing is a reality as well as a phenomenon of growing proportions. Therefore, as we argue in the third section, it is important to analyse it so that it can be addressed explicitly – in an effective, efficient, and equitable way.

As the concluding section underlines, like – or perhaps even more so than – the provision of public goods nationally, that of global public goods is a political issue. Politics matters more internationally, because at the international level there exists no equivalent of the institution of the state nationally; therefore, international cooperation has to succeed voluntarily. This means, it has to be a fair bargain and provide significant net benefits for all.

INTRODUCING (GLOBAL) PUBLIC GOODS

Economic growth, development, and, ultimately, people's well-being depend on two main types of goods: private goods and public goods.[2] As Box 4.1 explains, private goods are goods with excludable benefits. Therefore, individual actors can appropriate them and prevent others from enjoying the good. Public goods, by contrast, are in the public domain, available for all to consume, or affecting all.

Some may note that this definition differs from the standard economic theory, which states that public goods are marked by non-rivalry in consumption and non-excludability of benefits, with private goods having the opposite properties. Figure 4.1 presents a classification of select goods according to this definition. A brief examination of Figure 4.1 reveals that a number of the goods mentioned could also figure in another quadrant. For example, land exists as a private good and as a public good (such as a public park which can be enjoyed by all). Knowledge, too, can exist as a public good available to all. Or its use can be restricted through intellectual property rights.

It is important to distinguish between a good's potential publicness and its de facto publicness. De facto public goods are those that are actually in the public domain and affect all, often leaving little choice for the public. What

---

[2] The term 'good' is used here to denote things (such as bread), services (like the control of communicable diseases), or conditions (such as international financial stability). Thus, the term 'good' is value-neutral.

determines a good's publicness or privateness is the *nature of its actual benefits and costs* – not how the good is produced nor, in most cases, the intrinsic nature of the good. The very nature of these benefits and costs can be shaped in most instances by policy. Whether a good is public or not, therefore, is often a matter of choice.

Box 4.1 Defining private goods and public goods

People's consumption is composed of two basic types of goods: private goods and public goods.

Private goods have excludable benefits, meaning that the owner of the good can restrict its use. Goods that have this characteristic can be exchanged and traded in markets. They are also often viewed as things that individual actors ought to obtain by themselves. Examples are basic food items (such as milk and bread), means of information (for example, a radio), and shelter. Many private goods also tend to be rival in consumption, meaning that one person's consumption limits the availability of the good's benefits for others. Such rivalry not only implies a certain excludability of benefits but also encourages people to seek ownership protection.

Public goods are in the public domain, available for all to consume, and affecting all. Examples are the natural commons, such as moonlight or the high seas. It would be impossible, or at least very difficult and costly in economic and political terms, to prevent others from enjoying these goods. The same tends to apply to human-made public goods, such as a judiciary system. Public goods that have non-rival properties, such as scientific knowledge, especially lend themselves to being public in consumption, because one person's consumption of the good does not reduce its availability for others.

Transnational – global or regional – public goods are those goods whose benefits are public (as opposed to private) and extend across several countries (in the case of regional public goods) or countries in several regions as well as several generations (in the case of global public goods).

Figure 4.2 lists various goods according to their de facto properties, based on the observation of how they appear to us in reality. It shows that the actual characteristics of goods can be quite different from their original, basic properties. This means that publicness and privateness are malleable: these properties can be – and often are – changed, either as a result of deliberate public-policy decisions or owing to policy neglect and oversight. Figure 4.2 demonstrates the nature of 'common pool resources' and the policy choices that can be made to manage the potential overexploitation that tends to affect these goods. Fish stocks, for example, are non-excludable but rival goods.

Society can choose, though, to define quotas and assign harvesting entitlements to various actor groups (such as countries). Similarly, pollution entitlements or caps can be defined, creating an intangible barrier in front of the atmosphere and turning it from a non-exclusive good into a good to which individual actors have only limited access.

	RIVAL	NON-RIVAL
EXCLUDABLE	QUADRANT 1 Examples: • Milk • Land • Education	QUADRANT 2 Examples: • Research and development • Non-commercial knowledge (such as the Pythagoras' theorem) • Norms and standards • Property-rights regimes • Respect for human rights • Television signals
NON-EXCLUDABLE	QUADRANT 4 Examples: • Atmosphere • Wildlife	QUADRANT 3 Examples: • Moonlight • Peace and security/conflict • Law and order/anarchy • Financial stability/excessive financial volatility • Economic stability/flagging growth • Growth and development potential (such as an educated workforce) • Efficient/inefficient markets • Spread/control or eradication of communicable diseases

Figure 4.1 Conventional approach to public goods[3]

The benefits and costs of public goods can be different in reach. Some may just have a local span, for example, a street sign. Others, such as judiciary systems, can have national reach. And yet others may have transnational (regional or global) scope. An example of a regional public good is the control of river blindness, since this disease is endemic mostly in certain parts of Africa; an example of a global public good is moonlight. Whether local or

[3] Adapted from Kaul et al., 2003: 82.

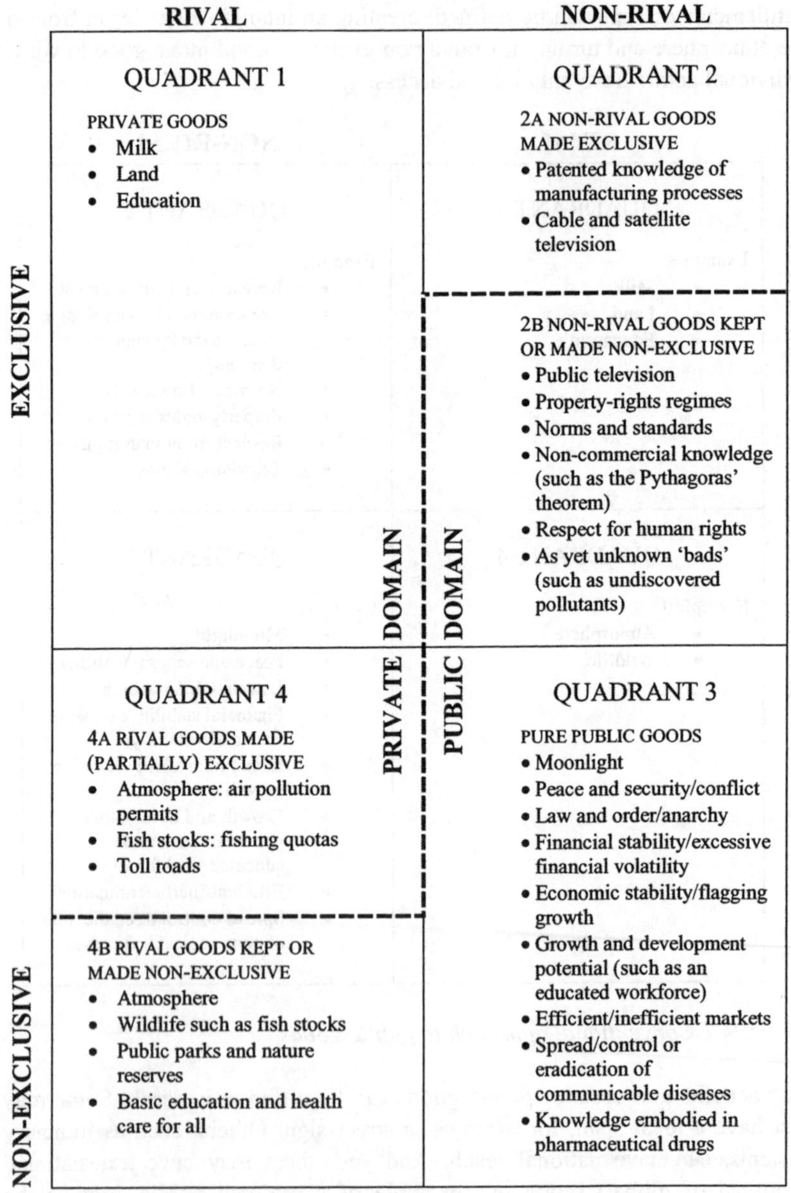

	RIVAL	NON-RIVAL
EXCLUSIVE	QUADRANT 1 PRIVATE GOODS • Milk • Land • Education	QUADRANT 2 2A NON-RIVAL GOODS MADE EXCLUSIVE • Patented knowledge of manufacturing processes • Cable and satellite television
		2B NON-RIVAL GOODS KEPT OR MADE NON-EXCLUSIVE • Public television • Property-rights regimes • Norms and standards • Non-commercial knowledge (such as the Pythagoras' theorem) • Respect for human rights • As yet unknown 'bads' (such as undiscovered pollutants)
	QUADRANT 4 4A RIVAL GOODS MADE (PARTIALLY) EXCLUSIVE • Atmosphere: air pollution permits • Fish stocks: fishing quotas • Toll roads	QUADRANT 3 PURE PUBLIC GOODS • Moonlight • Peace and security/conflict • Law and order/anarchy • Financial stability/excessive financial volatility • Economic stability/flagging growth • Growth and development potential (such as an educated workforce) • Efficient/inefficient markets • Spread/control or eradication of communicable diseases • Knowledge embodied in pharmaceutical drugs
NON-EXCLUSIVE	4B RIVAL GOODS KEPT OR MADE NON-EXCLUSIVE • Atmosphere • Wildlife such as fish stocks • Public parks and nature reserves • Basic education and health care for all	

PRIVATE DOMAIN | PUBLIC DOMAIN

Figure 4.2 De facto properties of goods[4]

[4] Adapted from Kaul et al., 2003: 83.

global in reach, public goods matter to people primarily as components of their consumption basket, or put differently, as ingredients of their well-being. Just as people have varying preferences for private goods, their preferences for public goods, including global public goods, vary. For example, fostering international financial stability would have higher priority for those participating in capital markets than for those eking out an existence on one dollar a day.

The fact of varying preferences for public goods is particularly important considering that public goods are often not only public in consumption but also public in production: they are dependent on cooperative efforts. This means that states play an important role in the provision of public goods, but also that these are not necessarily state-provided. In fact, the provision of public goods requires a number of actors, as indicated in the previous section. The role of states is crucial, but pertains to complementing individual-action inputs with CaCs.

Put differently, the fact that markets fail to generate an optimal provision level, for example, of greenhouse gas emission reductions and climate stability, does not mean they become superfluous and do not matter to public good provision. It is often better for governments to focus interventions on the definition and assignment of new types of property rights (such as transferable quotas) and then leave it up to markets – with their price mechanisms and ability to process large volumes of information – to establish the distribution of these entitlements. Figure 4.3 summarises graphically the provision of public goods in a generic way. Its main message is that few public goods are only state-provided. States may often play an important role, but many public goods also require private input.

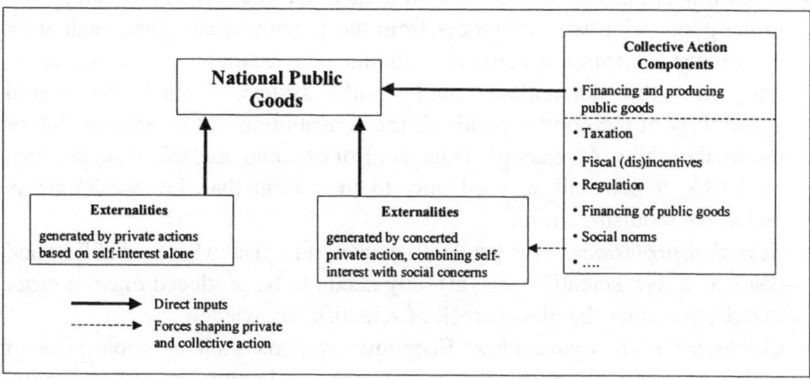

Figure 4.3 The provision of national public goods

Few goods are immutably global and public. Rather, they represent national public goods that in the wake of increasingly open borders have become *globalised.* Many forms of air pollution, of course, have never respected national borders. Other types of public effects now spill across borders much more easily than before because at-the-border controls have been reduced and cross-border activities have increased, with adverse effects on the environment accumulating. Thus, many global public goods can also be described as national public goods that require for their adequate provision international cooperation and an international-level CaC.

The provision of global public goods, as summarised in Figure 4.4, results from external private effects, the contributions of national public goods, and international CaCs.

Who provides the international CaC? The state cannot perform the same role internationally that it does nationally. The institution of the nation state lacks an international-level equivalent. States behave internationally much like private actors nationally: they are guided in their actions primarily by national (private) self-interest. The specific role played by the state, and the relative importance of different groups of actors and of coordination, depends on the specific nature of the production path that leads to the provision of the global public good. Four different types of production paths can be distinguished:

- *Summation processes.* All actors – countries, firms, households, and individuals – have to behave in a certain way in order for a particular good (e.g., global climate stability) to emerge. Often, all actors have to make the same contribution (e.g., reduce greenhouse gas emissions).
- *Weak-link situations.* In this case, it is also necessary for all to contribute to the good, with two differences from the previous case. First, each actor must make a location-specific contribution; for example, each government must strengthen its national public-health system. Second, the overall availability of the good depends on the contribution of the weakest link or the weaker links. An example is the control of communicable diseases such as SARS, which will succeed only to the extent that the 'weak' actors undertake requisite efforts.
- *Best-shot provisions.* This production scenario exists where a public good (such as a new scientific insight) only needs to be produced once in order to exist and enter the global stock of scientific knowledge.
- *Common-facility approaches.* Economic reasons such as economies of scale and scope, or political considerations of legitimacy and neutrality, sometimes make it desirable for individual actors to create a common facility (such as the United Nations, UN) or infrastructure (such as a satellite system). The actual production or maintenance of the good is

often delegated to a particular handling agent (for instance, the UN Secretariat).

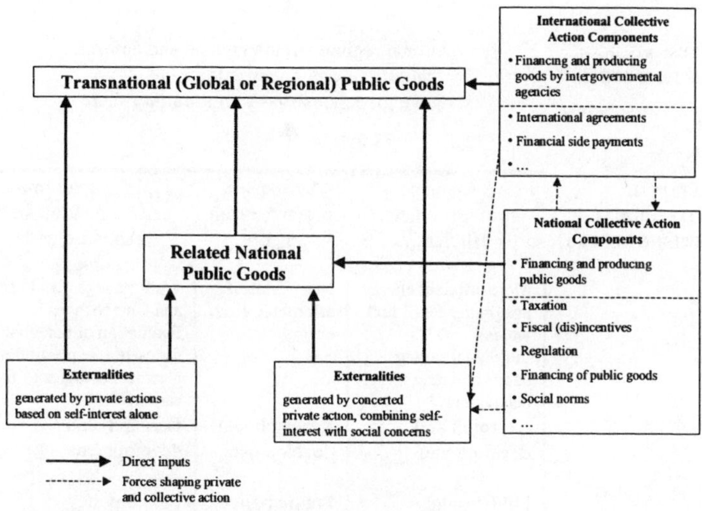

Figure 4.4 The provision of global public goods

Thus, when seen from the consumption side, global public goods can be defined as public goods whose benefits and costs span national borders and regions, and, sometimes, even generations. When seen from the production side, global public goods – with the exception of some rare best-shot goods – can be described as public goods that cannot be adequately produced through domestic policy action alone but require international cooperation, even if this cooperation only consists of national-level policy adjustments to discourage or encourage cross-border spillovers in response to outside pressures or international commitments into which countries might have entered.

Put more simply, global public goods tend to be composed of national public goods plus an element of international cooperation. This fact is also evident in Figure 4.5, which depicts the building blocks of the global public good 'climate stability'. Both the consumption and the production characteristics of global public goods have important implications for their financing, as we show below.

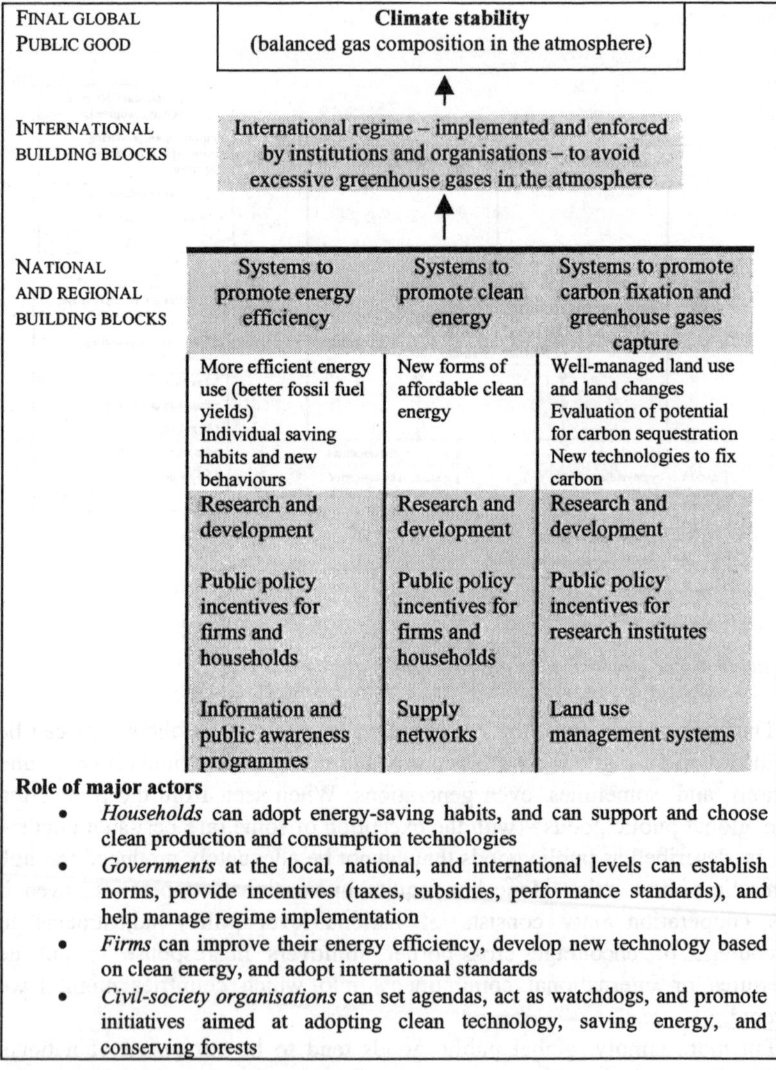

FINAL GLOBAL PUBLIC GOOD
Climate stability
(balanced gas composition in the atmosphere)
INTERNATIONAL BUILDING BLOCKS
International regime – implemented and enforced by institutions and organisations – to avoid excessive greenhouse gases in the atmosphere
NATIONAL AND REGIONAL BUILDING BLOCKS
Systems to promote energy efficiency
Systems to promote clean energy
Systems to promote carbon fixation and greenhouse gases capture
More efficient energy use (better fossil fuel yields)
Individual saving habits and new behaviours
New forms of affordable clean energy
Well-managed land use and land changes
Evaluation of potential for carbon sequestration
New technologies to fix carbon
Research and development
Research and development
Research and development
Public policy incentives for firms and households
Public policy incentives for firms and households
Public policy incentives for research institutes
Information and public awareness programmes
Supply networks
Land use management systems
Role of major actors
Households can adopt energy-saving habits, and can support and choose clean production and consumption technologies
Governments at the local, national, and international levels can establish norms, provide incentives (taxes, subsidies, performance standards), and help manage regime implementation
Firms can improve their energy efficiency, develop new technology based on clean energy, and adopt international standards
Civil-society organisations can set agendas, act as watchdogs, and promote initiatives aimed at adopting clean technology, saving energy, and conserving forests

Figure 4.5 Producing the global public good 'climate stability'[5]

[5] Adapted from Kaul et al., 2003: 375.

DIFFERENT GOODS, DIFFERENT FINANCING

It is clear from the previous section that the financing of global public goods may require action at the national as well as the international level. The objective is to ensure through financial policy tools (for example, taxes, subsidies, and other fiscal measures), as well as non-financial instruments (such as regulation), that all actors concerned allocate appropriate resources to these goods. In other words, the financing of global public goods is seen not just as a question of paying for a good; and it is certainly not seen just as a matter of allocating public revenue. Rather, the term 'financing' as it is used here means ensuring a desired level of public and/or private spending on a particular good as a means of influencing its provision.

In some instances, achieving an appropriate allocation may entail increasing the level of both public and private spending on the good. In other cases, it may mean decreasing current resource allocations. The most desirable policy option may also be to increase (or decrease) the costs of certain actions so that actors, at a given expenditure level, consume or produce less (or more) of a particular good. Clearly, given the varied nature of public goods (global and otherwise), notably the differences in their production paths, there exists no single, ideal formula for financing them. Financing arrangements have to be good-specific.

Nevertheless, when some of the financing measures for global public goods that exist at present are examined, the following main sets of measures can be distinguished: national-level financing arrangements to discourage negative cross-border spillovers and encourage positive cross-border spillovers; sharing the operating costs of common international-level facilities; resourcing international-level financial intermediaries; managing global scarcities through the creation of new property rights and new markets; levying user fees for the services of international organisations; and levying user fees for global services provided nationally or privately. In the remainder of this section, the six sets of measures listed above are described.

National-Level Arrangements to Influence Cross-Border Spillovers

The measures within this category serve to reduce negative cross-border spillovers and promote positive externalities. As such, they pertain primarily – but not exclusively – to global public goods whose production path follows a summation process. Some best-shot global public goods also figure in this category. For example, a number of countries have – as a result of national policy choices or in response to international agreements – adopted a variety of policy instruments that contribute to preventing negative externalities from spilling across national borders and adversely affecting actors in other countries. Examples are national carbon or energy taxes. Mention can also be made of improvements in human-rights conditions that contribute to easing international flows of asylum seekers, and of public investments in adequate banking supervision that may help prevent financial crises and potential international contagion effects. In some instances, the primary reason for introducing these measures may not have been to contribute to the provision of a global public good. Rather, the primary concern may have been to improve the national public domain. Thus, as pointed out before, national public goods are important ingredients of global public goods. Where national public goods are adequately provided, there is often no need for special corrective international-level interventions.

Similarly, national-level policy measures can foster the production of positive cross-border spillovers. For example, national support for scientific research may benefit researchers worldwide. Similarly, many technological innovations have, at least in the longer run, positive global spillover effects. The rapid progress that developing countries have achieved in such areas as the reduction of infant mortality is no doubt attributable to medical and pharmaceutical advances in industrial countries, notably the diffusion of knowledge and information on the mechanisms of the spread of diseases. Likewise, the national defence forces of a country or group of countries can deter nations from attacking other nations. While negative cross-border spillovers commonly receive considerable policy attention, positive spillovers are often just quietly – and free of charge – enjoyed by the global public. The unilateral provision of a global public good, such as pharmaceutical knowledge, is not necessarily appreciated. It may entail technological dependence; therefore, countries may prefer to create multiple production centres rather than to be dependent on best-shot providers, however benevolent their intentions may be.

There is a shortage of data on national and international-level spending on global public goods. It seems that national-level expenditure on reducing negative spillovers or encouraging positive ones constitutes the bulk of global-public-goods financing. Even when only public financing is considered, the ratio of national-level financing to international-level financing, according to some estimates, ranges between 200:1 and 400:1.[6] Of course, private financing also occurs nationally. Moreover, as states withdraw from the direct provision of public goods, focusing on creating incentives for private actors and employing such methods as marketisation (e.g., user fees for state-provided services), private spending is likely to be a growing part of the total spending on these goods.

---

6 Kaul et al., 2003.

Cost-Sharing of Common International-Level Facilities

International venues for state and non-state actors to come together and negotiate international cooperation issues constitute, in a way, core global public goods: without such facilities, international cooperation and the provision of global public goods would suffer. It is perhaps also for this reason that the few firmly binding financial commitments on which the international community has so far agreed primarily concern nation states' contributions to the regular budget of international organisations, such as the UN and its specialised agencies. Contributions to the Bretton Woods institutions are also obligatory in nature.

In most instances, the cost-sharing formula applied follows the ability-to-pay principle, sometimes modified by other criteria, such as a country's volume of civil-aviation traffic, postal mail, or telecommunications.

Resourcing International-Level Financial Intermediaries

International-level financial intermediaries are created for several purposes. The following are among the most important:

Supporting weak-link providers

As noted, such support can be desirable when the provision of a global public good follows a summation process but providers differ sharply in terms of their ability to pay or their technical capacities. In such situations, it may benefit all for the financially or technically stronger providers to assist the weakest link or the weaker providers. For example, civil-aviation safety calls for vigilance and proper screening processes at all airports. Yet, some countries may simply be too poor to make the necessary investments. For other countries to enjoy a particular level of protection, it is necessary and rewarding to extend support for this purpose to the poorer countries.

Providing compensation for global public services

The incremental cost payments effected within the context of the GEF are a case in point. Particular actors – mostly, but not necessarily, countries – undertake efforts (for example, aimed at carbon sequestration or biodiversity conservation) beyond those that they would make if they were guided solely by national self-interest. Since these efforts benefit the world as a whole, the GEF pays compensation to the providing actors in the form of incremental cost reimbursement.

Offering pooled incentives to best-shot providers

Governments may provide national incentives for actors to contribute to particular global public goods (such as the fight against global communicable diseases). Often, however, it is more efficient for countries to join forces and to pool resources internationally in order to encourage best-shot actors to provide these goods. Such encouragement could, for example, take the form of purchase commitments: guarantees to buy a certain quantity of the goods concerned so as to allow the providers to recoup related research and development costs.

Facilitating risk management and financing development

In effect, these are the main purposes of some of the most powerful international financial intermediaries, including the International Monetary Fund and the World Bank. Internationally agreed-upon insurance schemes such as those required for international oil shipments can also be mentioned.

Managing Global Scarcities

An important group of global public goods are the natural commons that are non-excludable (i.e., in the global public domain) but have rival consumption properties, and are, therefore, at risk of being overused and becoming depleted. Some of the aforementioned national measures (for instance, national carbon taxes) and some of the international-level financing arrangements (such as the incremental cost payments of the GEF) could help avert these risks. Additional measures to support such purposes include the assignment of new property rights such as tradable fishing quotas or pollution permits. This makes the scarcity, which is 'hidden' behind non-excludability, visible and explicit.

These measures combine a 'control-and-command' approach with the advantages of markets by making the defined allowances tradable. They are preferable where the exploitation of a particular global common resource has reached a critical threshold.

Levying User Fees for Services of International Organisations

Organisations such as the World Intellectual Property Organization (WIPO) help in the provision of a global public good (viz., that of patent protection and global knowledge management). At the same time, the services of the WIPO generate important private benefits for the patent holders, who are, therefore, willing to pay a patent registration fee. The revenue generated from these fees constitutes a major part of the WIPO's income.

For similar reasons, it is possible for the International Treaty on Plant Genetic Resources for Food and Agriculture, which constitutes and regulates a sort of global gene pool, to charge the users of the gene pool and to use these charges to pay for the preservation of genetic resources and to support the providers, many of whom are farmers in developing countries.

Table 4.1 Purposes and tools for financing public goods

Purposes	Tools National (dis)incentive measures	National incentive measures	Cost-sharing of common international facilities	Resourcing international financial intermediaries	Assignment of new property rights/ markets	User fees for global services
To reduce negative cross-border spillovers	X					
To encourage positive spillovers		X				
To exploit economies of scale			X			
To pool incentives				X		
To manage global scarcity					X	
To build/maintain national components of global networks						X
To recoup costs of services by international organisations						X

Levying User Fees for Global Services Provided Nationally or Privately

The global systems of civil aviation or postal services would not be possible without all countries contributing, through state or private provision, components of harmonised national infrastructure. Given that, in these cases, individual actors such as airlines or private individuals find it convenient and in their interest to use these systems, a variety of user charges exist in these fields. For example, air travellers pay airport fees and thus help finance airport facilities; and aircraft pay overflying charges that help countries recoup some of their civil-aviation infrastructure costs. In the case of international mail, the costs of postage are not only paid by mail users, but are shared on the basis of intergovernmental agreements between the sending and receiving countries.

Table 4.1 presents the above-mentioned financing arrangements for global public goods in summary form. It clearly shows that different goods and financing purposes call for different financing tools.

FINANCING INTERNATIONAL COOPERATION

An intriguing aspect of international policy debates today is that the financing issue is often narrowed down to just one modality: direct money flows between governments; in addition, it is often couched in terms of foreign aid. It would as yet be rare in international venues to find that financing global issues is discussed in terms of public finance, as outlined above. Since, in the minds of many, the issue at stake is one of aid, the international cooperation part of global-public-goods financing today is often paid out of ODA resources. It has been estimated that at present some 30 per cent of ODA resources flow into global purposes.[7]

However, as Table 4.2 shows, there are important differences between the financial aid modality and the financing of global public goods. Most important is the difference between the rationale for each of the activities, which can be related to the difference that is made at the national level between the distribution (for financial aid) and allocation (for financing global public goods) branches of public finance.[8]

Table 4.2 Financial aid versus financing global public goods[9]

Issue	Financial aid	Financing global public goods
Rationale	Equity	Efficiency
Branch	Distribution	Allocation
Policy tool	Transfer of resources	Panoply of instruments
Policy focus	Country	Issue
Main net beneficiary	Developing countries	Potentially all countries and all generations

[7] Raffer, 1999; Te Velde et al., 2002; World Bank, 2001.
[8] Kaul et al., 2003.

The basic motivation for providing foreign aid, as set forth in many official policy statements, is to assist poorer countries in their development endeavours. Thus, aid is a financial transfer from richer to poorer countries. As indicated above, it constitutes the international component of the distribution branch of public finance. The national components of this branch include such transfers as support for the unemployed or for vulnerable population groups. The motivation for the transfers is primarily equity; and equity has conventionally also been the stated motivating concern for development assistance.[10]

Global-public-goods financing, by contrast, can be viewed as the international component of the allocation branch of public finance. The primary motivation in this case is to enhance efficiency, to improve the provision level of a public good because this makes economic sense. The question, however, is, 'economic sense from whose perspective'? The answer to this question today tends to be, 'from the richer, industrial countries' perspective'. The reason for this is that, as noted in the first section, preferences for (global) public goods vary (for example, across income groups and regions). And the richer countries have the financial means to back their policy preferences with money and to finance policy reforms in developing countries. Yet, judging from some of the current global policy controversies on such issues as climate stability, the multilateral trade regime, and financial-architecture matters such as debt management, a number of these reforms are not (yet) based on a firm global consensus.

As also shown in Table 4.2, global-public-goods financing focuses on particular goods or issues. Countries come into the picture only to the extent that they can play a pivotal role in helping enhance the provision of desired goods. Using ODA resources to finance such global public goods as financial crisis prevention and management, or climate stability, may thus divert resources from poor countries and from poverty reduction. Or, if ODA funds are channelled to poor countries for these purposes, they may be inefficiently used, and may shift an undue burden of correcting global public 'bads' to the poorer countries.[11]

---

[9] Kaul et al., 2003: 337.

[10] It is a well-known fact that some foreign aid has also been siphoned off for non-aid purposes, such as promoting the donor countries' export products or fostering political alliances; this diversion of aid continues today. However, since ODA is limited, amounting to 0.2-0.3 per cent of the donor countries' income, further diversions (such as those for global-public-good purposes) have a high impact and are felt intensely by developing countries.

[11] ODA resources are also sometimes used for so-called 'offsetting arrangements': in these cases, donor countries provide financial support to developing countries for implementing, for example, projects aimed at carbon sequestration. The purpose is to lower for the donor countries the cost of achieving certain pollution-reduction targets. Financing such schemes is not foreign aid, nor does it constitute financing global public goods. It can perhaps best be classified as an international trade activity.

The development of poor countries also requires certain priority global public goods, such as a system of global knowledge management to complement current policies for intellectual property rights with measures to facilitate more proactively the application of knowledge to development. For global purposes that reflect developing countries' priorities, it would, of course, be appropriate to use ODA. Which type of knowledge production to encourage and at what price should preferably be a decision of the developing countries themselves. Where a global public good, for example, international peace and security, is of mutual interest to both industrial and developing countries, a 50:50 formula can perhaps be applied: meeting financing needs using partly ODA and partly non-ODA resources.

CONCLUSION: SYSTEMATISING GLOBAL FINANCING

The financing of global public goods faces a number of important challenges. Yet, it already exists as a 'hidden reality'. The additional changes required appear to be quite doable. They primarily consist of approaching the issue of global-public-goods financing in a more explicit as well as more coherent and integrated manner. Some reform steps that would lead to this include:

- Recognising that cooperation often happens within national borders and global challenges can, at least in part, be met by adequate national policy-making.
- Recognising the dual agenda of international cooperation financing (viz., the 'aid + global public goods' agenda) to avoid the current confusion between the two modalities, which appears to be detrimental to both.
- Tapping new sources of financing global-public-goods purposes to replace, as necessary, the current ODA resources. An obvious source for this purpose is the budget of the concerned national government agency. Environment ministries could, for example, support not only the national building blocks of 'their' global public goods but also the international cooperation components that might be required.
- Formulating a theory of global-public-goods financing to help re-equip policy-makers with new, tested, and reliable instruments that help them in meeting today's policy challenges, using the measures described above.
- Reducing information problems, notably on such issues as whether taking action to reduce a global public 'bad' is a good investment compared with financing the provision of private goods or other public goods. Without enhanced information on this point, many global public goods are likely to continue being addressed only when their underprovision has assumed crisis proportions. Table 4.3 presents a preliminary assessment, showing

for select global public goods the costs of current underprovision and the costs of possible corrective action. The message is clear: corrective action – rather than inaction – appears in most instances to be the less costly option.

- Improving participation in international negotiations, in line with the nationally well-established principle of fiscal equivalence.[12] This may require enhanced participation by various national constituencies that may at present not feel fully represented by their countries' international negotiating teams. It would also require giving full voice to all governmental teams, notably those from developing countries. Lack of participation tends to translate into lack of policy ownership and, consequently, a decrease in willingness to contribute and cooperate, financially or otherwise. Especially if supported by the foregoing measures, more participatory decision-making on global-public-goods issues could lead not only to more balanced agenda-setting but also to the identification of more opportunities for designing production paths and financing schemes so that all can gain from cooperation.

Table 4.3 Annual costs of global inaction and corrective action[13]

Type of costs	International financial stability	Multilateral trade regime	Reducing the excessive disease burden	Climate stability	Peace and security[14]
Inaction	50[15]	260[16]	1,138[17]	780[18]	358
Corrective action	0.3[19]	20[20]	93[21]	125[22]	71

[12] This principle calls for matching the jurisdiction (i.e., the circle of those taking decisions on the provision of public goods, with circles of stakeholders). This means that local public goods would be decided locally, national public goods nationally, and, by implication, global public goods in venues where all concerned parts of the global public can be heard and fairly represented.

[13] Adapted from Kaul et al., 2003: 159; amounts in US dollars.

[14] Refers to just nine conflicts in the 1990s for the full duration of the conflicts.

[15] Includes only banking crises in developing and transition countries; excludes currency and twin crises.

[16] Net benefits from removing distortions in goods markets of industrial and developing countries.

[17] Refers only to Africa's excessive burden of communicable diseases (relative to the burdens in Europe and North and South America) in 2000 in purchasing power parity exchange rates.

[18] Indicates the midrange potential reduction in global gross domestic product if the atmospheric concentration of carbon dioxide reaches twice the level of the pre-industrial era.

[19] Includes only technical assistance spending by the International Monetary Fund.

[20] Involves mostly one-time costs associated with capacity building.

[21] Estimated funding required by 2007 for the interventions proposed in CMH (2001), including commitments by both industrial and developing countries, to scale up existing interventions.

Nationally, it is often states that, based on their special coercive powers including their taxation authority, nudge or compel actors to band together and help provide public goods. The institution of the state lacks an international equivalent. Thus, international cooperation mostly happens voluntarily, which requires that the cooperative effort make sense for all. Hence, the financing of global public goods is not just a technocratic issue; it is also – and perhaps most importantly – a political one. But clarifying the technical aspects and making matters more transparent can help in revealing the distribution of net benefits, and thus help with the politics.

REFERENCES

CMH (2001), *Investing in Health for Economic Development*, Geneva: World Health Organization WHO.

Kaul, I., P. Conceição, K. Le Goulven, and R. Mendoza (eds) (2003), *Providing Global Public Goods: Managing Globalisation,* New York: Oxford University Press.

Raffer, K. (1999), *ODA and the Provision of Global Public Goods: A Trend Analysis of Past and Present Spending Patterns*, ODS Background Paper, New York: United Nations Development Programme.

Te Velde, D., O. Morrissey, and A. Hewitt (2002), 'Allocating aid to international public goods: An empirical analysis by donor and sector', in M. Ferroni and A. Moody (eds), *International Public Goods: Incentives, Measurement and Financing*, Dordrecht and Washington, DC: Kluwer/IBRD.

World Bank (2001), *Global Development Finance 2001*, Washington, DC.

[22] Annual costs to industrial countries, over 10 years, of meeting Kyoto Protocol targets for carbon dioxide emissions.

5. National Environmental Policies and Multilateral Trade Rules

Marion Jansen and Alexander Keck[1]

SUMMARY

We provide an overview of institutional, economic, and legal aspects of the relationship between national environmental policies and the multilateral trading system. Some of the difficulties are analysed that the Dispute Settlement System of the World Trade Organization (WTO) faces when evaluating disputes on national environmental policies that have an impact on international trade. From an economist's point of view, it would be desirable that optimal environmental policies, correcting for market failures, be ruled consistent with multilateral trade law. In theory, WTO law provides appropriate tools to ensure rulings that are consistent with economic thinking. Yet, the precise welfare effects of different types of environmental policies are insufficiently known. In practice, therefore, it is questionable whether economists are able to give adequate guidance to legal experts when it comes to the evaluation of national environmental policies. This is one of the reasons why there continues to be some uncertainty as to the possible interpretations of certain WTO rules in the context of environmental disputes.

INTRODUCTION

In the last decades, the intensity of commercial links between countries around the globe has increased significantly. Owing to lower transportation and communication costs, technological innovation, and lower trade barriers,

[1] The opinions expressed in this chapter should be attributed to the authors. They are not meant to represent the positions or opinions of the WTO and its members, and are without prejudice to members' rights and obligations under the WTO. The authors would like to thank Robert Teh and Frank Wijen for their comments on earlier drafts of this chapter.

the ratio of trade to gross domestic product (GDP) has more than doubled for the world as a whole in the last thirty years. Over the same period, concern about the deterioration of the environment – such as air and water pollution, acid rain, deforestation – has become an increasingly prominent policy issue. This raises the question of the links between trade and trade policy, on the one hand, and the environment and environmental policies, on the other hand. There are numerous linkages of different types between trade and the environment. Trade flows may have an effect on the environment through different channels. Trade, for instance, involves transport between countries and transport may be polluting. More trade is then potentially harmful to the environment as it increases transportation. Trade can also be expected to have a positive impact on growth, and several environmental indicators have been found to depend on a society's level of development. Trade, therefore, has an indirect impact on the environment through its effect on development. The activities of the WTO are relevant to this link to the extent that the WTO favours trade liberalisation and thus trade.[2]

In the Preamble to the 1994 Marrakesh Agreement, countries adhering to the General Agreement on Tariffs and Trade (GATT), recognised that 'their relations in the field of trade and economic endeavour' should be such that they allow for 'the optimal use of the world's resources in accordance with the objective of sustainable development, seeking both to protect and preserve the environment and to enhance the means for doing so in a manner consistent with their respective needs and concerns at different levels of development'.[3] In the Marrakesh Decision on Trade and Environment (1994), GATT members also noted that it should not be contradictory to safeguard the multilateral trading system (MTS), on the one hand, and act for the protection of the environment and the promotion of sustainable development, on the other hand. Even so, trade disputes about national environmental policies have arisen in recent years, which shows that conflicts sometimes exist between environmental policies and multilateral trade law. Some of these disputes have received much public attention, and have led some to conclude that WTO activities systematically undermine environmental standards.[4]

National environmental policies may have an effect on trade mainly through two mechanisms: they may affect companies' international competitiveness and they may affect the competitiveness of foreign companies in the national market. The first effect has received much attention in the public debate. It has been argued that national environmental policies represent additional

[2] This link is not discussed here.
[3] GATT, 1994a: 6.
[4] See, for instance, Public Citizen, 2003.

costs for domestic companies, thus giving them a competitive disadvantage in international markets. As a result, polluting industries move from countries with stringent environmental policies to countries with laxer regulations (the 'pollution haven' effect). In order to avoid massive re-allocations of companies, national governments feel under pressure to lower environmental standards, and standards, therefore, drop in accordance with the 'race to the bottom' hypothesis. Evidence of the existence of such a race to the bottom has so far been weak. National policies that give domestic companies a competitive advantage in foreign markets may conflict with WTO provisions.[5] Yet, the environmental policies discussed above tend to give domestic companies a competitive disadvantage and are, therefore, not in conflict with international trade rules. The question of whether policy-makers have incentives to implement more lenient environmental policies in order not to hurt the international competitiveness of domestic companies is not treated in detail here.

National policies that give foreign companies a disadvantage in the domestic market may be incompatible with WTO law. This may be the case in particular if such policies imply the ban of imports from foreign countries. An example is the ban of the European Union (EU) on imports of beef containing hormones. It may also be the case if governments apply taxes that mainly affect imported products, for instance, if a country decides to levy higher taxes on fuel-inefficient cars than on fuel-efficient ones and if the majority of imported cars fall into the fuel-inefficient category as defined by the environmental policy. In addition, national standards or regulations can lead to a competitive disadvantage for foreign companies if compliance with those standards or regulations implies higher costs for foreign companies than for national ones.[6] Standards and regulations can in these cases act as 'technical barriers to trade' (TBT).

As mentioned before, national environmental policies have in the past led to disputes at the WTO. Consider, for instance, the US–Shrimp case, in which a policy of the United States (US) restricted the imports of shrimps that had been captured using techniques endangering the lives of turtles. Some WTO members complained that this policy was incompatible with multilateral trade law and a first panel decision ruled that this was indeed the case.[7] This decision led to outrage among parts of the environmental community. Our intention is to give some insights into the decision-making process and to

---

5 This is especially the case for subsidies that affect exported products. These policies are not treated here.

6 According to common usage, regulations are mandatory, while standards are voluntary.

7 This decision was later reversed after the US revised its policy. See the section on legal provisions for more details on this case.

explain why certain national environmental policies may be considered incompatible with WTO law and others not. In this chapter, we address the links between multilateral trade law and domestic environmental policies, and in particular the potential conflicts that may arise between them. Our findings reveal that it is incorrect, if not naive, to assume from the outset that all environmental policies have been designed for purely environmental reasons and that it is unjustified to conclude that WTO case law shows a systematic bias 'against the rights of sovereign states to enact and effectively enforce environmental laws'.[8] At the same time, however, we point to certain difficulties the WTO faces when dealing with the interface between multilateral trade law and domestic environmental policies.

We focus on the linkages between institutional, economic, and legal aspects of the relationship between national environmental policies and the MTS. We do not pretend to provide a comprehensive discussion of any of the three domains individually. The legal discussion, for instance, focuses on the GATT Agreement, which we consider to be of fundamental importance to any environmental matter. For analytical purposes, we frequently refer to the TBT Agreement, but we do not discuss this Agreement in detail. Because of its narrow focus on certain aspects of life and health protection, the Agreement on the Application of Sanitary and Phytosanitary Measures (SPS) is not considered, although this Agreement may be relevant to certain environmental measures.[9] We focus on trade in goods and do not, therefore, deal with environmental provisions in the General Agreement on Trade in Services (GATS).

The reader should keep in mind the following. Although there is a formal obligation for WTO members to ensure that all laws and regulations are compatible with WTO law, there is no institutional policing or control of members' laws. In practice, therefore, this policing is left to individual members' complaints. Only in case of such complaints may disputes arise in which WTO compatibility is evaluated. In this process, legal experts have to base their arguments on existing legal texts, some of which were written more than 50 years ago. Ideally, such texts should give clear guidelines as to how to assess whether environmental policies are justified and thus compatible with WTO law. Justified should, in turn, ideally refer to the economic concept of optimal intervention (i.e., policies that effectively resolve existing market failures related to the environment). There are several reasons why disputes may often not lead to such an ideal outcome.

First, as we argue, it is often difficult to know exactly what the optimal policy instrument would be. Economists can generally give only rough

---

8 Public Citizen, 2003.

9 For a more detailed discussion of the legal relationship between GATT, TBT, and SPS, see Marceau and Trachtman, 2002.

indications of how certain policy interventions should be evaluated. Concrete answers can be given only on a case-by-case basis, and even then they are likely to be imprecise.[10] Second, laws have not necessarily been written on the basis of economic arguments, as imprecise as the latter may be. This is likely to be the case for certain WTO provisions that are relevant to environmental policies and that were written decades ago when, in most WTO member countries, environmental concerns were not high on the political agenda.

The remainder of this chapter proceeds according to the following structure. Section 2 gives an overview of the discussions and negotiations on trade and environment that have taken place within the GATT/WTO since the creation of the GATT. In the next section, our analysis of the economic linkages between environmental policies, on the one hand, and trade and trade policy, on the other hand, is described. This is followed by a discussion of key GATT provisions that form the legal basis for the analysis of conflicts between national environmental policies and the MTS, including a discussion of how Dispute Settlement Reports have interpreted these provisions in cases that are relevant to the trade and environment debate. Finally, we present conclusions.

DISCUSSIONS WITHIN THE GATT/WTO

When the MTS was reconstructed after the Second World War, the environmental consequences of economic integration were not a primary concern for policy-makers. This may explain why references to the environment were only indirect in the original GATT. The issue was first put on the agenda in the early 1970s, but it was only in the 1990s that a real discussion on the relationship between trade and the environment started within the GATT. This discussion is ongoing in the WTO, including in the WTO's Committee on Trade and Environment (CTE). Aware of the importance of policy coordination at the national level, the CTE actively seeks an exchange of ideas with environmental and trade experts from member governments. Although this exchange of ideas has certainly helped to enhance our understanding of the linkages between trade and the environment, progress has been slow with regard to mitigating the potential tensions between multilateral trade law and national environmental policies.

During the preparatory work for the Conference on the Human Environment (Stockholm, 1972), the GATT Secretariat was requested by the

[10] What is the optimal level of an environmental tax? When should a labelling policy be preferred over a price-based policy?

Secretary-General of the Conference to make a contribution. In the debate around this contribution, the idea came up to create within the GATT a flexible mechanism to observe the problems that could be created for international trade by anti-pollution measures concerning industrial processes. In 1971, a Group of Environmental Measures and International Trade was established for this purpose. It was set up as a standby machinery which would be ready to act at the request of a contracting party. In nearly twenty years, however, no request was made to convene a meeting of the Group.[11]

A request was made for the first time in 1990 by the countries of the European Free Trade Association (EFTA). Among the reasons they gave for their request, they explained that:

> [t]he approach to environmental policy making varied considerably from country to country due to differing geographical settings, economic conditions, stages of development and environmental problems. Accordingly, governments' priorities on these problems differed as well. The important point here was that the resulting differences in actual policies could set the stage for trade disputes. The EFTA countries' prime concern was to ensure that GATT's framework of rules worked, provided clear guidance to both trade and environment policy makers and that its dispute settlement system was not faced with issues it was not equipped to tackle...[12]

As a result of this request, a meeting of the Group of Environmental Measures and International Trade was finally convened, and the links between trade and environmental policies have been a subject of discussion within the GATT/WTO ever since. Even so, the concern expressed by the EFTA at the time is probably still valid today.

The debate on trade and the environment was further institutionalised within the GATT through the Marrakesh Decision on Trade and Environment in 1994. In this decision, GATT members noted that it should not be contradictory to safeguard the MTS, on the one hand, and act for the protection of the environment and the promotion of sustainable development, on the other hand. They further noted their desire to coordinate policies in the field of trade and the environment, 'but without exceeding the competence of the MTS, which is limited to trade policies and those trade related aspects of environmental policies which may result in significant trade effects'.[13] The Marrakesh Decision also foresaw the establishment of the CTE, which took place in January 1995. The tasks allocated to this Committee were to 'identify the relationship between trade measures and environmental measures, in

[11] See Annex I of Nordström and Vaughan, 1999.
[12] Annex I of Nordström and Vaughan, 1999: 68.
[13] GATT, 1994a: 470.

order to promote sustainable development' and 'to make appropriate recommendations on whether any modifications of the provisions in the MTS are required, compatible with the open, equitable and non-discriminatory nature of the system'.[14] The Marrakesh Decision lists ten items of relevance, including:

- 'The relationship between the provisions of the MTS and trade measures for environmental purposes, including those pursuant to multilateral environmental agreements'
- 'The relationship between environmental policies relevant to trade and environmental measures with significant trade effects and the provisions of the MTS'
- 'The relationship between the provisions of the MTS and:
  a) charges and taxes for environmental purposes
  b) requirements for environmental purposes relating to products, including standards and technical regulations, packaging, labelling and recycling'

The 1994 Marrakesh Agreement Establishing the World Trade Organization (the 'WTO Agreement') also introduced new elements to the legal texts that are of relevance to the trade and environment debate. Its Preamble included, for the first time in the context of the MTS, reference to the objective of sustainable development and to the need to protect and preserve the environment. During the negotiations preceding the Agreement (the 'Uruguay Round'), it was proposed to alter GATT Article XX and to explicitly mention the protection of the environment as a valid reason for a member to depart under certain conditions from its obligations under the Agreement. Although no effect was given to this proposal, Article XX is of relevance to all legal disputes concerning the compatibility of environmental policies with the MTS. This is also true of the 1994 WTO Agreements on TBT and SPS. Below, we provide a more detailed discussion of these Agreements and on how existing WTO Dispute Settlement Reports have interpreted WTO law in disputes that are relevant to the trade and environment debate.

The CTE continued to meet and pursue its mandate after the Marrakesh Conference. It focused its activities on the preparation of a report to the 1996 Ministerial Conference in Singapore. In this report, the CTE pointed out a number of concerns that are still of relevance today:

- On the issue of the relationship between trade measures in multilateral environmental agreements (MEAs) and the MTS, the report expressed concerns regarding the measures applied by MEA signatories to WTO

[14] Nordström and Vaughan, 1999: 72.

members that were not a party to the MEA in question.

- The report expressed concerns about the possible trade effect of eco-labels: the multiplication of eco-labelling schemes with different criteria and requirements, or the fact that they could reflect the environmental conditions, preferences, and priorities prevailing in the domestic market, might have the effect of limiting market access for overseas suppliers.
- More generally, the report also stressed that further discussion was needed on how criteria based on non-product-related processes and production methods should be treated under the TBT Agreement. This point is also relevant to the eco-labelling debate as eco-labels tend to be based on such criteria.

The Doha Declaration reflected the same concerns and, to a certain extent, stressed the need for further discussions.[15] Yet, its text also reflected the lack of real advances in the debate so far. For the first time, negotiations as opposed to 'discussions' on trade and the environment were initiated, as the Doha Declaration instructed members to negotiate on 'the relationship between existing WTO rules and specific trade obligations set out in multilateral environmental agreements'. Yet, the Declaration also stipulated that 'the negotiations shall be limited in scope to the applicability of such existing WTO rules as among parties to the MEA in question. The negotiations shall not prejudice the WTO rights of any member that is not a party to the MEA in question.'[16] This implies that the point of concern expressed in the CTE's Report to the Ministerial Conference in Singapore was explicitly excluded from the negotiations.

The Doha Declaration also instructed the CTE to pursue its work on all items on its existing agenda and to give particular attention to, among other things:[17]

- 'The effect of environmental measures on market access, especially in relation to developing countries, in particular the least-developed among them, and those situations in which the elimination or reduction of trade restrictions and distortions would benefit trade, the environment and development.'
- 'Labelling requirements for environmental purposes.'

Discussions so far reveal that members have espoused quite different positions, with some members stressing the importance of environmental protection and others being mainly concerned with avoiding the misuse of

---

[15] WTO, 2001d.
[16] WTO, 2001d: paragraph 31 (i).
[17] WTO, 2001d: paragraph 32.

environmental arguments to impose unjustifiable trade barriers. This may be one of the reasons for the lack of concrete results in the discussions so far. Nevertheless, it must be conceded that the problem at hand is complex and members seem to be aware of this. The different trade-offs are reflected in a CTE Report, which states:[18]

> In striking the appropriate balance between safeguarding market access and protecting the environment, it was felt that there was a need to examine *how* environmental measures could be designed by importing countries in a manner that (i) was consistent with WTO rules; (ii) was inclusive; (iii) took into account capabilities of developing countries; and, (iv) met the legitimate objectives of the importing country.

Although labelling requirements could in principle be considered to be one of many environmental measures affecting market access, they were considered to be a separate agenda item in the Doha Declaration. The CTE Report once again reflected the existing differences between members and the trade-offs that are inherent in the trade and environment debate. Most members agreed that 'voluntary, participatory, market-based and transparent environmental labelling schemes were potentially *efficient economic instruments* in order to inform consumers about environmentally friendly products. As such they could help move consumption on to a more sustainable footing. Moreover, they tended, generally, to be less trade restrictive than other instruments. It was also noted, however, that environmental labelling schemes could be misused for the protection of domestic markets.'[19] On the other hand, some members noted that 'for developing countries, and their SMEs in particular, difficulties arose along with the growing complexity and diversity of environmental labelling schemes in export markets'.[20] For developing countries, 'the recognition of the *equivalency* of their own certification systems was an area of particular concern'.[21] Regarding the *basis* for environmental labelling schemes, familiar differences of views remained on what was characterized by one member as 'the root cause of controversy surrounding the labelling debate: the WTO compatibility of measures based on non product-related processes and production methods (NPR-PPMs)'.[22]

In sum, an active debate on the issue of trade and the environment was initiated when EFTA countries pointed out in 1991 that it was desirable for the GATT's framework of rules to work, to provide clear guidance to both

---

[18] WTO, 2003: paragraph 6.
[19] WTO, 2003: paragraph 30.
[20] WTO, 2003: paragraph 31.
[21] WTO, 2003: paragraph 32.
[22] WTO, 2003: paragraph 34.

trade and environmental policy-makers, and to prevent its dispute settlement system from being faced with issues it was not equipped to tackle. Today, GATT Articles III and XX (from 1947) and, to a certain extent, the 1994 WTO Agreement on TBT form the legal basis for the analysis of conflicts between national environmental policies and the MTS. In the following sections, we argue that these provisions and resulting WTO jurisprudence give an imperfect reflection of potentially complex economic linkages between environmental policies and trade flows. As a consequence, and notwithstanding the intensive debate within the GATT/WTO, the MTS still does not seem to provide full certainty as to how environment-related trade disputes are dealt with.

Lack of progress in the debate and negotiations on trade and the environment can be explained by a number of factors. First, it should be noted that any tension at the international level between the two policy areas arises because countries differ and, therefore, have different priorities when it comes to environmental policies. It is because of these differences that countries also defend different interests when discussing or negotiating issues relating to trade and environmental policy, which sometimes makes it difficult to find common ground. Second, developing countries are particularly sceptical about the trade and environment debate, as they fear that developed countries press the issue at the WTO with a protectionist intent. Third, the relationship between environmental policies, on the one hand, and trade and trade policies, on the other hand, is a complex one. In the next section, we argue that even from a purely economic point of view (i.e., ignoring political and legal issues), it is not always clear how environmental policies should be designed when countries with different priorities start trading. This is a major explanation of the above-mentioned shortcomings in the legal situation, and is further discussed below.

ENVIRONMENTAL POLICIES IN OPEN ECONOMIES

The Role of Product Differentiation in Trade and Environmental Policy

National environmental policies typically aim at making economic activity more environment-friendly. Governments may, therefore, try to encourage the use of certain environment-friendly products and discourage the use of environment-unfriendly products. In order to reduce the emission of greenhouse gases, a government may, for instance, decide to tax cars according to their fuel efficiency. Such an environmental policy would de facto discriminate between varieties belonging to the same product category. It is from this discrimination that the potential conflict with international trade law arises. The aim of WTO provisions is to avoid protectionism (i.e., discrimination against foreign products with the aim of favouring domestic producers). Protectionist policies imply discrimination between varieties in the same product category on the basis of their origin. A problem may arise if an environmental policy discriminates between varieties of the same product category and if mainly varieties produced abroad suffer the negative consequences of this discrimination. It may be difficult in such cases to determine whether the relevant policy is an environmental policy measure that happens to discriminate against foreign producers or whether it is a protectionist measure disguised as an environmental policy.

The notion of products appearing in different varieties is common in the economic literature. Numerous economic models contain elements of product differentiation, some of which are popular in the trade literature.[23] This concept does not seem to be adequately reflected in the WTO legal texts. Article III of GATT 1994 plays an important role in determining whether a policy is protective. Paragraph 2 of this article states that foreign products should not be treated less favourably than 'like domestic products' once imported into the territory of another country. This paragraph thus creates the impression that national policies conflict with WTO law if they treat foreign products less favourably than domestic products belonging to the same product group. The fact that the foreign and the domestic products may belong to different varieties within the same product group is thus not adequately reflected in the legal text. Cars with different levels of petrol consumption per kilometre and different impacts on the environment may thus be considered to be goods belonging to the same product group. A policy imposing higher taxes on cars with higher petrol consumption could be considered an infringement of Article III if the country imposing the tax happened to be an importer of cars with high petrol consumption. Below, we provide a more detailed discussion of WTO law and the way it has been applied to cases relevant to the environment debate. In this section, we focus on why the concept of product varieties is important in relation to environmental policies.

Economists distinguish between two types of product varieties: varieties that differ in a 'vertical' way and varieties that differ in a 'horizontal' way. In the case of vertical differentiation, different varieties can be ordered according to a certain scale. One variety is better than another, larger than another, noisier than another, etc. Examples of such varieties are chocolates with higher or lower contents of butter, or cars that use more or less petrol per kilometre. The characteristic that differentiation is based on is the content of butter in the first case and petrol use in the second case. This characteristic

---

[23] Consider the literature on intra-industry trade (e.g., Krugman, 1980).

appears to a greater or lesser extent in the different product varieties. One characteristic of vertical differentiation is that it often leads to price differences among varieties. Consumers may, for instance, be willing to pay a somewhat higher price for a car that consumes less petrol because it will in the long run be cheaper to drive such a car. In the case of horizontally differentiated products, the characteristics that are responsible for the differentiation cannot be ranked. Colour and flavour are examples of such a characteristic. A red t-shirt is different from a blue t-shirt, but the two varieties cannot be ranked according to an objective scale. The same holds for strawberry ice-cream and vanilla ice-cream. Horizontal differentiation typically does not lead to price differences. Many products are differentiated along both lines. Cars, for instance, appear in different colours and differ in their use of petrol. But for the purpose of this study, we draw a clear distinction between the two types of differentiation. Environmental policies in general aim at characteristics leading to vertical product differentiation.

Consumers differ and they appreciate the characteristics of products in different ways. The availability of different varieties of products in the market should, therefore, be welcomed. In general, it can be presumed that markets provide those varieties demanded by consumers and that they provide them in the appropriate quantities. However, this is not always the case. In some cases, consumers may be better off if governments or private institutions enforce the supply of only one product variety in the market (harmonisation). In other cases, there is an undersupply of varieties in the market, or certain varieties are not supplied in optimal quantities.

Harmonisation can mostly be justified in the case of certain types of horizontally differentiated goods (for instance, plugs). It is more efficient and less costly for consumers if they know that any plug they buy fits in all sockets available. For efficiency reasons, it makes sense to harmonise the design of plugs and sockets sold in the market. Note that it does not matter for the functioning of the market what the plug and socket look like (e.g., wide or narrow pegs). What matters is that the plugs and sockets supplied fit together.

An undersupply of varieties occurs mainly in the case of vertically differentiated products and tends to be the result of information asymmetries in the market. Information asymmetries occur when producers have information about the characteristics of the goods they produce that consumers do not have when they purchase these products. In many cases, consumers discover those characteristics once they use the products. Economists refer to 'experience goods' in these cases. Consumers may not see that a washing machine is broken when buying it, but if the machine starts losing water upon use, they know that something is not in good order. Some product characteristics, however, may never be discovered. This can happen in the case of 'credence goods'. Consumers do not know how many calories a chocolate bar has, even after eating it, and they will never know the flour content of their toothpaste. It is when consumers do not know the characteristics of the products they use, even after a long time span, that undersupply of varieties may occur.[24] This is because all product varieties are the same for consumers when buying them. They will, therefore, go for the cheapest products supplied as they cannot be sure that the more expensive products are of a 'higher quality', which would in this context correspond to more environment-friendly varieties. If producing varieties that are more environment-friendly is more expensive, producers of these varieties may not be able to break even, as less environment-friendly varieties are supplied more cheaply in the market and consumers are only willing to pay the lowest price they perceive. As a consequence, and depending on a number of market characteristics, environment-unfriendly varieties may push more environment-friendly products out of the market.

Externalities are another type of market failure that may justify government intervention in the supply of vertically differentiated goods, as they tend to lead to the over- or undersupply of particular product varieties. Sometimes, in the process of producing or consuming certain goods, harmful or beneficial side-effects called 'externalities' are borne by people not directly involved in the market exchange. An often-used example of an externality is that of a chemical firm polluting the river on the banks of which it is based. Fish in the river die as a consequence of the pollution and fishermen living on the banks of the same river are unable to make a living. The chemical firm's activity thus has a negative impact on the fishermen's activity, but without government intervention, the firm does not have to make any (compensation) payments to the fishermen. In the absence of government intervention, the chemical firm, therefore, produces more than is optimal from the point of view of the whole economy. Intervention is, therefore, desirable to reduce production and thus reduce the pollution of the river.

Why Markets Fail in Relation to the Environment

Environmental policies are typically related to problems of externalities. It is useful to distinguish two types of externalities: externalities that occur when consuming a good and those that occur when producing a good. An example

[24] In the case of experience goods (i.e., if consumers can discover product characteristics upon use), producers can offer guarantees to signal high product quality or have the possibility of building a reputation for offering a certain type of product (Shapiro, 1983). As a result, a large range of product varieties will be supplied in the situation of market equilibrium. If consumers do not discover product characteristics upon use, only a restricted number of varieties may be supplied in equilibrium, or markets may completely disappear (Akerlof, 1970).

of the first category is driving a car. The carbon dioxide emissions involved deteriorate the quality of the air and also contribute to the problem of global climate change. Driving a car negatively affects the well-being of others, because of the air pollution this activity creates. Externalities related to consumption can be local or global in nature, this example being one with a global character. Externalities related to the production process can also be local or global in nature. Many production processes, for instance, involve the burning of fossil fuels and thus have a global effect. The previous example of a chemical plant polluting a river is one of a local nature.

Economists' suggestion for solving this kind of problem is to use taxes or subsidies in order to ensure that market prices reflect the real value of a product or activity. The chemical company would, for instance, have to pay taxes for polluting the river. To achieve a reduction in the emission of greenhouse gases, a government would, for instance, need to impose a tax on the carbon content of fuels. A tax on cars related to their fuel efficiency would be another option, which is lower in the hierarchy of possible policy instruments, but superior to a tax on cars based on size.

In some cases of production externalities, prices can reflect the environmental concerns of consumers without government-imposed taxes or subsidies, although only to a certain extent. Take the example of salmon. It has been argued that salmon farms pollute the marine environment they are situated in. Some consumers of salmon may be concerned about the marine environment and, therefore, decide no longer to buy farmed salmon but to switch to wild salmon instead. They may even be willing to pay a higher price for wild salmon than they paid for farmed salmon. The mark-up consumers are willing to pay is thus a reflection of the value of the externality, i.e., of how much consumers value the more environment-friendly production method. It is, however, unlikely that the externality is completely controlled for. There may be plenty of people who care for the marine environment, but do not like to eat salmon. Those people have no opportunity to express their disapproval through the market. The externality is still at work, but only partly.

Another market failure may lead to additional complications. Information asymmetries may make it difficult for consumers to distinguish between farmed and wild salmon when buying the product. Farmed salmon is normally grey and thus easily distinguishable from the pink wild salmon. But producers add chemical additives to make farmed salmon look similar to wild salmon. It was explained above what happens if consumers lack important information about a relevant product characteristic. They will consider farmed salmon and wild salmon to be the same product and will pay the same price for both. Prices will, therefore, not reflect the environmental concerns of consumers, because of the information asymmetry. It cannot be excluded that wild salmon would slowly but surely disappear from the market if its 'production' were more costly than that of farmed salmon. This is because catching wild salmon would no longer be a profitable activity at the prices prevailing in the market.

The introduction of a labelling policy can help to overcome the problems caused by an information asymmetry. A label indicating whether salmon is farmed or wild would allow consumers to distinguish between the two product varieties. Wild salmon would then be able to obtain the mark-up in the market reflecting the willingness of consumers to pay for the environment. Note that this mark-up would not reflect the full extent of the externality at work, as the opinion of those not consuming salmon would not be reflected in this mark-up. This is why, in theory at least, a government-imposed tax on farmed salmon would be a preferable instrument, as the government would take into account all individuals in the economy when evaluating the value of a cleaner marine environment and the resulting tax level. But what works well in theory does not necessarily work well in practice. Indeed, in practice, it is virtually impossible for a government to know the value each individual attaches to the marine environment and to calculate the optimal tax to be applied in this particular case. Excessive taxation would introduce new distortions to the market, while too-moderate taxation would only partly account for the externality. A price-based policy may completely miss the target and, besides, expose the government to the criticism that its policy does not reflect consumer interests. A labelling policy will lead to market prices that reflect the interests of at least a section of the consumers. This makes labelling policies attractive, notwithstanding their above-mentioned shortcomings.

Note that, in the above-mentioned case, the labelling policy corresponds to what is referred to as a 'standard' in the WTO context. A standard defines a set of criteria that producers can choose to satisfy or not. In the case of credence goods, a label is needed in order to indicate to consumers which standard the relevant producers have chosen (e.g., whether salmon has been raised or captured). Labels in such cases merely reveal information regarding the application of a predefined standard. Labels can also serve the purpose of giving information on product characteristics, like the indication of calorie content on food, fruit content in juices, etc. Which labelling policy is more appropriate is a complex question that is not dealt with here.

Another policy a government may need to impose in order to overcome environmental externalities is a regulation imposing the circulation of only a limited number of product varieties in the market. Product varieties not satisfying the specified criteria are thus banned from the market. If the relevant good is traded, a regulation may thus result in an import ban, to the extent that product varieties not satisfying the criteria cannot be imported. Market mechanisms are no longer at work. Such a policy indicates a strong preference for the environment in the trade-off between environment and price, as cheaper but less environment-friendly products are no longer supplied.

Different Concerns, Different Market Failures, Different Policies

In general, there would be no potential for conflict between environmental policies and international trade law if environmental policies were equal across countries. So why is this not the case? Why does one country ban a certain product, while another country does not? Why do the environmental standards introduced in one country not correspond to those in another country? There are basically two reasons for such differences. The first reason is that countries pursue different environmental objectives. The second reason is that countries prefer to use different policy measures to achieve a given environmental objective.[25]

In principle, it is perfectly legitimate for governments to pursue different environmental objectives. Countries differ in many aspects, such as size, climate, population density, and geography. And people differ across countries and within countries. They have different tastes, face different financial restrictions, and have different attitudes towards risk. All these aspects determine whether and to what extent certain environmental concerns affect their well-being. Waste reduction is likely to be a greater concern in densely populated countries than in countries with scarce population. The scarcer a country's water supply, the more likely people are to be worried about water pollution. In general, concern for the environment seems to increase with wealth. It is easier for consumers to pay a higher price for free-range eggs, bio milk, and wild salmon when they do not have to worry about having enough money to cover their basic expenses for food, shelter, and clothing.

Different objectives also lead to different policies across countries, and this may have an effect on trade.[26] Looking at price-based policy instruments, assume that a country decides to impose a tax on a product variety that it considers to be less environment-friendly than other product varieties, and assume that the country imports a certain quantity of the less environment-friendly variety. Foreign producers of the relevant variety would be hit equally hard by the tax as would domestic producers. A distinction should be

---

[25] Differences in conformity assessment procedures can also represent a barrier to trade; this, however, is not discussed in detail here. See, for instance, Sykes (1995) for more information on this issue.

[26] See also Vaughan, 2001.

made in this case between a local environmental externality and a global externality. If the relevant externality is local in character (remember the example of the chemical firm), it is not obvious why taxes should affect imported products at all. If the externality is global, however, a tax is, in principle, perfectly legitimate from an economic point of view, even if the outcome of this policy is that mainly imported product varieties are affected by the tax. Certainly, the relevant tax may appear suspicious in such a case and give the impression of being used for protective reasons, but it is possible that this is just the unfortunate outcome of an otherwise legitimate environmental policy.

It is often difficult to come to a clear-cut judgement of such situations. Governments typically have some room when deciding which varieties will be affected by the tax, for instance, when deciding the maximum level of fuel consumption at which cars will not be taxed. Governments should take into account a number of factors when taking such decisions, including consumer preferences, production costs, and the extent of the externality. Yet, as pointed out above, information on those factors is far from perfect and the decisions are, therefore, likely to be imperfect from an economic point of view. Government decisions can, of course, also be influenced by protectionist purposes, leading to outcomes that favour domestically produced varieties over foreign-produced varieties. In practice, therefore, government decisions may be determined by a combination of three factors: legitimate environmental purposes, imperfect information about the policy instrument to use, and protectionist intents. In cases of trade disputes, it is difficult for external observers to disentangle the three factors. Yet, this is exactly what the WTO Dispute Settlement System is supposed to do.

If governments choose a non-price-based policy to discourage the use of environment-unfriendly products, the situation becomes more complex. Consider the case of a label indicating the use of a standard, such as an eco-label. Product varieties carrying the eco-label signal to consumers that the varieties have been produced using certain environment-friendly production methods. The government decides which production methods qualify for the use of the label.[27] As discussed above, the typical outcome of such a label is that labelled products are sold at a higher price as consumers are willing to pay a mark-up for more environment-friendly production methods.

---

[27] Labelling policies can also be based on private initiatives. Private entities (such as companies or non-governmental organisations) in those cases determine the labelling policy and guarantee that labelled varieties are indeed produced according to the specified methods. Yet, in the case of credence goods, it is difficult for consumers to verify the assertions made by the private entities and there are significant incentives for the private entities to cheat. See, for instance, Brown (1999) on the implementation problems of labelling programmes related to the use of child labour.

Labelling policies have the potential to have stronger trade effects than do price-based policies. One important reason for this is that the effect of the labelling policy can be different for foreign producers than for domestic producers if foreign producers face a different labelling policy at home. Assume that the foreign labelling policy is less restrictive than the domestic one in the sense that varieties qualify for the eco-label in the foreign market that would not qualify in the home market. Foreign producers now have the option of upgrading their production methods, which is costly. Besides, the upgraded methods may be too expensive for the foreign market as it is not necessary to apply these methods in order to qualify for the label there. As a consequence, foreign producers may be forced to apply different production methods for the home market and for the foreign market, which is also costly. Labelling policies, therefore, bear an element of discrimination against foreign producers when those policies differ across countries, which is not necessarily the case for price-based policies. Labelling policies, however, keep the advantage of bearing a stronger element of consumer sovereignty than do price-based policies, in the sense that consumers decide whether and to what extent labelled varieties can command higher prices in the market.

It would be interesting to know in which situations price-based policies should be preferred over regulatory policies, such as labelling, and vice versa. Economists, however, tend to analyse the two policy tools in different types of models, which makes a direct comparison of the measures difficult.[28] Currently, economic analysis does not, therefore, seem to permit the giving of clear indications as to which policy tool to apply when.[29]

In order to overcome the problems inherent in labelling policies, trading partners have a number of policy options. They may agree to mutually recognise their labelling policies, in the sense that product varieties carrying an eco-label in the foreign market can automatically qualify for carrying the label in the home market. This seems to be an attractive option, in particular where local externalities exist. It should be noted, however, that this approach carries the risk of the home label being undermined. If consumers cannot distinguish between the foreign and the domestic label, they will no longer be willing to pay a higher price for the domestic one, even though it is more

[28] Taxes tend to be examined in models with homogeneous consumers, whereas regulatory measures tend to be analysed in models with heterogeneous consumers. This makes a direct comparison of the effects of both measures difficult.

[29] The approach towards regulatory policies taken in WTO Agreements also seems to reflect a certain dichotomy in that the TBT Agreement contains references to the design of regulatory policies and to international standards, but not to price-based policies such as tariffs or taxes. It could, therefore, be argued that, in cases of disputes about a measure such as an eco-label, legal experts are unlikely to evaluate whether an eco-tax would be a less discriminatory policy in the relevant situation.

environment-friendly.[30] The problem is, again, that consumers do not know this is the case. Mutual recognition, therefore, seems to be an option only for similar countries that pursue similar environmental objectives.[31] Accepting a trading partner's labelling policy in such cases would not have the effect of putting the national policy objectives at risk. Article 2.7 of the TBT Agreement, on the equivalence of foreign technical regulations, seems to reflect a similar line of thinking. This Article states that, if policies of trading partners differ from domestic policies but 'adequately fulfil domestic policy objectives', members should consider accepting these policies as equivalent.[32] Article 2.7 thus seems to encourage mutual recognition of regulatory policies, but only to the extent that these policies fulfil the same objectives.

Harmonisation of labelling policies is another option; trading partners may consider agreeing on common standards or applying already existing international standards.[33] It may be quite cumbersome to reach agreement on such standards given the differences in priorities. It was mentioned above that, from an economic point of view, it is not clear whether harmonisation is a desirable outcome at all. In principle, each country is better off applying the policy that best corresponds to its own setting. Harmonising policies thus implies a welfare loss for each country. On the other hand, it may lead to welfare gains from increased trade. Harmonisation, therefore, makes sense if the gains from trade exceed the losses from harmonisation.[34] To a certain extent, the TBT Agreement seems to take this trade-off into account, as it reflects the idea that harmonisation may not be desirable if countries pursue significantly different objectives. Article 2.4 of the TBT Agreement states that members shall use relevant international standards if they exist, except when such international standards or relevant parts of them are ineffective or inappropriate to fulfil the 'legitimate objective' pursued by the member. Deviation from international standards is thus discouraged,[35] but only to the extent that these standards allow member countries to fulfil their policy

---

[30] Jansen and Lince de Faria, 2002.

[31] The EU's approach to the treatment of different national standards within the EU contains a strong element of mutual recognition (Messerlin and Zarrouk, 2000).

[32] GATT, 1994a: 140.

[33] The size of countries may affect their bargaining power in such harmonisation attempts. Baldwin (2000) pointed out that hegemonic harmonisation is de facto the default option for many small nations that are heavily dependent on large trading partners, in the sense that the large trading partners impose their own policies on the smaller trading partners.

[34] Compare this with the previous discussion on horizontally differentiated goods, in which such a trade-off is not present as harmonisation is efficiency-enhancing. Strictly speaking, there might be an adjustment cost for one of the trading partners, but the loss from harmonisation in the case of vertically differentiated goods is a permanent one.

[35] Paragraph 2.5 of the TBT Agreement further indicates that technical regulations that are in accordance with international standards are presumed to be WTO-compatible.

objectives.[36] It could, therefore, be argued that harmonisation in the TBT Agreement should be interpreted as the harmonisation of policy measures for a given policy objective and not as harmonisation of policy objectives. This approach is in line with economic thinking on this question. Geographical and climatic conditions differ across countries, as do consumer preferences and other aspects that are relevant to the definition of environmental policy objectives. It would, therefore, not be desirable to harmonise those objectives at the international level. Yet, where similar objectives are pursued, it is desirable to harmonise policies in order to minimise economic inefficiencies introduced by the policies themselves.

Another important problem occurs when labels refer to NPR-PPMs, because they do not leave any easily measurable impact on the characteristics of the products when they come on the market. The problem that arises is one of control, and it applies to our example of eco-labels. The use of more environment-friendly production methods in the production of, for instance, carrots has no systematic, measurable effect on carrots. Their size and taste may be influenced, but the former is not affected in a systematic way and the latter is not something that can be measured according to an objective criterion. The question then arises as to how importing countries determine whether the imported product varieties qualify for labelling. In order to take such decisions, the production sites themselves would need to be controlled, but they are located abroad. Countries may be reluctant to allow foreign entities to control their production sites, as they may interpret this as a loss of sovereignty.[37] This is probably the single most important reason why measures based on NPR-PPMs have so far been incompatible with WTO rules, and why it is so difficult to change this. The extensive but so far fruitless debates within the WTO on the issue of eco-labelling are a reflection of the relevance of this problem, but also of the difficulties of solving it.

How to Deal with Trade Disputes on Environmental Policies?

A number of lessons can be learned from the above with regard to evaluating a trade dispute in which one trading partner accuses the other of applying a protectionist policy that is disguised as an environmental policy. The task is to identify whether and to what extent three different factors have determined the specific design of the particular environmental policy: legitimate environmental purposes, imperfect information, and protectionist intents. The

[36] The definition of policy objectives is, therefore, of some relevance. The Preamble to the TBT Agreement, for instance, indicates that each member is free to define these objectives as it wishes. Each member can thus choose to protect the environment at the level it considers appropriate. See Marceau and Trachtman, 2002, for more on this issue.

[37] Although in the settled US–Shrimp case, the US now offers to certify imports on a vessel-by-vessel basis.

first conclusion to be drawn from our discussion is probably that there are not many straightforward rules to be applied and that there is still much room for economic research to improve our understanding of the linkages between environment and trade policies.

The fact that a policy differentiates between product varieties within a product category is not an indication of an illegitimate policy. On the contrary, environmental policies typically aim at making such a differentiation. The fact that the policy is applied by one country but not by another is also not an indication of protectionist intentions. It is perfectly legitimate for countries to have different preferences with respect to environmental quality, and differences in policies may reflect such differences in preferences. It is also not always clear how to rank different types of policies. The discussion above shows, for instance, that the use of labelling policies has both advantages and disadvantages when compared with price-based policies.

If a policy predominantly affects product varieties supplied by foreign producers, this could be a reason for suspecting that the policy has been installed with the intention of protecting domestic producers. But it can at best be a reason for suspicion; it does not constitute a proof that the policy is unacceptable. Perfectly legitimate environmental policies may happen to affect foreign producers harder than domestic ones. Economic theory gives several reasons why this may happen. Foreign producers may, for instance, have a competitive advantage in the production of less environment-friendly products; or domestic producers may have a better knowledge of the domestic market, and have already adapted to consumer preferences before the introduction of the policy.

Evaluation of a case should focus on the characteristics that determine the differentiation between varieties relevant to the environmental policy. This characteristic should be related in an evident way to the environmental objective that is pursued. A car's level of fuel consumption can, for instance, easily be related to the objective of reducing air pollution.

The 'cut-off point' that determines whether a variety is affected by the policy or not can also give some indications as to whether an element of protection is present in the design of the policy. If a minor change of the cut-off point (e.g., the level of petrol consumption that determines whether a vehicle is taxed or not) has no major impact on the environmental effect of the policy, but does reduce imports significantly, there are reasons for suspicion that the design of the policy is not based purely on environmental concerns.

The consistency with which a policy is applied can also play a role in evaluating a trade dispute. The policy should be applied consistently across varieties in the same product group, and maybe also in related product groups, where varieties can be distinguished on the basis of the same characteristics of relevance to the environmental policy. It could, for instance, be considered inconsistent to tax fuel-inefficient passenger cars, while fuel-inefficient sports vehicles are not taxed. It could also be considered inconsistent to ban the use of hormones in beef production but not in pig rearing. In such cases, it can make sense for the WTO Dispute Settlement to require, as has happened in the past, greater consistency in order for the relevant policy to be considered WTO-compatible.

It is possible to be a bit more specific with reference to externalities with a local effect. One conclusion that may be drawn from the above analysis is that foreign producers should not be affected by policies aiming at local production externalities. Regarding local consumption externalities, it may be argued that price-based policies should be preferred over regulations or standards. This is because a tax would affect foreign and domestic suppliers equally. Standards or regulations, however, have the potential to inflict higher costs on foreign producers than on domestic producers. This is the case if norms in the foreign suppliers' country of origin are different and if foreign suppliers are, therefore, forced either to use two production processes simultaneously or to abandon one of the two markets.

RELEVANT GATT PROVISIONS AND JURISPRUDENCE

As stated above, the focus of this chapter is on national environmental policies that encourage the use of environment-friendly products and discourage the use of products seen as being relatively less environmentally benign. Thus, a regulatory distinction is made between product varieties belonging to the same product category. It must be assumed that usually both varieties are produced both domestically and by foreign producers. A trade problem may arise if the product variety disfavoured by an environmental policy is imported to any significant extent and/or is of major export interest to another country.[38] It needs to be examined whether the measure simply happens to affect foreign producers comparatively harder or whether it is de facto protectionist and disguised as an environmental policy. WTO rules clearly aim at preventing discrimination against foreigners and disguised

[38] It is interesting to note that merely 'different' treatment of like products, including on the basis of origin, cannot automatically be considered a violation of GATT Article III. In Korea–Beef, the Appellate Body clarified that '[a] formal difference in treatment between imported and like domestic products is thus neither necessary, nor sufficient, to show a violation of Article III: 4' (WTO, 2001a: paragraph 137). What mattered instead was a determination of 'less favourable' treatment. See the Korea–Beef case (WTO, 2001a: especially paragraph 138), the EC–Asbestos case (WTO, 2001b: especially paragraph 100), and our discussion of less favourable treatment below.

restrictions on international trade, while, at the same time, affording sufficient regulatory autonomy to governments to pursue policies genuinely targeted at environmental concerns, and this perhaps irrespective of the relative incidence of foreign and domestic producers. One of the central questions posed by domestic policy-makers may, therefore, be how genuine national environmental policies may be designed in order not to be seen as protectionist, even in the absence of explicit references to origin.

This section presents key legal provisions and WTO jurisprudence that may be relevant to national policy-makers in forming their judgement on the potential WTO compatibility of environmental policies. It is important to note that there is no obligation for national governments to undergo a 'compatibility test' with multilateral trade law a priori. Governments are free to set environmental policies according to their own preferences. Nevertheless, these policies can be challenged by other WTO members if those have the impression that they are incompatible with WTO law. This typically happens only when the relevant members believe that their own exports to the country are negatively affected by the policy (i.e., if the policy has a trade impact). In order to avoid a resulting trade dispute, policy-makers may want to ensure that their policy is WTO-compatible. As we show below, no clear guidelines exist as to when this is the case and when it is not. The only guidelines for policy-makers are the relevant legal texts – many of which were formulated more than 50 years ago – and the interpretation given to these texts by WTO case law.

We discuss how relevant WTO provisions have been interpreted, and a number of unresolved or as yet unspecified issues become apparent. The analysis has the following structure: first, the GATT national treatment provision is considered and it is examined how products are determined to be 'like'. Second, recent approaches to interpreting 'less favourable treatment' of 'like' foreign products are reviewed. This includes exploration of possible shortcomings of the chosen method and a comparison with the earlier aims-and-effects approach. Third, we examine the possibility of justifying environmental policies under the GATT's general exception clause in Article XX when a national environmental policy measure was found to violate the 'national treatment' obligation. Finally, some concluding remarks are offered on the ambiguous and incomplete picture that emerges from a legal analysis of the potential WTO (in)compatibilities of national environmental policies.

'Likeness' as a Precondition for Possible Discrimination

In our discussion of national environmental policies that appear to favour producers in the home country over those in a foreign country, we do not cover all legal aspects that may be relevant. We leave aside, for instance, a discussion of the most-favoured-nation principle contained in GATT Article I and the prohibition of quantitative restrictions in GATT Article XI. Because of its central importance, we limit our analysis to the national treatment obligation contained in GATT Article III.[39] We also assume that the environmental policies in question do not contain an explicit differentiation based on origin. It is often implied that the disfavoured variety is not produced domestically but is exclusively imported. This may hold true in exceptional cases, but it is assumed here that, in general, both varieties are produced in both countries.

In order for violations of Article III to occur, the two product varieties first need to be found to be 'like'. The concept of like products is probably meant to correspond to what economists view as varieties within one product group. In other words, national policies that treat different product categories differently (e.g., cars and bicycles, alcoholic beverages and soft drinks) do not pose a problem for WTO legislation. Policies that differentiate between varieties within one product group (e.g., between cars with different fuel efficiency) may pose a problem if they are considered to treat like products differently.

The definition of like products is, therefore, of considerable importance to policy-makers. In the EC–Asbestos case, the Appellate Body concluded that 'a determination of 'likeness' under Article III:4 is, fundamentally, a determination about the nature and extent of a competitive relationship between and among products.'[40] Recognising that there is 'a spectrum of degrees of 'competitiveness' or 'substitutability' of products in the marketplace',[41] the Appellate Body recommended as an appropriate starting point the use of four categories of characteristics that the products involved might share: (i) the physical properties of the products; (ii) the extent to which the products are capable of serving the same or similar end-uses; (iii) the extent to which consumers perceive and treat the products as alternative means of performing particular functions in order to satisfy a particular want

---

[39] Because of space limitations, we do not deal specifically with Agreements other than the GATT. We may make reference to certain aspects, especially the TBT Agreement, but only cursorily for analytical purposes and not in any consistent manner. We assume that the national environmental policies hypothetically referred to are not, for instance, TBT measures in accordance with relevant international standards, which would, pursuant to TBT Article 2.5, 'be rebuttably presumed not to create an unnecessary obstacle to international trade'. For the present purposes, it is not necessary either to delve deeper into the differences between GATT Articles III: 2 (relating to taxation) and III: 4 (relating to other domestic policy measures). We also skip the discussion on how broadly 'like' products are defined in Article III: 4 as opposed to the spectrum of 'like' products in the first sentence of Article III: 2 and as opposed to 'directly competitive and substitutable' products referred to in the interpretative note to Article III: 2 in connection with its second sentence. For more, see WTO, 2001b: paragraph 100.

[40] WTO, 2001b: paragraph 99.

[41] WTO, 2001b: paragraph 99.

or demand; and (iv) the international classification of the products for tariff purposes.[42]

Competition law has made intensive use of cross-price elasticities in order to define 'relevant markets'. Similarly, the Appellate Body confirmed that common end-uses of two products may, inter alia, be shown by their elasticity of substitution,[43] but cautioned that such quantitative measures could not be the decisive criterion for determining whether products are 'directly competitive or substitutable'.[44] In the Japan – Taxes on Alcoholic Beverages Panel Report, many contradictory arguments were presented by the parties with regard to an appropriate method for the estimation of elasticities of substitution, relating, in particular, to the necessary statistical adjustments in time series analysis versus panel data approaches on the basis of household surveys.[45] In the Korea–Taxes on Alcoholic Beverages case, the Appellate Body confirmed the panel's sceptical attitude to the quantification of the competitive relationship between imported and domestic products.[46] At the same time – and also recognising that under national antitrust law regimes elasticities are widely used to measure the extent to which products directly compete – it was acknowledged that studies of cross-price elasticity may nevertheless provide some useful insights in assessing latent demand that is suppressed by regulatory barriers but crucial to gauge the degree of *potential* competitiveness between substitutable products. In view of the goal of protecting expectations of equal competitive relationships, it was held that any examination under Article III needed to go beyond the static concept of situations in which consumers already regarded products as alternatives. In order to effect the necessary calculations, it was noted that 'evidence from other markets may be pertinent to the examination of the market at issue'.[47] The utility of quantitative factors as possible inputs in the examination of the nature and degree of competitiveness was also confirmed in the EC–Asbestos case.[48] Yet, in the overall consideration of likeness, the Appellate Body recognised that '[n]o one approach … will be appropriate for all cases. Rather, an assessment utilizing 'an unavoidable element of individual, discretionary judgement' has to be made on a case-by-case basis'.[49]

---

[42] These criteria were first mentioned in the Report of the Working Party on Border Tax Adjustments (GATT, 1972).

[43] It should be noted that cross-price elasticities and elasticities of substitution are related but different concepts, the latter being more widely used in production theory and the determination of conditional factor demand.

[44] WTO, 1996b: 32.

[45] WTO, 1996b.

[46] WTO, 1999.

[47] WTO, 1999: paragraph 137.

[48] WTO, 2001b.

[49] WTO, 2001b: paragraph 101.

Concerning the four criteria endorsed by the Appellate Body, it is not at all clear whether the fulfilment of one or even of all four of them would suffice to make two products like, nor what would happen if two of the Appellate Body criteria, or even different aspects of the same criterion, were to suggest contradictory outcomes as to the likeness of the products in question.[50] For each criterion, a decision on whether it supports likeness or not would need to be reached, as it has been made clear that, in all cases, all elements of the fourfold framework must be examined,[51] and that any piece of evidence that could make the case for or against likeness may be adduced and may not be ignored by the panel.[52] Given the emphasis on the non-exhaustive nature of the fourfold framework, the question remains under which circumstances other criteria may be added or may shape the consideration of the four criteria (e.g., of the third criterion concerning consumer perceptions of alternative means of satisfying a particular want) to support (un)likeness. This question is particularly relevant in the discussion on whether differences in NPR-PPMs – which do not leave any visible or otherwise detectable impact on the final products – can render two product varieties unlike.

Hence, indications as to the relative importance of the four criteria seem unavoidable. However, on this account, the Appellate Body does not go into much detail, hinting solely – in the EC–Asbestos case – that the physical differences (including toxicity) between asbestos and substitute products already placed a great burden on the complaining party to show that there was a competitive relationship 'between the products such that, all of the evidence, taken together, demonstrates that the products are 'like''.[53] Yet, this approach also raises questions. If two products are considered to be unlike because they have different levels of toxicity owing to their physical disparities (for example, as regards one molecule), could this also be said about similarly different products that have widely divergent environmental impacts? Governments would then be free to design their environmental policies in more or less trade distorting ways and would presumably go unchallenged, as long as those goods were considered unlike. The analogy is probably defective. It seems unlikely that anybody would consider a toxic, potentially lethal, product a substitute for a good that is innocuous. In other words, even though the relevant product varieties may have, for instance, similar end uses, the difference in level of toxicity is such that they can be considered to fall into different product groups. This is an extreme example

[50] This issue is acknowledged by the Appellate Body to be relevant, probably in many cases. See WTO, 2001b.

[51] Each criterion should also be examined separately in order to prevent different, and distinct, elements of the analysis from becoming entwined along the way. See WTO, 2001b.

[52] WTO, 2001b.

[53] WTO, 2001b: paragraph 136.

and it would, therefore, not apply to most of the environmental policies of relevance to our discussion. This seems to be confirmed by what was acknowledged, on an individual basis, by one of the Appellate Body members in stressing the possible 'supremacy' of the physical properties and qualities of a product:

> It is difficult for me to imagine what evidence relating to economic competitive relationships as reflected in end-uses and consumers' tastes and habits could outweigh and set at naught the undisputed deadly nature of chrysotile asbestos fibres, compared with PCG fibres, when inhaled by humans, and thereby compel a characterization of 'likeness' of chrysotile asbestos and PCG fibres. [... This is not meant to suggest] that any kind or degree of health risk, associated with a particular product, would a priori negate a finding of the 'likeness' of a product with another product, under Article III:4 of the GATT 1994, [... but it is simply to express a certain doubt] as to the 'fundamental', perhaps decisive, role of economic competitive relationships in the determination of the 'likeness' of products under Article III:4.[54]

In two GATT cases (United States – Taxes on Automobiles and US – Malt Beverages), the issue of likeness was approached differently.[55] The so-called 'aims-and-effects test' was applied in such a way that the examination of likeness and less favourable treatment to afford protection were collapsed into one single test. As a result, the panel affirmed the right of countries to define likeness in terms of the regulatory objective being pursued (in this particular case, fuel efficiency). The defendant's policy was found not to be in breach of Article III, because the regulatory distinction was not aimed at protecting the domestic industries. Hence, the question of determining the likeness of fuel-inefficient and fuel-efficient vehicles did not arise. The danger of such an approach would be that, in a somewhat circular manner, products would be considered unlike as long as the regulating government defined them as such. Consequently, an inquiry into likeness would be rendered meaningless.[56]

Less Favourable Treatment of Like Products and Protection

Once two products or product varieties are considered like for the purposes of GATT Article III, it needs to be determined whether the importing country accords less favourable treatment to the group of like imported products than to the group of like domestic products. In this regard, the Appellate Body

---

[54] WTO, 2001b: paragraphs 152-153.

[55] See GATT (1994b) and GATT (1992), respectively.

[56] It should be recalled that this report was never adopted and that an inquiry into subjective intent (such as the reasons stated by the legislator) was explicitly rejected by the Appellate Body (WTO, 1996b and WTO, 1997).

explained in the Korea–Beef case that mere distinctions between like products did not necessarily imply less favourable treatment.[57] The latter would take place only if the conditions of competition were modified to the detriment of imports. Going further in this direction, the Appellate Body, in the EC–Asbestos case, recalled its ruling in the Japan–Taxes on Alcoholic Beverages case[58] that less favourable treatment expressed the general principle in GATT Article III:1 that 'internal measures 'not be applied to imported and domestic products so as to afford protection to domestic production.' ... Article III protects expectations not of any particular trade volume but rather of the equal competitive relationship between imported and domestic products'.[59]

Two important conclusions may be drawn from this: first, the Appellate Body clarified that the mere distinction between like products drawn by an environmental regulation is not sufficient to conclude that the group of like imported products is being accorded less favourable treatment than the group of like domestic products. It insisted that protectionism, the central theme of Article III, must be proven to exist. Second, the right of importers not to be treated less favourably by a regulation does not imply that actual trade volumes are maintained. This appears to suggest that import volumes of the like product may decline following the introduction of environmental legislation as long as this is not the consequence of protectionism and the equal competitive relationship is maintained. From an economic point of view, it makes sense not to use trade volumes as a criterion when evaluating an environmental policy. Yet, the protection of an equal competitive relationship does not seem to be a workable concept either. A policy implemented for purely environmental reasons (be it a tax, subsidy, or a norm), that negatively affects a larger proportion of foreign than domestic products, is likely to have an impact on both the competitive relationship and trade flows.

The question, therefore, remains as to which criterion should be used to decide whether there is less favourable treatment of foreign goods.[60] As noted earlier, it cannot normally be assumed that foreign countries exclusively produce the disfavoured product varieties and home countries only the

---

[57] WTO, 2001a.

[58] More precisely, the Appellate Body recognised the importance of applying the principles contained in GATT Article III:1 also to Article III:4, and not only to Article III:2 second sentence, as implied in previous decisions.

[59] WTO, 2001b: paragraph 97.

[60] A 'before-after exercise' to detect a change in competitive opportunities to the detriment of imports is rarely as straightforward as in the Korea–Beef case, where beef was distinguished according to origin: imported and domestic beef were distributed through the same channels, when, in 1990, retailers had to choose between imported and domestic beef. Not surprisingly, most chose domestic beef. For more, see WTO, 2001a: in particular paragraph 145.

favoured ones. With an environmental regulation, home countries discriminate against at least some of their own producers. It would be absurd, and most likely economically inefficient, if national treatment was to be understood such that foreign producers automatically receive the best possible treatment, no matter which product varieties they produce.[61] It would, however, be equally unfair if discrimination was rejected out of hand because at least some domestic products are disfavoured as well. This raises the questions of how the group of imports comprising both varieties should be compared to the group of domestically produced like varieties. No explicit answer can be found in the WTO jurisprudence. Ehring advocated a point-in-time comparison between the domestic proportion of favoured to disfavoured products with the same ratio for imported varieties.[62] A larger difference between favoured and disfavoured products on the domestic side could be considered to buttress a suspicion of protectionist motives, which would need to be examined further. It could mean that the aim of home-country governments was to provide an advantage to domestic producers, knowing that a majority or all of them were or would be more competitive in producing the favoured varieties than were their foreign competitors. Ehring drew on analogies in the areas of tax and gender discrimination under European Community Law, in which asymmetric effect is a necessary condition for a case not to be dismissed.

The proportion of favoured to disfavoured products should not represent the ultimate yardstick in any examination of less favourable treatment. There may be numerous reasons why environmental measures affect foreign products in a disproportionate way, even in the absence of protectionist intent. It is, for instance, likely that domestic producers are more aware of changes in consumer preferences or in environmental legislation than are foreign producers. They are, therefore, more likely to anticipate the legislation and switch to the favoured varieties. They may also have stronger incentives to do so than have foreign producers, if domestic markets absorb a higher share of the output of domestic producers compared with the output share of foreign producers. Foreign producers may have decided, for a variety of reasons, not to supply the favoured variety to the country in question, despite their equal ability to do so. It is conceivable that they were able to sell the favoured varieties at a higher margin elsewhere owing to different demand functions and contented themselves with exporting predominantly the disfavoured varieties to the countries initiating the new environmental regulation. Foreign producers would be even less inclined to adapt their production if they expected that, in the context of a WTO dispute, the panel

61 Some would argue that this notion was, however, seriously considered in GATT, 1989.
62 Ehring, 2002.

would take a comparison of the favoured-disfavoured ratio between domestic and foreign products as evidence of protectionism.

Thus, for an investigation of protectionism and a purposeful modification of competitive opportunities, it may be more relevant to ask whether environmental regulation inevitably makes it more burdensome for suppliers in foreign countries to produce the favoured product varieties owing, for instance, to circumstances that cannot be changed, at least in the short term. To some extent, this would amount to an evaluation of reasons why the production of the favoured varieties cannot be conducted in an equally competitive manner as that of the disfavoured ones. For instance, in the US–Malt Beverages case, wine made from grapes growing only in one region of the US and the Mediterranean received more favourable tax treatment than other wines.[63] In view of the limited geographical dispersion and, presumably, considerable difficulty or unfeasibility of growing the specific grapes in new locations, the panel's observation that the tax break was put in place so as to afford protection to domestic producers seems well-founded.[64] Thus, in this case, while the effect of the regulation was taken as an important piece of evidence for possible protectionism, it was further examined whether there were good reasons to believe that the disproportionate impact may also have been its aim.

There appears to be no cogent reason why double-checking the aims and objectives of new regulation should be foregone, given that it is always conceivable that neutrally designed and purpose-oriented legislative measures lead to disproportionate impacts on different groups of people (countries) that are to be treated equally.[65] For instance, if the possibility of serving in the

---

[63] GATT, 1992.

[64] It may be argued that this is the 'pure' aims-and-effects test. In contrast, the aims-and-effects test in the US–Taxes on Automobiles case (GATT, 1994b, not adopted) – later explicitly rejected by the Appellate Body – seemed to have equated the regulatory aim with ascertaining a regulation's subjective intent (see, for instance, paragraph 5.10). Some have argued that it is generally impossible to identify the intent of states. It would, therefore, be impossible to know whether changes in competitive opportunities were the desired outcomes of laws that may have gone through laborious parliamentary and other debates and been modified a number of times to accommodate special interests. They may have had initial aims that became blurred, changed, or amplified into several aims. Yet, it seems useful to evaluate alternative regulatory schemes and scopes in the light of the stated environmental objective and to look for evidence of ostensibly inconsistent, and hence possibly purposeful, regulatory design that modifies competitive opportunities. We agree with Mattoo and Subramanian (1998) that if, for instance, in the US–Taxes on Automobiles case, it had been found that, at some stage in the legislative process, the exemption for sports vehicles was introduced in order to cater to specific constituencies or lobby groups, this may provide an important proof that something else, beyond a – possibly – genuine environmental concern, may have been pursued.

[65] In fact, economists would probably emphasise the need of policies to meet the relevant environmental purposes, rather than focusing on the burden these policies represent for foreign producers.

military was limited to men, this could be seen as discriminating against women. Yet, if the ability to lift X kg was indispensable in order to be able to hold a firearm for a sustained period of time, this could be seen as a reasonable requirement for entering to a military career that may be fulfilled by, say, only a quarter of all women as compared to every other man. Considerations of this kind may call for a more far-reaching approach, such as the aims-and-effects test, which, besides in the US–Malt Beverages case, has been applied once more albeit in an unadopted GATT panel and later been discarded by the Appellate Body. Its revival was advocated by several legal experts, most prominently Bob Hudec.[66] Ehring, again highlighting possible model cases in the jurisprudence of the European Court of Justice, noted that the Court, 'in the area of sex discrimination, ruled against sex-neutral differentiations only when they hit predominantly one sex and do not service a legitimate objective'.[67] Interestingly, the requirement to take into account the aims of regulation is not unfamiliar to the WTO's TBT and SPS Agreements. The TBT Agreement holds that technical regulations are not to be more trade-restrictive than is necessary to fulfil a legitimate objective, taking account of the risks non-fulfilment would create. This appears to suggest that, even if regulation places a considerable (or even disproportionate) burden on foreign producers, the objectives may be too important and their non-fulfilment represent too great risks to tinker with it.

In the two instances in which an aims-and-effects test was applied in the GATT, the examination of likeness and of less favourable treatment to afford protection were collapsed into one single test. This implied that the defendant's logic in defining likeness on regulatory grounds was followed, and the panel did not embark upon establishing likeness pursuant to a pre-determined framework. Instead, the intent and effect of the regulatory distinction between products were examined. As long as it was assured that protection was neither intended nor afforded to domestic products, any regulatory distinction between product varieties seemed acceptable. In the US–Taxes on Automobiles case, a luxury tax was levied on cars selling above a specified dollar threshold, as was an environmental tax on cars not achieving a certain fuel efficiency.[68] The panel declined to find protectionism despite an initially larger incidence of tax on imports and a more extensive use of exemptions by US manufacturers. Putting a dynamic spin on the concept of competitive opportunities, the panel observed that foreign manufacturers had the capability of producing small cars for other markets. The various threshold levels were, therefore, not inherently more easily missed by foreigners than by domestic producers. Similarly, in examining a

---

[66] Hudec, 1998.
[67] Ehring, 2002: 950.
[68] GATT, 1994b (not adopted).

measure that treated beer according to its alcohol content (the US–Malt Beverages case), the panel noted that there was no reason to believe that the resulting burden was inherently more difficult to bear by foreign than by domestic manufacturers.[69]

One criticism levelled at the aims-and-effects approach was that this inherence test was impossible to fail, as mere production possibilities are rarely confined to national boundaries, except where nature sets definite geographical limits (such as for the above-cited grape species).[70] It may be that many technologies are available across boundaries such that the production of specific goods *could* take place anywhere in the world. Yet, in search of evidence for protectionist motives, it may still be of legal interest to know whether or not the favoured varieties are *already* being produced in the foreign countries to any significant extent (for instance, for export to third countries). Existing production and export of the favoured varieties by the complainant to third countries with economic characteristics similar to those of the regulating countries may provide at least some evidence that the defendants may not have sought to protect domestic producers, as they must have expected production of the favoured varieties by the complainants to pick up easily and undermine such protection. Besides, in the case US–Taxes on Automobiles, the domestic policy was not applied consistently by the defendant in the domestic market. Sports vehicles were exempted from the various tax measures, although these products are closely related to other passenger vehicles that were being taxed depending on their fuel efficiency. Thus, it may be suspected that at least part of the problem with the aims-and-effects test is grounded in the fact that it was applied in an unsatisfactory way in the past. This may not be enough reason to conclude that it cannot be an appropriate tool in future cases, especially if combined with and preceded by an inquiry into likeness.

It is often argued that, since the Japan–Taxes on Alcoholic Beverages case, the Appellate Body has rejected the aims-and-effects test in declaring that 'it is irrelevant that protectionism was not an intended objective if the ... measure in question is nevertheless ... *applied* to imported or domestic products so as to afford protection to domestic production'.[71] This seems little more than stating the obvious, namely, that it is not enough for the defendants to invoke innocent regulatory objectives. Indeed, the Appellate Body continued to see the need to examine the way a regulation was designed in order to gather evidence to sustain a hidden protectionist purpose. It conceded that '[a]lthough it is true that the aim of a measure may not be

---

[69] GATT, 1992.

[70] Mattoo and Subramanian, 1998.

[71] WTO, 1996b: 32 (emphasis in original). See also WTO (1996b) and WTO (1997).

easily ascertained, nevertheless its protective application can most often be discerned from the design, the architecture, and the revealing structure of a measure'.[72] The major interpretational change should be considered to lie in the separation of the likeness and protection determinations into two successive steps. The emphasis of protective application revealed by the design of measures instead of protective aims which can also be discerned in scrutinizing the structure and context of measures may be more a clarification of a self-evident legal nuance than the drastic reversal of what can simply be called a common-sense approach.

The procedure suggested in the Japan–Taxes on Alcoholic Beverages case is geared towards collecting a maximum of evidence for alleged protectionism. This was probably also the initial goal of the aims-and-effects test. Thus, by turning against an inquiry into aims-and-effects, the Appellate Body simply rejected both the uncritical acceptance of the stated purposes of legislation, on the one hand, and the second-guessing of the intention behind it, from its mere letter, on the other. Apparently, the Appellate Body did not want to reduce the analysis to a mere investigation of protective effects on competitive conditions with no aims test. This cannot be better explained than in Hudec's words:

> The Appellate Body ruled that ... something more [is required] than just an analysis of protective effect. The Appellate Body took pains to make clear that it was not talking about the analysis of regulatory purpose called for by the 'aim' in 'aim and effects.' However, the 'aim' analysis it seemed most concerned with rejecting was a search for the actual motivation behind a measure, repeatedly stressing that the intent of legislators or regulators was irrelevant. The additional element the Appellate Body called for was an investigation of something called 'protective application,' a concept that for all the world looked like an objective analysis of regulatory purpose. ... The [quotation of the respective Appellate Body report] makes a great deal more sense if one substitutes the word 'purpose' for 'application.' Indeed, neither the Appellate Body's insistence on different words nor its insistence on objective analysis serve to mark a clear distinction between its 'protective application' concept and the 'aim and effects' analysis.[73]

Hudec's advice was seemingly heeded in the case Chile–Taxes on Alcoholic Beverages.[74] In recalling its ruling in the Japan–Taxes on Alcoholic Beverages case, the Appellate Body reconfirmed that '[t]he subjective intentions inhabiting the minds of individual legislators or regulators do not bear upon the inquiry, if only because they are not accessible to treaty interpreters. It does not follow, however, that the statutory purposes or objectives – that is, the purpose or objectives of a member's legislature and

---

[72] WTO, 1996b: 33.
[73] Hudec, 1998: 19-20.
[74] WTO, 2000.

government as a whole – to the extent that they are given objective expression in the statute itself, are not pertinent'.[75] In addition, it made the link between the concept of protective application and the purposes of measures explicit: 'We called for examination of the design, architecture and structure of a tax measure precisely to permit identification of a measure's objectives or purposes as revealed or objectified in the measure itself. Thus, we consider that a measure's purposes, objectively manifested in the design, architecture and structure of the measure, are intensely pertinent to the task of evaluating whether or not that measure is applied so as to afford protection to domestic production'.[76] The examination of the design, architecture, and structure of the new Chilean system ultimately revealed that the application of dissimilar taxation '*will* afford protection to domestic production'.[77] In arriving at this conclusion, the Appellate Body also took into account that most domestic beverages were found in the lower tax bracket and that the (considerably) higher tax bracket began at the point at which most imports, by volume, were found.[78] In that regard, the Chile–Taxes on Alcoholic Beverages case also provides an illustration of how the evidence of a differing distribution of favoured and disfavoured products between domestic and imported varieties can usefully be married with the investigation into protective application, without making such a proportions approach a hard and fast let alone automatic litmus test.

The way in which the examination under Article III is conducted determines the likelihood of violations occurring and, thus, how often national environmental policy-makers need to take recourse to the general exceptions contained in GATT Article XX. Article XX comprises ten policy objectives that can be invoked in order to justify policy measures contravening other GATT provisions. Two of those relate to the environment and, although environmental protection is not explicitly listed as an objective, they are commonly viewed as being worded broadly enough for national environmental policies to be covered. It should be mentioned that there is a host of other policy goals (for instance, in the social or consumer policy areas), which may not fall within the range of those ten objectives and for which a justification under Article XX is, therefore, not available. Hence, the identification of objectives or purposes of measures as revealed or objectified in the measures themselves, as part of the process of establishing compliance with GATT Article III, may prove an important safety net for policies other than those listed in Article XX.

---

[75] WTO, 2000: paragraph 62.
[76] WTO, 2000: paragraph 71.
[77] WTO, 2000: paragraph 66 (emphasis added).
[78] WTO, 2000.

Lawyers continue to struggle with the limited scope of Article XX and the question of how to dismiss its possible redundancy if an examination of aims is allowed under Article III.[79] Interestingly, the Appellate Body pointed out in the EC–Asbestos case that 'the fact that an interpretation of Article III:4, under those rules, implies a less frequent recourse to Article XX(b) does not deprive the exception in Article XX(b) of effet utile'.[80] In the same paragraph, it is also noted that the same or similar evidence serves different but related purposes under both provisions. Under Article III:4, it serves to establish whether the competitive relationship in the marketplace between allegedly like products has changed to the detriment of foreign producers, whereas it is used in Article XX to determine whether measures in pursuit of one of the objectives listed in that provision nevertheless represent a disguised restriction on international trade. This indicates that the analysis of the design, architecture, and revealing structure of measures, first, allows verification of whether there is protective application. If there is protective application, and hence a violation of GATT Article III, the measures may still provisionally qualify under one of the general exceptions in GATT Article XX. The design, architecture, and structure would again be looked at in order to, secondly, test the good faith of members in applying those measures (i.e., to prevent abuse of the exceptions).

The formal procedure purports to neatly separate protectionism (prohibited under Article III) from protectionist effects if implementation of measures in good faith provisionally qualifying under one of the subparagraphs of Article XX is ascertained under the chapeau (i.e., the introductory clause). We have shown that, in practice, the determination of protectionist effects may not be easy – to a large extent owing to the fact that favoured and disfavoured products exist both domestically and in the foreign country – and, hence, clear separability may not be on the cards. If de facto discrimination is at issue, the same evidence is likely to be reviewed in both determinations. Environmental policy-makers may, therefore, be more concerned with how the design, architecture, and structure of their measures is likely to be perceived than with legal strategy. It is not surprising that Hudec, in the aftermath of the Japan–Taxes on Alcoholic Beverages case, was able to trace the aims-and-effects test throughout major cases, and this, he argued, was 'merely the tip of an iceberg'.[81]

---

[79] Such redundancy would be in breach of the Vienna Convention on the Law of Treaties, Article 31.

[80] WTO, 2001b: paragraph 115. By the same token, the Appellate Body stated in a previous report, 'Nor may Article III:4 be given so broad a reach as effectively to emasculate Article XX(g) and the policies and interests it embodies' (WTO, 1996a: 18).

[81] Hudec, 1998: 27.

It should be recalled at this point that Article XX was drafted more than 50 years ago when little if any experience in these matters could be drawn on.[82] It is worth noting that the later TBT Agreement, for instance, seems to reflect an approach better suited to deal with de facto discrimination.[83] Three aspects are particularly noteworthy for our purposes. First, the TBT Agreement, in Article 2.1, formally prohibits discriminatory regulations without containing an equivalent to the exceptions contained in GATT Article XX. At the same time, TBT Article 2.2 'explicitly calls for attention to the protective 'aim and effects' of product standards. Under Article 2.2, the issues of trade effects and regulatory justification are considered on the same level, without any conclusion as to violation until both sides of the equation have been fully considered'.[84] Thus, Article 2.1, in connection with Article 2.2, seems to call for the kind of consideration required in cases of de facto discrimination under GATT Article III. Secondly, Article 2.2 contains elements to be analysed in the GATT under Article XX, namely, inquiry into whether the degree of trade restrictiveness is indeed necessary to accomplish the regulatory objectives. As discussed above, the type of cost-benefit analysis that should be carried out under TBT Article 2.2 can be seen as similar to a proper balancing test under GATT Article XX. Finally, the TBT Agreement enumerates a non-closed illustrative range of standard legitimate objectives, whereas GATT Article XX features a narrow list of eligible policies. This must clearly be seen within the historical context. It may also be legitimate to assume that the Article XX objectives are of such importance or frequency that is has been made explicit that exceptions are to be made, even in the face of blatant, albeit justifiable, discrimination. The exemplary objectives in the TBT Agreement also have the advantage of being clearly identified as legitimate, whereas the Appellate Body highlighted in the EC–Sardines case that, in all other cases, 'there must be an examination and a determination on the legitimacy of the objectives of the measure'.[85]

---

[82] In a related context, this line of thought was confirmed by the Appellate Body in the case US–Import Prohibition of Certain Shrimp and Shrimp Products (WTO, 1998: paragraph 129), where it was stated that Article XX(g) must be read 'in the light of contemporary concerns of the community of nations about the protection and conservation of the environment'. Particular reference was made to the WTO Preamble, which explicitly acknowledges the objective of sustainable development as informing the GATT and all other agreements.

[83] Marceau and Trachtman (2002) summarised and juxtaposed all disciplines of the GATT, TBT, and SPS Agreements that may be interesting to compare.

[84] Hudec, 1998: 36-37. Article 2.2 reads, 'Members shall ensure that technical regulations are not prepared, adopted or applied *with a view to or with the effect* of creating unnecessary obstacles to international trade. For this purpose, technical regulations shall not be more trade restrictive than necessary to fulfil a legitimate objective, taking account of the risks non-fulfilment would create. Such legitimate objectives are, inter alia: national security requirements; the prevention of deceptive practices; protection of human health or safety, animal or plant life or health, or the environment' (emphasis added).

[85] WTO, 2002b: paragraph 286.

These and other issues of compatibility between disciplines in different WTO Agreements may be puzzling from a legal perspective, but many are analytically less interesting. Suffice it to say that it is hard to imagine that the TBT Agreement, in particular Articles 2.1 and 2.2, and GATT Articles III and XX were meant to be incongruent, had they been drafted at the same time. Some would go as far as to say that later Agreements (such as TBT and SPS) have simply merged the requirements of GATT Articles III and XX.[86] The relevant jurisprudence on Article XX concerning national environmental policies is briefly reviewed below.

General GATT Exceptions for Environmental Policies

In order to defend environmental measures found to be in breach of other GATT provisions under Article XX, the policies pursued through the measures need to fall within the range of policies designed to protect human, animal, or plant life or health (Article XX(b)), or to conserve exhaustible natural resources (Article XX(g)). For instance, policies to protect dolphins, clean air, or petroleum were considered under either one (or both) of the two paragraphs in the past. Most notably, other bodies of international law (such as the Convention on International Trade in Endangered Species of Wild Fauna and Flora, CITES) were referred to in the Appellate Body's deliberation on whether sea turtles constituted an exhaustible natural resource for the purposes of Article XX(g). Further specific requirements need to be fulfilled in order for the environmental measure to provisionally qualify under Articles XX(b) or (g), in particular the elements of necessity in paragraph (b) and of relatedness in paragraph (g). Finally, none of the requirements contained in the introductory clause to Article XX must be violated. Each of these elements is discussed in turn below.

The necessity test in paragraph (b) was interpreted by a number of GATT panels such that measures could only qualify if there were no GATT-consistent measures reasonably available to governments and if, among those measures reasonably available, they were the ones with the least degree of inconsistency with other GATT provisions.[87] This approach became commonly known as the requirement of least trade-restrictiveness; and it was pointed out that the goal of least trade-restrictiveness does not correspond to the one of welfare maximisation, which is the objective preferred by

---

[86] It may be added that, legally speaking, this merger was in favour of the environment. This argument may be deduced from the Appellate Body's ruling in the EC–Sardines case under TBT Article 2.4 in connection with Article 2.2 that the complainants must demonstrate that measures taken in pursuit of legitimate objectives, such as the environment, are not WTO-compatible. Conversely, under GATT Article XX, the burden of proof rests with the respondents in invoking the environmental provisions in their defence.

[87] See, for instance, GATT, 1990.

economists.[88] Decisions in the past critically depended on what measures were considered to be reasonably available to governments. In the US–Gasoline case, the panel clarified that an alternative measure did not cease to be reasonably available simply because it involved administrative difficulties;[89] and, in the US–Malt Beverages case, the panel even provided examples of less trade-restrictive alternatives applied by other US states.[90] In the Korea–Beef case, the Appellate Body elucidated further that the 'determination of whether a measure, which is not 'indispensable', may nevertheless be 'necessary' ... involves in every case a process of *weighing and balancing a series of factors* which prominently include the contribution made by the compliance measure to the enforcement of the law or regulation at issue, the importance of the common interests or values protected by that law or regulation, and the accompanying impact of the law or regulation on imports or exports'.[91] It has been observed that, following the Korea–Beef case, a more adequate terminology may be to refer to the necessity test as an inquiry into less trade-restrictiveness supplemented with an evaluation of proportionality.[92] This interpretation would be closer to the concept of welfare maximisation, where the welfare losses due to imposed restrictions on trade are weighed against the welfare benefits from correcting market distortions linked to the pursued policy objectives which would be environmental in this context.

The exact relationship between the suggested factors remains elusive in the jurisprudence, although the principle direction is clearly expressed in statements such as '[t]he more vital or important those common interests or values are, the easier it would be to accept as 'necessary' a measure designed as an enforcement instrument'.[93] TBT Article 2.2 seems to refer to a similar test of less trade restrictiveness in connection with proportionality. In the context of TBT, the trade costs of measures are weighed against the benefits inherent in the importance of the objectives pursued and the degree to which the risk of non-fulfilment is minimised. If the latter is assumed to mean the same as the contribution made by the measure to achieving a given policy objective, the two approaches seem similar.

The relatedness requirement in Article XX(g) has commonly been seen as being less difficult to fulfil, as it clearly covers a wider range of measures

---

[88] Mattoo and Subramanian, 1998.
[89] WTO, 1996a.
[90] GATT, 1992.
[91] WTO, 2001a: paragraph 164 (emphasis added).
[92] WTO, 2002a.
[93] WTO, 2001a: paragraph 162. Some may argue that the value at stake can be such that trade restrictiveness loses all relevance.

than merely those necessary.[94] In the Canada–Herring case, the panel clarified that measures for the purposes of Article XX(g) had to be primarily aimed at the conservation of exhaustible natural resources. Subsequent panels followed this interpretation and determined that measures primarily aimed at conservation could not be based on unpredictable conditions, as would, for instance, be the case if the measures were effective only if other countries changed their policies. A key observation was made by the Appellate Body in the case US–Gasoline, highlighting that it was the *measure* itself (i.e., the fact of establishing baselines for the quality of gasoline) that needed to be primarily aimed at environmental conservation. It found that the baselines were primarily aimed at achieving certain clean air standards, since without such baselines the stabilisation and subsequent reduction of air pollution was not feasible. In so doing, it reversed the decision of the panel, which had wrongly assumed that it was the less favourable treatment of the imported product which needed to aim primarily at the policy objective of environmental conservation.

Further clarification was provided in the US–Shrimp case that it was the relationship between the general structure or design of the measures and the policy goals that mattered: 'The means [must be] ... reasonably related to the ends. The means and ends relationship between [the US measure] and the legitimate policy of conserving an exhaustible, and, in fact, endangered species, is observably a close and real one, a relationship that is every bit as substantial as that which we found in the US–Gasoline case between the EPA baseline establishment rules and the conservation of clean air in the United States'.[95] In addition, GATT Article XX(g) specifies that the measure must be 'made effective in conjunction with restrictions on domestic production or consumption'. In the US–Gasoline and US–Shrimps cases, the Appellate Body established the need for 'even-handedness' in the imposition of restrictions, meaning that a measure may not be imposed on imports only, but, in the name of the conservation of exhaustible natural resources, must also curb domestic production and consumption.[96] This evokes the concept of consistency, emphasised above.

Finally, environmental measures must not violate any one of three requirements contained in the introductory clause to Article XX: they may not be a means of unjustifiable or arbitrary discrimination between countries where the same conditions prevail, nor a disguised restriction on international trade. This chapeau essentially fulfils the function of preventing abuse of Article XX by keeping a balance between the right of members to build their

---

[94] GATT, 1988.
[95] WTO, 1998: paragraph 141.
[96] WTO, 1996a; WTO, 1998.

defence on one of the general exceptions contained therein while safeguarding the rights of other members under the GATT. The Appellate Body in the US–Shrimp case endorsed earlier rulings that such a determination must address not only 'the detailed operating provisions of the measure' but also the manner in which the measure 'is actually applied'.[97]

The first two elements clearly imply that Article XX measures may discriminate between countries, but not in an unjustifiable or arbitrary manner. In the US–Gasoline case, unjustifiable discrimination was considered to be discrimination that could be 'foreseen' and that was not 'merely inadvertent or unavoidable'.[98] This was translated into two concrete requirements in the US–Shrimp case. First, the US was expected before resorting to unilateral measures to make serious efforts to negotiate a conservation agreement with all relevant parties, and not only with some. Second, in the way it was implemented, the original US measure lacked flexibility to take into account the different situations in different countries.[99] The Appellate Body further elaborated that an inquiry into whether measures required essentially the same regulatory programmes of exporting members as those adopted by the importing members was a useful tool in identifying measures that do not meet the requirements of the chapeau of Article XX,[100] whereas 'conditioning market access on the adoption of a programme *comparable in effectiveness*, allows for sufficient flexibility in the application of the measure'.[101]

The same rigidity of the original measure that required foreign countries to adopt essentially the same policies and enforcement practices as those applied to domestic shrimp trawlers in the US, without inquiring into the appropriateness of that programme to the conditions prevailing in the exporting countries, also led to the finding of arbitrary discrimination. Conversely, the modified implementation measure allowed for the possibility of demonstrating that an alternative programme was comparable to that of the US. As it was no longer based on the application of certain methods, but on the achievement of given objectives, the measure was considered flexible enough to qualify under Article XX. [102]

Finally, the Appellate Body, to a large extent, equated the third requirement with the other two conditions listed in the chapeau in deciding that ''disguised restriction', whatever else it covers, may properly be read as embracing restrictions amounting to arbitrary or unjustifiable discrimination

---

97 WTO, 1998: paragraph 160.
98 WTO, 1996a: 27.
99 WTO, 1998.
100 WTO, 2001c.
101 WTO, 2001c: paragraph 144 (emphasis in original).
102 WTO, 1998.

in international trade taken under the guise of a measure formally within the terms of an exception listed in Article XX'.[103] The implementation panel in the US–Shrimps case decided that the design, architecture, and revealing structure of the measure, as actually applied by the US authorities, might help in determining whether it constituted a disguised restriction on international trade. In view of the flexibility allowing the use of alternative programmes for the protection of sea turtles and the offers made by the US to provide technical assistance to develop the technology required, the panel was confident that there was no disguised restriction on international trade. This view was upheld by the Appellate Body.

Evaluating National Environmental Policies under the GATT

There appear to be three major lines of defence under the GATT for national environmental policies that may be seen as making an implicit regulatory distinction between domestic and imported product varieties. All three feature considerable uncertainty as to their possible interpretations. First, if regulating countries can show that the disfavoured product varieties are unlike the favoured ones, any negative effects on the trade of other countries may supposedly be ignored. Yet, environmental policy-makers have little ex ante guidance as to quantitative measures, such as thresholds of cross-price elasticities (e.g., a precise non-negative value), that can show that consumers are not prepared to substitute one product for another. The four qualitative criteria advanced by the Appellate Body are hazily defined as well, both individually and regarding their relationships to each other. Considerable room for interpretation remains for other elements to be adduced, and the defending parties seem entitled to insist on having additional evidence considered in support of unlikeness. Nevertheless, while there appear to be limitations, it is not clear whether, for instance, NPR-PPMs as part of consumer preferences may be a basis for such a distinction. The legal uncertainty and judicial discretion regarding likeness is readily admitted by the Appellate Body, which stated that '[t]he concept of 'likeness' is a relative one that evokes the image of an accordion. The accordion of 'likeness' stretches and squeezes in different places as different provisions of the WTO Agreement are applied. The width of the accordion in any one of those places must be determined by the particular provision in which the term 'like' is encountered as well as by the context and the circumstances that prevail in any given case to which that provision may apply'.[104]

Second, environmental policy-makers may be in the position to

[103] WTO, 1996a: 23.
[104] WTO, 1996b: 26.

demonstrate that pieces of legislation regulating like product varieties may not necessarily afford less favourable treatment to the group of like imports. It must be shown by complainants that the application of the policy measures protects domestic producers. This is not a straightforward exercise. It involves collecting a maximum of evidence without the ex ante certainty that one particular piece of evidence may be powerful enough to make or break the case. It cannot be excluded that the defending parties did have innocent legislative purposes that, at the moment of the regulation's coming into being, affected imports comparatively harder. Should this still be regarded as protectionism, even if it was not intended and the measures are excellent policy instruments, albeit not optimal in economic terms? If the approach suggested in the previous sections were followed, this would probably not be the outcome. Despite a strong negative impact on importers, the measures would be found in breach of Article III obligations only if their design, architecture, and revealing structure suggested a protective application. The suggestion to inquire into what the optimal policy instruments to pursue may have been given environmental objectives in order to evaluate the genuine environmental character of measures, may form part of this examination.[105] In other words, while the analysis of less favourable treatment boils down to an investigation of the protective effect of measures, this, in all likelihood, cannot be determined without asking whether the measures have been designed in such a way that they have the potential to protect domestic producers. This seems to blur the major distinction between GATT Article III and Article XX, namely, that less favourable treatment does not presuppose the intention of protectionism whereas the non-fulfilment of the Article XX chapeau does. This may be because less favourable treatment is not easily ascertained, let alone amenable to quantitative assessments. Thus, in the absence of hard (quantitative) evidence, an inquiry into the regulatory design, etc. is necessary in order to find indications of wilful acceptance of less favourable treatment of importers, although this analysis should have been reserved for Article XX.

Third, and finally, national environmental measures may be found to provide less favourable treatment to importers. The measures may still be defensible if they qualify for one of the general exceptions contained in GATT Article XX. Here, it is worth noting that the factors that must be balanced against each other in judging whether environmental measures are necessary would be similar if both the *aims* of the environmental measures (including how well they achieve their purposes) and their *effects* were to be considered under Article III: the importance of the policy objectives pursued is linked to the degree to which the measures really contribute to achieving

[105] Mattoo and Subramanian, 1998.

their objectives and to their impact on trade. Again, if aims-and-effects considerations were to be precluded from examinations under Article III, then this approach, supposedly providing additional leeway for policy measures, would be reserved for the ten objectives listed under Article XX. If they are not precluded, it begs the question what additional examination can be conducted under Article XX, except for a hardening of the finding that the measures were put in place with a protectionist *intent* by looking at the chapeau (*disguised* restriction on international trade). It still seems strange that the design, architecture, and structure may be analysed twice for allegedly separate purposes.

In sum, in anticipating a possible defence strategy under the GATT, environmental policy-makers continue to face a significant amount of legal uncertainty, especially as regards the definition of like products and the way the determination of less favourable treatment is handled. Conversely, Article XX seems to provide good protection for many genuine environmental policies, as an *intention* to protect domestic producers must be adduced if justification under Article XX is to be rejected. Following the US–Shrimp case, even a possible role for NPR-PPMs in characterising goods may have been established, given that a measure was justified under Article XX that distinguished products on the basis of how they were produced, namely, in banning shrimp harvested in a way that adversely affected certain sea turtles. The safeguards against abuse of these exceptions built in the chapeau of Article XX also seemed to have worked well, as testified in the Shrimp DSU Article 21.5 Implementation Panel and Appellate Body reports.

CONCLUSIONS

In this study, we analysed the WTO's activities which are of relevance to national environmental policies. Two types of activities can be distinguished. First are discussions and negotiations within the WTO that aim at enhancing members' understanding of the linkages between environmental policies, on the one hand, and trade and trade policy, on the other hand. These discussions and negotiations may at some point have an impact on the formulation of WTO rules. The second type of activities concerns trade disputes, in particular those about environmental policies. Existing WTO rules form the basis of such disputes and an analysis of the relevant legal texts and the existing case law can give domestic policy-makers an idea of potential incompatibilities between domestic environmental policy and multilateral trade law.

When the GATT was established after the Second World War, the environmental consequences of economic integration were not a primary concern. This may explain why references to the environment were indirect in the original GATT. The issue was first put on the agenda in the early 1970s, but it was only in the 1990s that a real discussion on the relationship between trade and the environment started within the GATT. This discussion is ongoing in the WTO. But although the ensuing exchange of ideas has helped to enhance our understanding of the linkages between trade and the environment, progress has been slow in relation to mitigating the potential tensions between multilateral trade law and national environmental policies.

Trade disputes arise when one or several WTO members believe that policy measures of other members discriminate against the producers of the former. Such policies may be claimed to be environmental measures by the defending governments. In such cases, the WTO dispute settlement system is supposed to distinguish between genuine environmental policies that happen to be discriminative and protectionist measures disguised as environmental policies. Only the latter should be considered to be in conflict with multilateral trade law. The experts charged with the cases have to base their arguments on existing case law and existing legal texts, some of which were written more than 50 years ago, when the environment was not an issue for most member governments. Their task is rendered more complex by the fact that the policies they have to evaluate may be the combined outcome of genuine environmental purposes and some level of protectionist intent.

Ideally, the evaluation of the environmental policies of members should take the optimal environmental policy as a point of reference.[106] Yet, we have argued that economists have an imperfect knowledge of the precise welfare effects of different types of environmental policies. It is, for instance, not clear in many cases whether price-based measures should be preferred over norms (i.e., regulations or standards, potentially combined with a labelling policy). In addition, governments and economists are unlikely to have complete information about the extent of the relevant environmental distortion, which makes it difficult to pin down the precise design of policy measures. But economic analysis does provide an idea of which aspects of the design of policies indicate environmental purposes and which indicate protectionist intent.

Our discussion of WTO provisions and jurisprudence revealed that there continues to be considerable uncertainty as to the possible interpretations of existing rules. We also found that existing WTO law provides appropriate tools to evaluate disputes about environmental policies. In particular, the approach of analysing the design, architecture, and revealing structure of the relevant measures seems promising. We have argued that certain elements of

---

[106] Optimal is considered from the point of view of economic analysis. See also Mattoo and Subramanian (1988) on this point.

such a policy will at the same time give indications of potential protective aims and the resulting trade effects, on the one hand, and of the environmental purposes of policies, on the other hand. Ultimately, any decision should be based on a process that weighs these three elements against each other, in an attempt to make maximum use of available evidence in order to approximate an economist's decision criterion: the trade-off between the losses from trade distortions created by environmental policies and the gains from attenuating market distortions that are harmful to the environment.

REFERENCES

Akerlof, G. (1970), 'The market for 'lemons': Quality uncertainty and the market mechanism', *Quarterly Journal of Economics*, **84**(August), 488-500.

Baldwin, R. (2000), *Regulatory Protectionism, Developing Nations and a Two-Tier World Trade System*, CEPR discussion paper 2574, London: Centre for Economic Policy Research.

Brown, D. (1999), *Can Consumer Product Labels Deter Foreign Child Labor Exploitation?*, Medford, MA: Department of Economics Tufts University.

Ehring, L. (2002), 'De facto discrimination in world trade law: National and most-favoured-nation treatment – Or equal treatment?', *Journal of World Trade*, **36**(5), 921-977.

GATT (1972), *Report of Working Party on Border Tax Adjustments*, L/3464, BISD 18S/97, 2 December, Geneva: GATT Secretariat.

GATT (1988), *Canada – Measures Affecting Exports of Unprocessed Herring and Salmon (Canada – Herring)*, Panel Report, BISD 35S/98, adopted on 22 March, Geneva: GATT Secretariat.

GATT (1989), *United States – Section 337 of the Tariff Act of 1930*, Panel Report, BISD 36S/345, adopted on 7 November, Geneva: GATT Secretariat.

GATT (1990), *Thailand – Restrictions on Importation of and Internal Taxes on Cigarettes*, Panel Report, BISD 37S/200, adopted on 7 November, Geneva: GATT Secretariat.

GATT (1992), *United States – Measures Affecting Alcoholic and Malt Beverages (US–Malt Beverages)*, Panel Report, BISD 39S/206, adopted on 19 June, Geneva: GATT Secretariat.

GATT (1994a), *The Results of the Uruguay Round of Multilateral Trade Negotiations: The Legal Texts*, Geneva: GATT Secretariat.

GATT (1994b), *United States – Taxes on Automobiles, Panel Report*, DS31/R, circulated on 11 October, not adopted, Geneva: GATT Secretariat.

Hudec, R. (1998), 'GATT/WTO constraints on national regulation: Requiem for an 'aim and effects' test', available at: *http://www.worldtradelaw.net/articles/hudecrequiem.pdf*.

Jansen, M. and A. Lince de Faria (2002), *Product Labelling, Quality and International Trade*, CEPR discussion paper 3552, London: Centre for Economic Policy Research.

Krugman, P. (1980), 'Scale economies, product differentiation and the pattern of trade', *American Economic Review*, **70**, 950-959.

Marceau, G. and J. Trachtman (2002), 'The Technical Barriers to Trade Agreement, the Sanitary and Phytosanitary Measures Agreement, and the General Agreement on Tariffs and Trade: A map of the World Trade Organization law of domestic regulation of goods', *Journal of World Trade*, **36**(5), 811-881.

Mattoo, A. and A. Subramanian (1998), 'Regulatory autonomy and multilateral disciplines: The dilemma and a possible resolution', *Journal of International Economic Law*, **1**, 303-322.

Messerlin, P. and J. Zarrouk (2000), 'Trade facilitation: Technical regulations and customs procedures', *The World Economy*, **23**(4), 577-593.

Nordström, H. and S. Vaughan (1999), *Trade and Environment,* special study 4, Geneva: WTO.

Public Citizen (2003), *http://www.citizen.org/trade/wto/environment.*

Shapiro, C. (1983), 'Premiums for High Quality Products as Returns to Reputations', *Quarterly Journal of Economics*, **98**(4), 659-680.

Sykes, A. (1995), *Product Standards for Internationally Integrated Goods Markets,* Washington, DC: The Brookings Institution.

Vaughan, S. (2001), 'Reforming environmental policy: Harmonisation and the limitation of diverging environmental policies: The role of trade policy', in Z. Drabek (ed.), *Globalization under Threat: The Stability of Trade Policy and Multilateral Agreements*, Cheltenham: Edward Elgar.

WTO (1996a), *United States – Standards for Reformulated and Conventional Gasoline (US – Gasoline)*, Panel and Appellate Body Reports, WT/DS2/R and WT/DS2/AB/R, adopted on 20 May, Geneva: World Trade Organization.

WTO (1996b), *Japan – Taxes on Alcoholic Beverages*, Panel and Appellate Body Reports, WT/DS8/R, WT/DS10/R, WT/DS11/R and WT/DS8/AB/R, WT/DS10/AB/R, and WT/DS11/AB/R, adopted on 1 November, Geneva: World Trade Organization.

WTO (1997), *European Communities – Regime for the Importation, Sale and Distribution of Bananas (EC – Bananas)*, Panel and Appellate Body Reports, WT/DS27/R and WT/DS27/AB/R, adopted on 25 September, Geneva: World Trade Organization.

WTO (1998), *United States – Import Prohibition of Certain Shrimp and Shrimp Products (US – Shrimp)*, Panel and Appellate Body Reports, WT/DS58/R and WT/DS58/AB/R, adopted on 6 November, Geneva: World Trade Organization.

WTO (1999), *Korea – Taxes on Alcoholic Beverages*, Panel and Appellate Body Reports, WT/DS75/R, WT/DS84/R and WT/DS75/AB/R, WT/DS84/AB/R, adopted on 17 February, Geneva: World Trade Organization.

WTO (2000), *Chile – Taxes on Alcoholic Beverages*, Panel and Appellate Body Reports, WT/DS87/R, WT/DS110/R, and WT/DS87/AB/R and WT/DS110/AB/R, adopted on 12 January, Geneva: World Trade Organization.

WTO (2001a), *Korea – Measures Affecting Imports of Fresh, Chilled and Frozen Beef (Korea – Beef)*, Panel and Appellate Body Reports, WT/DS161/R and WT/DS169/R, and WT/DS161/AB/R and WT/DS169/AB/R, adopted on 10 January, Geneva: World Trade Organization.

WTO (2001b), *European Communities – Measures Affecting Asbestos and Asbestos-Containing Products (EC – Asbestos)*, Panel and Appellate Body Reports, WT/DS135/R and WT/DS135/AB/R, adopted on 5 April, Geneva: World Trade Organization.

WTO (2001c), *United States – Import Prohibition of Certain Shrimp and Shrimp Products, Recourse to Article 21.5 by Malaysia (US – Shrimp (Article 21.5))*,

Panel and Appellate Body Reports, WT/DS58/RW and WT/DS58/AB/RW, adopted on 21 November, Geneva: World Trade Organization.

WTO (2001d), Ministerial Declaration, WT/MIN(01)/DEC/1, Geneva, WTO, adopted on 14 November, Geneva: World Trade Organization.

WTO (2002a), *GATT/WTO Dispute Settlement Practice Relating to GATT Article XX, Paragraphs (b), (d) and (g) – Note by the WTO Secretariat*, WT/CTE/W/203, Geneva: World Trade Organization.

WTO (2002b), *European Communities – Trade Description of Sardines (EC – Sardines)*, Panel and Appellate Body Reports, WT/DS231/R and WT/DS231/AB/R, adopted on 23 October, Geneva: World Trade Organization.

WTO (2003), *CTE Report to the 5th Session of the WTO Ministerial Conference in Cancún*, WT/CTE/8, Geneva: World Trade Organization.

6. Towards an Effective Eco-Innovation Policy in a Globalised Setting

René Kemp, Luc Soete, and Rifka Weehuizen

Two things are unlimited: the number of generations we should feel responsible for and our inventiveness.
Jan Tinbergen

SUMMARY

Technological innovations can contribute to more environment-friendly production and consumption practices. In order to realise such eco-innovations, national governments should tailor their policy instruments to the prevailing circumstances. We discuss the technology impacts of national environmental policies in developed countries. Next, the tenets, purposes, and context-specificity of different environmental policy instruments (standards, taxes, tradable permits, subsidies, communication, and covenants) are analysed. While national policies may be suitable to address local environmental problems, they are insufficient to cope with global, especially common-resource, issues. To fill this governance gap, we propose a more prominent role for international non-governmental organisations (NGOs), using the analytical framework of exit, voice, and loyalty. We conclude that since markets and governments are not capable of bringing about environmental sustainability at the global level, international NGOs can and should fulfil an important role in raising the intrinsic motivation of producers and consumers.

INTRODUCTION

One of the major policy questions of our time, literally a 'matter of life and death', is whether growth in population, industrialisation, and mobility around the world can be and will be compatible with sustainable development. More people have been added to the world's population in the past five decades than in all previous millennia of human existence. The overall world economy has been growing over the last decade at an historically unprecedented rate. Combined, these developments have had severe consequences for the world's environment in terms of pollution and exhaustion of natural resources. The world seems to be on a track of 'unsustainable development', and the major concern is that the limits of this course will only be acknowledged in terms of global policy when the world has already crossed them and feels the global consequences. Since these consequences are partly irreversible, the challenge is to change course before they occur. Technological development is part of both the problem and the answer. In this chapter, we examine the possibilities for governments to steer the process of technological and social innovation so as to achieve environmentally sustainable development at the global level.

Traditionally, science and technology have been viewed as the 'deus ex machina' solution to people's unsustainable activities in terms of production and consumption. Technological development and innovation could be geared towards environment-friendly methods of production, distribution, and consumption. Often, it is not clear how science and technology are expected to achieve this; it is a 'black box' – or rather, a 'green box' – out of which will emerge the secret of long-term sustainable development.[1] In order to understand how technology and innovation can contribute to sustainable development, we need to understand what is going on in this green box. An important driver of environmentally benign innovation ('eco-innovation') is government policy. With the growing experience of a variety of national environmental policies and innovation policies, much more can be said about the effectiveness of governments' attempts at steering development and growth in a more sustainable direction.

As highlighted in the next section, most advanced countries today have sophisticated national environmental policies. These policies focus increasingly on what may be called the 'greening of technology': the design of a wide variety of policies addressing particular features of the innovation process which may be central in reducing some of the negative environmental aspects of industrial production and consumption. Such policies range from the development of new process or product technologies, and the particular role of technology in enforcing standards therein, to technology diffusion with various schemes of subsidies and/or taxes aimed at an accelerated uptake of more environment-friendly technologies by industry as well as consumers. The conclusion we draw, as elaborated in section 3, from examination of the by now quite voluminous set of specific policy instruments in operation in different countries is that such policies have often been instrumental in

---

[1] Kemp and Soete, 1992.

bringing about a significant recognition of the importance of environmentally benign production and consumption in many advanced, high-income countries, and in some newly industrialising countries, but that their impact has been limited in many areas with international implications. Improvements in the quality of some aspects of the local environment (air, water, or soil pollution) have sometimes been impressive. Environmental unsustainability at the local level is now increasingly identified as a sign of lower levels of development; or more broadly, of a lower level of the quality of life of citizens living in that environment.

From this perspective, globalisation is a complex process in which there are many interactions between population growth and poverty, urbanisation, deforestation, and global environmental degradation. Addressing those on the basis of national eco-innovation policies does not seem to be a fruitful route. Furthermore, from a global perspective, technological change appears, to some extent, to be as much part of the problem as part of the solution. Thus, the primarily technology-induced process of rapid industrialisation at the global level often leads – at least in its early take-off phase – to increased exploitation of unique, sometimes common, natural resources (such as rain forests and oceans). The continuous improvements in transport efficiency and the resulting reductions in transport costs have enlarged beyond imagination the opportunities for the trading of physical goods by sea, truck, or air, with carriers generally not incorporating most of the negative, environmentally damaging externalities associated with such transport. Revolutionary digital technologies have facilitated information and communication exchanges at the global level, broadening the consumption expectations and anticipations of world citizens living in rich and poor contexts. Those developments have improved global welfare, but the failure to address the environmental implications of such developments raises questions about the sustainability of the present pattern of global development.

The focus of this chapter is on innovation-related incentive structures with respect to long-term sustainable changes in the behaviour of individuals (whether consumers or producers), private and not-for-profit organisations, and countries or groups of countries. Government policy in the form of regulations, taxes, and subsidies has been important historically; therefore we start with an overview of national policies in sections 2 and 3. Changes in behaviour can be induced in a number of ways: through extrinsic incentive structures such as direct regulation and law; through self-regulating market incentives such as prices, taxes, subsidies, and labels; and through intrinsic incentives such as the internalisation of goals in organisations and people's internal utility perceptions, as a result of which people behave in more environment-friendly ways, not so much because of the fear of being fined but rather inspired by internal beliefs.

Within a global context, the first strategy will quickly reach its natural limits, as is argued in section 4. When environmental problems are global in nature, direct international regulation often is – and will probably for a long time remain – a vulnerable and weak policy instrument in terms of effectiveness. Its lack of effectiveness will discourage national policy-makers from participating in this type of regulation, thereby further weakening it. The second strategy, providing market incentives, is likely to be more effective in bringing about changes in behaviour when dealing with private goods, so-called 'excludable' goods.[2] From this perspective, the development of new international markets (for instance, in the area of international trade in emission rights, such as the recent Directive of the European Union (EU) on the Internal Trading of Carbon Dioxide Emissions), represents an interesting attempt at enlarging the market incentives structure both geographically and in the direction of new, excludable, private environmental goods, to include such emissions. However, sustainable development at the global level is also crucially dependent on changes in behaviour with respect to so-called non-excludable goods: common or global public goods. This type of goods represents a problem area for governance: regulation will be ineffective, and market incentives can be applied only to a limited extent.

This is where, as we argue in section 5, international NGOs provide a response aimed at the third strategy of changing individual behaviour, emerging as citizen-based institutions, between the state and the market. NGOs typically focus on awareness and action, on mobilising the intrinsic motivation to behave in environment-friendly ways of individuals and organisations, and on reducing the impression of individual helplessness by showing how concrete action is feasible and, in some cases, even effective. International NGOs operate at the global level, yet, because of their local focus, appear to be effective in bringing about behavioural change. Apart from direct influence such as in buying a piece of rain forest, they are also capable of influencing international law and regulation, and market behaviour. NGOs have acquired, we argue, the function of a global 'collective conscience', to use the term of the sociologist Emile Durkheim, which has an important regulatory function. Using Hirschmann's 'exit, voice, loyalty' framework, we argue that international NGOs provide today the 'voice' strategy of changing human behavioural patterns to enable long-term sustainable global development.[3]

---

[2] See also the chapter by Alkuin Kölliker in this book.

[3] This is based on Soete and Weehuizen, 2003.

An effective eco-innovation policy at the global level must consist of a variety of incentive-inducing mechanisms: global regulatory mechanisms setting limits to process emissions, international research policies of the EU aimed at the development of eco-innovation, and a host of other policies focusing on the more rapid uptake of environment-friendly technologies; and behavioural mechanisms motivating people across the globe to reduce unsustainable consumption and production practices.

TECHNOLOGY IMPACT OF ENVIRONMENTAL POLICY

In this section, we provide a brief description of the technology responses to the various environmental challenges which have emerged over the last thirty years in most developed countries. These technology responses range from attempts at steering the development of eco-innovations, product substitution, and the development of new processes to the diffusion of existing technology. The most common response to environmental regulation has been incremental innovation in processes and products and the diffusion of existing technology (in the form of 'end-of-pipe' solutions and non-innovative substitutions of existing substances). The stringency of regulations appears to be an important determinant of the degree of innovation, with stringent regulations such as product bans being most effective for radical technology responses.[4] Often, technology-forcing standards appear to be a necessary condition for bringing about innovative compliance responses. They do not 'pull' innovation in a simple way: long before the regulations are promulgated, there is a search process to find solutions to the problems, by the regulated industry (mostly for defensive reasons), its suppliers, and outsiders. This happened in the case of PCBs and CFCs, where firms both in and outside the chemical industry sought substitutes ten years before the use of PCBs and CFCs was actually banned.[5] In this case, the certainty that a product or activity would be subject to regulations was an important factor.

As to the nature (incremental or radical, product- or process-related) and the source of technological solutions, the following stylised facts have been established:[6]

---

4 There exists some literature on the impact of the stringency of environmental regulations on compliance innovation and clean technology. This literature consists of the work by Ashford and others in the United States (Ashford, 1993; Ashford et al., 1979, 1985), and a number of German studies (Hartje, 1985; Hemmelskamp, 1997; Klemmer et al., 1999). The focus of most of these studies is on technical innovation, rather than on organisational innovation.

5 Ashford et al., 1985.

6 See note 4 and Kemp, 1997; Strasser, 1997.

- High-volume, mature sectors were more resistant to change, although very amenable to environmental monitoring and process controls that improve efficiency. This fits with the Abernathy-Utterback product life cycle model, according to which sectors become rigid during the lifetime of products, especially those sectors that are capital-intensive. The problem here is often one of the sheer size of the sunk costs, the initial fixed capital costs.
- Significant process innovations occurred in response to stringent regulations that gave firms in the regulated industry enough time to develop comprehensive strategies. There is a clear trade-off between achieving quick results, on the one hand, and more long-term and more radical change, on the other hand.
- Smaller firms and potential new entrants tended to develop more innovative responses. A possible explanation for this is that incumbent firms, especially large ones, are vested in old technologies – both economically and mentally. Environmental change has, in other words, also much to do with organisational change.
- The environmental-goods-and-services industry provided compliance strategies that were at best only incrementally innovative but which diffused quickly owing to their lack of disruption and their acceptability to regulators.
- Regulatory flexibility towards the means of compliance, variation in the requirements imposed on different sectors, and compliance time periods were aspects of performance standards that contributed to the development of superior technological responses.

Technology responses are not always a simple response to regulatory pressure.[7] Apart from the regulatory stimulus, many other factors had an influence. This suggests that the stimulus-response model popular in the economics literature is too simplistic. For one thing, it assumes that social innovation starts with regulation, which is often not the case. Regulation is not the be-all and end-all of social innovation. The knowledge required for such innovations is often already available, with regulations providing the leverage or some extra stimulus for their exploitation. Regulation is, as we argue at greater length in section 4, only one of many stimuli. It may, in some cases, even not be needed for environmental innovation. Many technologies producing environmental benefits have been adopted for normal business reasons of reducing costs and enhancing product quality. These options have been referred to as eco-efficiency options.[8] However, even for

[7] Kemp, 1997; Klemmer et al., 1999.

[8] The term eco-efficiency was coined by the World Business Council for Sustainable Development (Schmidheiny, 1992).

environmentally beneficial technologies that do not combine environmental gains with economic gains, regulation may not be needed. In the case of environmentally harmful products, there will often be local pressures to reduce the harm. These pressures come from a range of actors: insurance companies, banks, customers, employees, environmentalist groups, and consumer organisations through product tests that include environmental aspects.[9] We come back to the role of such 'motivating' organisations in the international context in section 5.

An example of the above is the following. When the early synthetic detergents of the 1960s created visible environmental problems (foam in surface water), the detergent companies and especially their suppliers developed new processes – leading to biodegradable synthetic detergents – without government regulation, although future regulation was expected. The local voicing of concern and the threat of regulation may be enough to induce industry to look for alternative solutions. This does not remove the need for regulation. But care must be taken in using regulation for promoting innovation. Given the information problems of policy-makers, the threat of regulation might actually be a better means of stimulating technological innovation than actual regulation.[10] At the international level, regulation is likely to undermine the international competitiveness of the firm or industry. It is difficult to formulate regulations that are not disruptive in some sense.

Environmental innovations, like normal innovations, must meet a number of requirements: they should be expendable; it should be possible to fit them into existing processes; and, in the case of products, they should meet user requirements in terms of performance characteristics. Water-saving shower heads should be comfortable (have sufficient stream power), and environmentally improved detergents should have good washing performance. User benefits and social-performance benefits must be balanced and co-optimised. It is the need for co-optimisation that creates a problem for innovators and environmental regulators. For example, it proved difficult to develop phosphate-free detergents with washing power equal to that of the phosphate-based ones. In the search for a phosphate substitute, detergent companies spent more than 250 million US dollar (USD). The actual regulations on phosphate content co-evolved with the results obtained from product tests (both toxicological tests and tests of washing performance).[11]

---

[9] According to an environmental programme manager at Sony, environmental aspects account for about 20 per cent of the scores obtained by consumer products in consumer organisations' product tests. A discussion of the factors that led to the use of ONO installations for the control of metal discharges is offered in Kemp, 1997.

[10] Rip and Kemp, 1998.

[11] Hartje, 1985.

The above example shows that innovations cannot simply be 'elicited by legal fiat'.[12] This fits with insights from technology studies, according to which technology cannot be moulded in a pre-defined, socially desirable shape. This is why emission limits are based on assessments of what is technologically possible and economically affordable, and why environmental permits are often based on the concept of 'Best Available Technology' (BAT) or 'Best Practicable Means' that are specified in BAT lists or guidance notes. There is a dynamic interplay between innovation and regulation, with innovations often paving the way for regulations. The stimulus-response model fails to appreciate this dynamic interplay and circular causality.

The obvious implication of all this is that the macro management of technical change is not a simple matter. It is difficult to design instruments that do the job and do it well, in the sense that society as a whole is better off. Evaluation studies of environmental policy instruments show that the instruments in themselves are either ineffective in achieving a set goal or outcome, or inefficient in terms of costs or choice of technology. An example is the ONO technology used in the Dutch metal-plating industry to control the release of metals in waste water, which led to the production of toxic sludge containing heavy metals that had to be treated.

MERITS AND LIMITS OF ECO-INNOVATION POLICIES

Given the underlying complexity in designing appropriate and effective policy instruments, what can be learned from the actual experiences of countries with alternative eco-innovation policy instruments? Are there 'best-practice' environmental policy instruments to encourage technological innovation and diffusion? Clearly, there is not a single best-practice policy instrument that can stimulate the use of clean technology. All instruments have a role to play, depending on the context in which they are to be used. Not surprisingly, from an innovation perspective, the experiences with environmental policies are very mixed. All we can do here is to offer suggestions as to the purposes for which specific instruments may be used to obtain environmental protection benefits through the use of technology. We limit the discussion to the most common alternative eco-innovation policy instruments.

---

[12] Heaton, 1990.

Standards

Emission standards were often based on available end-of-pipe technologies and provided little incentive for the development of new, more effective technologies; they merely stimulated the diffusion of existing technologies. This demonstrates the danger of using technology-based standards and the importance of taking a long-term view towards environmental protection. Technology-forcing standards that require the development of new technologies are a better way of encouraging technological innovation, as the regulatory experiences in the United States (US) demonstrate. However, these may impose high costs on companies unless regulators are willing to soften and delay standards, which, in its turn, will have a negative effect on the willingness of suppliers to develop innovations. Technology-forcing standards should only be used when technological opportunities are available that can be developed at sufficiently low cost.

In using standards, it is important that regulators give industry enough time to develop solutions that are environmentally benign and meet important user requirements. Time may also be needed to examine whether solutions are environmentally benign and do not pose other hazards. One way of dealing with the problem of compliance time is to give firms innovation 'waivers' that exempt them from regulations during a certain period. If innovation waivers are used, it is important to give firms sufficiently long time allowances and to make sure that eligibility criteria are clear. Another strategy is to set long-term standards that require the development of new technology.

Tradable Permits

Decentralised incentive systems (such as taxes and tradable pollution rights) are an alternative to command-and-control policies. They are generally favoured by economists. The theoretical benefits of incentive-based approaches to reducing pollutant emissions are many. First, effluent fees (or charges and taxes) and tradable quotas are more efficient because polluters are given a choice between compliance and paying the polluter's bill. Polluting firms cannot be forced to undertake emissions control whose marginal costs would be higher than the effluent fees. This means that environmental benefits are achieved at the lowest abatement costs. Second, there is a financial incentive to diminish all pollution; such an incentive is absent under emission limit values (ELVs), where companies have no incentive to reduce their emissions below the ELVs. Thus, they create a constant demand for innovation.[13] The economic belief that incentive-based

[13] Stewart, 1981.

approaches provide a greater inducement to innovation is based on this argument. Third, such a system depends less on the availability of pollution-control technology than standards-based policies and, therefore, can be introduced more quickly at lower decisional costs by reducing demands on the regulatory process to take decisions on complex, detailed engineering and economic questions.[14] Fourth, the danger that polluting industries fail to develop new technologies for strategic reasons is lower under an incentive-based regime because they are not dependent on available solutions. Finally, economic instruments tend to stimulate process-integrated solutions (including recycling technology), rather than the end-of-pipe technologies that have been commonly applied in the past.

A disadvantage of effluent charges is the uncertainty about polluters' responses. Another disadvantage is that the total environmental costs – abatement costs plus tax payments – are likely to be high, which lowers their political attractiveness and may induce regulators to set low taxes, as has happened in the countries in which they are being used. Since freely distributed tradable pollution quotas do not have this drawback, they may work better than taxes or charges in stimulating environmental innovations. However, there are also other disadvantages to economic instruments. First, in order to be effective, polluters must be responsive to price signals, which is not always the case. For instance, two evaluation studies in the Netherlands showed that price considerations played a limited role in the timing of investments in thermal home improvements. This suggests that price incentives are probably better suited to changing the behaviour of firms than the purchasing decisions of consumers. Second, the price incentive must be sufficiently high to induce firms to develop and implement environmentally beneficial technologies. This was not the case for most environmental policies in which economic instruments were used. And, third, in dealing with transnational environmental problems (such as global warming), taxes should be used unilaterally only if their introduction does not put national industries at a serious competitive disadvantage. They should be introduced in those sectors where environmental costs are a small part of total costs or in sectors sheltered from international competition.

Subsidies

Uncertainty about the demand for cleaner technologies, partly related to unpredictable government policy, may call for the use of research and development (R&D) subsidies or loans. However, agencies responsible for subsidy programmes should take care not to stimulate second-rate

[14] Stewart, 1981.

technologies. The use of subsidies should be restricted to environmentally beneficial technologies for which there are as yet no markets, for example, technologies with long development times (such as energy technologies) or technologies where appropriating the benefits of innovation by innovators is problematic (such as technologies that are easy to imitate). R&D programmes may also be used to increase the number of technological solutions if there is uncertainty about environmental solutions. Subsidies for investments in pollution-control technology are less useful. They clash with the polluter-pays principle and are expensive; in addition, evaluation research in the Netherlands has proved they are only minimally effective. There is a great risk that such subsidies provide windfall gains for the firms and consumers receiving them. They should be used only if a switch to cleaner technology involves high costs and produces competitive disadvantages owing to less strict regulations in other countries.

Communication

Communication instruments can be useful policy tools for addressing information problems related to products and processes. Environmental management and auditing systems in business, demonstration projects, and information campaigns can be useful to ensure that firms make better use of their possibilities for emission reduction, especially cost-reducing environmental measures. Information disclosure requirements, such as those in the US, which force firms to communicate environment-related product information are also believed to be useful. They increase pressure on firms to improve their environmental record while enhancing their environmental awareness. Eco-labels are very important for green purchases. They make the market for green products more transparent. Eco-labels offer a stimulus for producers to innovate, but the requirements are often not technologically challenging.[15] Information instruments are believed to be useful as additional mechanisms, not as substitutes for environmental regulations or taxes.

Covenants

Covenants are a new environmental policy instrument in the EU and the US. Covenants are negotiated agreements between industry or an industrial sector and government in which industry promises to reduce the environmental burden of its activities progressively within a certain period (often five to ten years) according to certain targets. They are sometimes referred to as voluntary agreements, if firms belonging to certain sectors are free to enter

[15] Hansen et al., 2000.

into sectoral agreements (if they do not, they will be subject to regular licensing procedures). Covenants are attractive to industry because they provide more freedom as to the methods and moments of compliance, thus lowering the so-called regulatory burden. By handing over responsibility for achieving environmental improvements to industry, covenants may stimulate environmental responsibility in firms, which is important for the wider integration of environmental concerns to company decisions. From the perspective of environmental control agencies, covenants are attractive because they lower administrative burdens and help to establish better, more cooperative relationships with business.

A clear disadvantage of the use of covenants is the danger of strategic exploitation of such agreements by firms who may engage in free-rider behaviour, or, more likely, may underexploit opportunities for innovation by claiming that it is impossible for them to meet targets through compliance technology that fulfils important user requirements. Such behaviour may jeopardise the fulfilment of environmental agreements. Further, the softness of covenants – or of voluntary agreements in general – means there is little incentive for third-party suppliers to develop compliance technologies as the market for the new technologies is insufficiently secured. If covenants continue to be used in the future, as they probably will be, they should be more oriented towards innovation. One way of doing so is by negotiating technology compacts between public authorities and private firms to implement long-term technological change.[16] In such compacts, industry commits itself to performance goals that require new and advanced technology in exchange for enforcement flexibility and guaranteed acceptance of the new technology. The system of technology compacts looks attractive but, as with covenants, it could be exploited by industries (including the 'environmental industry') which have superior knowledge of what is technologically possible.[17]

The list of various eco-innovation policies is summarised in Table 6.1. The table describes the effectiveness and efficiency characteristics of various policy instruments, the purposes for which they can be used (to stimulate technological innovation or diffusion), and the contexts in which they may be applied, based on experiences with environmental policies and studies of environmentally benign technical change.

The discussion above makes clear that all policies have certain advantages and drawbacks. Generally, policy instruments should be combined with one another to benefit from synergistic effects. A combination of standards with

---

[16] Heaton and Banks, 1999.

[17] Aggeri (1999) offered a discussion of the usefulness of cooperative approaches to promote innovation, and provided useful suggestions for managing the process of collective learning.

Table 6.1 Eco-innovation policy instruments

Policy instruments	Characteristics	Purposes	Application
Technology-based environmental standards	• Effective in most cases (if they are adequately enforced) • Uniform standards give rise to inefficiencies in case of heterogeneous polluters	• Technological diffusion and incremental innovation	• When differences in the marginal costs of pollution abatement are small and economically feasible solutions to environmental problems are available
Technology-forcing standards	• Effective (in focusing industry's minds on environmental problems) • Danger of forcing industry to invest in overly expensive and suboptimal technologies • Problem of credibility	• Technological innovation	• When technological opportunities are available that can be developed at low costs • When there is consensus about the appropriate compliance technology
Innovation waivers	• Same as technology-forcing standards	• Technological innovation	• When technological opportunities are available and when there is uncertainty about the best solution
Taxes	• Efficient • Uncertainty about industry response • Danger that they provide a weak and indirect stimulus • Total environmental costs for industry are likely to be high • Limited political attractiveness	• For recycling, and material and energy saving • Technological diffusion and incremental innovation	• In cases of heterogeneous polluters that respond to price signals • When there are many different technologies for achieving environmental benefits

Table 6.1, continued

Policy instruments	Characteristics	Purposes	Application
Tradable permits	• Technically effective • Cost-effective (i.e., environmental benefits are achieved at lowest costs)	• Technological innovation and diffusion	• Same as taxes • Costs of monitoring and transaction should not be prohibitive
R&D subsidies	• Danger of funding second-rate projects • Danger of providing windfall gains to recipients	• Technological innovation	• When markets for environmental technology do not yet exist and when there is uncertainty about future policies • When there are problems of appropriating the benefits from innovation • When there exist important knowledge spillovers • In cases of large social benefits and insufficient private benefits
Investment subsidies	• In conflict with polluter-pays principle • Danger of windfall gains • Politically expedient	• Technological diffusion	• When industry suffers a competitive disadvantage due to less strict regulations abroad
Communication	• Helps to focus attention on environmental problems and available solutions • Little coercive power	• Technological diffusion and innovation (in the case of eco-lables)	• When there is a lack of environmental consciousness • When there are information failures
Covenants and technology compacts	• Uncertainty about whether industry will meet agreements (should be supplemented with penalty for non-compliance) • Low administrative costs	• Technological diffusion	• In cases of many polluters and many technological solutions • When monitoring environmental performance is expensive

economic instruments is particularly useful since this combines effectiveness with efficiency. A good example of an effective and economically efficient environmental policy is the US corporate automobile fuel economy (CAFE) standards, which set progressive fuel economy targets for automobile manufacturers in the 1979-1985 period under penalty of a 50 USD fine for each mile per gallon of shortfall per car sold. Tradable pollution permits also deserve to be used more, as they also combine effectiveness with efficiency. At present, there is a nation-wide market for sulphur dioxide in the US where utilities can trade sulphur dioxide rights at the Chicago Board of Trade. Early results suggest that the tradable permits for sulphur dioxide emissions will reduce the costs of the 1990 acid rain programme by 50 per cent or more.[18] In April 2004, the European Union (EU) introduced rules and regulations on the internal EU trading of carbon dioxide emissions, so as to bring the EU closer to the emission-reduction targets such as established in the Kyoto Protocol.

In short, despite the complexity involved in using the appropriate mix of alternative eco-innovation policies and the context-specificity in assessing the effectiveness of such policies, significant progress seems to have been made in most developed countries in designing effective national environmental policies that have been successful in reducing local environmental concerns.

The progress made in achieving local sustainability in many rich countries, where growing environmental concerns have triggered policy interventions and where awareness of environmental quality and/or degradation has been integrated into local concepts of quality of life, contrasts, however, sharply with global environmental concerns, where much less progress has been achieved, as we will discuss in the next sections.

THE LACK OF GLOBAL GOVERNANCE

The 'greening of technology' policies reviewed in the previous section appear, by and large, to be reflections of the growing environmental concerns of citizens in high-income countries. As income and welfare have grown, demands for environmental quality have also grown. This argument can be expanded to the international context: the high environmental standards and the market forces driven by high-income domestic consumer demands for 'green' products will ultimately give high-income countries an international edge in further developing and applying new environmental production and consumption technologies. Companies will recognise the importance and potential of environmental technologies for increasing their international competitiveness. The EU, for example, in its 2003 Environmental

---

[18] Palmer et al., 1995.

Technologies Action Plan (ETAP), claims to have become a leading producer and exporter of key environmental technologies and services such as photovoltaics, wind energy, and water supply and services, thanks to a whole set of measures (the Directives on Integrated Pollution Prevention and Control,[19] Environmental Management Systems,[20] Eco-Labelling,[21] and the Community's Greenhouse Gas Emissions-Trading Scheme).[22] Thanks to the international technology transfer associated with the export of and foreign investment in such environmentally benign production technologies, goods and services, national and European eco-innovation policies have a much broader, international impact. Effective eco-innovation policies will ultimately lead to competitive advantage.

Alkuin Kölliker, in his chapter elsewhere in this volume, pays greater attention to the trade protection and international trade regulation features of national environmental policy measures, and in particular to the way multilateral trade rules, as in the context of the WTO, might take precedence over national environmental legislation. Here, we are less interested in the emergence of competitive advantages from technology-related green trade or foreign direct investment in high-income countries. The analysis, apart from the specific implications of transport costs,[23] is broadly in line with most 'North-South' technology trade models.[24] We are more interested in the question whether eco-innovation policies aiming to establish international technological leadership in environmental technologies, as in the case of the EU, will provide sufficient incentives to producers and consumers in the South to curtail the environmental degradation associated with their rapid industrialisation.

Apart from local, circumscribed impacts, the slowness of diffusion and uptake of environment-friendly technologies in the South also has direct global environmental implications for global common or public goods: impacts on climate change, biodiversity, ocean fishing, etc. In this global-impact perspective, it is clear that national eco-innovation policies are rarely effective: they hardly push for a more rapid diffusion of new products and processes with global environmental benefits beyond national borders (for example, by providing subsidies). On the contrary: from a broader historical perspective, the spread of technology worldwide and the global market expansion which followed in the wake of cheaper transport technologies have undoubtedly contributed to the more rapid global environmental degradation

[19] European Commission, 1996.
[20] European Commission, 2001.
[21] European Commission, 2000.
[22] European Commission, 2003.
[23] Soete and Ziesemer, 1997.
[24] Dosi et al., 1990; Krugman, 1979.

of (global) public goods. The rapid take-off growth which has taken place in the emerging economies over the last few decades on the basis of the old, heavy-energy and emission-based industrialisation path has contributed significantly to global environmental degradation and the disappearance of remaining, sometimes unique, natural resources. It is the current success of global industrialisation, with emerging economies such as China and India contributing to a rapidly growing world economy,[25] which is challenging global environmental sustainability most directly. Adjustment towards more environment-friendly technologies in those rapidly growing countries is undoubtedly high – for example, New Delhi rapidly transformed its public bus transport system, using natural gas – but is primarily aiming to improve local environmental conditions, in line with the rise in income and wealth.

The more fundamental question, however, remains whether overall industrial adjustment towards a more environmentally benign production structure will be sufficiently fast to offset the high output growth rates in those countries and the rapidly rising degradation of global common goods, such as the climate change accompanying it. At the same time, and at the high-income consumption end, there is similar concern that consumption adjustment towards less fossil-fuel-intensive consumption patterns is also too slow to offset the global catching up to such high-income and high-energy-consumption patterns with their adverse global environmental impact. In short, both poverty and affluence contribute to environmentally wasteful behaviour. This points to the other dark side of the globalisation process: emerging global environmental problems (such as global warming, ozone depletion, and, in some cases, natural disasters) are beyond national remedy, linking the fates of nations, policy-makers, producers, and consumers more closely together, but with nobody feeling responsible.

In this perspective, the process of globalisation poses global challenges, and dealing with these ultimately requires some form of global governance. The term 'governance' refers to the whole structure of institutions regulating human behaviour: the triad of state institutions, market institutions, and community institutions. 'Global' governance resembles other governance structures, but has several unique features, such as its geographical and organisational scale and size, and the diversity of issues it faces. More fundamental, however, is the lack of a (meaningful) external environment. Other forms of governance always deal with insiders and outsiders: the function of governance structures be they governments, firms, or gangs is to enhance the well-being of the insiders, while in several ways taking into account the outsiders: as allies or enemies, as partners or competitors, as

---

[25] In 2003, we witnessed the world's highest output growth ever.

relevant or irrelevant. This is the clearest in formal government structures, which, because of the distinct nature of their authority, always need explicit justification – more than the market or the community. Ancient city-states, medieval feudal kingdoms, and modern nation states generally came into being as a result of internal association (shared language, culture, and history), on the one hand, and external pressures (wars, foreign oppression, and natural threats), on the other hand. They derived an important part of their legitimacy – and thereby their authority and effectiveness – from the fact that they represented and protected the interests of their subjects or citizens against an outside world. In global governance, this element is missing: by definition, the world has no 'outside world'.

The reason why we deal with the notion of governance and global governance in such detail is because of its particular relevance to the effectiveness of environmental policies. Here too, the three components of governance can be considered separately: governments or states with their portfolios of direct regulation measures, markets with prices, taxes, and subsidies as self-regulation tools, and, finally, community institutions focusing to some extent on the 'activation' of intrinsic incentives with consumers and citizens.

As we discussed in sections 2 and 3, national eco-innovation policies focusing on what could be called extrinsic incentive structures (such as the setting of technology-enforcing standards) might ultimately result in local success but are unlikely to be effective in dealing with the environmental degradation of global common environmental goods. As argued by Kölliker, it is also unlikely that international cooperation among countries will naturally emerge in these last cases. On the contrary, global governance is likely to fail with respect to global common environmental goods. Market-driven eco-innovation policies are likely to be more effective in bringing about changes in consumer and producer behaviour, particularly if the negative environmental features of production or consumption can be effectively internalised. The organisation of such 'privatised' environmental features, such as opening up carbon dioxide emissions to international trade, can be an effective way of addressing global common environmental problems such as climate change. Even if national governments refuse to participate in the global governance of such markets, they may well come under increasing pressure from both domestic firms and citizens to do so anyway. A third set of eco-innovation policies, for example, results directly from community- or citizens-based pressure groups. The focus here is on the internalisation of goals in organisations and on the internal utility perceptions of environment-friendly ways in people, not out of economic self-interest but out of internal conviction and intrinsic motivation.

This is where one of the most interesting but perhaps least expected actors appears on the scene of effective eco-innovation policy: the international

NGOs. The definition of NGO is not commonly agreed upon. In the widest sense, NGOs are self-governing, private, not-for-profit organisations aiming to improve quality of life in a general sense. NGOs are different from grassroots organisations or voluntary organisations, or public service contractors.[26] The term 'NGO' is generally used for organisations with a clear formal organisational structure. Though they often work with volunteers, NGOs have a core of people working as paid employees. And though NGOs often provide public services, they do so in a self-assigned capacity, not as contractors. International NGOs are taken here to refer to those organisations that are officially established, run by employed staff (often urban professionals or expatriates), well-supported (by international funding), and often relatively large and well-resourced. In the following section, we discuss the role of international NGOs in global governance, using the triad of state, market, and community.

NGOS: VOICE FOR GLOBAL SUSTAINABILITY?

NGOs share many important characteristics with governments: they are not for profit and work for the public good. Relations between NGOs and governments are close. An important part of NGO work consists of influencing governments, lobbying, and putting issues on the policy agenda. NGOs need governments to get things done, to change policy and legal frameworks, and to fund their activities: governments are major donors to NGOs, especially in the area of development aid. NGOs do not only work closely with governments, but they also share characteristics of governmental organisations in terms of organisation structure. With increasing size, NGOs – originally devised for their transparency, efficiency, responsibility, and accountability – tend to develop bureaucratic structures resembling those of governments.

NGOs are more prominent on the international stage today than they were twenty years ago. Yet, international prominence or mere activity is not sufficient proof of influence. It is clear that NGOs are constrained in the degree and type of influence they have on national governments and supranational organisations. NGOs appear to have the greatest influence if issues have high salience and low policy. 'High salience' is important because it is easier to influence issues that are already on people's minds and on policy-makers' agendas. Of course, such issues need to be put on the policy agenda first or be brought to public awareness; NGO activity is not always

---

[26] Some other abbreviations often used in this context are INGOs (international NGOs) and IGOs (international governing organisations).

suitable for doing so. Grassroots actions and large demonstrations are the means of drawing attention to focal issues: 'Low policy' encompasses those areas that do not threaten national security, are fairly cheap and easy to manage bureaucratically, and are unknown or broadly uncontroversial. In these policy areas, there is often a policy vacuum, and NGOs are welcomed for their expertise to generate policy. In addition, low-policy issues are generally undertaken by the weaker parts of bureaucracy. These are easier to influence because they are more open, and NGOs are not infrequently seen as the allies of governments in the face of public or intragovernmental conflicts. Another key element in influence and success is early and privileged access to decision-making bodies. As the rising legitimacy and popularity of NGOs gave them greater access, there has also been a rise in NGOs' use of insider tactics.

With respect to eco-innovation policy, international NGOs are essential in raising awareness of the shared nature of global public goods and the vulnerability of such shared resources to overexploitation and environmental degradation. In this sense, international NGOs contribute to the missing dimension of a global 'collective conscience' or 'global loyalty'. From this perspective, international NGOs are the 'voice' outlet for the human need for global sustainable development.

Albert Hirschmann developed the concepts of exit and voice to explain the behaviour of firms,[27] and these concepts are also applicable here. In economics, exit may be an effective way of signalling consumer dissatisfaction with particular products or services in highly competitive markets, but, in situations of monopoly supply, there is ultimately nowhere else for consumers to go. In the case of the environmental degradation of shared resources, exit is not an option. Voice – complaints or expressions of dissatisfaction with particular products or services – will often require only some alert citizens to have an impact. In the case of sustainable development, voice has raised awareness and policy concerns. However, while exit is a fairly crude, binary response, voice responses are more subtle: consumers may express more clearly what they desire and can verbalise the intensity of their dissatisfaction more easily. In practice, this means that global governance tends to be based on voice. Loyalty is the most difficult concept in Hirschman's framework. Hirschman introduced the concept of loyalty to try and understand why exit is virtually ruled out in certain contexts such as the family, tribe, church, and state. Loyalty clearly serves as a brake on exit, preventing it from occurring when this would otherwise be rational. Hence, loyalty often tends to activate voice.

[27] Hirschmann, 1970.

With respect to global sustainability, the exit, voice, and loyalty framework can be related in a relatively straightforward way to the governance triad discussed in section 4: the market, the state, and the community. Exit can be associated with the type of rationale that is dominant in markets (choices: A or B, yes or no). To the extent that alternative markets exist, exit can be an effective strategy for inducing firms to shift towards more environmentally benign production methods and products. For many products in rich countries (cars, domestic appliances, etc.) as well as in poor states (sewerage systems, clean water, etc.), however, no such alternatives exist. Voice can be associated with political structures but primarily in terms of national public goods (such as elections in democratic states). At the global level, the rules governing global voice are, unfortunately, not the result of democratic voice. Loyalty is generally linked to community since it involves socio-psychological attachments. Enhancing global loyalty is, to some extent, the raison d'être of international NGOs.

International NGOs have been a major force, arguably the most important force, behind corporate social responsibility (CSR), a notion closely related to corporate sustainability. Ewa Charkiewicz even spoke of an avalanche in CSR, which began in 1999 when Kofi Annan proposed to the World Economic Forum in Davos to establish the Global Compact: a set of voluntary guidelines which include principles of human rights, core labour standards, and environmental management principles.[28] A year later, also in Davos, James Wolfensohn announced the World Bank Program on Corporate Social Responsibility and Competitiveness.[29] In 2000, following an exceptionally fast negotiating process with stakeholders, the Organisation for Economic Co-operation and Development (OECD) issued revised Guidelines for Multinational Corporations. In 2001, the European Commission decided to develop a CSR framework for business.[30] And in 2002, a Minister for CSR was appointed in the United Kingdom.[31] In the same year, Chevron signed an agreement with the United Nations Development Programme (UNDP) for community development, health and education programmes in the Niger Delta.[32] McDonald's signed an agreement with UNICEF, the United Nation's Children Fund.[33] Companies want to be seen as responsible actors. Image is important. There are many indications that presentational innovation will be increasingly important in the new economy. Not all companies are equally prone to reputation pressures. One should not expect all companies to shift to

---

[28] United Nations, 2004.
[29] World Bank, 2004.
[30] European Commission, 2004a, 2004b.
[31] UK Government, 2004.
[32] UNDP, 2002.
[33] EarthRights International, 2004.

CSR, but CSR brings benefits in the absence of regulations. We cannot do without regulations for controlling global problems and managing global common goods. But we are not completely dependent on regulations either.

CONCLUSIONS

Over the past decades, most developed countries have devoted an increasingly large number of resources to the development of environment-friendly technologies. Combined with various incentives and regulations, such technology policies – and in particular eco-innovation policies – have contributed significantly to the reduction of local emissions and industrial waste output, improvements in surface water and river quality, environmental conservation, and many other features of the local environment in many rich countries. At the policy level, such local achievements have sometimes even been translated into claims that environmental issues were problems of the past.

At the global level, progress has been much more limited. Using the triad concept of state, market, and community in describing the essential elements of governance, we have argued that, in addition to the development of global regulations and global markets, global governance needs to take much more seriously the particular role international NGOs can play in changing intrinsic motivation in favour of sustainable development. Given the failure of national governments to effectively address issues of long-term sustainability with respect to common-resource goods, world citizens can only express their growing concern through such NGOs. The role of international NGOs on the global scene has become crucial in a very short time. Nation states and multinational corporations – the most important market players at the global level – have quickly grasped the particular role NGOs can play, given the unique requirements of global governance, in dealing with the size, scale, and diversity of the world's environmental problems. To fulfil their intermediary position and their main function of 'voice', NGOs should have the competence to speak different languages, not only literally but also figuratively: they must speak the language of different sectors, spheres, and cultures in order to engage across diverse institutional boundaries and foster interorganisational linkages.

Whether all this will be sufficient for a global sustainable development path remains an open question. Today, national governments appear to be using an exit strategy in dealing with global environmental problems: their focus remains strongly dominated by national concerns. Yet, the global aspects of environmental degradation and unsustainable development are likely to have the same global devastation effects, without making any distinction between advanced and less developed countries. In this sense, global warming is like a modern version of the plague. In its global impact, it resembles the local devastation in Italian cities centuries ago: it will act on rich and poor alike.

REFERENCES

Aggeri, F. (1999), 'Environmental policy and innovation: A knowledge-based perspective on cooperative approaches', *Research Policy*, **28**(7), 699-717.

Ashford, N. (1993), 'Understanding technological responses of industrial firms to environmental problems: implications for government policy', in K. Fischer and J. Schot (eds), *Environmental Strategies for Industry: International Perspectives on Research Needs and Policy Implications*, Washington, DC: Island Press.

Ashford, N., C. Ayers, and R. Stone (1985), 'Using regulation to change the market for innovation', *Harvard Environmental Law Review*, **9**(2), 419-466.

Ashford, N., G. Heaton, and W. Priest (1979), 'Environmental, health, and safety regulation and technical innovation', in J. Utterback and C. Hill (eds), *Technological Innovation for a Dynamic Economy*, New York: Pergamon Press.

Dosi, G., K. Pavitt, and L. Soete (1990), *The Economics of Technical Change and International Trade*, Brighton, Wheatsheaf, and New York: New York University Press.

EarthRights International (2004), *http://www.earthrights.org/news/mcunicefpr.shtml.*

European Commission (1996), *Directive 96/61/EC,* Brussels: European Commission.

European Commission (2000), *Regulation 1980/2000*, Brussels: European Commission.

European Commission (2001), *Regulation 761/2001*, Brussels: European Commission.

European Commission (2003), *Directive 2003/87/EC*, Brussels: European Commission.

European Commission (2004a), *http://europa.eu.int/comm/enterprise/services/Social_policies/index.htm.*

European Commission (2004b), *http://www.eldis.org/static/DOC12824.htm.*

Hansen, O., J. Holm, and B. Søndergård (2000), *Construction of a Selective Milieu of Environmental Development in the Danish Textile Industry*, paper presented at the 3rd International Conference of the European Society for Ecological Economics, Vienna, 3-6 May.

Hartje, V. (1985), *Environmental Product Regulation and Innovation: Limiting Phosphates in Detergents in Germany*, discussion paper 85-5, Berlin: International Institute for Environment and Society.

Heaton, G. (1990), *Regulation and Technological Change*, paper presented at the WRI/OECD Symposium Toward 2000: Environment, Technology and the New Century, Annapolis, Maryland, 13-15 June.

Heaton, G. and R. Darryl Banks (1999), 'Toward a new generation of environmental technology', in L. Branscomb and J. Keller (eds), *Investing in Innovation: Creating a Research and Innovation Policy that Works*, Cambridge: MIT Press.

Hemmelskamp, J. (1997), 'Environmental policy instruments and their effects on innovation', *European Planning Studies*, **5**(2), 177-193.

Hirschmann, A. (1970), *Exit, Voice and Loyalty: Responses to Decline in Firms, Organizations and States*, London: Harvard University Press.

Kemp, R. (1997), *Environmental Policy and Technical Change; A Comparison of the Technological Impact of Policy Instruments*, Cheltenham: Edward Elgar.

Kemp, R. and L. Soete (1992), 'The greening of technological progress: An evolutionary perspective', *Futures*, **24**(5), 437-457.

Klemmer, P., U. Lehr, and K. Löbbe (1999), *Environmental Innovation*, Berlin: Analytica.

Krugman, P. (1979), 'A model of innovation, technology transfer, and the world distribution of income', *Journal of Political Economy*, **87**(2), 253-266.

Palmer, K., W. Oates, and P. Portney (1995), 'Tightening environmental standards: The benefit-cost or the no-cost paradigm?', *Journal of Economic Perspectives*, **9**(4), 119-132.

Rip, A. and R. Kemp (1998), 'Technological change', in S. Rayner and E. Malone (eds), *Human Choice and Climate Change; An International Assessment*, Washington, DC: Batelle Press.

Schmidheiny, S. (1992), *Changing Course: A Global Business Perspective on Development and the Environment*, Cambridge: MIT Press.

Soete, L. and R. Weehuizen (2003), *No Exit: Reflections on Global Governance*, paper presented at the First Globelics Conference, Rio de Janeiro, October.

Soete, L. and T. Ziesemer (1997), 'Gains from trade and environmental policy under imperfect competition and pollution from transport', in H. Feser and M. von Hauff (eds), *Neuere Entwicklungen in der Umweltökonomie und –politik*, Regensburg: Universität Kaiserslautern.

Stewart, R. (1981), 'Regulation, innovation and administrative law: A conceptual framework', *California Law Review*, **69**(5), 1256-1270.

Strasser, K. (1997), 'Cleaner technology, pollution prevention, and environmental regulation', *Fordham Environmental Law Journal*, **9**(1), 1-106.

UK Government (2004), *http://www.businessandsociety.gov.uk*.

UNDP (2002), *http://www/undp.org/dpa/frontpagearchive/2002/january/18jan02/*.

United Nations (2004), *http://www.unglobalcompact.org*.

World Bank (2004), *http://www.wbicsr.org*.

7. Collaboration of National Governments and Global Corporations in Environmental Management

Kees Zoeteman and Eric Harkink

SUMMARY

Globalisation is transforming the power of nation states to manage environmental problems and sustainability issues. States are losing their traditional power in certain areas while, at the same time, doors are being opened for the use of new policy instruments. To identify possible opportunities for such new instruments, attitudes of actors towards sustainability were assessed and analysed. We developed a quantitative model of sustainability attitudes, which we tested on nation states, national governments, and two business sectors (the oil and gas industry and the dairy industry). The results of the study show that growing parallels in the attitudes of multinational businesses and their countries of origin are evolving simultaneously, especially in North-Western Europe. Key factors in stimulating these parallels are the vulnerability to global consumer choices and the impact of home-country culture in the global operations of multinational companies.

INTRODUCTION

Globalisation has positive and negative impacts on the sustainability levels of nation states and the possibilities of governments of safeguarding environmental quality. Extra economic growth that results in a higher standard of living for large groups in society is generally seen as a positive national effect of globalisation. However, this does not occur automatically.

Only when nation states have a good governance structure, implement laws well, have free press, and have democratised institutions is globalisation likely to benefit a country. Friedman saw the nation states in this respect as the plugs that connect countries with their potential global systems.[1] Economic globalisation may also result in the magnification of risk, such as the increase of pollution or the depletion of resources. Furthermore, globalisation makes all countries, both developing and developed, more vulnerable to external influences. Developing nation states experience the risk of being excluded from the global system; developed countries experience the externalities that result from their global interconnectedness. The more nation states are connected with global markets, the more they will be affected by economic fluctuations.

Nation states that promote free trade and open borders (for example, the United States (US) and the United Kingdom (UK)), do not necessarily possess the most globalised economies, as Figure 7.1 illustrates. Small but open economies such as those of Belgium, Singapore, and the Netherlands are highly globalised and are generally more sensitive to global market changes than are the economies of the US (index value for economic globalisation: 0.37), China (0.38), and Japan (0.33).

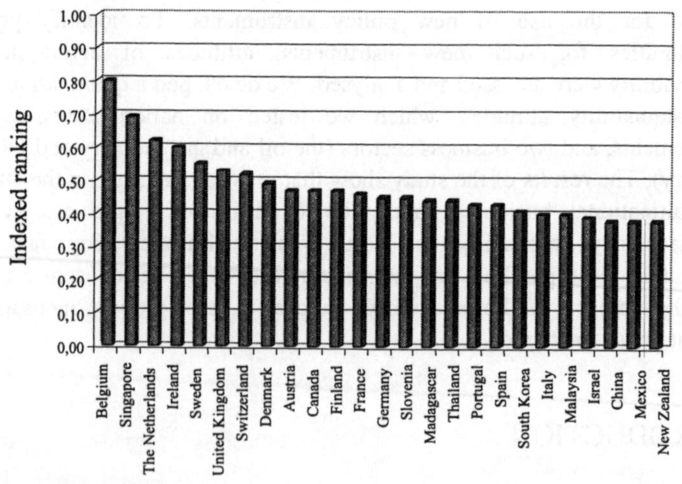

Figure 7.1 The globalisation degree of national economies[2]

---

1 Friedman, 2002.

2 Foreign direct investment as a percentage of gross domestic product (GDP), trade as a percentage of GDP and export concentration, indexed to the highest values of each aspect (CIA World Factbook, 1997; World Bank, 2002).

Environmental policies over the past 25 years have increasingly expressed the need to regulate and reduce polluting activities, restrict the exhaustion of natural resources, and restore the global habitat. Globalisation has spread these concepts to all nation states, and the fruit of this can be found in international agreements and environmental institutions. Yet, it is unclear whether widespread global environmental progress is likely to continue. Globalisation may jeopardise established frameworks for the management of environmental problems. When pressing economic or safety issues hit the media, attention for environmental concerns can diminish, even when the scientific urgency to take action is greater than before. At the national level, a major downgrading of the political priority of environmental issues has taken place since the year 2000 in countries such as Denmark, the Netherlands, and the US. How can environmental policy be sustained under completely different social circumstances than those prevailing during the last decades of the 20th century? And what kinds of roles should nation states develop? Policy-makers should ask themselves these questions before reality takes them by surprise.

In this context of a globalising society, it is of great interest to analyse how national governments cope with their changing power positions. The general view is that states are losing part of their power and sovereignty.[3] Can states compensate for this loss by developing new policy tools and ways to implement policies that cross old borders between governments and societal actors such as multinational corporations? Corporations are opening up to sustainable development as a result of external and internal factors. Internal factors are increased corporate awareness of sustainable development, stakeholder integration, and continuous improvement and innovation.[4] Other internal factors include the need to be attractive as an employer and to secure long-term supply and demand. External factors include pressure from government (particularly in home countries) and international society, the preference of certain sales markets, and the preferences of business partners and critical consumers. Through external factors, governments have the possibility of influencing corporate attitudes.

In this chapter, we address the changing roles of nation states and the prospects of closer collaboration between governments and multinational companies in pursuing global sustainable development. For this purpose, the oil and gas industry and the dairy industry were investigated as examples. A model of five sustainability attitudes, quantifying and comparing the attitudes of actors, passes in review. Finally, new environmental policy instruments for nation states and transnational corporations are discussed.

---

3 Held et al., 1999.
4 Hart, 1995; Sharma and Vredenburg, 1998.

GLOBALISATION AND NATIONAL ROLES

Many authors have recently described how globalisation has reduced the role of national governments to exert sovereignty in their territory (see Figure 7.2).[5] Proactive transnational companies have taken over roles that previously belonged solely to national governments, such as the management of natural habitats and the worldwide implementation of codes of conduct. The economic size of several multinational companies has grown far beyond the economic size of many states,[6] which have undergone a steady process of subdivision and left the global community with three times as many states of smaller size than a century ago.

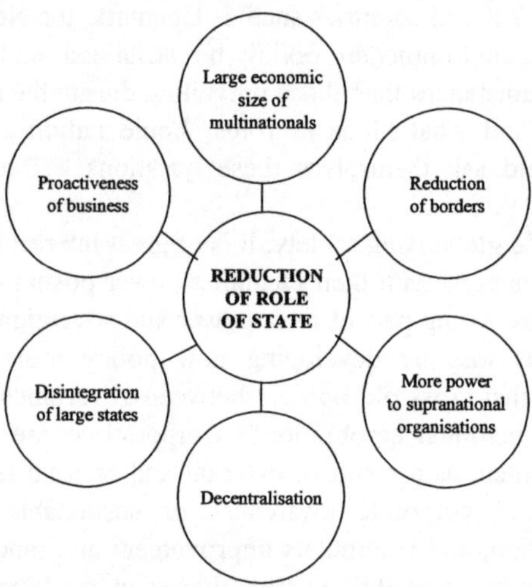

Figure 7.2 Factors reducing the role of states[7]

It is not only the growth of the economic size of multinational companies relative to the size of nation states that takes governance influence away from national governments. More than 40 per cent of international trade takes place within multinational companies, which have most functions centralised in a small number of global mega-cities such as New York, London, Paris, Tokyo, and Hong Kong.

Furthermore, many financial transactions are carried out through the internet, which has no territorial barriers, and is designed and mainly controlled by multinational businesses. Virtual currency trade, for example,

---

5 Beck, 1997; Held et al., 1999; Keohane and Nye, 2000; Sassen, 1996.
6 De Grauwe and Camerman, 2002.
7 Adapted from Zoeteman, 2002: 28.

had already exceeded a volume of 1 billion US dollar (USD) per day in 1995. The power of national central banks over monetary key variables has decreased and their main concern is to satisfy the desires of the financial markets rather than to implement domestic economic and social targets. Speed, synchronicity, and the increased interconnectedness of economic, environmental, and social networks makes it possible to affect individual nation states in ways difficult to predict, as was demonstrated by the Asian financial crisis in 1997, the aftermath of 11 September 2001, and the global panic caused by the SARS outbreak in China in 2002. Sassen also pointed at the special development zones that have been created in developing countries to attract foreign direct investment (FDI) by exempting businesses from local taxes and legislation.[8] She called these areas 'denationalised zones'.

Three mechanisms have facilitated the transfer of power to transnational levels and changed the authority of the nation state: relatively new supranational organisations such as the World Trade Organization (WTO), the International Monetary Fund (IMF), the World Bank (WB), the Global Environment Facility (GEF), and United Nations (UN) organisations such as the World Health Organization (WHO), the Food and Agriculture Organization (FAO), and the UN Environment Programme (UNEP); recent international treaties (for example, climate, biodiversity) and guidelines; and new transnational rules in the private sector (for instance, codes of conduct of the Organisation for Economic Co-operation and Development, OECD).

These three developments are the result of the active involvement of nation states and give the states new roles in which they share sovereignty with each other but also have to accept the judgment of international organisations in case of dispute settlements. An example is the WTO. Because of loss of national sovereignty in the framework of the WTO, a large group in the US has long opposed its conception. The governments of many developing countries have perceived the set of conditions for financial support applied by the IMF, known as structural adjustments, very negatively. Another example is the UN Framework Convention on Climate Change, which will penalise nation states that do not meet their greenhouse gas reduction obligations before 2012. This transfer of sovereignty to international bodies and conventions, with their secretariats, has led to what Sassen called the 'transformation' of nation states. As a result of these processes, nation states obtain a 'de-concentrated sovereignty' and, in different respects, a 'denationalised territory'.

Keohane and Nye argued that national governments nowadays compete for influence in the international arena, whereas they previously possessed

---

[8] Sassen, 1996.

sovereignty within their own national territory.[9] They concluded, however, that nation states will remain central in global governance. Nation states have been remarkably successful in repositioning themselves in this globalisation era. One reason for this is that even the neo-liberal philosophy of the so-called 'Washington consensus' – promoting globalisation by deregulation, expanding free markets, and reducing the impact of national borders – has to recognise the need for rules set by states to protect basic commercial interests and prevent the development of business monopolies, as they ultimately end the functioning of free markets. Another argument is that economic integration still has a long way to go. Toronto, for example, trades ten times more with Vancouver than with Seattle, showing the influence of national borders on a continent where one would least expect it. Furthermore, globalisation does not always favour lower social classes; it does not reduce social inequalities, making corrective actions from national governments still necessary. In more open markets, national social-safety nets to safeguard minimal living standards remain essential.

This brief overview shows that the role of nation states is affected by globalisation, but not in such a way that they become obsolete. On the contrary, national governments have two additional roles to play following their transformation:

- Interacting abroad in the international governance structures, a highly competitive arena, and further developing their own de-concentrated sovereignty.
- Protecting their citizens and domestic functions from the negative effects of globalisation, such as denationalising of their territories and decreasing sustainability, and at the same time utilising the opportunities that come with globalisation.

As long as the step towards a real global government is not feasible, nation states and the regional institutions (the European Union (EU), the North-American Free Trade Agreement (NAFTA), etc.) through which they are represented are the only democratic forms of government available to shape as good a global governance as possible.

A similar picture of transformation could be described from the points of view of other actors, such as transnational businesses, international non-governmental organisations (NGOs), and individual citizens. In this chapter, we focus in particular on the new opportunities globalisation offers to transformed nation states in promoting sustainable development.

---

[9] Keohane and Nye, 2000.

GLOBALISATION AND NATIONAL POLICY AREAS

The impacts of globalisation can also be described from the more detailed point of view of the specific functions of the state. Figure 7.3 gives a general overview of these functions for medium-sized nation states and an indication of the degree to which they are influenced by globalisation.

Public environmental management is positioned in this figure as being strongly influenced by economic, social, and environmental globalisation, but also as being manageable using national measures. Functions such as the content of the information provided by the media and the power over production are much less in the hands of medium-sized governments. The right part of Figure 7.3 shows functions of national governments that can be managed without much outside interference. Functions such as physical planning and nature conservation are mainly dependent on national and local processes. A similar argument can be developed for the specific areas of environmental policy, varying from air policies, which are strongly affected by globalisation, to soil policies, which are less influenced by the global ecosystem, although the global economy can lead to chemical waste being dumped in the soil. Such an analysis would identify the environmental policy areas most vulnerable to the impacts of globalisation and most promising for developing new policy instruments.

People expect their governments to eliminate the undesirable effects of globalisation. These may be environmental effects resulting from actions

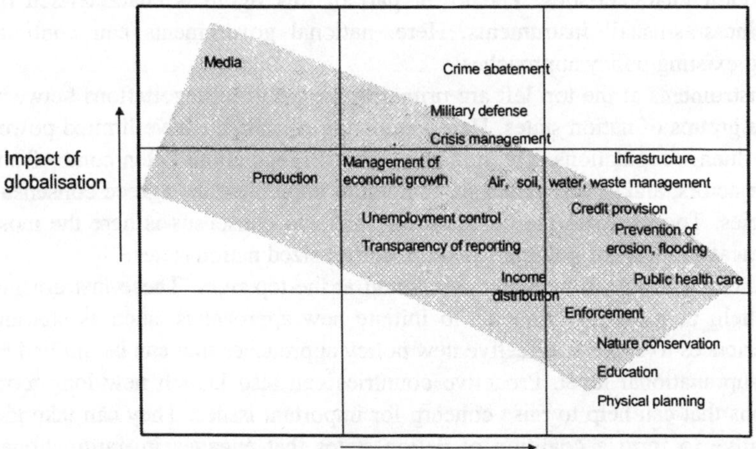

Figure 7.3 Globalisation and national policy areas

abroad, such as climate change, depletion of the ozone layer, the as yet hardly noticed pollution of outer space, as well as the import and export movements of waste, protected species, etc. Governments often lack effective national policy instruments to cope with these problems and are forced to look to international agreements and cooperation frameworks. Once governments have agreed to these, and accommodated wishes and approaches preferred by the other participants, they have lost part of their sovereignty.

Nation states, and particularly the smaller ones, have to acquire new capabilities in order to influence global governance institutions and rules, such as multi-layered governance approaches and innovative, interactive policy instruments. Modern nation states need to develop an improved awareness of the motives of all players in the global interconnected network, from culturally diverse states to the global civil society, from transnational companies to individual citizens, whose power is growing.

POSSIBLE NEW ROLES OF STATES

In our globalised world, national governments could explore new ways to function properly. This requires a closer look at both existing and new policy instruments. These instruments, like the functions of states, differ depending on their link with globalisation and their manageability by the national government. Figure 7.4 gives an indicative overview of policy instruments and their characteristics. The lower part of this figure is characterised by 'business-as-usual' instruments. Here, national governments can continue using existing policy approaches.

Instruments at the top left are primarily the result of negotiations between large groups of nation states. Here, national governments have limited power to influence regulations. The initiative for such regulations often comes from other actors, and governments are obliged to implement the agreed consensus policies. To anticipate the common international consensus is here the most practical strategy for governments of medium-sized nation states.

Another group of instruments is found at the top right. These instruments can help national governments to initiate new approaches, such as sharing experiences to come to effective new policy approaches that can be applied at the supranational level. Proactive countries can also launch new long-term visions that can help to raise concern for important issues. They can take the initiative to form a coalition of nation states that engages in multinational agreements with businesses to achieve long-term environmental goals. A new method of promoting global corporations operating in an environment-friendly way is the use of the internet as a means of informing the public about corporate performance. Governments may actively promote the

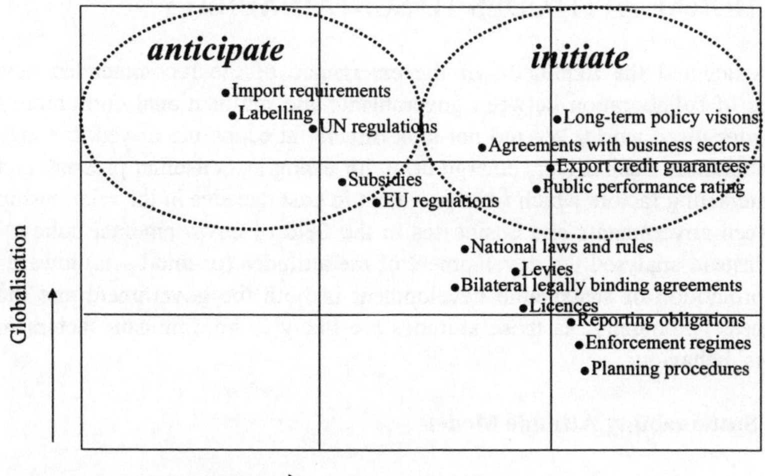

Figure 7.4 Globalisation and national policy roles

development and operation of sustainability rating mechanisms, which inform the public and which reward corporations that perform well.

Zarsky and Daly pointed out that globalisation exerts a downward pressure through the functioning of the market (race to the bottom) on the level of ambition of national environmental and other policies.[10] States must be more competitive than before, which reduces their willingness to accept additional costs for social and environmental measures. In this way, the global market could become a strong driving force for change in environmental and social policies. This force is offset by other upward-acting forces, such as the adoption of voluntary codes of conduct by transnational companies to avoid consumer boycotts.[11] Furthermore, the bonds of states with a sustainable track record may become more attractive. These bonds have lower expected financial risks and are generally expected to possess a more certain future prosperity.[12]

Through the new forms of collaboration discussed, governments may overcome constraints to their ambitions and multinational companies may obtain more control over future operating conditions in their markets.

---

[10] Daly, 2000; Zarsky, 1997.
[11] Hettige et al., 1996.
[12] Van Gompel and Bastiaans, 2002.

NATIONAL ATTITUDES TO SUSTAINABILITY

We examined the likelihood of the emergence of the recommended new forms of collaboration between governments and multinational corporations in a globalised world. We did not look directly at economic drivers for such collaboration. Such drivers emerge from, for example, consumer pressure and counteracting factors which were prevalent in past decades in the relationship between governments and businesses in the field of environmental policy.[13] We instead analysed the development of the attitudes (or mind-sets) towards the promotion of sustainable development in both the government and the business community, as these attitudes are likely to be dominant factors in actors' behaviour.

The Sustainability Attitude Model

We looked for a method that could be used to uniformly measure the current state of sustainable development within different types of organisations. We did not look for methods based on indicators of sustainable development that use flow parameters such as emissions, or stock parameters such as volume of forest.[14] These parameters result in a wide range of data which are hard to integrate, particularly if the aim is final integration into one representative figure for the sustainability of an organisation or a community. Furthermore, it is difficult to compare different sectors objectively because activities differ greatly across sectors.

Instead, we developed a sustainability-attitude index,[15] and used this to measure the attitudes of organisations on the basis of five mind-sets and their corresponding types of activities. These five attitude levels are related to the normative presumption that underlies the notion of sustainable development,[16] which aims at societies with growing integration of sectors, mutual voluntary responsibilities, and consideration of the present and the future needs of all. The resulting figure indicates the dominant attitude of actors investigated and aids understanding of the differences and similarities in the mental states of actors.

---

[13] Such factors include the presupposed conflicting interests of short-term profit-making and long-term environmental protection, and the 'enemy image' of business people towards environmental bureaucrats and vice versa (Walley and Whitehead, 1994).

[14] Bossel, 1997, 1999.

[15] Zoeteman, 2001, 2002.

[16] Brundtland, 1987.

Definition of Attitude

Attitudes of actors can be defined as the revealed mind-sets or mental codes that steer all physical behaviour of individuals or organisations. Attitudes represent sets of normative presumptions and other motives that actors or groups of actors use – consciously or unconsciously – in making choices. They are, in other words, the manifested willpower to apply normative principles while satisfying needs. Beck and Cowan called such mind-sets value systems, which, as organising principles, pervade decision structures and cultural expression.[17]

Attitudes have four characteristics that make them valuable as overall indicators for sustainable development. Firstly, attitudes are more pervasive than other indicators of the mental conditions of persons. This has been verified by the ease with which people report evaluative reactions to a wide variety of objects, the difficulty of identifying categories of objects within which evaluative distinctions are not made, and the pervasiveness of an evaluative component in judgments of meaning.[18]

Secondly, attitudes are a selective force in perception and memory. People selectively seek information that agrees with their attitudes while avoiding disagreeing information,[19] and remember agreeable information in preference to disagreeable information.[20] The focus of persons on a subject that is positively associated with the attitude is thus not likely to change in the future.

Thirdly, attitudes predict behaviour. Attitude and behaviour are correlated when the observed behaviour is judged to be relevant to the attitude, the attitude and behaviour are observed at comparable levels of specificity, and mediation of the attitude-behaviour relationship by behavioural intentions is taken into account.[21] Attitude and behaviour are correlated when the attitude is based on direct experience with the attitude object, and to the extent that the attitude is cognitively accessible.[22] Attitude strength is an important determinant for the attitude-behaviour relationship. The strength of the attitude depends on factors such as attitude certainty, importance, accessibility, and ambivalence. Additionally, attitude retrieval is crucial to attitude strength.[23] Strong attitudes are based on past knowledge and may be retrieved. Weak attitudes tend to be constructed on the spot. The latter are likely to be shaped by behaviour, as attitudes are not based on vast

---

[17] Beck and Cowan, 1996.
[18] Osgood et al., 1957.
[19] Festinger, 1957.
[20] Levine and Murphy, 1943.
[21] Fishbein and Ajzen, 1974.
[22] Fazio, 1986.
[23] Holland et al., 2002.

knowledge. Compared to weak attitudes, strong attitudes have more impact on behaviour, are less susceptible to self-perception effects, and are more stable over time. By measuring the attitude to sustainability over a diverse number of subjects, we aimed to address the strong part of the sustainability attitude.

The fourth reason to choose attitudes as indicators is that they serve various psychic functions. The attitude of persons has a function in utilitarianism, social adjustment, object appraisal, knowledge, value expression, and ego-defence.[24]

Attitude-conform behaviour is realised when attitudes are strong enough, practical opportunities for the realisation of the intended behaviour are available, and resistance to the behaviour (such as old habits) is overcome. For these reasons, attitude-conform behaviour may not be realised in all circumstances. Lack of practical opportunity may result in underestimation of the real attitude level.

The attitude index represents implicit values, which can be deduced from social or legal behaviour, demonstrated by, for example, the implementation of international agreements by governments or voluntary membership of organisations with ethical codes of conduct by citizens or corporations.

To avoid the risk of adopting carefully orchestrated public images as corporate sustainability attitudes, we broadened the set of indicators so that the focus of companies on certain sustainability programmes or omissions in their overall approach to sustainability would show up. We also compared the information provided by organisations with information from published newspapers and respected internet news websites. All information on the environmental, social, and economic sustainability of companies was derived from publicly available annual social and environmental reports from the year 2000, and from news sources and company websites.

In Table 7.1, the five sustainability attitudes are characterised for three main social actors. The five levels of sustainability range from very unsustainable (ignorant) and unsustainable (reactive) to nearly sustainable (proactive), sustainable, and, ultimately, post-sustainable. At the sustainable level, a static and dynamic equilibrium is reached with and between economic, environmental, and social flows and stocks. The post-sustainable level represents a mental state where people are willing to do more than is necessary to attain a balance with and between the three pillars of sustainable development. Actors at level 5 strive to improve the economic, environmental, and social states beyond the system of the past, and to overcome the scarcity dilemma by adopting new ways of value creation.

---

24 Katz, 1960; Smith et al., 1956.

Table 7.1 Levels of sustainability and societal attitudes[25]

Level of sustainability	Attitude of societal actors		
	Businesses	Governments	NGOs
1. Very unsustainable	Exhaustion of resources	Ad hoc use of power	Getting organised
2. Un-sustainable	Resistance to law-making	Top-down law-making	Hard action
3. Nearly sustainable	Anticipation of laws	Voluntary agreements	Joining licensing process
4. Sustainable	Anticipation of global consumer needs	Broad consensus policies	Initiation of sustainable enterprises
5. Post-sustainable	Management of commons	Facilitation of private initiatives	Joining business initiatives

An example is the transformation of deserts to fertile natural areas by the introduction of saltwater agriculture. The post-sustainability phase is a dynamic level that evolves over time.

When governments change their overall governance philosophy, they move from one sustainability level to another. Governments typified by command and control that change their philosophy to a more cooperative one that accepts positive outside interventions in law-making shift from sustainability level 2 to level 3. Such enhancement of the sustainability attitude had to be traced back using the selected indicators that are presented below.

[25] Zoeteman, 2001, 2002.

Measurement of National Sustainability Attitudes

The sustainability-attitude index of nation states was calculated on the basis of publicly available indicators (see Table 7.2). In the construction of this index, we did not include other well-established overall environmental indices from the CIA World Factbook or the Environmental Sustainability Index developed by the World Economic Forum, as our purpose was specifically to form a picture of the attitudes of actors to sustainability. To uphold a high level of transparency and reliability, we decided to select sub-indicators from these established sources. Because we aimed at discovering the psychological attitudes of actors towards sustainable development, we included indicators from Ronald Inglehart and Geert Hofstede, as well as the corruption perception index from Transparency International.

The national sustainability index consists of indices for government, business, and citizens, as well as the intensity of collaboration among the three actors. For each of the three actors, indicators for environmental, social, and economic sustainability were collected.[26] Owing to limitations of available data, indicators of the environmental dimension of sustainability dominate the picture. For each sub-indicator, values were defined that correspond with the five sustainability levels.[27] The four indices composing the overall national sustainability-attitude index were given equal weight in calculating the national average. The sustainability attitude was calculated for 85 countries, as shown in Table 7.3.

North-West European countries and Australia were found to be the most progressive in striving for sustainable development, with attitude levels of 3.5 and more. Here, sustainable development has become a core value of society.[28] The Netherlands' top position was achieved mainly as a result of its high score for the citizens' attitude. The cluster of countries, with attitude values between 3.0 and 3.5, consists of countries such as Germany, the UK, Canada, and New Zealand, where sustainability policies are proactively aimed at but not yet fully internalised. In the third group of countries, with attitude values between 2.5 and 3.0, are the US, France, Costa Rica, Spain, and Japan. Costa Rica has a remarkably high sustainability-attitude ranking in a region where the giving of priority to economic growth is dominant. Japan leads in the Asian region.

---

26 Zoeteman and Harkink, 2003b.

27 Zoeteman and Hamelink, 2002. For example, for the sub-indicator Membership of the World Business Council for Sustainable Development (WBCSD), scores ranged from less than 1.5 members per one billion USD (attitude level 1) to over 50 members per one billion USD (attitude level 5).

28 The results were based on information available in 2002.

Table 7.2 National sustainability-attitude indicators

	Sustainability indices	Sustainability indicators
Sustainability attitude of country	Attitude of government	Participation in environmental treaties[29] Membership of international organisations[30] Participation in Montreal fund[31] Priority enforcement of environmental treaties[32]
	Attitude of businesses	Corruption perception index[33] ISO 14001 certified companies[34] WBCSD membership[35]
	Attitude of citizens	Participation in voluntary work index[36] Individuality index[37] Subjective well-being index[38]
	General governance aspects	Interpersonal trust[39] IUCN membership[40] Corruption of governance index[41] Power distance index[42]

[29] CIA World Factbook, 1997.
[30] World Economic Forum, 2002.
[31] World Economic Forum, 2002.
[32] World Economic Forum, 2002.
[33] Hodess et al., 2001.
[34] World Economic Forum, 2002.
[35] World Economic Forum, 2002.
[36] Inglehart, 1997.
[37] Hofstede, 2001.
[38] Inglehart, 1997.
[39] Inglehart, 1997.
[40] World Economic Forum, 2002.
[41] World Economic Forum, 2002.
[42] Hofstede, 2001.

Table 7.3 Overall sustainability-attitude level of 85 nation states

The Netherlands	4.17	South Africa	2.13	Sudan	1.58
Denmark	4.06	Czech Republic	2.10	Peru	1.54
Sweden	4.02	Chile	2.08	Latvia	1.52
Finland	4.00	Argentina	2.06	Cameroon	1.50
Norway	3.81	Jordan	2.06	Colombia	1.50
Australia	3.69	South Korea	2.02	Moldova	1.50
Switzerland	3.60	Uruguay	2.02	Turkey	1.50
Germany	3.46	Brazil	2.00	Ecuador	1.46
Canada	3.44	Malaysia	2.00	Romania	1.46
United Kingdom	3.42	Morocco	2.00	Zambia	1.44
Iceland	3.38	Slovakia	1.98	Nigeria	1.44
New Zealand	3.38	Tunisia	1.94	Ethiopia	1.42
United States	2.94	Slovenia	1.92	Pakistan	1.42
Austria	2.88	Bolivia	1.89	Paraguay	1.42
Ireland	2.85	Thailand	1.85	Syria	1.42
Belgium	2.85	Senegal	1.83	El Salvador	1.42
France	2.75	Algeria	1.83	Cuba	1.39
Spain	2.75	Egypt	1.78	Central Afr. Rep.	1.33
Italy	2.67	Bulgaria	1.77	Libya	1.33
Japan	2.63	China	1.75	Saudi Arabia	1.33
Israel	2.48	Ghana	1.75	Venezuela	1.27
Hungary	2.46	India	1.75	Iraq	1.25
Poland	2.42	Lithuania	1.69	Indonesia	1.21
Costa Rica	2.42	Kenya	1.67	Madagascar	1.17
Greece	2.42	Macedonia	1.67	Belarus	1.17
Estonia	2.40	Mexico	1.65	Ukraine	1.08
Singapore	2.25	Russia	1.63	Albania	1.00
Portugal	2.21	Panama	1.60		
Croatia	2.17	Kuwait	1.58		

The wide range in attitudes in the EU member states after the expansion in 2004 shows the great differences in sustainability attitudes, which have to be bridged during the coming decades. Turkey is at the extreme bottom of the scale with an attitude value of 1.5. Uniform policies are not likely to be successful in these circumstances.

Figure 7.5 shows a sub-set of the data that is particularly relevant here: the environmental sustainability attitudes of the governments of the 25 best-scoring nation states. The data are based on the four sub-indicators for governments shown in Table 7.2.

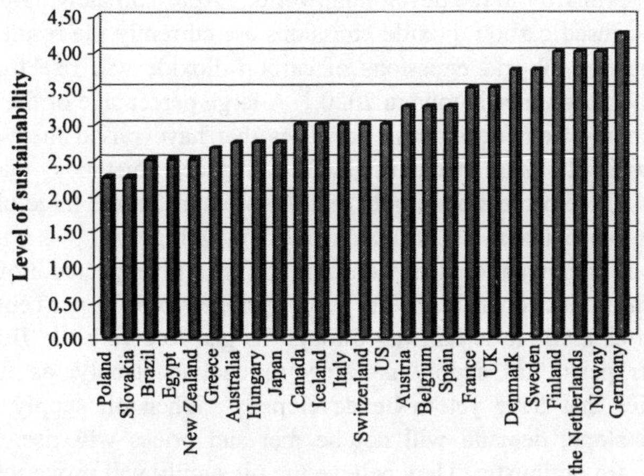

Figure 7.5 Environmental attitudes of national governments

An earlier study indicated that economic growth may often overshadow sustainability-attitude effects in reducing environmental impacts such as carbon dioxide emissions and household water use, but that examples exist where the attitude to sustainability correlates positively with improved eco-efficiency, such as reduced water use in agriculture.[43] A more detailed study of these relations is needed.

Nation states with high sustainability-attitude scores are not necessarily the least polluted ones and those which exhibit the most sustainable behaviour. But they are the most conscious of the need to act, and also the most prepared to do so. The attitude to sustainability is, therefore, a good indicator for countries that are likely to create opportunities to adopt innovative approaches to environmental management.

SUSTAINABILITY ATTITUDES OF TWO SECTORS

The Oil and Gas Industry

In the oil and gas industry, a global presence is important for long-term competitiveness and growth. Global competition is important for corporate learning, innovation, and development. The influence of the major oil and gas companies is bound to increase as the consumption of energy is still

[43] Zoeteman and Hamelink, 2002.

increasing, primarily in the developing world.[44] Approximately 80 per cent of all human-caused carbon dioxide emissions are currently the result of fossil-fuel combustion. World emissions of carbon dioxide will rise from 6.1 in 1999 to 9.9 billion metric tons in 2020.[45] A large percentage of the projected increase comes from developing countries that have raised their economic activity without using technologically advanced methods that reduce emissions. These projections imply that the pressure on the large oil and gas companies to produce in a more sustainable way will grow.

The peak in conventional oil consumption is to be expected around 2010, after which production will decline at an average rate of 3 per cent a year.[46] The world as a whole is halfway through its gas reserve peak. But when it reaches its peak, gas supply is likely to decline steeply, as it requires technologies that have yet to be developed.[47] When oil supply enters its downward slope, demand will not be met and prices will rise. Many oil companies are optimistic. They believe the oil supply will prove to be larger than is now expected, and hope technological advances will reveal more and new oil wells.[48] This signifies that these oil and gas companies are not yet fully convinced of the need to become more sustainable.

In order to assess the corporate sustainability attitude of oil and gas companies, we selected 17 indicators.[49] They are presented in Table 7.4.[50] Each higher attitude level represents a higher level of corporate commitment. In this way, worldwide-accepted sustainability-commitment programmes were included in the criteria for measuring corporate attitudes. A company's 14001 certification, for example, represents an attitude of level three for the environmental orientation of production facilities.

Using these criteria the following 26 companies were selected and assessed: Agip ENI (Italy), BG Group (UK), BP (UK), Chevron (US), Conoco (US), Esso (Imperial Oil) (US), ExxonMobil (US), Gazprom (Russia), LUKoil (Russia), Marathon (US), Nexen (Canada), Norsk Hydro (Norway), Petrobras (Brazil), Phillips 66 (US), Repsol YPF (Spain/Argentina), Santos (Australia), Sasol (South Africa), Sibneft (Russia), Shell (the Netherlands/UK), Suncor (Canada), Tatneft (Russia), Texaco (US),

44 Energy Information Administration, 2002.
45 Energy Information Administration, 2003.
46 Campbell and Laherrère, 1995.
47 Bentley, 2002.
48 Bentley, 2002; Campbell and Laherrère, 1995.
49 See Harkink (2002) for the rationale of these indicators.
50 As an example, the indicator for human rights and child labour ranged from level 1 (complaints; no care for human rights) to level 5 (integrated responsibilities for social and regional development with NGOs). An indicator of environmental orientation is ISO14001 certification of production facilities (level 3).

TotalFinaElf (France), Unoca (US), Woodside (Australia), and Yukos (Russia).

Table 7.4 Indicators of corporate sustainability attitudes

	Sustainability indices	Sustainability indicators
	Economic sustainability attitude	Risk and crisis management Nature of investments Strategic strengths Official codes of conduct Integration of sustainability within organisation
Overall oil and gas company sustainability attitude	Environmental sustainability attitude	Environmental NGO relationships Communication of environmental impact Environmental reporting and transparency Environmental orientation of production facilities Environmental product design and development Environmental community care Position with respect to environmental legislation
	Social sustainability attitude	Employee facilities Employee development possibilities Product health and safety care Human rights and child labour care Social community care

The sample included the largest and the fastest-growing oil and gas companies. The regions that were not represented in our sample are the

Middle East and Asia, where oil and gas companies are national property and do not provide public information. Q8 is the most important company that was not included in the sample. Although more and more companies voluntarily disclose environmental information, some companies investigated did not provide separate social and environmental reports. The quantity and quality of environmental information disclosed by companies is still limited. Therefore, it signals some level of commitment when companies report on environmental progress. Firms which publish social reports tend to outperform firms that do not.[51] Environmental disclosures are significantly correlated with external pollution-control indices in the oil industry.[52]

The results of the study of oil and gas companies are represented in Figure 7.6. Shell and BP were found to be the sustainability leaders in the sector, with overall attitudes of almost 4. The average score for overall sustainability of the investigated companies was 2.84. Hence, the industry is about to enter a stage of proactive action, but still needs a transformation in order to become sustainable. Social sustainability (average level: 3.06) is better developed than environmental sustainability (average level: 2.56), in particular in the North-American companies.

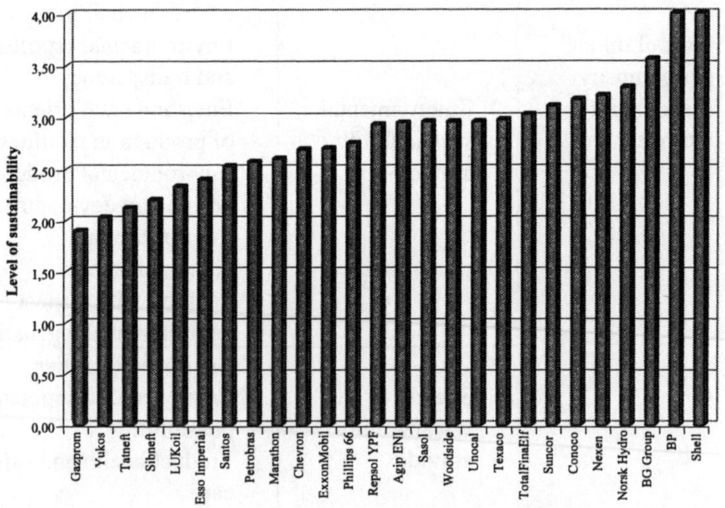

Figure 7.6 Overall attitudes of oil and gas companies

[51] Anderson and Frankle, 1980.
[52] Freedman and Wasley, 1990.

This shows that most companies are driven only by the need to comply with regulations.

Five companies were found to have reached environmental sustainability (level 3, which is where companies start to behave proactively). These companies are Repsol YPF, Agip ENI, Suncor, BP, and Shell. The analysis of the environmental strategies showed that their approaches were still under development and thus slowly evolving in a more sustainable direction. Five major oil and gas companies (Gazprom, LUKoil, Petronas, Tatneft, and Yukos) lacked a consistent environmental programme. Only eight developed sustainable energy solutions. Figure 7.7 specifies the environmental sustainability attitudes of these companies.

In conclusion, only three oil and gas companies (BG, BP, and Shell) had taken up the challenge of sustainable development seriously by building a corporate culture and structure that can create a sustainable mode of doing business. These three companies scored higher than 3.5.

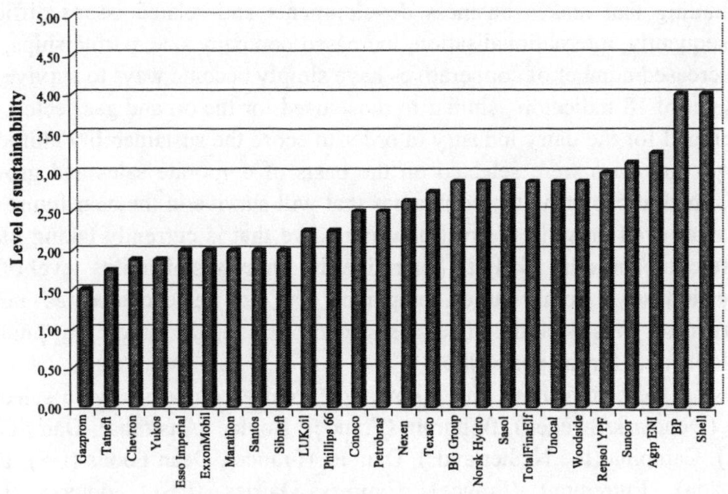

Figure 7.7 Environmental attitudes of oil and gas companies

The Dairy Industry

The dairy industry is specific because of four distinguishing factors.[53] Milk is a highly perishable product that depends heavily on farm management for product quality. Moreover, milk-producing dairy farmers and dairy cooperatives are in a vulnerable position when they cannot offer differentiated dairy products and are only in a limited way able to adjust their products to market demands. Thirdly, large diversified cooperatives and multinationals have a strong bargaining position in the milk-processing industry. Fourthly, milk is a valuable and relatively expensive raw material. The processing industry is important for the dairy-farming sector.

The dairy industry is currently in a state of rapid change. Important restructuring operations have taken place over the past ten years and are set to continue in the future. The dairy industry is becoming an increasingly global industry. The main share of restructuring took place in the cooperative sector, where margins are especially low as pressure is exerted to pay higher milk prices than is justified under market conditions.[54] This has resulted in a lack of funding that makes business developments and related issues difficult. Consequently, internationalisation, increased company size, partnerships, and an increased number of cooperatives have simply become ways to survive.

A set of 18 indicators, similar to those used for the oil and gas sector, was developed for the dairy industry in order to score the sustainability attitude.[55] Dairy companies were selected on the basis of corporate sales and growth. We aimed at capturing the companies that will survive in the near future and that may even grow in the restructuring wave that is currently taking place. These companies are likely to determine the future sustainability level of the dairy industry. Some large companies had to be excluded as public information was lacking. The assessment was again based on publicly available data for the year 2001.

Using these indicators, the following 25 dairy companies were selected: Alra (Denmark/Sweden), Bongrain (France), Bonlac (Australia), Dairy Crest (UK), Campina (the Netherlands), Danone (France), Dean Foods (US), DFO (Canada), Entremont (France), Express Dairies (US), Fonterra (New Zealand), Friesland Coberco (the Netherlands), Glanbia (Ireland), Kraft (US), Land O'Lakes (US), Leerdammer (the Netherlands), Meiji Dairies (Japan), Morinaga (Japan), Muller (Germany), Nestlé (Switzerland), Parmalat (Italy), Snow Brand (Japan), Tatura (Australia), Unilever (the Netherlands/UK), and Valio (Finland). The resulting sustainability attitudes of the dairy companies are shown in Figure 7.8.

---

[53] Schelhaas, 1999.
[54] Mauser, 2001.
[55] Zoeteman and Harkink, 2003a.

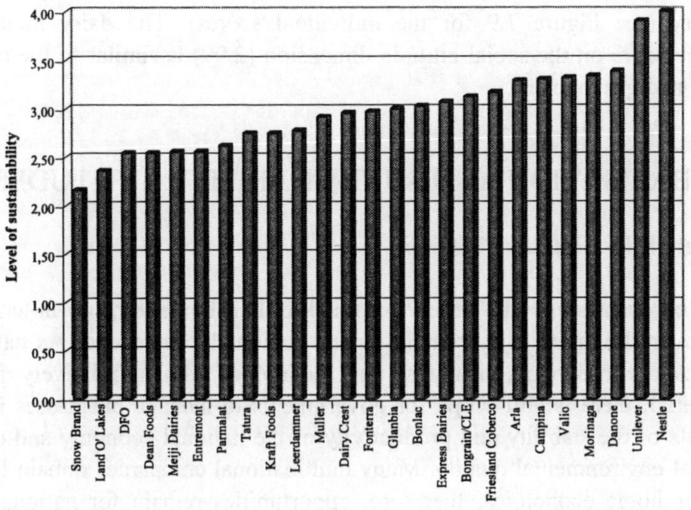

Figure 7.8 Overall attitudes of dairy companies

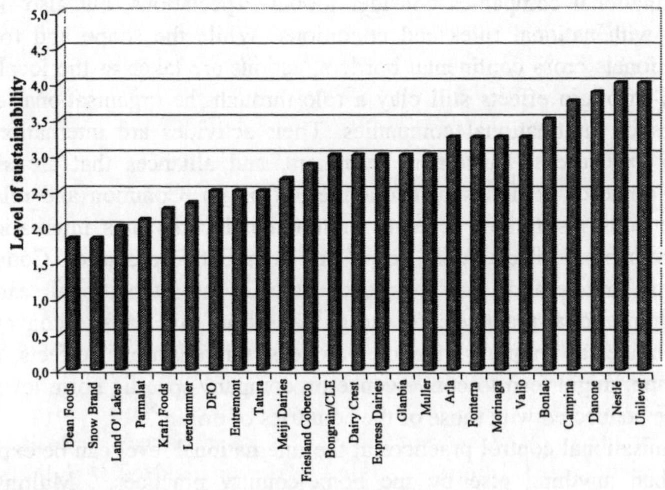

Figure 7.9 Environmental attitudes of dairy companies

As in the case of the oil and gas industry, two European multinationals (Nestlé and Unilever) lead the sector with an attitude level of almost 4. It is especially in the environmental dimension of sustainability (with an average level of 2.95) that the dairy industry scored better than the oil and gas industry (see Figure 7.9 for the individual scores). The dairy industry's average score on the social-attitude dimension (2.99) is similar to the one of the oil and gas industry

GLOBALISATION AND SUSTAINABILITY ATTITUDES

Impact of the Country of Origin

Many companies are still largely confined to their home territory in terms of their overall activity; they remain largely nationally embedded. As national systems of production, business, and technology remain relatively firmly influential, there is still scope for governments to manage businesses in the interests of the stability and productivity of the national economy and of the national environmental quality. Many multinational companies remain linked to their home economies; therefore, opportunities remain for national and regional bodies to specifically monitor and regulate them. The ability of nation states to regulate multinational companies goes together with increased expectations of global consumers, which bring increased behavioural scrutiny and expectations, and make sustainability issues a higher priority.

Multinational companies consider global expectations, but also need to comply with national rules and conditions. While the scope and focus of multinationals cross continental borders, actions are taken at the local level. Country-of-origin effects still play a role through the organisational control practices of multinational companies. Their activities are internationalised through exposure to customers, suppliers, and alliances that are situated outside domestic societies or economies of origin. Common and relatively homogeneous institutions regulate multinational companies internationally. They are subject to regulatory norms and competitive pressures. Companies constantly imitate and learn from each other in the international search for best organisational practices. Due to globalisation, this leads to convergence on the universal template of good practices. Organisational effects, on the other hand, might lead to convergence on company-specific norm levels that might be connected with those of the countries of origin.[56]

Organisational control practices at the international level can be explained more than anything else by the home-country practices.[57] Multinational companies have practices of corporate control geared to facing the specific problems that their subsidiaries in different societies face, even though they

---

[56] Mueller, 1994.
[57] Harzing and Sorge, 2003.

are embedded in different societal contexts. Moreover, the internationalisation of activities leads to the reproduction of specific institutions and cultural habits at the international level. Multinationals are, therefore, likely to be strongly influenced by institutions and predispositions arising in the countries of origin.

Additionally, companies that are affected by changed environmental legislation traditionally maintain close bonds with local regulators. As multinational companies affect the local socio-economic and environmental well-being, regulators might like to cooperate with corporate boards from their countries of origin. In this way, governments can influence (the attitudes and strategies of) companies, and the latter keep up to date regarding the latest developments in environmental legislation. By anticipating and learning from home-country regulators, companies might be able to adopt international strategies that improve their long-term corporate image, profitability, and sustainability internationally.

To investigate the outcome of this relationship empirically, the sustainability attitudes of the selected companies and those of their countries of origin were compared. Regression analyses relating national sustainability attitudes and corporate sustainability attitudes showed a correlation (i.e. explained variance) of 56 per cent (significant at the 1 per cent level) for the oil and gas sector and 16 per cent (significant at the 5 per cent level) for the dairy industry. National sustainability-attitude values thus correspond better with those of the oil and gas companies than with those of dairy companies. This shows that the home country's sustainability attitude is better controlled and transferred to subsidiaries in the oil and gas industry than in the dairy industry. Oil and gas companies seem to be better able to anticipate on the policies implemented and the environmental direction taken by the home government. Or global market expectations and media pressures may play a greater role in the oil and gas industry when companies originate from nation states in which practices are highly sustainable.

Impact of Globalisation

Companies, once part of the global market, have become vulnerable to damage to the image of their corporate brands in the eyes of global consumers. A high-quality profile and good environmental management is then required for survival in global markets. Additionally, a high-quality image and a positive sustainability attitude can be utilised globally to claim market access and consumer demand. Not many multinational companies have reached this state. When a critical number of businesses in the global market have adopted a more sustainable strategy, others will follow suit. The pace at which companies adopt sustainable development is likely to increase.

Once more companies become sustainable, the chances of late resisters meeting with trouble increase as well.

When these considerations are dominant, it can be expected that the globalisation of companies creates stronger impulses to adopt higher corporate sustainability attitudes. To investigate the influence of an increased global presence on a company's attitude, we constructed a globalisation index consisting of the number of countries in which a company produces, the number of countries in which a firm sells, and the percentage of corporate sales obtained outside the home country. Each indicator was scored on a relative scale from 0 to 1, where 1 represents the case of complete globalisation. Data applying to 2002 were used for several variables of the index.[58]

The impact of globalisation on the corporate sustainability attitude seems to be stronger in the dairy sector than in the oil and gas industry. The findings with regard to the influence of globalisation in the oil and gas sector were less conclusive than we expected: a statistically insignificant correlation of 15 per cent. The industry has always had to deal with global pressure. This is a natural result of the importance of this energy-supplying industry. Nevertheless, the oil- and gas-producing companies have not felt the urge to comply with consumer expectations completely. This may be because consumers cannot easily distinguish the more environment-friendly products. Oil and gas products are merely commodities that are bought primarily on the basis of arguments of price and availability.

The relationship between corporate sustainability attitude and levels of globalisation in the oil and gas industry is also affected by the Russian companies (Gazprom, LUKoil, Sibneft, Tatneft, and Yukos), which have reached a high level of corporate globalisation but still possess low sustainability attitudes. Market pressure on these companies to act according to the laws of sustainable development is relatively low. Although they sell their products in a large number of countries, their main market is Russia, where the pressure to operate sustainably is relatively low.

The effect of globalisation is larger in the dairy sector, with a correlation of 32 per cent (significant at the 5 per cent level). An explanation may be that global food brands are expected to display high standards of sustainability in order to maintain credibility.

[58] Harkink, 2002.

NEW INSTRUMENTS FOR STATE POLICIES

Conditions for New Partnerships

We have shown above how economic globalisation and home-country sustainability attitudes may coincide with corporate sustainability attitudes. The data indicate that national environmental authorities could make more use of the developing sustainability attitudes of businesses as national and corporate attitudes coincide. It is an opening window of opportunities that have not been fully utilised to the benefit of both parties.

Private enterprises are at a crossroads in deciding whether to take the lead and enter a new phase on their own account or to wait for governments to prescribe conditions for future production. Where private enterprises take the lead, they require governments to create proper conditions for sustainable business. Governments also have a role to play in facilitating a social debate in order to compose clear and tight objectives for achieving sustainable management of natural capital, both internationally and nationally.[59] This means that both businesses and governments have an interest in cooperating in order to promote higher levels of sustainable development at reduced risks.

It may be questioned whether this opportunity is sufficiently valued or if there is a structural reluctance and disbelief among actors to follow this path. Noreena Hertz, for example, listed a number of arguments to stress the downside of giving companies more responsibilities:[60]

- Corporations' contribution to societal needs will always remain marginal as their motives are commercial rather than moral.
- The priorities of corporations will change when customers' priorities change during an economic downturn and profit-making starts to diverge with ethical business.
- The growing dependence of governments on businesses will blur the boundaries between the public and the private sector; excessive corporate power may not be curbed sufficiently.

In summary, according to Hertz, ethics and business sometimes coincide, but as businesses are morally ambivalent and governments are supposed to be the custodians of society, their tasks should be well separated. Otherwise, governments risk losing the support of the people. Recent waves of protests are signals that states should reclaim the support of the people.

[59] Keijzers, 2003.
[60] Hertz, 2001.

We come to a different conclusion. We think that the position that anti-globalists take and defend is characterised by polarisation, which typically represents a sustainability attitude level of 2. This level is characteristic of the common attitude in the global arena of negotiations on environmental issues. At this time, it is of crucial importance for governments and multinational companies to move to the next sustainability level in the global arena.

North-West European governments, for example, promote policies that may move national parties from sustainability level 3 to level 4, using cooperative arrangements and instruments. Instead of polarising approaches, they want to build trust, increase transparency, and promote long-term societal common goals to governments, businesses, and NGOs. Businesses which have roots in these countries cannot neglect or ignore these attitudes. The critical points Hertz mentioned merit concern, but do not encourage collaborative arrangements between governments and progressive corporations when sustainability attitude levels of 3 and higher prevail.

New Instruments Supporting Mutual Arrangements of Actors

To give an indication of how governmental environmental regulations and policies have paralleled the environmental strategies of multinationals, we looked at the correlation between the environmental sustainability attitudes of multinational companies and those of their home-country governments.

In the group of multinationals investigated, we found a large spread in environmental attitudes among companies from the same home countries. To provide a representative picture, we examined only those countries from which at least three companies were included in our sample of companies. For each country, we calculated the average environmental company attitude. Regression analysis showed a correlation of 69 per cent (significant at the 5 per cent level) between the environmental attitudes of governments and those of multinationals.

There are many examples of new forms of collaboration between governments and multinational corporations. One third of the approved partnerships submitted at the 2002 World Summit on Sustainable Development in Johannesburg had governments as the leading partners. Many more governments are involved as partners in initiatives launched by non-governmental actors. The Global Compact launched by Kofi Annan in 1999 is another good example.[61] Other actions may include the promotion of collective actions of states to avoid competition distortions, of

[61] In an address to The World Economic Forum in Davos on 31 January 1999, UN Secretary-General Kofi Annan challenged business leaders to join the Global Compact initiative to bring companies together with UN agencies, labour, and civil society by supporting nine principles in the areas of human rights, labour, and the environment.

communication on consumer trends that influence the performance of multinational companies, and of self-regulation of the global business community.[62] Alliances between governments, businesses, and NGOs also help create the stable conditions needed for innovations leading to sustainability.[63]

Governments as well as multinational corporations could explore means of collaboration beyond the national borders, for example, by agreeing on arrangements between proactive companies in one or more sectors and like-minded nation states. The arrangements could include export-credit guarantees, co-financing of private environmental-sustainability initiatives in developing countries, rewarding sustainable corporations by abolishing policy measures such as 'perverse' subsidies that support companies which perform less well, co-financing of breakthrough technologies, labelling of and providing public information on ethical entrepreneurship and sustainable products, elimination of financial and administrative drawbacks for financial institutions which promote sustainable investments abroad, promotion of more positive sustainability impact using official development assistance funds, and granting sustainable management contracts to qualified companies for public environmental goods such as forests and coastal waters.

The globalisation of companies with positive corporate cultures and good attitudes to sustainable development can be beneficial to worldwide economic competition and to governments that want to promote sustainable development nationally and globally. Companies that possess the technical know-how and enjoy a respectable sustainability reputation probably do not want to lose this edge in the highly competitive global marketplace. A race to the bottom is unlikely in such a situation, as it would damage the corporate reputation and its perceived product quality. It is more likely that such companies will propagate their sustainability attitudes to boost competitive advantage and reduce societal critique. Moreover, the country-of-origin effect, together with the globalisation effect, will create conditions in which companies are held accountable and need a licence to operate. Integrated chain management to improve the sustainability performance will expand the upward drive to companies that do not sell directly to consumers.

The concept of arrangements between multinational companies and national governments is not new, but is still poorly developed. As proclaimed in 2002 in Johannesburg, the spirit to overcome boundaries between nation states and actors has to be ignited to make the steps reality.

---

[62] Zarsky, 1997.
[63] Faucheux et al., 1997.

REFERENCES

Anderson, J. and A. Frankle (1980), 'Voluntary social reporting: An iso beta portfolio analysis', *The Accounting Review,* **55**(3), 467-479.

Beck, U. (1997), *Was ist Globalisierung? (What is Globalisation?)*, Frankfurt am Main: Suhrkamp Verlag.

Beck, D. and C. Cowan (1996), *Spiral Dynamics: Mastering Values, Leadership, and Change*, Malden: Blackwell.

Bentley, R. (2002), 'Global oil and gas depletion: An overview', *Energy Policy,* **30**(3), 189-205.

Bossel, H. (1997), 'Finding a comprehensive set of indicators of sustainable development by application of orientation theory', in: B. Moldan and S. Billharz (eds), *Sustainability Indicators: A Report on the Project on Indicators of Sustainable Development (SCOPE 58)*, Chichester and New York: John Wiley.

Bossel, H. (1999), *Indicators for Sustainable Development: Theory, Method, Applications*, Canada: International Institute for Sustainable Development.

Brundtland, G. (1987), *Our Common Future,* Oxford: Oxford University Press.

Campbell, C. and J. Laherrère (1995), *The World's Supply of Oil 1930-2050*, Geneva: Report Petroconsultants.

Central Intelligence Agency (1997), *The World Factbook 1997*, Dulles: Brassey's Inc.

Daly, H. (2000), *Globalization*, paper presented at the 50th Anniversary Conference of the Aspen Institute, Aspen, Colorado.

De Grauwe, P. and F. Camerman (2002), 'How big are the big multinational companies?', *Tijdschrift voor Economie en Management*, **47**(3), 311-326.

Energy Information Administration (2002), *International Energy Outlook 2002*, Washington: US Department of Energy.

Energy Information Administration (2003) *Emissions of Greenhouse Gases in the United States 2002*, Washington: US Department of Energy.

Faucheux, S., M. O'Connor, and I. Nicolaï (1997), 'Economic globalisation, competitiveness, and environment', *Globalisation and Environment,* Paris: OECD.

Fazio, R. (1986), 'How do attitudes guide behavior?' in A. Pratkanis, S. Breckler, and A. Greenwald (eds), *Attitude Structure and Function*, Hillsdale: Lawrence Erlbaum Associates Publishers.

Festinger, L. (1957), 'Theory of cognitive dissonance', in A. Pratkanis, S. Breckler, and A. Greenwald (eds), *Attitude Structure and Function*, Hillsdale: Lawrence Erlbaum Associates Publishers.

Fishbein, M. and I. Ajzen (1974), 'Attitudes towards objects as predictors of single and multiple behavioral criteria', *Psychological Review,* **81**(1), 29-74.

Freedman, M. and C. Wasley (1990), 'The association between environmental performance and environmental disclosure in annual reports and 10K's', *Advances in Public Interest Accounting*, **3**, 183-193.

Friedman, T. (2002), 'States of discord', *Foreign Policy,* **March/April**, 65-66.

Harkink, E. (2002), *Sustainable Development and the Oil and Gas Industry*, report no. 02.04, Tilburg: Globus, Tilburg University.

Hart, S. (1995), 'A natural-resource-based view of the firm', *Academy of Management Review,* **20**(4), 986-1014.

Harzing, A. and A. Sorge (2003), 'The relative impact of origin and universal contingencies on international strategies and corporate control in multinational enterprises: Worldwide and European perspectives', *Organization Studies*, **24**(2), 187-214.

Held, D., A. McGrew, D. Goldblatt, and J. Perraton (1999), *Global Transformations – Politics, Economics and Culture*, Cambridge: Polity Press.

Hertz, N. (2001), 'Better to shop than to vote?', *Business Ethics*, **10**(3), 190-193.

Hettige, H., M. Huq, and D. Wheeler (1996), 'Determinants of pollution abatement in developing countries: Evidence from South and Southeast Asia', *World Development*, **24**(12), 1891-1904.

Hodess, R., J. Banfield, T. Wolfe and P. Eigen (2001), *Global Corruption Report 2001: Transparency International*, Berlin: Intl Specialized Book Service.

Hofstede, G. (2001), *Culture's Consequences. Comparing Values, Behaviors, Institutions, and Organizations across Nations*, 2nd ed., Thousand Oaks: Sage Publications.

Holland, R., B. Verplanken, and A. Knippenberg (2002), 'On the nature of attitude-behavior relations: The strong guide, the weak follow', *European Journal of Social Psychology*, **32**(6), 869-876.

Inglehart, R. (1997), *Modernization and Postmodernization: Cultural, Economic, and Political Change in 43 Societies*, Princeton: Princeton University Press.

Katz, K. (1960), 'The functional approach to the studies of attitudes', in A. Pratkanis, S. Breckler, and A. Greenwald (eds), *Attitude Structure and Function*, Hillsdale: Lawrence Erlbaum Associates.

Keijzers, G. (2003), *Creating Sustainable Directions: A Collaborative Stakeholder Approach of Governments and Businesses*, dissertation, Rotterdam: Erasmus University.

Keohane, R. and J. Nye (2000), *Governance in a Globalizing World*, Washington, DC: Brookings Institution Press.

Levine, L. and G. Murphy (1943), 'The learning and forgetting of controversial material', *Journal of Abnormal and Social Psychology*, 38, 507-517.

Mauser, A. (2001), *The Greening of Business: Environmental Management and Performance Evaluation: An Empirical Study in the Dutch Dairy Industry*, Delft: Eburon.

Mueller, F. (1994), 'Societal effect, organizational effect and globalization', *Organization Studies*, **15**(3), 407-428.

Osgood, C., G. Suci, and P. Tannenbaum (1957), 'The measurement of meaning', in A. Pratkanis, S. Breckler, and A. Greenwald (eds), *Attitude Structure and Function*, Hillsdale: Lawrence Erlbaum Associates.

Sassen, S. (1996), *Losing Control? Sovereignty in an Age of Globalisation*, New York: Columbia University Press.

Schelhaas, H. (1999), 'The dairy industry in a changing world', in L. Falvey and C. Chantalakhana (eds), *Smallholder Dairying in the Tropics,* Nairobi: International Livestock Research Institute.

Sharma, S. and H. Vredenburg (1998), 'Proactive corporate environmental strategy and the development of competitively valuable organizational capabilities', *Strategic Management Journal*, **19**(8), 729-753.

Smith, M., J. Bruner, and R. White (1956), 'Opinions and personality', in A. Pratkanis, S. Breckler, and A. Greenwald (eds), *Attitude Structure and Function*, Hillsdale: Lawrence Erlbaum Associates.

Van Gompel, J. and E. Bastiaans (2002), 'Scandinavische landen meest duurzaam' ('Scandinavian countries most sustainable'), *Economisch Statistische Berichten,* **87**(4379), 752-754.

Walley, W. and B. Whitehead (1994), 'It's not easy being green', *Harvard Business Review*, **May/June**, 46-52.

World Bank (2002), *World Development Indicators 2002*, Washington, DC: World Bank.

World Economic Forum (2002), '2002 environmental sustainability index', *http://www.ciesin.columbia.edu/indicators/ESI.*

Zarsky, L. (1997), *Stuck in the Mud? Nation States, Globalization, and the Environment*, The Hague: OECD Economics Division.

Zoeteman, K. (2001), 'Sustainability of nation states', *International Journal of Sustainable Development and World Ecology*, **8**(2), 93-109.

Zoeteman, K. (2002), *Globalisation and Sustainability: On Governance and the Power(lessness) of the Nation State*, Tilburg: Tilburg University Press.

Zoeteman, K. and S. Hamelink (2002), *Op Zoek naar de Bronnen van Nationale Duurzaamheid: Een Berekening van de Duurzaamheidshouding van Landen (In Search of Sources of National Sustainability; A Calculation of the Sustainability Attitudes of Countries)*, report no. 02.05, Tilburg: Globus, Tilburg University.

Zoeteman, K. and E. Harkink (2003a), *Sustainable Development in the Dairy Industry*, report no. 03.02, Tilburg: Globus, Tilburg University.

Zoeteman, K. and E. Harkink (2003b), 'Samenwerken voor duurzaamheid; Belangen Noordwest-Europese overheden en hun progressieve multinationals lopen steeds meer parallel' ('Cooperating for sustainability; Interests of North-Western European governments and their progressive multinationals parallel increasingly'), *Arena*, **9**(September), 88-92.

8. Globalisation and the Role of Citizen-Consumers in Environmental Politics

Gert Spaargaren and Susan Martens

SUMMARY

We address the environmental behaviour of citizen-consumers in a globalised setting. National government should focus on the consumption patterns of citizen-consumers for four environmental reasons and one generic motive. We review the literatures on ecological citizenship (highlighting political roles of consumers), sustainable consumption patterns (concerned with the environmental impact of consumption), and global civil society (relating the roles of consumers and citizens in a global context). The results from four Dutch projects aimed at reducing the environmental impact of consumption are presented. They show that the impact can be reduced if the technical, social, and political aspects of citizen-consumer behaviour are considered simultaneously. National environmental policy should be geared to attuning supply and demand. Besides, government should bridge the growing distance between international developments and domestic commitment. We conclude that national environmental policies of citizen-consumer behaviour require an integrative policy approach, a challenge which globalisation has enhanced.

INTRODUCTION AND OUTLINE OF THE ARGUMENT

The connections between 'environmental' and 'global' change are well established and documented in the social science literature. Environmental problems are among the most frequently used examples to illustrate the new quality or dimension which social problems display when they manifest themselves primarily at the global level. Global warming, for example, was discussed by Anthony Giddens as a High Consequence Risk (HCR) to illustrate the new, risky dynamics of global modernity. The term 'global modernity' refers to the current radicalised and universalised phase of post-traditional modern society, in which locales are penetrated and shaped by social influences distant from them.[1] Trans-boundary pollution by nuclear radiation in the aftermath of the Chernobyl disaster was used by Ulrich Beck as an early and convincing example of the emergence of the risk society, where it is impossible for citizens to escape from environmental risks that threaten large parts of the population.[2]

The globalisation of environmental policies and politics has been reinforced by the emergence of global problems. However, the development of these policies cannot be explained exclusively by the fact that the most pressing environmental problems (global warming, ozone-layer depletion) have obtained a global dimension. The management and control of environmental problems have developed into a transnational and global affair for other reasons as well, the main reason being the emergence of a global network society.[3] In the global network society, politics in general take on a different characteristic when compared with earlier phases of modernity. In order to be effective, environmental policies have to be developed at different levels, the nation state just being one – yet still very important – level.

As a result, we witnessed over the past decade a qualitative and quantitative growth in arrangements for the management of the global commons, for world-wide control of carbon dioxide emissions at levels set in the Kyoto Protocol, and for managing problems like desertification, biodiversity, and water scarcity beyond the regional levels. The globalisation of environmental politics is not restricted to official, governmental policies, however. The agendas of global civil-society actors, and especially those of the international non-governmental organisations (NGOs), are fuelled to a considerable extent by environmental issues, and the influence of NGOs such as WWF, Greenpeace, and Friends of the Earth International is judged to be substantial by most policy actors involved.

Although the globalisation of environmental problems and politics triggered a vital and still growing discourse in the environmental sciences, at least one theme has not been given proper attention in the literature: the changing role of citizen-consumers in the globalisation of environmental politics. Do citizens participate in, or at least feel some commitment to, the Rio, Kyoto, and Cancún processes? When confronted with new regulations, eco-labels, or eco-taxes, do citizen-consumers know which national, transnational, or global state actors and institutions can legitimately be held responsible for them? What are the consequences of globalisation for national environmental policies that aim to address the lifestyles, consumption routines, and political commitments of citizen-consumers? Do they play the

1 Giddens, 1990.
2 Beck, 1986.
3 Castells, 1996.

consumption-card, the citizen-card, or a combination of both? These questions, referring to the impacts of globalisation on both the political roles of people as citizens and the market roles of individuals as consumers, deserve careful treatment and detailed discussion.

In this study, we examine the changing roles of citizen-consumers in globalising environmental politics, and look in detail at the potential impacts of globalisation on national environmental politics that aim to address the consumption patterns and lifestyles of 'their' citizen-consumers. We start by posing the question why citizen-consumers matter in environmental politics in the first place. In section one, we present five arguments for the need for a stronger orientation towards the role of citizen-consumers in environmental policy-making, both at the national and the transnational level. In section two, we specify the role of citizen-consumers in a theoretical way, by presenting a selective review of the social science literature on ecological citizenship and sustainable consumption behaviours. Using the recent literature on 'global civil society', we investigate the hybridisation of consumer roles and citizenship roles as they are seen to result from the globalisation of society. Having considered citizen-consumer roles in a theoretical perspective, we then explore citizen-consumer issues in the context of applied (environmental) politics. We do so by examining four Dutch case studies, deriving from them some key questions and principles for citizen-consumer-oriented policies. This is described in section three. In the fourth and final section, we discuss a few key questions which could help to further develop the policy and research agendas at the global or transnational level with respect to citizen-consumer involvement in environmental policies.

THE NEED FOR A CITIZEN-CONSUMER ORIENTATION

In the environmental field, policy experts are frequently joined by scientific experts who are specialised in the physical aspects of environmental flows. There is much technical knowledge involved in understanding the dynamics of environmental flows such as energy, waste, clouds, or species. As a result, in many segments of the environmental policy field, technical or system rationalities tend to prevail. This dominance of technical scientific rationalities generates a number of 'translation' problems when policy-makers are confronted with the task of organising a dialogue with citizen-consumers about the goals and general principles of environmental policies. The technical goals of emission reduction or the percentages of material recycling set in official policy documents have to be translated and adapted to the rationalities of ordinary daily life.[4] Adding to this translation problem the fact that citizen-consumers are a heterogeneous, loosely organised, and modestly informed target group of environmental policies, it is understandable that policy experts in the environmental field usually prefer to do business with the more powerful, well-informed, and well-organised target groups such as transnational companies, farmers' organisations, professional NGOs, and utility companies. For reasons of effectiveness and efficiency, so it is argued, it is better to avoid investing too much effort and money in policy arrangements that require the active involvement of citizen-consumers in bringing about environmental change.

In the Netherlands, the role of citizen-consumers in national environmental politics fits the pattern described above to a considerable extent. In the national environmental policy plans (NEPPs), appreciated and praised by friends and enemies alike for their thorough, consistent, and well-elaborated views of environmental change, the role of citizen-consumers is discussed as a complex and complicated issue.[5] Consumers are regarded as a target group that is difficult to reach and hold. The only thing that seems feasible for national environmental policy-makers is to inform, with the help of so-called social or soft policy-instruments, citizens about the reasons behind policy measures resulting from the negotiations between national government and its counterparts in industries and professional NGOs. Citizen-consumers' direct engagement with and commitment to environmental policies, however, was avoided until recent times. At the international level, the situation does not seem to be much better. According to Cohen, Dutch environmental policy-makers are among the frontrunners when it comes to developing strategies to promote sustainable consumption patterns.[6] When judged against the background of the overall 'environmental capacity' as it was developed by nation states in advanced societies from the 1970s onwards,[7] the potential control of citizen-consumers and their direct contribution to environmental pollution is poorly understood and under-theorised when compared with the politics developed for institutional actors.

With the apparent neglect of the important role of citizen-consumers in environmental policy-making, policy experts are swimming against the tide of social science studies emphasising the need to take a citizen-consumer-oriented perspective on social change in contemporary modernity, also when environmental change or transition processes are at stake. The need to take into account the perspective of citizen-consumers when designing and implementing environmental policies can be argued for on a number of

4 Spaargaren, 2001.
5 VROM, 1993.
6 Cohen, forthcoming.
7 Jänicke, 1991.

grounds, the most important of which can be summarised in 4 + 1 main points. The first four arguments are 'internal' to the field, since they are rooted in the need to make environmental politics more legitimate and effective. The +1 argument refers to the changing character of politics in global modernity: affecting environmental politics but not being restricted to the environmental field.

Four Arguments

First, the need to develop a consumer orientation in environmental policy is obvious considering the environmental impacts of modern consumption patterns. The relative and absolute contribution of citizen-consumers to total national environmental impacts is significant and still increasing in many advanced societies. According to a recent study by the national environmental research institute (RIVM) in the Netherlands, environmental policy goals cannot be realised without a further dematerialisation of consumption.[8]

Second, the contemporary neglect of citizen-consumers owing to an emphasis on more 'profitable' target groups in the earlier stages of environmental policy-making is difficult to defend in the present-day situation, where we are confronted with mature, fully developed environmental policies. With the advancement of environmental policy over the years, the fruits on the lower branches of the trees have been harvested. What remain are the more complex, persistent problems that need special treatment and new, integrated policy strategies. Viewing the problems that are labelled by the Dutch government as 'persistent' problems (such as problems that are in need of specific forms of transition management), it can be concluded that they all have a strong 'consumer-behaviour dimension'. For example, the energy consumption related to inhabiting a house and driving a car is significant in terms of environmental impacts, but difficult to change 'behind the backs' of citizen-consumers. Although it is acknowledged by many policy-makers that consumer behaviours are crucial in this respect, it is indicated at the same time that the existing policy instruments are not suitable for this job and that we need new, more sophisticated tools and equipment.[9]

Third, when compared with the early days of environmental policy-making, the 'material foundations' for citizen-consumer-oriented politics have become much broader and more developed. For a long time, consumer empowerment in the environmental field was restricted to lifestyle groups that were able to combine high environmental awareness with high incomes. This can be explained by the fact that, in the early days of environmental policy-

---

[8] Vringer et al., 2001.

[9] VROM, 1993.

making, the provisioning of citizen-consumers with more sustainable goods and services occurred at low levels and also tended to be poorly organised. As a result, sustainable options in consumption domains such as food, housing, travel, and recreation were inaccessible to the majority of the population. Now that the process of ecological modernisation of production and consumption results in more green products and services being offered in more accessible and explicit ways to a greater diversity of lifestyle groups, this argument no longer holds true.[10] Environmental (product) choices or (lifestyle) policies are no longer restricted to top consumers in niche markets or to well-informed citizens who are active in environmental organisations.

Fourth, not only have the material foundations of consumer-oriented environmental policies improved, but the 'social foundations' have also become more solid and better established. Over the past thirty years, the process of internalisation of environmental policy goals and strategies has not been restricted to companies and governments exclusively; at present, it extends also to our daily lives and the consumption routines implied in them. Environmental learning has not been restricted to the well-organised actors and institutions in governments and markets; civil-society actors have become better informed about and engaged with the 'environmental dimension' of their daily behavioural routines. Citizen-consumers 'know about' waste-management principles, green electricity schemes, green product labels, and the environmental risks associated with daily routines such as sunbathing or driving a car. They not only 'know about' these products and processes, but they increasingly know how to handle them in the context of recursively organised social practices. On the demand side, citizen-consumer empowerment was fuelled over the last decades by world-wide processes such as 'local Agenda 21', the sustainable-cities dialogues, the emergence of sustainable-housing associations, and a variety of consumer organisations. Consumer organisations often take the form of loosely organised groups, associations, consumer tribes, or virtual communities, ranging from cyclists and users of public transport to vegetarians, internet eco-shoppers, and eco-travellers. Therefore, next to the articulation of green provisioning by governments and producers, articulation has emerged on the demand side of the process as well.

Plus One

The arguments discussed so far are internal to the environmental policy field and refer exclusively to environmental politics. From the broader perspective of politics in general, a fifth and crucial argument can be added in favour of a

[10] Mol and Sonnenfeld, 2000; Spaargaren, 2000.

stronger orientation towards the roles and behaviours of citizen-consumers. The role of classical or 'emancipatory' politics in modern societies seems to be under great pressures.[11] The general public lacks interest in supporting the core business of traditional political parties, with the memberships of official political parties in decline. The numbers of people making use of their right to participate in democratic elections at different levels of policy-making – local, national, and the European Union (EU) – seem to be diminishing, as is the willingness of people to participate in any way in formal processes of political decision-making at local and national levels.

Political scientists refer to these developments in terms of the 'erosion of politics' and are involved in lively debates on the 'withering away of the nation state', on the legitimacy crisis of state-based powers, and on the negative image of political organisations compared with their counterparts in the market.[12] Many commentators agree that these developments point to the need to rethink and reformulate politics 'from the viewpoint of the citizen-consumer'. They argue that, in order to regain the trust of citizens in politics and to bridge the gap between politics and everyday life, the existing policies have to become better adjusted to the life-worlds of citizen-consumers.[13] Environmental policies are part of this broader problem of the crisis of modern politics and the lack of support from the general citizenry cannot be dealt with by simply moving away from local politics or by neglecting the everyday views and concerns of citizen-consumers. Environmental politics need to come to terms with citizen-consumers.

SOME THEORETICAL CONSIDERATIONS

Although the general arguments on the changing roles of citizen-consumers in globalising environmental policies are underdeveloped, there are some narratives in the making in this respect. Three debates pass in review. First, there is an emerging debate on ecological citizenship, exploring mainly the political roles of citizens in globalising (environmental) politics.[14] Second, especially since the Rio Conference in the early 1990s and the Rio+10 gathering in Johannesburg, a world-wide debate emerges on the need for sustainable consumption patterns, especially in the overdeveloped parts of the

---

[11] Giddens, 1991.

[12] Held et al., 1999.

[13] The lack of trust in politics and the failure of official politics to take popular views into account more often and more seriously were discussed in the Netherlands as important factors contributing to the upsurge of the populist political movement headed by the late Pim Fortuyn. After Fortuyn was murdered shortly before the national elections, his party, LPF, temporarily gained a strong position in Dutch parliament.

[14] Held et al., 1999; Holemans, 1999; Urry, 2000; Van Steenbergen, 1994.

world. Third, a series of round-table discussions on the Global Civil Society was organised by the London School of Economics (LSE), focusing on the positions of civil-society actors and organisations with respect to globalisation and the global network society.

Ecological Citizenship

When, in advanced societies, the electricity grid breaks down or the drinking water system is temporarily out of order, people tend to get angry, not just because it is unpleasant to see daily life being disrupted but also because people feel they are entitled to an uninterrupted supply of these necessary underpinnings of their daily routines. This basic attitude of having the 'environmental right' of access to clean water, proper waste-management services, and a guaranteed supply of electricity is the result of a history of more than a hundred years of utility provisioning in Europe; a history in which the role of the nation state has been of overriding importance. The state guaranteed all citizens access to these goods and services under a set of fixed, state-determined circumstances.[15] Only when monopolistic public provisioning began to give way to liberalisation and privatisation did the established and taken-for-granted character of these ecological citizenship rights come to be discussed and, in some respects, (re)discovered by a generation that grew up regarding these rights as basic and customary. When European citizens in the future have the right of choice in utility provisioning, when they are free to buy, for example, their green electricity from 'foreign' utility companies or even international environmental NGOs, how will these ecological citizenship rights be guaranteed and what will be the role of nation states in this? If citizens develop into consumers, and when more than 80 per cent of the regulation governing the provisioning of water and electricity is set at the EU level, who will be held responsible when the grid breaks down or when providers do not live up to their contractual obligations? Will consumers phone the municipality, the police, the private utility companies' local or international offices, or just the newspapers to air their complaints?

This example illustrates some of the changes in citizenship and politics from a sociological point of view.[16] The interesting aspects of ecological citizenship do not concern 'voting green' or becoming a member of a green party or NGO. Neither is the most important or interesting aspect of the ecological citizenship debate to be found in 'ecological rights' as the fourth and most recent set of comprehensive civil rights following earlier generations of (civil, political, and socio-economic) rights.[17] Ecological

[15] Van Vliet, 2002.
[16] Held, 1995.
[17] Marshall, 1973.

citizenship is about the *societal frameworks* of individual rights and responsibilities with respect to sustainable development.[18] As Held argued, citizenship rights in Europe emerged historically in the context of the debate on (the sovereignty of) the emerging nation states, which 'granted' rights to individuals who were assigned a series of rights and responsibilities they could legitimately claim or articulate in their relationship with the state. Therefore, the framework to make the rights 'work' was defined along the individual–nation-state axis.

Against this background, we should ask ourselves with the help of what kind of framework citizenship rights can best be organised now that globalisation threatens the classic 'modern' framework provided by the nation and the nation state. Nation states have become enmeshed in many regional and global networks of environmental governance, and are confronted with threats to the 'ecological safety' of their citizens that are beyond the control of the individual nation state. The discussion that Beck set in motion in the social sciences about the 'ecological risks of reflexive modernity' – using Chernobyl, the BSE disease, and genetically modified organisms (GMOs) as examples, next to the well-known cases of ozone-layer depletion and global warming – was meant not only to illustrate the changing role and authority of modern science and technology but also, and especially, to show that these risks are beyond the control of individual nation states. The nation state is no longer an exclusively 'national' phenomenon, and what an 'individual citizen' is must be reconsidered in examining the kind of citizenship rights people articulate when voting for or against the EU (elections), Brent Spar politics (consumer boycotts), or the World Trade Organization (WTO, demonstrations).

As we argue in more detail in section four, the challenge for the environmental social (political) sciences is to develop new frameworks for ecological citizenship rights and responsibilities. While the classic framework of national, emancipatory politics will have an important role to fulfil in future environmental policy, it needs to be complemented by the frameworks of life-politics, on the one hand, and arrangements for transnational and global environmental governance, on the other. Since the notion of life-politics is meant to emphasise the more direct relationships in global modernity between 'the personal and the planetary', this concept has an important role to fulfil as successor, or at least complement, of the citizen–nation-state framework of the classic, emancipatory politics.

[18] Although these rights and responsibilities are 'attributed to' individual human agents, it goes without saying that the formation and reproduction of these civil rights and responsibilities are not an individual affair and must be examined at the level of social practices.

Sustainable Consumption Behaviour

In some respects, the debate on sustainable consumption behaviour overlaps with issues of ecological citizenship rights. In the aftermath of the Rio Conference, attempts have been made to determine the 'amount of environmental space' or the 'ecological footprint' that regions and nations, and also (future) individual consumers, should be permitted or entitled to. The question 'how much is enough',[19] reasonable, or necessary in terms of overall levels of consumption cannot be decided upon with the help of ecological-technical knowledge alone. Notwithstanding the sociologically 'naive' solutions – one (wo)man, one resource or vote – offered by most environmentalist studies of footprints, these questions must be addressed at the political level as well.[20]

Although consumer roles and citizenship roles are interconnected, sustainable consumption issues are usually discussed in the context of markets and chains or cycles of production and consumption. For example, when talking about sustainable food consumption, the roles of different actors in production-consumption chains, from farmers to retailers to consumers to (organic) waste-managers, are at the centre of analyses. As consumption in this respect is correlated to production, *sustainable* consumption must be conceived of as the necessary complement to the ecological modernisation of production. If consumption processes at the downstream end of the production-consumption chain are ignored, the dematerialisation of production and consumption practices – by a factor of four or ten – will be understood only halfway. Or perhaps less than halfway, since, according to some commentators, the dynamics at the consumer end of the production-consumption chains are becoming by far the most decisive elements in modern consumer societies. Consumer behaviour is becoming important, not just because life cycle assessments (LCAs) tell us that major environmental leakages occur in the consumption (and waste) phase of the cycle. Consumption dynamics in the 'Age of Access' are worth investigating because they are decisive for the future organisation of production-consumption chains.[21] Therefore, the ecological modernisation of production presupposes knowledge of and strategies for the ecological modernisation of consumption.

Against the background of the increasing need to understand modern consumption dynamics, it is a pity that it took so long before the environmental social sciences developed in-depth analyses of (sustainable) consumption. Many environmentalists were content for a long time to adhere

---

[19] During, 1992.
[20] Wackernagel and Rees, 1996.
[21] Rifkin, 2000.

to the fundamental critique of consumerism and the consumer society developed by Frankfurter Schule sociologists some decades ago.[22] These contributions were formulated in general terms, addressing the fundamental values and overall structure of the capitalist consumer society, offering as a solution the downsizing of consumption and the development of radical alternatives to western consumerism. Questions as to how to design concrete politics for sustainable consumption were dismissed, since they were considered managerial strategies missing the fundamental points and misleading in their tendency to 'blame the consumer'.[23]

In the 1990s, consumption studies in the social sciences were boosted by the serious attention of a number of scholars from different disciplinary backgrounds who joined a Lancaster-based international network to investigate consumption, everyday life, and sustainability.[24] The major challenge was to develop a perspective on consumption behaviour beyond the existing economic and psychological explanations which looked only at individual parameters to explain the social phenomenon of consumption. In the *infrastructural perspective* on (sustainable) consumption put forward by this Lancaster network, the key elements are (1) the need to understand consumption behaviour as an ordinary, everyday-life concern that is (2) made possible by socio-technical *systems of provision*[25] which deliver the goods and services under specific conditions of access to citizen-consumers who (3) use these goods and services to organise their daily lives as knowledgeable and capable agents in a meaningful way, thereby (4) reproducing the culturally mediated levels of comfort, cleanliness, and convenience they have become accustomed to over the course of their lives.

In developing some basic principles of consumer-oriented environmental policies, we took this infrastructural perspective on consumption behaviour as our starting point. This is discussed in section three. First, one question on the concept 'citizen-consumers' must be addressed: how do the two dimensions of this hybrid relate to each other? Where do citizens evolve into consumers, and vice versa? How does the hybridisation of social roles relate to the process of globalisation?

Citizen-Consumers in Global Civil Society

A complaint often made by companies that offer sustainable products and services at prices which are slightly higher than those of the conventional products, is that consumers – notwithstanding their frequent and loudly

---

22 Marcuse, 1964.
23 Princen et al., 2002.
24 Shove, 2003.
25 Fine and Leopold, 1993.

voiced 'green preferences' – in the end opt for the cheap alternatives. From this, a general discrepancy is derived between citizenship roles and behaviour, on the one hand, and consumer behaviour, on the other.[26] As citizens, people express great concern about the environment, about food safety, about fair incomes for small coffee farmers, etc. As consumers, however, they refuse to act on these sentiments and always choose the cheapest products. How realistic are these complaints and how immune are consumers to socio-political considerations when shopping for their daily needs?

There are different aspects or dimensions to be distinguished when analysing the discrepancy between 'citizen norms' and 'consumer behaviour'. The gap between 'saying' (concerned citizens) and 'doing' (not-buying consumers) has been discussed by many researchers as representing a specific case of a broader phenomenon, referred to by social psychologists as the gap between attitudes and behaviour. In the so-called attitude-behaviour paradigm within the (environmental) social sciences, this individual attitude or behavioural intention has been used to 'predict' future behaviour.[27] However, this attitude-behaviour approach to consumer behaviour can be considered a cul-de-sac for several reasons, the main reasons being the isolation of behaviours from their social contexts and the fact that most behavioural routines are conducted at the level of 'practical consciousness' instead of being consciously or discursively organised at all times and all places.[28]

When working within this specific attitude-behaviour set of theoretical and methodological premises, the gap between opinions and behaviours is put at the centre of analyses and reproduced as a theoretical dilemma which first and foremost refers to individual norms and (lack of) responsibilities. Individual citizens say they care for the environment, but the same persons fail to live up to these norms as consumers. However, when, instead of the attitude-behaviour scheme, the infrastructural perspective of consumer behaviour is taken as a starting point, the gap is analysed not just in terms of a lack of concern on the part of citizen-consumers, but also, and primarily, in terms of the quality of the provisioning of consumer behaviour by social structures. The complaints of companies about 'unwilling customers' is investigated against the background of (the lack of) consumer empowerment in the organisation of the process of 'green provisioning'.

A second and more interesting aspect of the citizen-consumer discrepancy is the fact that companies which complain about the lack of green buying

---

[26] Spaargaren, 2003a.

[27] In a similar vein, one could argue, economists try to determine the 'value' of (public) environmental goods by asking individuals how much they would be willing to pay for the objects or services under investigation.

[28] Spaargaren, 1997.

power in fact argue that green political preferences *should* be expressed more frequently and consistently via consumption behaviour. With this plea for more fluid boundaries between citizenship roles and consumer roles, they find many social scientists at their side. In contemporary societies, the borders between citizen roles and consumer roles are 'under reconstruction', and are less rigid and recognisable than before, owing to the impact of globalisation.[29] The confusion about people acting as citizens and/or consumers is seen as the result of a concrete set of historical developments affecting the relationship between states, markets, and 'civil society'. In the discussions of 'global civil society' organised by the LSE, the roles of change agents at the grass-roots level, beyond or in states and markets, was put at the centre of analyses.[30] The fluid borders between the roles of citizens (in the context of state policies) and those of consumers (in the context of market-based policies) in global modernity was discussed against the background of the emergence of a global civil society .

The increased importance of the 'third sphere' of global civil society is described in the LSE yearbooks as civil-society-based responses to globalisation. Four types of responses are discussed in some detail. At the extremes, we find 'uncritical support' (transnational companies and other main carriers of globalisation) and 'outright rejection' (left-wing and right-wing radicals) of globalisation. In between these extremes are the positions of the 'reformists', trying to 'civilise' globalisation, and the 'alternatives', protesting against globalisation by organising their own, local, alternative, 'organic food'-type solutions.[31] As is clear from this categorisation, civil-society-based responses to globalisation were discussed in the LSE round-table conferences very much in terms of the new *political* roles, protests, and division lines in global modernity. While emphasising the political responses to globalisation, the yearbooks offer little reflection on and conceptual space for reactions to globalisation based on the role prescriptions and expectations of (global) *consumers*. Especially when reformist strategies are at stake,[32] we would argue that the 'eco-civilisation of globalisation' also requires the development of countervailing power in the context of the greening of chains of production and consumption which are increasingly organised over global levels of distance in time and space. As we discuss briefly in section four, the new media and the new dynamics of time and space in global modernity also

---

29 Urry, 2000.
30 Anheier et al., 2001; Glasius et al., 2002.
31 Anheier et al., 2001.
32 Those who are familiar with the debate on ecological modernisation in environmental social sciences will recognise the similarities between the distinctions made in the LSE books between reformists and alternatives, on the one hand, and the eco-modernist versus demodernisation strategies discussed in environmental sociology, on the other hand (Mol and Sonnenfeld, 2000; Spaargaren, 2000).

make possible new articulations of consumer roles and consumer power at the level of global production-consumption cycles or chains.

To make sense also of the new emerging *consumer* roles in the governance of global environmental flows, the political outlook of the global-civil-society discussions needs to be complemented by an economic or market-based strategy. To this purpose, a new vocabulary will have to be developed, breaking away from the classic schemes of multi-level multi-actor governance, and instead taking (global, local, or regional) environmental flows and the networks and socio-physical 'scapes' that go along with them as the main units of analysis. The recent work by John Urry, elaborating upon Castells' concepts of flows, scapes, and networks, seems to offer a promising starting point for this.[33]

The theoretical overview offered in this section provides the background to our discussion of the specific roles citizen-consumers may perform in concrete environmental politics.

FOUR NATIONAL CASE STUDIES

Some policy experiments and pilot studies were conducted in the Netherlands in recent years to explore the possibilities of developing a more citizen-consumer-oriented environmental policy. The pilot projects brought together policy-makers and scientific researchers in an attempt to reflect more systematically on this type of environmental policies. In this section, we briefly introduce and discuss some of the relevant projects issued by the Dutch Ministry of the Environment (VROM), in order to derive from them the most important aspects of a consumer-oriented policy. Four projects are discussed in some detail. The 'Domain Explorations', the 'Future Perspective', and the 'Ecological Modernisation of Consumption Practices' projects emphasised the consumption role of citizen-consumers, while the 'Citizen and the Environment' project stressed the political roles of citizen-consumers in the environmental policy field.[34]

Domain Explorations: Alternatives for High-Impact Consumption

With the initiation of a series of so-called 'Domain Explorations' from the mid-1990s onwards, the Dutch Ministry of the Environment acknowledged the increasing environmental pressures resulting from consumption practices. The aim of the Domain Explorations was to identify the environmental profile

---

[33] Mol and Spaargaren, 2003.

[34] This section is based on a comprehensive report on the pilot project by Beckers et al., 2000. See also: Martens and Spaargaren, 2002; Spaargaren et al., 2002; and Spaargaren, 2003b.

(including the most detrimental elements) of a number of consumption practices of special relevance to environmental policy-making. Five domains were selected: food, clothing, housing, recreation, and personal care.[35] Within each of these domains, the environmental deterioration caused by various items of consumption and consumption behaviour was assessed, mainly with the help of technical methods such as LCA. In addition to identifying the major causes of environmental deterioration, referred to as identifying the 'big fishes', technical and/or behavioural ways to lower the environmental impacts were sought for. The Domain Explorations consisted of a series of systematic scans of the production-consumption chains implied in a specific consumption domain, in order both to find major sources of environmental deterioration as well as ways to repair them. In a number of cases, chain actors were brought together by national policy-makers in order to discuss the (development of) more sustainable consumption alternatives in their specific domains.

Through the development of the Domain Explorations, an important step was taken in the foundation of consumer-oriented policy in the Netherlands. Well-defined 'ordinary' consumption domains were taken as the points of departure for policy-making, instead of starting from already-determined policy goals and instruments and subsequently arguing for the need to incorporate them into consumption practices. By mapping out the big fishes (i.e., the most rewarding interventions from the point of view of environmental policy), it became possible to compare the effectiveness and efficiency of past and future interventions, both within and between the consumption domains. This kind of knowledge is essential in order to legitimise consumer-oriented environmental policies in a sustained dialogue with citizen-consumers. In the Domain Explorations, the legitimacy of consumer-oriented environmental policy was established primarily on technical grounds, in terms of technical environmental efficiency only.

This exclusive emphasis on the technical dimensions of environmental policies can be regarded as the major weakness of the Domain Explorations, since the extent to which citizen-consumers know and have internalised technical objectives (such as x per cent carbon dioxide reductions, a specific BOD/COD[36] for water quality, or a certain nitrate and phosphate concentration in the soil) is always limited. Technical objectives in the context of citizen-consumer-oriented policies, in our opinion, must be connected with the social behaviours, the expectations, and the experiences of citizen-consumers in their everyday lives. A link must be established with the

---

[35] CREM, 2000; Schuttelaar & Partners, 2000; TNO, 1999.

[36] Biochemical Oxygen Demand (BOD) and Chemical Oxygen Demand (COD) are technical indicators of the quality of surface water, based on the oxygen consumed when some substances degrade.

consumption routines behind the environmental impacts in the different domains. Carbon dioxide targets, for example, have to be translated into buying A-labelled refrigerators, washers, and dryers, and into insulating the home. Carbon dioxide consumption should be related to using fuel-efficient cars and/or considering eco-efficiency factors when choosing certain combinations of transport modalities (plane, train, car, bike, etc.) for holidays or work. Another weakness of the Domain-Exploration approach was the fact that the stage, organised in cooperation with chain actors to discuss feasible routes for the innovation and diffusion of green alternatives, was dominated by providers and their predominantly technical outlook and rationality, with citizen-consumers hardly being represented in the workshops.

The Domain Explorations were concerned with consumption, but they were not 'citizen-consumer-oriented'. They helped to establish consumption politics by determining in technical terms the environmental impacts of a number of consumption domains that affect most people in their daily lives. The policy strategy behind the Domain Explorations was to develop the knowledge needed to put pressure on the providers within the different consumption domains to 'make available' more sustainable alternatives in the future. These experiments, in other words, were targeted at the further development of the material basis of consumption politics by improving 'green provisioning' by producers, retailers, and other institutional actors at the provider end of the production-consumption chains.

But will citizen-consumers make use of these newly developed, more sustainable products and services? Under what circumstances can households be 'seduced' to green their daily lives and lifestyles?[37] To investigate these questions in more detail, the Future Perspective Project was initiated.

Future Perspective: Reading Social Structures

The Future Perspective Project[38] was launched as the follow-up to a number of studies in which attempts were made to assess the environmental dimension of the lifestyles of Dutch citizen-consumers in a quantitative way.[39] The environmental impacts of domestic consumption routines were operationalised in terms of direct (heating, lighting) and indirect energy use, with indirect energy referring to the energy needed to produce a certain product (a car, a refrigerator, a piano). By focusing on the long-term development of the domestic direct and indirect energy consumption of different lifestyle groups, the project served to provide the Ministry of the Environment with insights into the (potential) role and position of citizen-

[37] Bauman, 1983.
[38] CEA, 1999.
[39] Blok and Vringer, 1995; Vringer and Blok, 1993.

consumers in carbon dioxide policies.[40] In the aftermath of the Kyoto Conference, climate policies were given high priority by the Dutch national government.

During a two-year period, a small number of participating households were challenged to change their consumption patterns in order to achieve a 30-40 per cent reduction of their total domestic consumption of indirect energy. Since a continued rise in household incomes was expected, the participating households received a supplement to their regular budgets of 20 per cent. Additionally, the householders were assigned a computer programme providing information on the indirect energy content of most products available in the Netherlands, as well as a personal coach to assist them in permanently monitoring their energy use and to provide advice and information on energy-saving alternatives.[41] When returning from local shops or supermarkets, householders had to register the environmental dimension of their shopping, thereby learning about the best ways to spend their money in an environmentally efficient way and about the possibilities of replacing some polluting products or packages with less energy-intensive alternatives. The use of alternative products and services was expected to lead to a substantial reduction of environmental impact, while avoiding negative consequences for the levels of comfort, cleanliness, and convenience the householders were accustomed to.[42]

At the end of the two-year period, most participating households had developed a personal, energy-extensive consumption pattern, with an average of a 31 per cent reduction. Equally interesting, however, was the conclusion that householders differed with respect to the ways in which the energy-extensive consumption patterns were realised. Householders showed their own specific 'webs of life' before and after the project period, where the webs were used to represent the ways in which environmental impacts were distributed over the different consumption domains that together make up a lifestyle (Figure 8.1).

A year-and-a-half after the end of the project, the consumption patterns and energy uses of the participating households were monitored and evaluated again. In most households, the changes in the patterns of indirect energy use turned out to be rather stable, especially in the fields of food and leisure. Long-distance travel related to summer holidays, however, proved more difficult to alter on a lasting basis. Constraining factors in maintaining the energy-extensive consumption patterns were the loss of supplementary

---

[40] The indirect energy use of households amounts to about 60 per cent of the total household energy consumption.

[41] CEA, 1999.

[42] Shove, 2003.

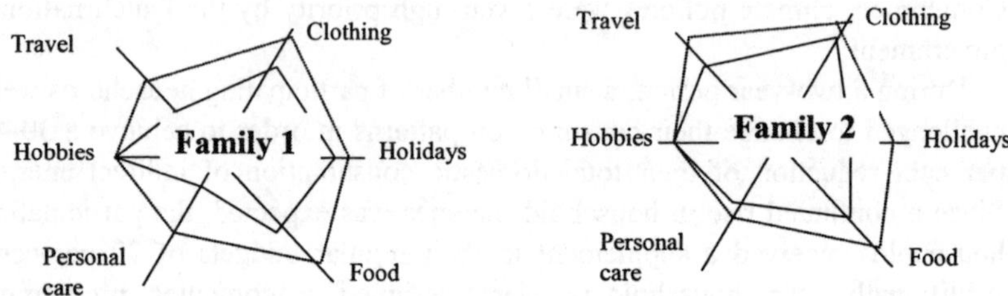

Figure 8.1 Webs of life(styles)[43]

budgets and personal feedback, in combination with overall high prices and the limited availability of sustainable products and services.

Factors enabling the new energy-extensive behaviour to become routine had to do with householders perceiving their new 'lifestyles' as healthier, more relaxed, and money-saving. Participating in the project was regarded by most householders as a positive experience, while some felt that they were judged by their social environments as being 'different' or 'special'.

The Future Perspective project did not focus on the environmental attitudes or opinions but on the actual behaviour of people in the context of their daily lives. We consider this focus on ordinary daily lives to be a fruitful starting point for exploring the possibilities of a politics of sustainable consumption. The project managed to provide insights into the relative feasibility of energy saving in different consumption domains. The potential for environmental change was found to be considerable, and related both to internal factors (income, composition, life-cycle phase, housing, etc.) and to external factors (such as the availability and accessibility of more sustainable alternatives in the different consumption domains).

With respect to the external factors, the Future Perspective project turned out to be an especially instructive exercise in the 'reading' of social structures. Actors were trained to constantly scan their social environments for the possibilities and constraints they offered with regard to realising more sustainable lifestyles. The fact that a special computer programme and a personal coach, in addition to the extra budgets, were needed to detect and evaluate the sustainable alternatives can be regarded as an indication of the fact that the ecological modernisation of consumption is not so well developed as the greening of production processes.

When discussing the results of the Future Perspective project in a number of workshops with scientific researchers, practitioners, and policy-makers, it

---

[43] Based on CEA, 1999.

was recognised that factors on the providers' side, determining both the level and the accessibility of sustainable alternatives, were not given proper attention in the project.[44] As a consequence, the task of greening everyday consumption patterns was to be performed almost exclusively by citizen-consumers themselves, 'independent' of actors on the supply side of the green economy. With respect to future pilot projects, it was suggested that especially the mediation between supply and demand (i.e., the facilitation of exchange between providers and consumers) should be enhanced at different levels, ranging from the national level (for example, with internet sites on green alternatives) to the local level (for example, with 'environmental brokers' operating at district or neighbourhood levels). If market actors and mechanisms alone cannot perform this task of mediation satisfactorily, it can, in principle, become an important task for environmental policy-makers to facilitate and stimulate this exchange, if not to organise themselves on the basis of a public-private partnership.

The mediation between supply and demand, in sociological terms the interrelation between consumption behaviours and their corresponding systems of provision, was the main focus of the third project.

Ecological Modernisation: Provisioning and Access

In the Social Practices approach to consumption, some of the strategic elements from the 'provider-oriented' Domain Explorations were combined with the 'end-user' orientation of the Future Perspective project, while at the same time an attempt was made to overcome some of the apparent weaknesses of both perspectives. We briefly summarise the basic assumptions of the Social Practices approach as a foundation of citizen-consumer-oriented policies, and then look into the results of two case studies conducted to empirically explore the feasibility of these assumptions.

In the Social Practices approach,[45] the daily behaviour of citizen-consumers is taken as the starting point. Unlike in the Future Perspective project, however, behaviour is not put forward as an individual (domestic) affair or a series of isolated behavioural strings (like buying product X or not performing behaviour Y). Instead, behaviour is analysed at the level of groups of citizen-consumers sharing some specific consumption routines that are reproduced over time and space. By taking social practices as the central unit of analysis and policy-making, the target group of citizen-consumers is specified and differentiated or segmented. This is not done according to the

[44] ResCon, 2000.

[45] For a more detailed introduction to and theoretical foundation of the Social Practices approach, also in relation to environmental policy-making, see: Beckers et al., 1999, Spaargaren, 1997; and Spaargaren et al., 2002.

opinions of the citizens about the environment, nor with respect to their overall lifestyle characteristics (such as income, education, or political preferences). Instead, the analysis focuses on the patterns of concrete behavioural routines such as shopping, gardening, housing, commuting, and playing soccer. Issues of lifestyles, personal motives, interests, and norms are discussed, not just as characteristics of individuals but also, and primarily, as characteristics of the social practices under investigation.

The Social Practices approach follows the Domain Explorations by looking at a specified number of consumption domains and the environmental impacts they represent. Unlike the Domain Explorations, however, these consumption practices or routines are analysed not only as the end-user aspects or elements of specific production-consumption chains. Consumption practices are conceived of primarily as the performance of skilful and knowledgeable human agents who, with the help of social structures, reproduce these consumption routines in a knowledgeable and skillful way.

In order to make environmental change 'work' at the level of social practices, policy definitions of sustainable consumption must address the way they fit or do not fit into the behavioural routines of groups of citizen-consumers. If the new, more sustainable routines affect the existing levels of comfort, convenience, and cleanliness[46] in a negative way, major obstacles will probably arise in the transition process. Also, when people do not recognise environmental policy goals as being legitimate (for example, because these goals are framed in an overly technical jargon), the transition process will probably not come about.

To deal with this problem, in the Social Practices approach, the need to develop environmental heuristics for each consumption domain is emphasised and taken into account. These environmental heuristics are thought to result from a dialogue or consultation between environmental policy experts and citizen-consumers. They specify the environmental policy goals in terms of life-world rationalities, and indicate how these goals can be approached in terms of behavioural changes.[47]

In the Social Practices approach, citizen-consumers' responsibilities for and commitments to the transition to more sustainable consumption practices are analysed and formulated with reference to the existing capacities for change in the domains under investigation. Capacities for change can so far be said to result from the concerted actions of (governmental and market-

---

[46] Shove, 2003.

[47] An example of an environmental heuristic used by citizen-consumers in the field of waste management is the principle of 'prevention and separation of wastes': if possible, the practices of shopping (bags, packages, etc.) and gardening (composting) should be performed in such a way that waste production is minimised and the different waste streams are kept separate in order to allow for a more efficient recycling and re-use of domestic wastes. For more examples, see Martens and Spaargaren, 2002.

based) providers and innovative groups of citizen-consumers. If the level of 'green provisioning' is high (both quantitatively and qualitatively) and when these environmental innovations are judged to be accessible to considerable numbers of lifestyle groups, then the responsibilities and 'to be expected commitments' are higher than in situations of under- or undeveloped capacities for environmental change.[48]

Concerning consumption practices that have good environmental performance records, the individual citizen-consumers can be asked to re-organise their daily lives in this respect (for example, for this specific segment or sector of the lifestyle). Therefore, in line with the Future Perspective project, the central objective and rationale of citizen-consumer-oriented environmental policies is the reduction (by a factor of four or more) of the net environmental impacts of consumption. This can and must be done with the help of high-quality green provisioning, enabling a greening of lifestyles without affecting the levels of comfort, cleanliness, and convenience in modern society. As shown by the Future Perspective project, the 'roads to reduction' are manifold and need not be specified by governments or NGOs. Sustainable lifestyles are the 'private affairs' of citizen-consumers; they determine the ways in which to realise the reduction of environmental impacts over the different segments or sectors of their lifestyles. There are many webs of life(styles) connected to the strategies householders might choose to reduce overall environmental impacts.

With the help of the assumptions formulated above, two case studies were conducted in close collaboration with policy-makers in the Dutch Ministry of the Environment. For two 'environment-relevant' consumption domains (inhabiting a house and going abroad for holidays), the levels of green provisioning were established. This was done by scanning the production-consumption chains related to the socio-technical innovations that were made available to citizen-consumers by providers higher up in production-consumption chains or cycles. By looking into the number of innovations, the way they were put into the market by providers, and the ways in which they were facilitated (or not) by governmental actors, the quality of the provisioning in these two consumption domains was established. A series of semi-structured interviews with actors at the provider end of the production-consumption chains was the main source of information in this respect, next to the information available in policy documents, on web sites, etc.

---

[48] Whatever the concrete instruments used (from regulatory eco-taxes to formal legislation to communication), the argument presented here implies that legitimate citizen-consumer-oriented environmental policies can be more strict and compulsory as the quality of provisioning with green alternatives increases.

With this information at hand, groups of citizen-consumers who showed themselves to be engaged with the consumption routines under investigation were invited to join a series of focus groups developing a citizen-consumer perspective on the greening of these specific consumption practices.[49] With the help of visual techniques (the construction of a series of posters representing different green alternatives available to citizen-consumers under different conditions of access), citizen-consumer definitions of the problem (and the corresponding environmental heuristics) were established, as were the (lifestyle-related) preferences for gaining access to the socio-technical innovations presently available. Finally, the 'responsibilities' of citizens, consumers, markets, and state providers were discussed against the background of the existing state of affairs in the specific consumption domains.[50]

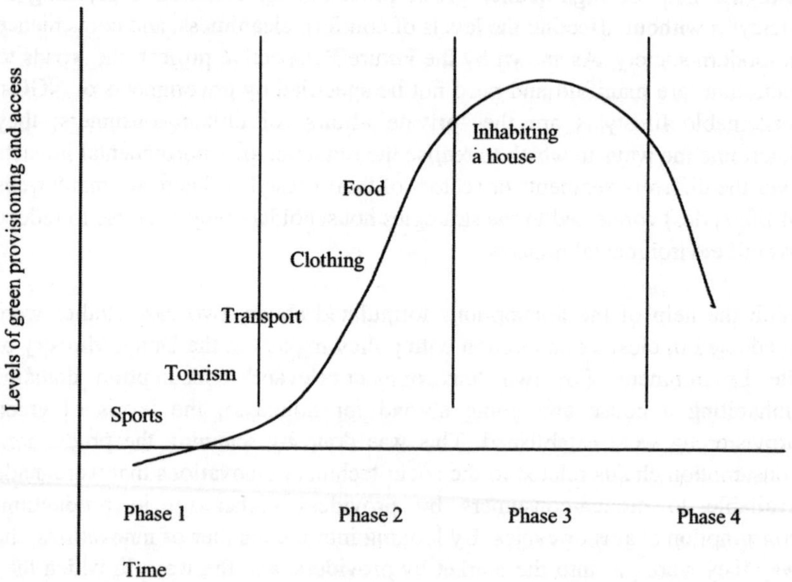

Figure 8.2 The ecological modernisation process

[49] Participants in the focus groups were selected with the help of 'on-site' short questionnaires. Two main criteria used in the selection process were being an active consumer and not having an aversion to (talking about) environmental aspects of these consumption practices. Active consumers for the Do-It-Yourself (DIY) focus group that was organised in the context of the consumption practice 'inhabiting a house' were selected while shopping for materials in a large DIY warehouse, while people for the focus group on tourism were addressed in travel agencies.

[50] Martens and Spaargaren, 2002; Spaargaren et al., 2002.

With the help of these research methods, it was concluded that the two consumption domains under investigation were in different phases of the policy process, with both qualitative and quantitative differences in green provisioning. Figure 8.2 illustrates the positions of the two consumption domains (inhabiting a house and tourism) in relation to each other, and also shows the estimated levels of green provisioning and access in some other domains in the Netherlands.

If more mature policies for sustainable consumption are to be established in the future, the kind of information provided by the Social Practices approach to consumption is of key importance. Only when the existing levels of green provisioning and access are established in a scientifically valid and socially legitimate way can the roles of governmental and non-governmental actors, including the roles of citizen-consumers, be specified in the transition process.

But do citizens feel the need to become involved in processes of environmental change? Do they regard environmental policies for sustainable consumption as something the (national) government should facilitate and promote? These questions of the legitimacy of national environmental (consumption) policies were addressed in the fourth and final project.

Citizen and Environment: Bothering People?

The Citizen and Environment project was instigated by the Dutch Ministry of the Environment to actively seek meaningful relations with citizen-consumers and to develop a dialogue on the (lack of) responsiveness of environmental policy-makers to the concerns of 'ordinary citizens'.[51] The project was aimed at moving away from the traditional emphasis on the technical system rationality dominating the environmental policy field. In the project, possible ways were sought to connect policy goals formulated by central governments with the needs of citizens, and their motivations and concerns in daily life. The focus was on the political roles of individuals as citizens. In what ways do people want or not want to address environmental problems and policies when they are busy organising their everyday lives? What definitions of environmental change do specific (lifestyle) groups of citizen-consumers use in this respect? What choices do they think the government or society should make in environmental matters and what courses of environmental actions do they think are relevant to themselves?

To find the answers to these questions, a series of small projects was conducted to explore the implications of a citizen-consumer orientation to agenda-setting, policy-making, and the division of roles and responsibilities

---

[51] B&A Groep, 2000.

in the implementation of environmental policies. Several methods and techniques (such as workshops, expert sessions, panel meetings, interviews with consumers, and focus groups) were employed to investigate the environmental motives and concerns of citizen-consumers in the context of different daily-life settings. Citizen-consumers were interviewed, for example, when visiting entertainment areas, day-care centres, or shopping malls. With the help of different research methods, an attempt was made to reconstruct how people encountered and viewed environmental issues in these settings and how they thought these issues should be addressed by environmental policies so that the interventions would fit their lifestyles and preferences.[52]

Several conclusions regarding citizen-consumer-oriented environmental policies were drawn from the results of the project. With respect to the ways in which citizens perceive the environmental agenda, it was concluded that environmental policies need societal embedding: problem definitions must not be restricted to technical ecological issues only. With regard to the division of labour between citizens and the (national and regional) governments in the context of environmental policy-making, it became clear that a distinction must be made between environmental problems for which consumers can reasonably be held (co)responsible, on the one hand, and problems that are the responsibility mainly of the (central) governments, on the other hand.

During the project, this problem was labelled as the 'bordjes-probleem', a Dutch saying referring to 'the need to serve meals on the proper plates'. Keeping in mind the social science literature on the prisoners' dilemma and the debate on the environment as a collective or a common good, it was concluded that the preconditions for bothering citizen-consumers with environmental problems and solutions (for example, regulations) must be specified in more detail than is done at the moment. Some problems should simply not be put on the plate of citizen-consumers in a specific manner during a specific phase of the policy cycle, since the next series of actions to be performed in that policy phase is to be undertaken by government and/or business actors and not by citizen-consumers. Also, when environmental problems are not 'visible' to citizen-consumers in their daily lives, and when citizens are not able to establish a meaningful link between the environmental problem and their daily behaviour, environmental policies are likely to fail. In these situations, new symbolic representations of environmental problems must be developed which match the daily life-worlds of citizen-consumers. These new representations ought to be generated in a direct dialogue between policy-makers and citizens.

One of the weaknesses of the Citizens and the Environment project was the

---

[52] B&A Groep, 2000.

strict focus on politics. The responsibilities of citizen-consumers for sustainable development were discussed exclusively in terms of the relationship between citizens and their (national) environmental state authorities. With regard to (the prime) responsibility for environmental problems, market- and civil-society actors should be included in the analyses as well. The prisoners' dilemma, in our view, is not restricted to citizens and states, but pertains to companies and consumers as well.

As we discuss in more detail in the next section, questions about the best way and time to confront citizen-consumers with political rights and responsibilities for environmental problems can be shown to be directly affected by the globalisation of environmental policies. The issue of putting the right problems on the right plates acquires a new dimension when national environmental policy-makers are joined by transnational and global policy-makers *and* market actors who operate most of the time outside the reach of national governmental politics.

CITIZEN-CONSUMERS AND GLOBALISING POLITICS

We explored the possible roles of citizen-consumers in environmental politics and the ways in which these roles are affected by the process of globalisation of both environmental problems and policies. Do not environmental politics 'move away' from citizen-consumers and their everyday lives, instead being formulated at ever higher levels of governance, up to the global level? Can environmental policies be said to be extra vulnerable to the 'erosion' of traditional nation-state-based policies, owing to the strong 'technical' nature of environmental problems and because their negative social impacts can often be said to be 'far away' in both time and space?

Our general response to these critical questions is that, indeed, the erosion of traditional nation-state policies does not halt at the environmental policy field. Perhaps even more than their colleagues in other policy domains, environmental experts at the national level are faced with a fast-growing body of transnational legislation and with the restructuring of the policy-making process at the national level, owing to the globalisation of social relations in the network society. Against this background, it is easy to understand why environmental problems are so often used by sociologists to illustrate the new dynamics and principles of policy-making in global modernity.

To face the challenges of environmental policies moving away from everyday life, a strong orientation towards the roles of citizen-consumers in environmental policy-making was put forward as an essential element of the strategies to be followed. Urging a stronger emphasis on citizen-consumers in (national) environmental policies, we used arguments both from within the field itself and from broader discussions on changing roles of nation states vis-à-vis citizen-consumers in global modernity. Having explored both the theoretical and practical implications of a stronger orientation towards the roles of citizen-consumers, we concluded that the agenda for citizen-consumer-oriented environmental politics is still very much in the making and that there are many research and policy questions yet to be answered. Among the questions that most urgently need to be addressed is that pertaining to the impacts of globalisation. We were not able to resolve this issue entirely. We looked into some of the key elements, first with respect to the impact of globalisation on the changing roles of *citizen*-consumers and then with respect to its impact on the roles of citizen-*consumers*.

Ecological Citizenship Beyond the Nation State

Ecological citizenship is a concept that has gained substance over the past thirty years. It refers to the rights and responsibilities which can be said to be connected with the overall process of ecological modernisation in contemporary societies. The concept of rights refers to both positive elements (like the right of access to clean water or waste services) and negative elements (such as the avoidance of exposure to environmental risks in the modern risk-society).

Although the Club of Rome reports[53] and the Stockholm Conference[54] in the early 1970s gave environmental politics its 'global' character right from the beginning, the overwhelming majority of environmental policy activities were organised at the (sub-)nation-state level in the first decades of environmental policy-making. The dominant framework for thinking about ecological citizenship rights and responsibilities was organised for that reason along the citizens vis-à-vis nation states axis, also in the environmental field.

In Europe, the development of environmental policies was influenced to a considerable extent by the strong position of the EU, which gained much (regulatory) power in this field from the 1980s onwards. As a result, EU environmental policies are, next to economic and agricultural politics, the most influential transnational set of policies in modern history. In the latest Dutch national environmental policy plan (NEPP-4), it is stated that about 80 per cent of all policy measures in the national environmental field are affected by, if they do not originate from, European environmental politics.

With EU influence growing stronger and stronger, the commitment of citizens to formal or official environmental policy-making at the EU level seems to be moving in the opposite direction. The lack of power of the

[53] Meadows, 1972.
[54] United Nations, 1972.

European Parliament and the information deficit at both national and European levels partly explains this, together with the apparent lack of interest in daily European environmental affairs of national mass-media in most member states. To us, however, the lack of a mediating role of the nation states between EU environmental politics and their national citizenry is among the most decisive factors in understanding why ecological citizenship is rarely discussed from a transnational or global perspective. Such a mediating role for the nation state – also in the process of ecological modernisation – was put forward in the debate on so-called 'third way' politics, although not unambiguously.[55] So far, only the vague contours of a framework to organise the ecological rights and responsibilities of citizen-consumers beyond the nation-state level have been developed.

The role of the EU in providing an alternative framework to organise citizenship rights beyond the nation state will be restricted primarily to official, emancipatory politics, with European citizens voting for European politics, politicians, and Parliament. To find frameworks that address globalisation in a more direct and encompassing way, we have to enter the debate on global civil society again.

Citizen-Consumers and Political Consumerism in Global Civil Society

Outside official policy-making circuits at the nation state and EU levels, new ways of organising citizenship involvement in environmental policies seem to be developing. Environmental NGOs have opened direct negotiations with market actors at different levels, using public pressures as their most important and sometimes effective sources of power. The Brent Spar case is discussed in the literature as a prototype example of 'sub-politics' at the transnational level.[56] Such politics are shaped in direct relation, or owing to confrontation, between multinational companies and multinational civil-society actors, both of which are outside the influence of official (state) politics. With nation states losing terrain in environmental affairs, these sub-political arrangements seem to be gaining influence, not least because they are much more flexible with respect to the (combination of) layers or levels of environmental policy-making.

The concepts of life politics and lifestyle politics, introduced to the social sciences by Giddens, represent the new relationships between the local and

---

[55] In the 'third way' debate, the renewal of social democracy in Europe (especially the United Kingdom (UK) and Germany) and the United States was at stake, and particular attention was given to the new roles of the nation state in an era of globalised markets. Both Tony Blair (UK) and Gerhard Schröder (Germany) made active contributions to this debate (Giddens, 1998, 2000).

[56] Beck et al., 1994.

the global, between the personal and the planetary. Life politics may refer to nature ('the eyes of the panda' or the 'tail of the whale'), as well as to the environment. An example of environmental life politics is provided by citizens who fulfil their Kyoto targets in a direct and personal way by fixing to their roofs a number of photovoltaic (solar) cells which roughly corresponds with the carbon dioxide reduction levels required by the Kyoto Protocol for individual households in the specific country or region.

As the examples in the field of sub-politics illustrate, it is difficult to decide where exactly in global civil society the citizen role develops into the consumer role and vice versa. Buying green will likely develop in the future into an umbrella category of environmental actions, covering both consciously organised forms of environmental behaviour shaped as political activism and routinised everyday consumption behaviour. However, instead of focusing on the lack of consistency between citizenship roles and consumer roles (see section 2), we prefer to emphasise the broadening of possible roles and welcome the new hybrids for involving and engaging citizen-consumers in environmental policies in different ways.

It is clear that, for all the different forms of engagement, the development of new communication and information technologies in the network society – the world of www, dot com, and mobile phones – very much enhances the possibilities of actors and institutions in the global civil society to become involved in processes of political consumerism. The internet offers opportunities for groups of citizen-consumers to exert more direct and effective control over producers (companies and governments alike), both in their negative roles as polluters or providers of environmental risks and in their positive roles as providers of eco-labels, green electricity schemes, and other green products and services.[57]

When the monitoring of environmental performance in production-consumption cycles is no longer organised exclusively along the established lines of providers' interests but also with the aim of increasing the empowerment of citizen-consumers, people are able to become engaged in more direct and diverse ways with processes of environmental change. The environmental scorecard, informing citizens in the public domain about sources of environmental risks in their direct living environments, would perhaps be at the political end of the spectrum, while interactive internet sites providing information on the best available green electricity schemes would be at the consumption end of the spectrum.

[57] Van den Burg et al., 2003.

Sustainable Consumption (Politics) in the Global Arena

Sustainable consumption is usually organised outside the direct influence of nation states and governmental bodies. In that strict sense, sustainable consumption is not a concern of environmental policy. It is primarily a market or chain concern, with all the relevant players struggling to make visible, and to buy and sell, the environmental qualities that (are claimed to) result from the improved environmental performance of producers and consumers. Although states do not usually organise sustainable production and consumption themselves, in thirty years of environmental policy-making on sustainable production, the role of states has been crucial in some phases and with respect to specific dimensions of the process. In section three, we discussed some pilot projects illustrating the involvement of the Dutch national government in the emerging consumption politics. Here, we discuss the role of transnational and globally operating state organisations in triggering sustainable consumption in the global arena.

In the year 2000, the Organisation for Economic Co-operation and Development (OECD) started a programme on sustainable consumption, in an attempt to find answers to questions similar to the ones discussed in section three. For a number of countries, the environmental impacts of household consumption was to be established. The programme focused not only on examination of the direct and indirect use of energy per household, but also on specification of environmental trends in key sectors of everyday consumption in the areas of, for example, food, tourism, travel, and housing.[58] The study drew on the experiences and results of a number of EU-funded international programmes such as Domus,[59] Homes,[60] and SusHouse,[61] in which sustainable domestic consumption was investigated from a comparative perspective. Next to the OECD, the United Nations (UN) Commission on Sustainable Development, the EU, the UN Environment Programme, and a number of proactive companies which joined the WBCSD[62] were also involved in making the first steps towards developing sustainable consumption policies in the global arena.[63]

This task seems difficult, however. When investigating the possibilities of developing sustainable consumption policies at the international level, we run into an extra set of complicating questions. If a selected number of everyday

---

[58] OECD, 2000.
[59] Chappels et al., 2000.
[60] Noorman and Schoot Uiterkamp, 1998.
[61] Knot, 2000.
[62] The World Business Council for Sustainable Development (WBCSD) is a coalition of 170 international companies united by a shared commitment to sustainable development via the three pillars of economic growth, ecological balance, and social progress.
[63] Cohen, forthcoming.

life-based, environmentally relevant consumption routines is to be chosen as the starting point for developing consumption policies in the global arena, how can the considerable differences between countries and social classes with respect to their overall consumption levels, the levels of comfort, cleanliness, and convenience adhered to, and the cultural meanings attached to sustainable consumption be confronted, and how must the different levels of green provisioning and access in the countries or regions under investigation be dealt with, to mention just a few of these complicating issues. Therefore, while not being an environmental policy concern in the strict sense, the *political dimension* of sustainable consumption programmes (i.e., the social distributional aspects of the greening of consumption at the global level) cannot be underestimated and manifests itself at the global level in a specific way.

An example may be helpful to illustrate this argument. Because EU and US consumers enjoy high levels of consumer protection in the area of food, producers in developing countries face a number of green barriers which result from the import restrictions imposed on them for reasons of environment and nature protection.[64] The greening of consumption at the global level cannot be envisaged without the concomitant greening of the international production, trade, and investment regimes regulated at the global level by the WTO. While NGOs push hard in this direction and many countries – also in less developed parts of the world – are theoretically in favour of such a greening of product flows, in the political reality of WTO negotiations there seems to be a 'fundamental clash of priorities between developed and developing countries' in this respect.[65] The greening of consumption that at the level of life(style) politics appears as the evident heuristic of buying cans of tuna with smiling dolphins on them, is discussed in global trade institutions in terms of Western protectionism, sold under the umbrella of nature conservation. The dolphin-free technologies in tuna production and the turtle-protective technologies in shrimp production illustrate the more confusing aspects of the relationship between the personal and the plenary in sustainable consumption.

The easiest way to deal with these nasty questions and complications is to do away with eco-labelling altogether and to discard the efforts to develop sustainable consumption programmes (also) in the global arena. The more difficult and more challenging approach is to try and tackle these issues in the context of internationally and globally discussed and coordinated research programmes and policy pilots for the life politics of global sustainable consumption.

---

64 Oosterveer, 2002.
65 Neumayer, 2001: 15.

CONCLUDING COMMENTS

Contrary to the tendency of environmental policy-makers to regard the (consumption) behaviour of citizen-consumers as too difficult to influence or assess from the viewpoint of environmental policy, we argued that a stronger orientation towards citizen-consumers can be supported on a number (four + one) of grounds. Both the material and social bases of citizen-consumer-oriented environmental policies and politics advanced over the past decennia, at least in the most developed parts of the world. The social context is more favourable now than it was in the early days of environmental policy-making, and the theoretical foundations of citizen-consumer-oriented politics have also developed and become more mature. With the help of four national (Dutch) case studies, we tried to show that if social practices in distinct consumption domains (food, travel, housing, etc.) are taken as the unit of analysis of policy and if they are approached with the help of the infrastructural perspective on consumption behaviour, a politics of sustainable consumption may result that combines democratic quality with environmental efficiency and does not blame the wrong actors for environmental deterioration.

While emphasising the need to rethink the role of citizen-consumers in environmental politics, we agree that globalisation makes this project more challenging and complicated. The findings obtained from the national case studies take on a different dimension when situated at the level beyond the nation state. We tried to map out the different ways in which globalisation might affect citizen-consumer-oriented environmental politics. We suggested that these impacts can be discussed and further investigated under at least three headings.

First, we discussed the ways in which globalisation disturbs the national formats that are used to organise ecological citizenship rights and responsibilities. Globalisation does not just result in more actors working on environmental policy arrangements at more levels. It leads to a new quality or category of policy-making, and these environmental politics of the space of flows need to be addressed differently.[66] We suggested that nation states have so far not lived up to their task of serving as mediator between global- and local-level environmental politics in the way suggested in the discussion of third way politics. We also suggested that the classic frames of emancipatory politics be complemented by the framework of life politics, making more direct connections between the planetary and the personal.

Second, we dealt with the emerging agenda for sustainable consumption politics and the complexities of situating this agenda in the global arena. Our

[66] Castells, 1996.

general argument that citizen-consumers' commitments to and responsibilities for domain-specific efforts to reduce the environmental impacts of consumption must be decided upon against the background of the quality of green provisioning seems to be valid in the global arena also. However, as the social inequalities in the global network society are greater and more complicated than those in the national environmental policy domains, we argued that the political dimension of sustainable consumption programmes needs to be taken seriously, especially when working in the global arena. We argued that, in this respect, the concepts offered by Urry and Castells of the space of flows versus the space of place deserve attention in future research, as they may be helpful in analysing and assessing the social distributional aspects of sustainable consumption politics at the transnational level.

Third, we discussed the concept of global civil society as put to the fore in a series of round-table meetings organised by the LSE. Global civil society refers to the realm of sub-politics beyond state and market, nourished to a great extent by the activities of (environmental) NGOs and supported by the (communication) technologies of our time. Notwithstanding the fact that its geographical scope is still restricted mainly to Western Europe, this global civil society and its main actors and institutions may in the future be of importance in setting the global agenda on issues of ecological citizenship and sustainable consumption. Like their predecessors in the 1970s, they fuel the global environmental debate with the pros and cons of reformist versus alternative strategies to cope with globalisation from the viewpoint of environmental policy. We contributed to this debate by arguing for the need to include in the predominantly political debate on the global civil society an analysis of the economic and consumption dimensions of global environmental change.

If an eco-civilised global modernity is to be realised, nation states and citizen-consumers have to re-invent their roles in environmental politics now that the traditional roles are partly immobilised by the extreme mobility of globalisation.

REFERENCES

Anheier H., M. Glasius, and M. Kaldor (eds) (2001), *Global Civil Society 2001*, Oxford: Oxford University Press.

Bauman, Z. (1983), 'Industrialism, consumerism and power', *Theory, Culture and Society*, **1**(3), 32-43.

B&A Groep (2000), *Burger en Milieu; Verslag van een Verkenning naar Potentie en Meerwaarde van 'Burger en Milieu' (Citizen and Environment; Report of an Exploration of the Potential and Added Value of 'Citizen and Environment')*, Hoofdrapport (Main report), The Hague: B&A Groep.

Beck, U. (1986), *Risikogesellschaft; Auf dem Weg in eine andere Moderne (Risk Society; Towards another Modernity)*, Frankfurt am Main: Suhrkamp.

Beck, U., A. Giddens, and S. Lash (1994), *Reflexive Modernisation; Politics, Tradition and Aesthetics in the Modern Social Order*, Cambridge: Polity Press.

Beckers, T., P. Ester, and G. Spaargaren (1999), *Verkenningen van Duurzame Consumptie (Exploration of Sustainable Consumption),* Publicatiereeks Milieustrategie, The Hague: Ministry of the Environment.

Beckers, T., G. Spaargaren, and B. Bargeman (2000), *Van Gedragspraktijk naar Beleidspraktijk; Een Analytisch Instrument voor een Consument-Georienteerd Milieubeleid (From Behavioural Practice to Policy Practice; An Analytical Instrument of a Consumer-Oriented Environmental Policy)*, The Hague: Ministry of the Environment.

Blok, K. and K. Vringer (1995), *Energie-Intensiteit van Levensstijlen (Energy Intensity of Life Styles),* Utrecht: Utrecht University.

Castells, M. (1996), *The Rise of the Network Society*, Cambridge: Blackwell.

CEA (1999), *Minder Energieverbruik door een Andere Leefstijl? Eindrapportage Project Perspectief (Lower Energy Consumption through a Different Lifestyle? Final Report of the Project Perspective)*, The Hague: Ministry of the Environment.

Chappells, H., M. Klintman, A. Linden, E. Shove, G. Spaargaren, and B. Van Vliet (2000), *Domestic Consumption, Utility Services and the Environment; Final Domus Report*, Wageningen: Wageningen University.

Cohen, M. (forthcoming), 'The politics of sustainable consumption', *International Journal of Contemporary Sociology.*

CREM (2000), *Domeinverkenning Recreëren; Milieuanalyse Recreatie en Toerisme in Nederland (Domain Exploration Recreation; An Environmental Profile of Recreation and Tourism in the Netherlands)*, Amsterdam: CREM.

During, A. (1992), *How Much is Enough*?, New York: Norton & Company.

Fine, B. and E. Leopold (1993), *The World of Consumption*, London: Routledge.

Giddens, A. (1990), *The Consequences of Modernity*, Cambridge: Polity Press.

Giddens, A. (1991), *Modernity and Self-Identity*, Cambridge: Polity Press.

Giddens, A. (1998), *The Third Way; The Renewal of Social Democracy*, Cambridge: Polity Press.

Giddens, A. (2000), *The Third Way and its Critics*, Cambridge: Polity Press.

Glasius, M., M. Kaldor, and H. Anheier (eds) (2002), *Global Civil Society 2002*, Oxford: Oxford University Press.

Held, D. (1995), *Democracy and the Global Order,* Cambridge: Polity Press.

Held, D., A. McGrew, D. Goldblatt, and J. Perraton (1999), *Global Transformations; Politics, Economics and Culture*, Stanford: Stanford University Press.

Holemans D. (1999), *Ecologie en Burgerschap (Ecology and Citizenship),* Antwerpen: Stichting Leefmilieu.

Jänicke, M. (1991), *The Political System's Capacity for Environmental Policy*, Berlin: Freie Universität Berlin.

Knot, M. (2000), *Sustainable Clothing Use and Care; Sushouse Project Final Report*, Delft: Technical University Delft.

Marcuse, H. (1964), *One Dimensional Man*, Boston: Beacon Press.

Marshall, T. (1973), *Class, Citizenship and Social Development*, Westport: Greenwood Press.

Martens, S. and G. Spaargaren (2002), *Het Gedragspraktijken Model Geïllustreerd aan de Casus 'Duurzaam Wonen' (The Social Practices Model Illustrated for the Case of 'Sustainable Dwelling')*, Publicatiereeks Milieustrategie 2002/2, The Hague: Ministry of the Environment.

Meadows, D., D. Meadows, J. Randers and W. Behrens (1972), *The Limits to Growth*, London: Earth Island Limited.

Mol, A. and D. Sonnenfeld (eds) (2000), *Ecological Modernisation around the World: Perspectives and Critical Debates*, London: Frank Cass.

Mol, A. and G. Spaargaren (2003), *Towards a Sociology of Environmental Flows: A New Agenda for 21st Century Environmental Sociology*, paper for the International Conference on 'Governing Environmental Flows', June 13-14, Wageningen.

Neumayer, E. (2001), *Greening Trade and Investment: Environmental Protection without Protectionism,* London: Earthscan.

Noorman, K. and T. Schoot Uiterkamp (eds) (1998), *Green Households? Domestic Consumers, Environment and Sustainability,* London: Earthscan.

OECD (2000), *Work Programme Environmental Directorate 1999-2000*, Paris: Organisation for Economic Co-operation and Development.

Oosterveer P. (2002), 'Reinventing risks politics: Reflexive modernity and the European BSE crisis', *Journal of Environmental Policy and Planning*, **4**(3), 215-229.

Princen, T., M. Maniates, and K. Conca (eds) (2002), *Confronting Consumption*, Cambridge: MIT Press.

ResCon (2000), *Het Perspectief Project Anderhalf Jaar Later (The Perspective Project a Year-and-a-Half Later)*, Utrecht: Novem.

Rifkin J. (2000), *The Age of Access*, London: Penguin Books.

Schuttelaar & Partners (2000), *Domeinverkenning Voeden; Ingrediënten voor een Gezond Milieu (Domain Exploration Food; Ingredients of a Healthy Environment)*, The Hague: S&P.

Shove, E. (2003), *Comfort, Cleanliness and Convenience; The Social Organization of Normality*, Oxford: Berg.

Spaargaren, G. (1997), *The Ecological Modernisation of Production and Consumption; Essays in Environmental Sociology*, dissertation, Wageningen: Wageningen University.

Spaargaren, G. (2000), 'Ecological modernization and the changing discourse on environment and modernity', in G. Spaargaren, A. Mol, and F. Buttel (eds), *Environment and Global Modernity*, London: Sage.

Spaargaren, G. (2001), *Milieuverandering en het Alledaagse Leven (Environmental Change and Everyday Life)*, inaugural address, Tilburg: Tilburg University.

Spaargaren, G. (2003a), 'Duurzaam consumeren of ecologisch burgerschap' ('Sustainable consumption or ecological citizenship'), in H. Dagevos and L. Sterrenberg (eds), *Burgers en Consumenten: Tussen Tweedeling en Twee-Eenheid (Citizens and Consumers: Between Split and Unity),* Wageningen: Wageningen Academic Publishers.

Spaargaren, G. (2003b), 'Sustainable consumption: A theoretical and environmental policy perspective', *Society and Natural Resources,* **16**, 1-15.

Spaargaren, G., T. Beckers, S. Martens, B. Bargeman and T. van Es (2002), *Gedragspraktijken in Transitie (Behavioural Practices in Transition),* Publicatiereeks Milieustrategie, The Hague: Ministry of the Environment.

TNO (1999), *Duurzame Consumptie: Verkenning Kleding (Sustainable Consumption: Exploration Clothing),* Delft: TNO.

United Nations (1972), *Report of the Conference on the Human Environment, Stockholm, 5-16 June,* New York: United Nations.

Urry, J. (2000), *Sociology beyond Societies*, London: Routledge.

Van den Burg, S., A. Mol, and G. Spaargaren (2003), 'Consumer-oriented monitoring and environmental reform', *Environment and Planning C: Government and Policy,* **21**, 371-388.

Van Steenbergen, B. (ed.) (1994), *The Condition of Citizenship*, London: Sage.

Van Vliet, B. (2002), *Greening the Grid; The Ecological Modernisation of Network-Bound Systems*, Wageningen: Wageningen University.

Vringer, K. and K. Blok (1993), *Energy Intensities of Dutch Houses*, Utrecht: NW&S/UU.

Vringer, K. et al. (2001), *Nederlandse Consumptie en Energiegebruik in 2030 (Dutch Consumption and Energy Use in 2030)*, Bilthoven: RIVM.

VROM (1993), *Nationaal Milieubeleidsplan 2: Milieu als Maatstaf (National Environmental Policy Plan 2: Environment as a Yardstick),* The Hague: SDU.

Wackernagel, M. and W. Rees (1996), *Our Ecological Footprint: Reducing Human Impact on the Earth*, Gabriola Island: New Society Publishers.

9. Environmental Federalism in the European Union and the United States

David Vogel, Michael Toffel, and Diahanna Post

SUMMARY

The United States (US) and the European Union (EU) are federal systems in which the responsibility for environmental policy-making is divided or shared between the central government and the (member) states. The attribution of decision-making power has important policy implications. This chapter compares the role of central and local authorities in the US and the EU in formulating environmental regulations in three areas: automotive emissions, packaging waste, and global climate change. Automotive emissions are relatively centralised in both political systems. In the cases of packaging waste and global climate change, regulatory policy-making is shared in the EU, but is primarily the responsibility of local governments in the US. Thus, in some important areas, regulatory policy-making is relatively centralised in the EU. The most important role local governments play in the regulatory process is to help diffuse stringent local standards through centralised regulations, a dynamic which has become more common in the EU than in the US.

INTRODUCTION

In the EU and the US, responsibility for the making of environmental policy is divided between EU and federal institutions, on the one hand, and local institutions, on the other. The former is comprised of the EU and the US federal government, while the latter consist of state and local governments in the US, and member states and subnational authorities in the EU.[1] Historically, environmental rules and regulations were primarily made at the state or local level on both sides of the Atlantic. However, the emergence of the contemporary environmental movement during the late 1960s and early 1970s led to greater centralisation of environmental policy-making in both the US and Europe.

In the US, this change occurred relatively rapidly. By the mid 1970s, federal standards had been established for virtually all forms of air and water pollution. By the end of the decade, federal regulations governed the protection of endangered species, drinking water quality, pesticide approval, the disposal of hazardous wastes, surface mining, and forest management, among other policy areas. The federalisation of US environmental policy was strongly supported by pressure from environmental activists, who believed that federal regulation was more likely to be effective than regulation at the state level.

In Europe, this change occurred more gradually, largely because the Treaty of Rome contained no provision providing for environmental regulation by the European Community (EC). Nonetheless, more than 70 environmental directives were adopted between 1973 and 1983. Following the enactment of the Single European Act in 1987, which provided a clear legal basis for EC environmental policy and eased the procedures for the approval of Community environmental directives, EC environmental policy-making accelerated. Originally primarily motivated by the need to prevent divergent national standards from undermining the single market, it became an increasingly important focus of EC/EU policy in its own right. Each successive treaty has strengthened the EU's commitment to and responsibility for improving environmental quality and promoting sustainable development throughout Europe. Thus, notwithstanding their different constitutional systems, in both the EU and the US, the locus of environmental policy-making has become increasingly centralised over the last three decades.

Nevertheless, state governments continue to play a critical role in environmental regulation on both sides of the Atlantic. Most importantly, states remain an important locus of policy innovation and agenda setting. In many cases, new areas of environmental policy are first addressed at the state level and subsequently adopted by the central authority. Many state regulations remain more stringent or comprehensive than those of the central authority; in some policy areas, states retain primary responsibility. In other cases, responsibility for environmental policy-making is shared by both levels of government. Not surprisingly, in both federal systems, there are

[1] For ease of presentation, we refer at times to both of the former as central authorities and both of the latter as states.

ongoing disputes about the relative competence of central and state authorities to regulate various dimensions of environmental policy.

We explored the dynamics of federal environmental policy-making in both the US and the EU. At what level of government are new standards initiated? Under what circumstances are state regulations diffused to other states and/or adopted by the central authority? Under what circumstances can or do states maintain regulations that are more stringent than those of other states? We conducted a comparative study of the development of US and EU regulatory policies in three areas: automobile emissions, packaging waste, and global climate change. Each policy area reflects a different stage in the evolution of environmental policy. These cases also demonstrate the differences and the similarities in the patterns of environmental policy-making in the US and the EU.

Automobile emissions typify the first generation of environmental regulation. A major source of air pollution, particularly in urban areas, automobiles were among the first targets of environmental regulation during the 1960s and 1970s and they remain an important component of environmental policy in every industrialized country. Packaging typifies the next generation of environmental regulation. Its emergence on the policy agenda during the 1980s reflected the increased public concern about the scarcity of landfills and the need to conserve natural resources. Unlike automobile regulation, which primarily affects only two industries, albeit critical ones (automotive manufacturers and the refiners of gasoline), packaging waste regulations affect virtually all manufacturers of consumer goods. The increased priority of reducing packaging waste and promoting re-use and recycling symbolises a shift in the focus of environmental regulation from reducing pollution to promoting eco-efficiency. Global climate change represents a relatively new dimension of environmental policy. It first surfaced during the mid-1980s, but it has become much more salient over the last decade. This policy area exemplifies the increasingly important international dimension of environmental regulation: global climate change both affects and is affected by the regulatory policies of virtually all countries. It also illustrates the growing economic scope of environmental regulation: few economic activities are likely to be unaffected by policies aimed at reducing the emission of carbon dioxide and other greenhouse gases.

These three policy areas provide a useful window on the changing dynamics of the relationship between state and central regulation in the US and the EU. Since the mid-1980s, automobile emissions standards have been more centralised in the EU than in the US. The US permits states to adopt more stringent standards, while the EU does not. However, both the EU and the US have progressively strengthened their regulations governing automotive emissions and fuel composition, though most US federal standards remain more stringent than EU ones. For its part, California, which is permitted its own emissions standards, has become a world leader in the effort to encourage the development and marketing of low- and zero-emission vehicles.

The dynamics of the regulation of packaging waste differs considerably. In the US, the federal government plays little or no role in setting standards for packaging waste: packaging, recycling, and waste disposal are all the responsibility of state or local governments. However, the lack of federal standards has neither prevented nor discouraged many state governments from adopting their own regulations. There has been considerable innovation at the state level: a number of local governments have developed ambitious programmes to reduce packaging waste and promote recycling. There has been little pressure for federal standards and the federal government has not attempted to limit state regulations with one important exception: federal courts have repeatedly found state restrictions on 'imports' of garbage to violate the interstate commerce clause of the US constitution.[2]

In the EU, the situation is more complex. Member states began to regulate packaging waste during the 1980s, while the EU became formally involved in this policy area in 1994. However, in contrast to automotive emissions, the responsibility for packaging regulation remains shared between central and state authorities. There is considerable diversity among state regulations, and member states continue to play an important role in policy innovation, often adopting regulations that are more stringent than those of the EU. State packaging waste regulations have been an ongoing source of conflict between central and local authorities, with the European Commission periodically challenging particular state regulations on the grounds of their incompatibility with the single market. In addition, the central government has imposed maximum as well as minimum standards for waste recovery, though this is likely to change soon. On balance, EU packaging standards are more stringent and comprehensive than those prevailing in the US. Europe's 'greener' member states have made more ambitious efforts to reduce packaging waste than have their American state counterparts, while the EU's Packaging Waste Directive provides a centralised floor on state standards which does not exist in the US. Nevertheless, there have been a number of important US state standards.

In the case of climate policy, important initiatives and commitments to reduce emissions of greenhouse gases have been undertaken in the EU at both the central and state levels with one often complementing and

---

[2] Berland, 1992.

reinforcing the other. In the US, by contrast, there are no federal climate change regulations. As in the case of packaging waste policies, there have been a number of state initiatives. But in contrast to the regulation of packaging waste, the lack of central regulation of climate policy has become politically salient; the nation's 'greener' states have strongly pressured for centralised standards, though their efforts to date have been unsuccessful. In addition, there has been conflict over the legal authority of states to establish policies in this area. The gap between US and EU regulatory policies regarding climate change is more substantial than the gaps in the other two policy areas. The EU and each member state have formally ratified the Kyoto Protocol, while the US has not. Since American states cannot enter into international environmental agreements, this means that no US regulatory authority is under any international obligation to regulate carbon dioxide emissions. While all EU member states have adopted climate change policies, many states in the US have not. Moreover, US state regulations tend to be weaker than those adopted or being adopted by the member states of the EU. The EU has established a regulatory regime based on emissions trading and shared targets to facilitate member states' carbon dioxide reduction programmes, while in the critical area of vehicle emissions, the US central government has become an obstacle to more stringent state regulations.

AUTOMOBILE EMISSIONS

United States

The regulation of automobile emissions in the US began in 1960 when the state of California enacted the Motor Vehicle Pollution Control Act. This statute established a state board to develop criteria to approve, test, and certify emission control devices.[3] Within two years, the board had certified seven devices that were bolt-on pollution controls, such as air pumps that improve combustion efficiency[4] and required their installation by 1965.[5] After opposing emissions standards in the mid-1960s, 'the automobile industry began to advocate federal emissions standards for automobiles [after] California had adopted state standards, and a number of other states were considering similar legislation.'[6] In 1965, Congress enacted the federal Motor Vehicle Air Pollution Control Act, which authorised the establishment

---

3 Percival et al., 1992.
4 California EPA, 2001.
5 Percival et al., 1992.
6 Revesz, 2001: 573.

of auto emissions standards. The first federal standards were imposed for 1968 model year vehicles for carbon monoxide and hydrocarbons.

Two years later, in 1967, Congress responded to the automobile industry's concerns about the difficulty of complying with different state standards by declaring that federal emission controls would preempt all state emission regulations. However, an exception was made for California, provided that the state afforded adequate lead time to permit development of the necessary technology, given the cost of compliance within that time.[7] The exemption was granted 'in recognition of the acute automobile pollution problems in California and the political power of the California delegation in the House of Representatives'.[8] One legal scholar noted, 'The legislative history of the 1967 waiver provision suggests two distinct rationales for its enactment: (1) providing California with the authority to address the pressing problem of smog within the state; and (2) the broader intention of enabling California to use its developing expertise in vehicle pollution to develop innovative regulatory programs.'[9]

In 1970, President Nixon asked Congress to pass more stringent standards based on the lowest pollution levels attainable using developing technology.[10] Congress responded by enacting the technology-forcing Clean Air Act Amendments of 1970, which required automakers to reduce their emissions of carbon monoxide and hydrocarbons by 90 per cent within five years and their emissions of nitrogen oxides by 90 per cent within six years.[11] These drastic reductions were intended to close the large gap between ambient urban air pollution concentrations and the federal health-based Nationally Uniform Ambient Air Quality Standards (NAAQS) established pursuant to the US Clean Air Act.[12] Once again, California was permitted to retain and/or enact more stringent standards, though these were specified in federal law.[13]

The 1977 amendments to the Clean Air Act established more stringent emissions standards for both automobiles and trucks and once again permitted California to adopt more stringent standards. In 1990, the Clean Air Act was again amended: 'the California Air Resources Board old tailpipe emissions standards for new cars and light duty trucks sold in that state were adopted by Congress . . . as the standard to be met by all new vehicles.'[14] In

7 US EPA, 1999.
8 Rehbinder and Stewart, 1985: 114.
9 Chanin, 2003: 699.
10 Percival et al., 1992.
11 Rehbinder and Stewart, 1985.
12 Congress based its 90 per cent reduction on 'the simple notion that since air pollution levels in major cities were approximately five times the expected levels of the NAAQSs, emissions would need to be reduced by at least 80 per cent, with an additional 10 per cent necessary to provide for growing vehicle use' (Percival et al., 1992: 834).
13 California EPA, 2001.
14 Bryner, 1993: 150.

addition to again waiving federal preemption for California, the 1990 legislation for the first time authorised any state that was not meeting NAAQS for automotive pollutants to adopt California's standards.[15] To date, Massachusetts, New York, Vermont, and Maine have chosen to do so. Thus, since 1990, the US has had a nationwide two-tiered system of automotive emission regulation: one based on federal standards and the other on California's. This regulatory policy 'reflects a compromise between two interests: the desire to protect the economies of scale in automobile production and the desire to accelerate the process for attainment of the NAAQS'.[16] Thus, while automotive emission standards are primarily shaped by federal legislation, the federal government provides states with the opportunity to choose between two sets of standards.

California continues to play a pioneering role in shaping automotive emissions policy. In 1990, California adopted a programme to encourage Low-Emission Vehicles (LEV). This included a Zero-Emission Vehicle (ZEV) programme meant to jump-start the market for these vehicles. The ZEV programme required that such vehicles comprise at least 2 per cent of new car sales by 1998, 5 per cent by 2001, and 10 per cent by 2003. When this requirement was approved, the only feasible technology that met ZEV standards were electric vehicles, whose emissions were over 90 per cent lower than those of the cleanest gasoline vehicles, even when including the emissions from the power plants generating the electricity required to recharge them.[17] Massachusetts and New York subsequently adopted the California LEV plan. However, in 1992, New York's decision was challenged in the courts by the automobile manufacturers on the grounds that it was sufficiently different from California's to constitute a third automotive emission requirement, which the Clean Air Act explicitly prohibits. Shortly afterwards, the manufacturers filed another suit against both states arguing that, since their standards were not identical with those of California, they were preempted by the Clean Air Act. As a result, both states were forced to modify their standards.

In 1998, California's Air Resources Board (California ARB) identified diesel particulate matter as a toxic air contaminant[18] and subsequently launched a Diesel Risk Reduction Plan in 2000 to reduce diesel particulate emissions by 75 per cent by 2010. It also established new requirements for low-sulphur diesel fuel and particulate standards for new diesel engines and vehicles, and required new filters to be put on existing engines.[19] More

---

15 Chanin, 2003; Revesz, 2001.
16 Revesz, 2001: 586.
17 California Air Resources Board, 2001.
18 California EPA, 2001.
19 California Air Resources Board, 2001.

recently, California's automotive emissions standards have become a source of conflict with the federal government, which has not enacted new automotive emissions standards since 1990. Two novel California regulations, which the state claims are designed to reduce automobile emissions, have been challenged by both the automotive industry and the federal government on the grounds that they indirectly regulate fuel efficiency, an area of regulation which Congress has assigned exclusively to the Federal government.[20]

The first case involves a modification California made to its ZEV programme in 2001 that allowed automakers to earn ZEV credits for manufacturing compressed natural gas, gasoline-electric hybrid, and methanol fuel cell vehicles.[21, 22] General Motors and DaimlerChrysler sued California's ARB over a provision that allowed manufacturers to earn ZEV credits by using technology such as that included in gasoline-electric hybrid vehicles, which were already being sold by their rivals Honda and Toyota. Because hybrids still use gasoline, General Motors and DaimlerChrysler argued that California's efforts were effectively regulating fuel economy.[23] The US Justice Department supported the auto manufacturers' claim on the grounds that the Energy Policy and Conservation Act provides that when a federal fuel-economy standard is in effect, a state or a political subdivision of a state may not adopt or enforce a regulation related to fuel-economy standards.[24] California responded by claiming that it was acting pursuant to its exemption under the US Clean Air Act to regulate auto emissions. In June 2002, a Federal District Court issued a preliminary injunction prohibiting the Air Resources Board from enforcing its regulation.[25] In response, the ARB modified the ZEV programme to provide two alternative routes for automakers to meet ZEV targets.[26] At the same time, California imposed new

---

[20] In the Energy Policy and Conservation Act of 1975, Congress established exclusive Federal authority to regulate automotive fuel Corporate Average Fuel Economy (CAFE) standards.

[21] At the same time, California extended ZEV market share requirements to range from 10 per cent in 2003 up to 16 per cent in 2018 (California Air Resources Board, 2001).

[22] The second dispute concerns climate change and is discussed below.

[23] Parker, 2003.

[24] Yost, 2002.

[25] California Air Resources Board, 2003.

[26] According to the California Air Resources Board (2003), 'Auto manufacturers can meet their ZEV obligations by meeting standards that are similar to the ZEV rule as it existed in 2001. This means using a formula allowing a vehicle mix of 2 per cent pure ZEVs, 2 per cent AT-PZEVs (vehicles earning advanced technology partial ZEV credits) and 6 per cent PZEVs (extremely clean conventional vehicles). Or manufacturers may choose a new alternative ZEV compliance strategy, meeting part of their ZEV requirement by producing their sales-weighted market share of approximately 250 fuel cell vehicles by 2008. The remainder of their ZEV requirements could be achieved by producing 4 per cent AT-PZEVs and 6 per cent PZEVs. The required number of fuel cell vehicles will increase to 2,500 from 2009-11, 25,000 from 2012-14 and 50,000 from 2015 through 2017. Automakers can substitute battery electric vehicles for up to 50 per cent of their fuel cell vehicle requirements'.

regulations which required that the auto industry sell increasing numbers of fuel-cell vehicles in the state over the next decade.[27] However, in the summer of 2003, both automobile firms dropped their suits against California after its regulatory authorities agreed to expand their credit system for hybrids to encompass a broader range of vehicles.[28] This will increase the availability of advanced-technology vehicles in states such as New York and Massachusetts, which have adopted regulations similar to California's. However, as there is no indication that federal standards will be similarly strengthened, tensions between federal and state requirements are likely to persist.

European Union

As in the US, in Europe, the regulations of state governments have been an important driver for centralised automotive emissions standards, with Germany typically playing the role in Europe that California has played in the US. The EU has progressively strengthened its automotive emissions standards, both to improve environmental quality and to maintain a single market for vehicles. However, European standards were strengthened at a much slower rate than were those in the US, and they were harmonised much later. Thus, in 1989, the EU imposed standards to be implemented in 1992 that were based on US standards implementing legislation enacted in 1970 and 1977, while the EU did not establish uniform automotive emissions requirements until 1987, although some fuel content standards were harmonised earlier. However, unlike in the US, which has continued to maintain a two-tiered system – and indeed extended it in 1977 by giving states the option of adopting either federal or California standards, in Europe, centralised standards for automobile emissions have existed since 1987. During the 1970s and 1980s, there was considerably more tension between central and state regulations in the EU than in the US. Recently, the opposite has been the case.

During the 1960s, France and Germany imposed limits on emissions of carbon monoxide and hydrocarbons for a wide range of vehicles, thus forcing the EC to issue its first automotive emissions standards in 1970 in order to prevent these limits from serving as obstacles to internal trade. Shortly afterwards, there was substantial public pressure to reduce levels of airborne lead, a significant portion of which came from motor vehicles. The most severe restrictions were imposed by Germany, which in 1972 announced a two-stage reduction: the maximum lead content for gasoline was defined at 0.4 grams per litre in 1972, and 0.15 in 1976. The United Kingdom (UK)

[27] Hakim, 2003a.
[28] Hakim, 2003b.

enacted less severe restrictions. No restrictions were imposed by any other member state. The resulting disparity in national rules and regulations represented an obstacle to the free movement of both fuel and motor vehicles within the EC. For not only did these divergent national product regulations limit intra-EC trade in gasoline, but since different car engines were designed to run on fuels containing different amounts of lead, they created a barrier to intra-Community trade in motor vehicles themselves. Accordingly, the EC had to move towards harmonised standards. After prolonged negotiations, the Community approved a Directive establishing both minimum and maximum lead standards (the latter being identical with Germany's standards). The EU subsequently enacted legislation urging all member states to reach the most stringent level as quickly as possible and to make at least some unleaded gasoline available for sale.

Unlike the lead standard, in the establishment of which the German regulations played an important role, the EC's standards for sulphur in fuel did not reflect the policy preferences of any member state. The sulphur standard adopted in 1975 required all countries, including France, Germany, and the UK, to reduce their sulphur emissions.[29] France, for instance, had already adopted standards for sulphur in diesel fuel in 1966, but the more stringent levels in the European-wide standard forced the French standards lower as well. Germany's standard was adopted at the same time and was similar to that of the EC.

In contrast to the fuel standards, the auto emissions standards adopted during the 1970s were not mandatory. In fact, until 1987, member states were permitted to have standards less stringent than the European-wide standards, although they could not refuse to register or sell a vehicle on their territory if it met EC maximum standards. In effect, these early standards were maximum or ceiling requirements. Indeed, they were not developed by the EC, but were based heavily on emissions standards of the United Nations Economic Council for Europe.

In 1985, the German minister responsible for environmental affairs announced, on his own initiative, that as of 1989 all cars marketed in Germany would be required to meet US automotive emissions standards, commonly referred to as 'US '83.' The adoption of these standards required the installation of catalytic converters, which could only use unleaded gasoline. This created two problems within Europe. Most importantly, it meant that automobiles produced in France and Italy, whose producers lacked the technology to incorporate the converters into their smaller vehicles, would be denied access to the German market. In addition, it meant that German tourists who drove their cars to southern Europe would be

[29] Bennett, 1991.

stranded, owing to the unavailability of unleaded gasoline in Greece and Italy. Germany's insistence on requiring stringent standards for vehicles registered in its country forced the EU to adopt uniform automobile emissions standards. This in turn led to a bitter debate over the content of these standards, pitting the EU's greener member states (Germany, Denmark, and the Netherlands) against the EU's (other) major automobile producers (the UK, France, and Italy), who favoured more flexible standards. The resulting Luxembourg Compromise of 1987 established different emissions standards for different sizes of vehicles with different timetables for compliance. It thus represented the first uniform set of automotive emissions standards within Europe. These standards have been subsequently strengthened several times, and some of the EU's most recent standards are more stringent than those of the US, which has not tightened its standards for more than a decade. In 1999, the EU adopted a labelling plan requiring that all new cars sold within the EU show their fuel efficiency and carbon dioxide emissions in a sticker on the product.

During the 1990s, the politics of automobile emissions standards became much less affected by member state differences or tensions between central and state standards. The most important initiative of this period, the Auto-Oil Programme, first adopted in 1996, was aimed at bringing together the Commission and the auto and oil industries to work on comprehensive ways to reduce pollution. After a series of negotiations, the programme ultimately tightened vehicle emission limits and fuel quality standards for sulphur and diesel, and introduced a complete phase-out of leaded gasoline.[30] In 2003, the EU approved a Directive requiring that all road vehicle fuels be sulphur-free by 2009. With the finalisation of Auto-Oil I and II, as the programmes are known, the shift from state to centralised automotive emission requirements appears to be complete. The debates and negotiations over proposals to regulate pollution from vehicles now take place between the auto and oil producers on the one hand, and the Commission and European Parliament (EP) on the other hand.

PACKAGING WASTE

United States

The regulation of packaging wastes is highly decentralised in the US. The role of the federal government remains modest and virtually all policy initiatives have taken place at the local level. While the 1976 Resource

[30] McCormick, 2001.

Conservation and Recovery Act (RCRA) established stringent requirements for the management of hazardous wastes, the RCRA also declared that the regulation of landfills accepting municipal solid waste (MSW) was to remain primarily the domain of state and local governments.[31] As a result, there is considerable disparity in the handling of packaging wastes throughout the US.

On balance, US standards tend to be considerably laxer than those in the EU. While many state legislatures have established recycling goals, few have prescribed mandatory targets.[32] The US generates more MSW per capita than any other industrialised country, and 50 per cent more than most European countries.[33] From 1995 to 1998, the percentage of the MSW generated that has been recovered for recycling remained steady at 44 in the US, while it rose from 55 to 69 in Germany, owing in part to Germany's Packaging Ordinance.[34]

State and local governments have implemented several policy mechanisms to reduce MSW, including packaging waste. Deposit-refund schemes, minimum recycling content requirements, community recycling programmes, and disposal bans are among the most common policy mechanisms designed to divert materials to recycling from waste streams destined for landfills or incinerators. Eleven states have developed deposit-refund schemes to encourage the recycling of beverage containers.[35] When Oregon passed the first bottle bill requiring refundable deposits on all beer and soft-drink containers in 1971, its objective was to control litter rather than to spur recycling. When the city of Columbia, Missouri, passed a bottle bill in 1977, it became the first local container-deposit ordinance in the US and remained the only local initiative until it was repealed in 2002.[36] In general, deposit-refund laws require consumers of soft drinks and beer packaged in glass, metal, and plastic containers to pay a deposit that is refundable when the container is returned.[37] These schemes typically do not require, however, that these containers be recycled or reused.[38] California recently expanded its

[31] US EPA, 2003a, 2003b, 2003c.

[32] American Forest & Paper Association, 2003.

[33] The latest OECD figures report that Americans generate 760 kg per capita, the French 510, the British 560, and Germans 540 (OECD, 2004).

[34] OECD, 2002.

[35] The eleven states with deposit-refund schemes on soft-drink containers are California, Connecticut, Delaware, Hawaii, Iowa, Maine, Massachusetts, Michigan, New York, Oregon, and Vermont. Hawaii's law takes effect in 2005 (Container Recycling Institute, 2003).

[36] Container Recycling Institute, 2003.

[37] Some deposit refunds are being expanded to include office products, while Maine and Rhode Island have created deposit-refund schemes for lead-acid/automobile batteries (US EPA, 1999).

[38] McCarthy, 1993.

programme to include non-carbonated beverages, which added roughly 2 billion containers, nearly 40 per cent of which are plastic.[39]

To reduce the burden on landfills and incinerators, whose construction and expansion are increasingly politically infeasible owing to community objections, many states and local governments have developed recycling programmes that enable or require the recycling of various materials. Such programmes remain exclusively the purview of state and local government because national laws do not allow EPA to establish federal regulations on recycling.[40] Virtually all New Yorkers, 80 per cent of the Massachusetts population, and 70 per cent of Californians have access to curbside recycling.[41] Recycling programmes typically include paper as well as metal and glass containers, while some programmes also include containers of particular plastic resins. In Oregon, container glass comprises nearly 4 per cent of that state's total solid waste stream, and its deposit-refund and collection schemes resulted in 55 per cent of this glass being collected and recycled.[42] Sixty per cent of Oregon's recycled container glass comes from its deposit-refund scheme, 25 per cent is collected from residential curbside programmes, and the remainder comes from commercial solid-waste hauler programmes, disposal sites, and other private recycling activities.

A few states have sought to facilitate recycling by banning packaging that is particularly difficult to recycle, such as aseptic drink boxes, which are made of paper, foil, and plastic layers that are difficult to separate. Connecticut banned plastic cans in anticipation of obstacles this product would pose to materials recovery. In 1989, Maine banned aseptic drink boxes because of a concern about their ability to be recycled, though this restriction was subsequently repealed. The Wisconsin Legislature considered imposing a ban on the sale of aseptic drink boxes and bimetal cans (drink cans with aluminium sides and bottom and a steel top). Instead, the state enacted an advisory process permitting it to review a new packaging design if the packaging proved difficult to recycle. In addition, a few states, including Wisconsin and South Dakota, have banned the disposal of some recyclable materials to bolster their recycling rates.[43]

Some states require certain types of packaging to contain some minimum amount of recycled material. Oregon's 1991 Recycling Act required that by 1995, 25 per cent of the rigid plastic packaging containers (containing eight ounces to five gallons) sold in that state must contain at least 25 per cent recycled content, be made of a plastic material that is recycled in Oregon at a

---

[39] US EPA, 2003a, 2003b, 2003c.
[40] Cotsworth, 2002.
[41] Dietly, 2001.
[42] Oregon Department of Environmental Quality, 2003.
[43] Thorman et al., 1996.

rate of at least 25 per cent, or be a reusable container made to be reused at least five times.[44] This law also requires glass containers to contain 35 per cent recycled content by 1995 and 50 per cent by 2000.[45] California requires manufacturers of newsprint, plastic bags, and rigid plastic containers to include minimum levels of recycled content in their products or to achieve minimum recycling rates. Manufacturers of plastic trash bags are required to include minimum percentages of recycled plastic post-consumer material in trash bags they sell in California. California's 1991 Rigid Plastic Packaging Container (RPPC) Act sought to reduce the amount of plastic being landfilled by requiring that containers offered for sale in the state meet criteria akin to those laid down in the Oregon law. These criteria 'were designed to encourage reuse and recycling of RPPCs, the use of more postconsumer resin in RPPCs and a reduction in the amount of virgin resin employed RPPCs'.[46] Wisconsin's Act on Recycling & Management of Solid Waste requires that products sold in the state must use a package made from at least 10 per cent recycled or remanufactured material by weight.[47] Industrial scrap, as well as pre- and post-consumer materials, count towards the 10 per cent requirement. Exemptions are provided for packaging for food, beverages, drugs, cosmetics, and medical devices that lack FDA approval. However, according to the President of the Environmental Packaging International, Wisconsin has done little enforcement of its 10 per cent recycled content law.[48]

Governments at the federal, state, county, and local levels have also promulgated policies prescribing government procurement of environmentally preferable products.[49] In 1976, Congress included in RCRA requirements that federal agencies, as well as state and local agencies that use appropriated federal funds, that spend over a threshold amount on particular items to purchase products with recycled content when their cost, availability, and quality are comparable to those of virgin products, though the RCRA does not authorise any federal agency to enforce this provision.[50] States requiring government agencies to purchase environmentally preferable products include California, Georgia, Oregon, and Texas. California's State Assistance for Recycling Markets Act of 1989 and Assembly Bill 11 of 1993 required government agencies to give purchasing preference to recycled

[44] All rigid plastic container manufacturers have been in compliance with the law since it entered into force a decade ago, because the aggregate recycling rate for rigid plastic containers has remained between 27-30 per cent since the law took effect (Oregon Department of Environmental Quality, 2003).
[45] Thorman et al., 1996.
[46] California Integrated Waste Management Board, 2003.
[47] Plastic Shipping Container Institute, 2003.
[48] Bell, 1998.
[49] California Integrated Waste Management Board, 2003; Center for Responsive Law, 2003.
[50] US EPA, 2003a, 2003b, 2003c.

products and mandated that increasing proportions of procurement budgets be spent on products with minimum levels of recycled content. Accordingly, the California Integrated Waste Management Board (CIWMB) developed the State Agency Buy Recycled Campaign, requiring that every State department, board, commission, office, agency-level office, and cabinet-level office purchase products that contain recycled materials whenever they are otherwise similar to virgin products.

Procurement represents one of the few areas in which there have been federal initiatives. A series of Presidential Executive Orders issued throughout the 1990s sought to stimulate markets for environmentally preferable products and to reduce the burden on landfills.[51] In 1991, President George Bush issued an Executive Order to increase the level of recycling and procurement of recycled-content products. In 1993, President Bill Clinton issued an Executive Order that required federal agencies to purchase paper products with at least 20 per cent post-consumer fibre and directed the US EPA to list environmentally preferable products, such as those with less cumbersome packaging. Clinton raised this recycled-content threshold to 30 per cent in a subsequent Executive Order in 1998.[52]

At the national level, several Congressional attempts to pass a National Bottle Bill between 1989 and 1994 were defeated, owing in part to successful industry lobbying. According to the non-profit Container Recycling Institute, a key reason why bottle bills have not spread to more states or become national law is 'the tremendous influence the well-funded, politically powerful beverage industry lobby wields'.[53] Thus, packaging waste policies remain primarily the responsibility of state and local governments.

European Union

The EU's efforts to control packaging waste contrast sharply with those of the US in two ways. First, with the enactment of the 1994 EU Directive on Packaging and Packaging Waste, central authorities have come to play a critical role in shaping politics to reduce packaging waste within Europe. Thus, in Europe, in marked contrast to the US, this area of environmental policy is shared between central and state governments. Second, unlike in the US, where federal authorities have generally been indifferent to state policies to promote the reduction of packaging waste, in Europe, such policies have frequently been challenged by Brussels (the Commission) on the grounds that they interfere with the single market. In addition, the EU's 1994 Packaging Directive established maximum as well as minimum recycling targets, while

[51] Lee, 1993.
[52] Barr, 1998.
[53] Container Recycling Institute, 2003.

maximums have never existed in the US. As a result, some member states have been forced by Brussels to limit the scope and severity of their regulations.

Historically, recycling policies were made exclusively by the member states. In 1981, Denmark enacted legislation requiring that manufacturers market all beer and soft drinks in reusable containers. Furthermore, all beverage retailers were required to take back all containers, regardless of where they had been purchased. To facilitate this recycling programme, only goods in containers that were approved in advance by the Danish environmental protection agency could be sold. Thus, a number of beverage containers produced in other member states could not be sold in Denmark. Foreign beverage producers complained to the European Commission that the Danish requirement constituted a 'qualitative restriction on trade', prohibited by the Treaty of Rome. The Commission agreed. When Denmark's modified regulation in 1984 failed to satisfy the Commission, the EC brought a complaint against Denmark to the European Court of Justice (ECJ). In its decision, the ECJ upheld most of the provisions of the Danish statute, noting that the Commission itself had no recycling programme. The Court held that since protecting the environment was 'one of the Community's central objectives', environmental protection constituted 'a mandatory requirement capable of limiting the application of Article 30 of the Treaty of Rome'.[54] This was the first time the Court had sanctioned an environmental regulation that clearly restricted trade.

The result of the ECJ's ruling was to give a green light to other national recycling initiatives. Irish authorities proceeded with a ban on non-refillable containers for beer and soft drinks, while a number of Southern member states promptly restricted the sale of beverages in plastic bottles in order to protect the environment, and, not coincidently, domestic glass producers. The Netherlands, Denmark, France, and Italy promptly introduced their own comprehensive recycling plans. The most far-reaching initiative to reduce recycling waste, however, was undertaken by Germany.

The 1991 German packaging law was a bold move towards a 'closed loop' economy in which products are reused instead of thrown away. It established very high mandatory targets, requiring that 90 per cent of all glass and metals, as well as 80 per cent of paper, board, and plastics be recycled. In addition, only 28 per cent of beer and soft drinks could be sold in disposable containers. The law also established 'take-back' requirements on manufacturers, making them responsible for the ultimate disposal of the packaging in which their products were sold and shipped. A quasi-public system was established to collect and recycle packaging, with the costs

[54] Vogel, 1995: 87.

shared by participating firms. In addition to making it more difficult for foreign producers to sell their products in Germany, the so-called Töpfer Law distorted the single market in another way. The plan's unexpected success in collecting packaging material strained the capacity of Germany's recycling system, thus forcing Germany to 'dump' its excess recycled materials throughout the rest of Europe. This had the effect of driving down prices for recycled materials in Europe, and led to the improper disposal of waste in landfills in other countries.[55] Yet, the ECJ's decision in the Danish Bottle Case, combined with its fear of being labelled 'anti-green', made it difficult for the Commission to file a legal challenge to the German regulation.

Accordingly, the promulgation of waste management policy now moved to the EU level. In 1994, following nearly three years of intense negotiations, a Directive on Packaging Waste was adopted by a qualified majority of member states with opposition from Germany, the Netherlands, Denmark, and Belgium. It required member states to recover at least half of their packaging waste and recycle at least one-quarter of it, within five years. Ireland, Greece, and Portugal were given slightly lower targets. More controversially, the Directive also established maximum standards: nations wishing to recycle more than 65 per cent of their packing waste could do so, but only if they had the facilities to use their recycled products. It was this provision which provoked opposition. The Packaging Waste Directive has played a critical role in strengthening packaging waste regulations and programmes throughout much of Europe, particularly in Great Britain and the South of Europe. As in the case of automobile emissions standards, it illustrates the role of the EU in diffusing the relatively stringent standards of some member states throughout Europe. Moreover, the decrease in some state standards as a result of the 1994 Directive was modest.[56]

Member states continue to innovate in this policy area and these innovations have on occasion sparked controversy within the EU. For example, in 1994, the European Commission began legal proceedings against Germany, claiming that a German requirement that 72 per cent of drink containers be refillable was interfering with efforts to integrate the internal market. Germany has proposed to do away with the requirement owing to pressure from the Commission, but it remains a pending legal issue. This packaging waste dispute tops the list of key single market disputes identified by the Commission in 2003, and the outcomes of numerous other cases hinge on its resolution.[57]

In 2001, Germany adopted a policy requiring deposits on non-refillable (one-way) glass and plastic bottles and metal cans in order to encourage the

---

[55] Comer, 1995.
[56] Haverland, 1999.
[57] Environment Daily, 2001a, 2003d.

use of refillable containers. This law, which went into effect in 2003, aroused considerable opposition from the German drinks industry, which held it responsible for a dramatic decline in sales of beer and soft drinks and the loss of thousands of jobs. In addition, the European Commission, acting in response to complaints from non-German beverage producers, questioned the legality of the German scheme. The Commission agreed that the refusal of major German retailers to sell one-way drink containers had disproportionately affected bottlers of imported drinks, a position which was also voiced by France, Italy, and Austria. However, after the German government promised to revise its plan in order to make it compliant with EU law, the Commission decided not to take legal action.

As occurred during the previous decade, the extent to which new packaging waste initiatives by member states threaten or are perceived to threaten the single market has put pressure on the EU to adopt harmonised standards. As the European Environmental Bureau noted in response to the Commission's decision to sue Germany over national rules protecting the market share of refillable drinks containers, 'national reuse systems will come under pressure if the Commission continues to legally attack them at the same time it fails to act at the European level'.[58]

In 2003, the Commission and the EP revised the 1994 Packaging Waste Directive by establishing new recycling targets. The EP attempted to extend the reach of the Directive to include, for example, a requirement that all packages contain an environmental indicator to show how 'green' they are, and to encourage packaging incineration and packaging reuse systems when they are preferable to recycling. But these negotiations dropped those proposals and the new Directive is focused on establishing new targets. While a key difference between the Council of Ministers and the EP concerns the date by which higher packaging recycling and recovery targets should be met, there is a broad consensus both that packaging recycling and recovery targets should be strengthened and that no limits should be placed on national waste recovery. Thus, the result of this Directive will further strengthen European standards.

CLIMATE CHANGE

United States

In the US, greenhouse gas (GHG) emissions remain largely unregulated by the federal government. In the 1990s, the Clinton administration participated

---

[58] *Environment Daily*, 2001b.

in the United Nations effort to establish a treaty governing GHG emissions. While the US signed the Kyoto Protocol, it was never submitted to the Senate for ratification. Soon after the Bush administration took office, it declared it would not support the Kyoto Protocol. Also refusing to propose any regulations for carbon dioxide emissions, it instead chose to encourage industry to adopt voluntary targets, through its Global Climate Change Initiative. Legislation to amend the Clean Air Act to encompass carbon dioxide emissions has been submitted in Congress, but has yet to be voted upon. The Congress has also persistently voted down proposals to strengthen fuel-economy standards.

The lack of federal regulation has created a policy vacuum which a number of states have filled. While 'some significant legislation to reduce greenhouse gases was enacted during the late 1990s, such as Oregon's pioneering 1997 law that established CO_2 standards for new electrical power plants . . . [state] efforts to contain involvement on climate change have been supplanted in more recent years with an unprecedented period of activity and innovation'. The US EPA has catalogued over 700 state policies to reduce GHG emissions.[59] Two reports describe various state-level initiatives that address climate change, either directly or indirectly.[60] 'New legislation and executive orders expressly intended to reduce greenhouse gases have been approved in approximately one-third of the states since January 2000, and many new legislative proposals are moving ahead in a large number of states'.[61]

New Jersey and California are undertaking measures that directly target climate change. In 1998, the Commissioner of New Jersey's Department of Environmental Protection (DEP) issued an Administrative Order that established a goal for the state to reduce GHG emissions to 3.5 per cent below the 1990 level by 2005, making New Jersey the first state to establish a GHG reduction target.[62] The DEP has received signed covenants from corporations, universities, and government agencies across the state pledging to reduce their GHG emissions, though nearly all are unenforceable. In an unusual move, the state's largest utility signed a covenant that includes a commitment to monetary penalties if it fails to attain its pledged reductions. Other states have employed air pollution control regulation and legislation to cap carbon dioxide emissions from large source emitters such as power plants. Massachusetts became the first state to impose a carbon dioxide emission cap on power plants when Governor Jane Swift established a multi-pollutant cap for six major facilities in 2001 that requires 'each plant to

---

59 US EPA, 2003c.

60 Bramley, 2002; Rabe, 2002.

61 Rabe, 2002: 7.

62 New Jersey Department of Environmental Protection, 1999.

achieve specified reduction levels for each of the pollutants, including a ten per cent reduction from 1997-1999 CO_2 levels by the middle-to-latter stages of the current decade'.[63] The New Hampshire Clean Power Act of 2002 requires the state's three fossil-fuel power plants to reduce their carbon dioxide emissions to 1990 levels by the end of 2006.[64] Oregon created the first formal standard in the US for carbon dioxide releases from new electricity generating facilities by requiring new or expanded power plants to emit no more than 0.675 pounds of carbon dioxide per kilowatt-hour, a rate 17 per cent below the most efficient natural-gas-fired plant currently in operation in the US.[65] In 2001, all six New England states pledged to reduce their emissions to 10 per cent below 1990 levels by 2020.[66]

Several states are pursuing indirect means to reduce GHG emissions.[67] For example, 16 states have enacted legislation that requires utilities to provide a certain percentage of electricity from renewable energy sources. Wisconsin is seeking to mimic the US EPA Toxic Release Inventory Program's success in spurring voluntary emission reductions by requiring public reporting of toxic releases by power plants. In 1993, the Wisconsin Air Contaminant Emission Inventory Reporting regulation began requiring any facility that emits more than 100,000 tons of carbon dioxide to report its emission levels. In 2002, 11 state Attorneys General wrote an open letter to President George W. Bush calling for expanded national efforts to reduce GHG emissions[68] and indicated their commitment to intensify state efforts if the federal government failed to act. In 2002, California passed legislation requiring its California Air Resources Board to develop and adopt GHG emission-reduction regulations by 2005 for passenger vehicles and light duty trucks, starting with vehicles manufactured in the 2009 model year. This made California the first and only legislative body in the US to enact legislation aimed at curbing global warming emissions from vehicles. As The New York Times pointed out, 'Though the law applies only to cars sold in California, it will force the manufacturers to develop fuel-efficient technologies that all cars can use. This ripple effect will be even greater if other states follow California's lead, as the Clean Air Act allows them to do.'[69] Indeed, a bill was introduced in April 2003 to the New York legislature calling for the adoption of California's automotive greenhouse gas standard.

[63] Rabe, 2002: 16.
[64] New Hampshire Department of Environmental Services, 2002.
[65] Rabe, 2002.
[66] New England Governors/Eastern Canadian Premiers, 2001.
[67] Rabe, 2002.
[68] The states are Alaska, New Jersey, New York, California, Maryland, and all six New England states (Sterngold, 2002).
[69] *The New York Times*, 2002.

The marked divergence between state and federal policies in this area has led to two lawsuits, one by automotive manufactures against a state and the other filed by states against the federal government. Stating its intention to challenge California's GHG standard in federal court, the president of the Alliance of Automobile Manufacturers argued that '[F]ederal law and common sense prohibit each state from developing its own fuel-economy standards'.[70] For its part, the EPA announced that it lacked the authority to regulate carbon dioxide emissions under the Clean Air Act. This cast an additional legal cloud over California's global warming initiative, which was taken pursuant to the state's authority under the Clean Air Act to adopt more stringent standards to control air pollution. In June 2003, Attorneys General from Connecticut, Maine, and Massachusetts filed a lawsuit against the federal government claiming that the EPA is required by the Clean Air Act to regulate carbon dioxide emissions as an air pollutant because these emissions contribute to global warming.[71] If this suit is successful, the EPA will be required to classify carbon dioxide as a pollutant and establish standards for permissible atmospheric levels. In October 2003, California announced that it planned to sue the EPA over the agency's decision that it lacks the authority to regulate GHG emissions from tailpipe and other sources. This lawsuit represents, in part, an effort to protect its 2002 statute from a challenge by the federal government. Nine other states, along with environmental groups, are expected to join this suit.

Thus, in contrast to developments in the area of packaging waste, the lack of federal regulations for GHG emissions has become a political issue in the US. Clearly, the issue of climate change is much more politically salient in the US than is the issue of packaging waste. Thus, proposals to address the former but not the latter frequently come before Congress. Finally, while packaging waste can be seen as a problem which can be effectively addressed at the local or state level, global climate change clearly cannot. Even the regulatory efforts of the most ambitious states will have little impact on global climate change in the absence of federal regulations that impose limits on carbon dioxide emissions throughout the US.

European Union

By contrast, both the EU and individual EU member states have been active in developing policies to mitigate climate change. In the early 1990s, several countries (including Finland, the Netherlands, Sweden, Denmark, and Germany) had adopted or were about to adopt taxes on either carbon dioxide specifically or energy more generally. Concerned that such taxes would

---

[70] Keating, 2002.

[71] Johnson, 2003.

undermine the single market, the EU attempted to establish a European energy tax.[72] The EU's 1992 proposal was for a combined tax on both carbon dioxide emissions and energy, with the goal of reducing overall EU emissions by the year 2000 to their 1990 levels. However, this proposal was vehemently opposed by the UK, which was against European-wide tax policies, and to a lesser extent by France, which wanted a tax on carbon dioxide only rather than the combined tax. By the end of 1994, the European Council abandoned its efforts and agreed to establish voluntary guidelines for countries that were interested in energy taxes.[73] In 1997, the Commission again proposed a directive to harmonise and, over time, increase taxes on energy within the EU; that proposal was finally approved in March 2003. It contained numerous loopholes for energy-intensive industry and transition periods for particular countries and economic sectors.[74] Thus, while the EU has had to retreat from its efforts to impose a carbon/energy tax, it has succeeded in establishing the political and legal basis to harmonise such taxes throughout the EU. Efforts to establish European energy-efficiency standards have also been largely stillborn, with the Commission relegated to setting general principles on which member states can base their own programmes.[75] There are no centralised targets or timescales.

In March 2002, the Council of Ministers unanimously adopted a legal instrument obliging each state to ratify the Kyoto Protocol, which they have subsequently done. Under the terms of this treaty, overall EU emissions must be reduced by at least 8 per cent of their 1990 levels by 2008-2012. The so-called 'EU bubble' in Article 4 of the Kyoto Protocol allows countries to band together in voluntary associations to have their emissions considered collectively. However, even before Kyoto was formally ratified, the EU had begun efforts to implement its provisions. In June 1998, a Burden Sharing Agreement gave each member state an emissions target which collectively was intended to reach the 8 per cent reduction target. In the spring of 2000, the EU officially launched the European Climate Change Program, which identified more than 40 emission-reduction measures. In 2001, the EU proposed a Directive for a system of emissions trading and harmonising domestic arrangements within the EU.[76] The Directive, approved by the EP in July 2003, calls on member states to prepare plans for allocating emissions by 2004. Under the Directive, governments are given the freedom to allocate permits as they see fit; the European Commission will not place limits on allowances, although governments are asked to keep the number of

[72] Zito, 2000.
[73] Collier, 1996.
[74] *Environment Daily*, 1997, 2003b.
[75] Collier, 1996.
[76] Smith and Chaumeil, 2002.

allowances low.[77] This trading scheme will initially cover 4,000 to 5,000 large factories, power stations, and similar installations, which are estimated to emit nearly all of Europe's carbon dioxide emissions. The first trade, between Shell Oil and Nuon, a power firm, has already been planned for when the Directive goes into effect.[78] There are plans to subsequently extend emissions trading to include additional GHG emissions and economic sectors.

The efforts at the European level have been paralleled by a number of member-state policy initiatives. Among the earliest efforts was an initiative by Germany in which a government commission established the goal of reducing carbon dioxide emissions by 25 per cent by 2005 and 80 per cent by 2050, though these targets were subsequently relaxed owing to concerns about costs. Germany subsequently enacted taxes on energy, electricity, building standards, and emissions. The German federal government has negotiated voluntary agreements to reduce carbon dioxide emissions with virtually every industrial sector. In 2001, the UK launched a comprehensive and pioneering gas-emissions trading scheme, involving nearly fifty industrial sectors. Participation is voluntary and the government has offered financial incentives to encourage industry participation. However, the British government has also levied a tax on energy use with rebates for firms and sectors that have met their emission-reduction targets. Like its German counterpart, the British government has officially endorsed very ambitious targets for the reduction of carbon dioxide emissions. This will require, among other policy changes, that a growing share of electricity be produced using renewable sources. While both Germany and the UK have reduced carbon dioxide emissions in the short run, their ability to meet the Kyoto targets to which they are now legally committed remains problematic. Other countries, such as France, Belgium, and the Netherlands, have established a complex range of policies, including financial incentives to purchase more fuel-efficient vehicles, investments in alternative energy, changes in transportation policies, voluntary agreements with industry, and the limited use of energy taxes. In 2002, Denmark approved legislation phasing out three industrial GHGs controlled by Kyoto.

ANALYSIS

The dynamics of the relationship between central and state authorities varies considerably across these six case studies. In three cases (automobile emissions in the EU and the US, and packaging waste policies in the EU),

[77] *Environment Daily*, 2003c, 2003e.

[78] *Environment Daily*, 2003a.

state governments have been an important source of policy innovation and diffusion. In these cases, state authorities were the first to regulate, and their regulations resulted in the adoption of more stringent regulatory standards by the central government. In the case of climate change policies, both EU and member state regulations have proceeded in tandem, with one reinforcing the other.

In the two remaining cases (packaging waste and climate change in the US), American states have been a source of policy innovation, but not of significant policy diffusion. To date, state initiatives in these policy areas have not prompted an expansion of federal regulation, though some state regulations have diffused to other states. The earlier US pattern of automotive emissions standards, in which California and other states helped ratchet up federal standards, has not applied to either of these policy areas. Nor has it recently applied to automobile emissions: federal standards have not been changed since 1990; California's more stringent standards have been adopted by few states. In short, in these cases, the 'California effect' (i.e., the diffusion of stringent environmental standards from more stringent to less stringent jurisdictions), has not occurred. Indeed, in the critical area of global climate change, state and central regulations appear to be moving in opposite directions rather than complementing or reinforcing one another. In contrast to its historic role of strengthening overall state standards, the federal government is trying to weaken some of them.

By contrast, in Europe, relatively stringent state environmental standards continue to drive or parallel the adoption of more stringent central standards. Thus, in the EU, the dynamics of the interaction between state and central authorities has become much more significant than in the US. Why has this occurred? Three factors are critical: two are structural and one is political. First, in the EU, states play a direct role in the policy-making process through their representation in the Council of Ministers, the EU's primary legislative body. This provides state governments with an important vehicle to shape EU policies. In fact, many European environmental standards originate at the national level; they reflect the successful effort of a member state to convert its national standards into European ones. In the US, by contrast, state governments are not formally represented in the federal government. While representatives and senators may reflect the policy preferences of the states from which they are elected, the states themselves enjoy no formal representation. Equally importantly, the separation-of-powers constitutional system in the US divides law-making authority between the legislative and executive branches. This means that the Congress has less power to shape central legislation than does the Council of Ministers. Consequently, for example, the senators and representatives from California enjoy less influence over US national environmental legislation than does Germany's representative in the Council of Ministers.

Second, the single market is more recent and more politically fragile in the EU than in the US. The federal government's legal supremacy over interstate commerce dates from the adoption of the US constitution, while the EU's constitutional authority and political commitment to create and maintain a single market is less than two decades old. Accordingly, the European central government appears more sensitive to the impact of divergent standards on its internal market than does the US central government. For example, the US federal government explicitly permits two different standards for automotive emissions, while the EU insists on a uniform one. Likewise, the US federal government appears relatively indifferent to the wide divergence in state packaging waste regulations; only state regulations restricting imports of hazardous wastes and garbage have been challenged by federal authorities.[79] By contrast, distinctive state packaging waste standards have been an important source of legal and political tension within the EU, prompting efforts to harmonise standards at the European level, as well as legal challenges to various state regulations by the Commission. There are numerous state standards for packaging waste in the US that would probably prompt a legal challenge by the Commission were they adopted by an EU member state. Significantly, the EU has established maximum state recovery and recycling goals, while the US central government has not. This means that when faced with divergent state standards, particularly with respect to products, the EU is likely to find itself under more pressure than the US central government to prevent them from interfering with the single market. Accordingly, they must be either challenged or harmonised. In principle, harmonisation need not result in more stringent standards. In fact, the EU's Packaging Directive imposes both a ceiling and a floor. But for the most part, coalitions of the EU's greener member states have been successful in pressuring the EU to adopt directives that generally strengthen European environmental standards. The political influence of these states has been further strengthened by the role of the European Commission, which has made an institutional and political commitment to improving European environmental quality; consequently, the Commission prefers to use its authority to force states to raise their standards rather than lower them. In addition, the increasingly influential role of the EP, in which green constituencies have been relatively strongly represented, has also contributed to strengthening EU environmental standards.

---

[79] Stone, 1990.

The third factor is a political one. During the 1960s and 1970s, there was a strong political push in the US for federal environmental standards. According to environmentalists and their supporters, federal regulation was essential if the US was to make effective progress in improving environmental quality. And environmentalists were influential enough to secure the enactment of numerous federal standards, which were generally more stringent than those at the state level. Thus, the centre of gravity of US environmental regulation shifted to Washington. But over the last decade, the national political strength of environmentalists and their supporters has diminished in the US, owing in part to the Republican Party's capture of both houses of Congress in 1994 and the election of a Republican president who was relatively indifferent to environmental concerns in 2000. As a result, the federal government has become less responsive to pressures for more stringent environmental standards, most notably in the critical area of global climate change.

Nevertheless, environmentalists and their supporters continue to be relatively influential in a number of American states. In part, this outburst of state activity is a response to their declining influence in Washington. Thus, in the US, a major discontinuity has emerged between the environmental policy preferences of many states and those of the federal government. This has meant that, unlike in the 1960s and 1970s, more stringent state standards are much less likely to ratchet up federal standards. Indeed, in marked contrast to two decades ago, when the automobile emissions standards of California and other states led to the progressive strengthening of federal standards in this critical area of environmental policy, California's recent policy efforts to regulate automobiles as part of a broader effort to reduce GHGs have produced the opposite effect: they have been legally challenged on the grounds that they violate federal fuel-economy standards, an area of regulatory policy in which the federal government has exclusive authority but which it has refused to strengthen in any meaningful way for more than two decades.

In the EU, the political dynamics of environmental regulation differ markedly. The 1990s witnessed both the increased political influence of pro-environmental constituencies within the EU – by the end of this decade, green parties had entered the governments of five Western European nations – and a decline in the influence of green pressure groups in the US federal government. During this period, a number of EU environmental policies became more centralised and stringent than those of the US.[80] Paradoxically, while the US federal government exercises far more extensive authority than

---

[80] Vogel, 2003.

the EU, in each of three cases we examined, EU environmental policy is now more centralised than that in the US.

CONCLUSION

The focal cases are summarised in Table 9.1. We conclude with general observations about the dynamics of environmental policy in the federal systems of the US and the EU. On one hand, the continued efforts of states in the US and member states of the EU to strengthen a broad range of environmental regulations suggest that fears of a regulatory race to the bottom may be misplaced. Clearly, concerns that strong regulations will make domestic producers vulnerable to competition from producers in political jurisdictions with less stringent standards have not prevented many states on both sides of the Atlantic from enacting many relatively stringent and ambitious environmental standards. On the other hand, the impact of such state policies remains limited, in part because not all states choose to adopt or vigorously enforce relatively stringent standards. Thus, in the long run, there is no substitute for centralised standards; they represent the most important mechanism of policy diffusion.

Accordingly, the most important role played by state standards is to prompt more stringent central ones. But unless this dynamic comes into play, the effectiveness of state environmental regulations will remain limited. In the areas of both global climate change and packaging waste, even the most stringent state regulations of the US are weaker than those of the EU. It is not coincidental that the case we examined in which EU and US standards are the most comparable – and relatively stringent – is automobile emissions, in which the US central government plays a critical role. By contrast, the lack of central regulations for both packaging waste and climate change clearly reflects and reinforces the relative laxity of US regulations in these policy

Table 9.1 Comparison of environmental regulations

Policy area	EU chronology	Status	US chronology	Status
Auto emissions	State to central	Centralised	State to central	Shared
Packaging waste	State to shared	Contested	State	Uncontested
Climate change	Shared	Uncontested	State	Contested

areas. The EU's more centralised policies in both areas reflect the greater vigour of its recent environmental efforts.

REFERENCES

American Forest & Paper Association (2003), 'State recycling goals and mandates', *http://www.afandpa.org/content/navigationmenu/environment_and_recycling/recycling/state_recycling_goals/state_recycling_goals.htm.*

Barr, S. (1998), 'Clinton orders more recycling; Government agencies face tougher requirements on paper', *The Washington Post*, September 16, A14.

Bell, V. (1998), 'President, Environmental Packaging International, environmental packaging compliance tips', *http://www.enviro-pac.com/pr02.htm*, August.

Bennett, G. (ed.) (1991), *Air Pollution Control in the European Community: Implementation of the EC Directives in the Twelve Member States*, London: Graham and Trotman.

Berland, R. (1992), 'State and local attempts to restrict the importation of solid and hazardous waste: Overcoming the dormant commerce clause', *University of Kansas Law Review*, **40**(2), 465-497.

Bramley, M. (2002), 'A comparison of current government action on climate change in the U.S. and Canada', Pembina Institute for Appropriate Development, *http://www.pembina.org/publications_item.asp?id=129.*

Bryner, G. (1993), 'Blue skies, green politics', *Washington, DC: Congressional Quarterly Press.*

California Air Resources Board (2000), 'California's diesel risk reduction program: Frequently asked questions (FAQ)', *http://www.arb.ca.gov/diesel/faq.htm.*

California Air Resources Board (2001), 'Fact sheet: California's Zero Emission Vehicle Program', *http://www.arb.ca.gov/msprog/zevprog/factsheets/evfacts.pdf.*

California Air Resources Board (2003), 'Staff report: Initial statement of reasons, 2003; Proposed amendments to the California Zero Emission Vehicle Program regulations', *http://www.arb.ca.gov/regact/zev2003/isor.pdf.*

California Environmental Protection Agency (EPA) (2001), 'History of the California Environmental Protection Agency', *http://www.calepa.ca.gov/about/history01/arb.htm.*

California Integrated Waste Management Board (2003), 'Buy recycled: Web resources', *http://www.ciwmb.ca.gov/buyrecycled/links.htm.*

Center for Responsive Law (2003), 'Government purchasing project "State government environmentally preferable purchasing policies"', *http://www.gpp.org/epp_states.html.*

Chanin, R. (2003), 'California's authority to regulate mobile source greenhouse gas emissions', *New York University Annual Survey of American Law*, **58**, 699-754.

Collier, U. (1996), 'The European Union's climate change policy: Limiting emissions or limiting powers?', *Journal of European Public Policy*, 3(March), 122-138.

Comer, C. (1995), 'Federalism and environmental quality: A case study of packaging waste rules in the European Union', *Fordham Environmental Law Journal*, **7**, 163-211.

Container Recycling Institute (2003), 'The Bottle Bill Resource Guide', *http://www.bottlebill.org.*

Cotsworth, E. (2002), 'Letter to Anna K. Maddela', *yosemite.epa.gov/osw/rcra.nsf/ea6e50dc6214725285256bf00063269d/290692727b7ebefb85256c6700700d50?opendocument.*

Dietly, K. (2001), *Research on Container Deposits and Competing Recycling Programs*, presentation to the Columbia, Missouri Beverage Container Deposit Ordinance Law Study Committee Meeting, 1 November.

Environment Daily (1997), March 13.

Environment Daily (2001a), March 29.

Environment Daily (2001b), October 4.

Environment Daily (2003a), February 28.

Environment Daily (2003b), March 21.

Environment Daily (2003c), April 2.

Environment Daily (2003d), May 5.

Environment Daily (2003e), July 2.

Hakim, D. (2003a), 'California regulators modify Auto Emissions Mandate', *The New York Times*, April 25, A24.

Hakim, D. (2003b), 'Automakers drop suits on air rules', *The New York Times*, August 12, A1, C3.

Haverland, M. (1999), *Regulation and Markets: National Autonomy, European Integration and the Politics of Packaging Waste*, Amsterdam: Thela Thesis.

Johnson, K. (2003), '3 States sue E.P.A. to regulate emissions of carbon dioxide', *The New York Times*, June 5.

Keating, G. (2002), 'Californian governor signs landmark Auto Emissions Law', Reuters, July 23, *http://www.enn.com/news/wire-stories/2002/07/07232002/s_47915. asp.*

Lee, G. (1993), 'Government purchasers told to seek recycled products; Clinton Executive Order revises standards for paper', *The Washington Post*, October 21, A29.

McCarthy, J. (1993), 'Bottle Bills and curbside recycling: Are they compatible?', Congressional Research Service (Report 93-114 ENR), *http://www.ncseonline org/nle/crsreports/pollution.*

McCormick, J. (2001), *Environmental Policy in the European Union*, New York: Palgrave.

New England Governors/Eastern Canadian Premiers (2001), 'Climate Change Action Plan 2001', *http://www.massclimateaction.org/pdf/necanadaclimateplan.pdf.*

New Hampshire Department of Environmental Services (2002), 'Overview of HB 284: The New Hampshire Clean Power Act, ground-breaking legislation to reduce multiple harmful pollutant from New Hampshire's electric power plants', *http://www.des.state.nh.us/ard/cleanpoweract.htm.*

New Jersey Department of Environmental Protection (1999), 'Sustainability Greenhouse Action Plan', *http://www.state.nj.us/dep/dsr/gcc/gcc.htm.*

New York Times, The (2002), 'California's message to George Pataki (Editorial)', July 24, A18.

OECD (2002), *Sustainable Development: Indicators to Measure Decoupling of Environmental Pressure from Economic Growth*, SG/SD(2002)1/FINAL, May, Paris: OECD.

OECD (2004), *Environmental Data Compendium: Selected Environmental Data*, OECD EPR/Second Cycle, February 9, Paris: OECD.

Oregon Department of Environmental Quality (2003), 'Oregon Container Glass Recycling Profile', *http://www.deq.state.or.us/wmc/solwaste/glass.html.*

Parker, J. (2003), 'California board's boundaries debated: Automakers say it oversees emissions, not fuel economy', *Detroit Free Press*, May 7.

Percival, R., A. Miller, C, Schroeder, and J. Leape (1992), *Environmental Regulation: Law, Science and Policy*. Boston: Little, Brown & Co.

Plastic Shipping Container Institute (2003), 'Wisconsin solid waste legislative update', *http://www.pscionline.org*.

Rabe, B. (2002), 'Greenhouse and statehouse: The evolving state government role in climate change', Pew Center on Global Climate Change, *http://www.pewclimate.org/projects/states_greenhouse.cfm*.

Rehbinder, E. and R. Stewart (1985), *Integration Through Law: Europe and American Federal Experience, vol. 2: Environmental Protection Policy*, New York: Walter de Gruyter.

Revesz, R. (2001), 'Federalism and environmental regulation: A public choice analysis', *Harvard Law Review*, **115**, 553-641.

Smith, M. and T. Chaumeil (2002), 'Greenhouse gas emissions trading within the European Union: An overview of the proposed European Directive', *Fordham Environmental Law Journal*, **13**(Spring), 207-225.

Sterngold, J. (2002), 'State officials ask Bush to act on global warming', *The New York Times*, July 17, A12.

Stone, J. (1990), 'Supremacy and commerce clause: Issues regarding state hazardous waste import bans', *Columbia Journal of Environmental Law*, **15**(1), 1-30.

Thorman, J., L. Nelson, D. Starkey, and D. Lovell (1996), 'Packaging and waste management; National Conference of state legislators', *http://www.ncsl.org/programs/esnr/rp-pack.htm*.

US EPA (1999), 'California State Motor Vehicle Pollution Control Standards; Waiver of federal preemption', *http://www.epa.gov/otaq/regs/ld-hwy/evap/waivevap.pdf*.

US EPA (2003a), 'Federal and California Exhaust and Evaporative Emission Standards for Light-Duty Vehicles and Light-Duty Trucks', Report EPA420-B-00-001, *http://www.epa.gov/otaq/stds-ld.htm*.

US EPA (2003b), 'Municipal Solid Waste (MSW): Basic facts', *http://www.epa.gov/apeoswer/non-hw/muncpl/facts.htm*.

US EPA (2003c), 'Global warming: State actions list', *yosemite.epa.gov/oar/globalwarming.nsf/content/actionsstate.html*.

Vogel, D. (1995), *Trading Up; Consumer and Environmental Regulation in a Global Economy*, Cambridge: Harvard University Press.

Vogel, D. (2003), 'The hare and the tortoise revisited: The new politics of consumer and environmental regulation in Europe', *British Journal of Political Science*, **33**(4), 557-580.

Yost, P. (2002), 'Bush administration is against California's Zero Emissions Requirement for Cars', *Environmental News Network*, *http://www.enn.com/news/wire-stories/2002/10/10102002/ap_48664.asp*.

Zito, A. (2000), *Creating Environmental Policy in the European Union*, New York: St. Martin's Press.

10. Globalisation and Policies/Politics towards Sustainable Development in Developing Countries

Hans Opschoor

SUMMARY

The complex interrelationships between regional and global economic integration, on the one hand, and the emergence of environmental and resource-oriented policies in developing countries, on the other hand, are addressed in this chapter. I discuss the notions and processes of sustainable development, globalisation, and institutions, including their interrelations. Three types of agents are identified: economic agents (national and transnational corporations), political bodies (national governments and supranational development organisations), and civil-society organisations (in particular, non-governmental organisations). Globalisation has had a positive effect on national environmental policies in developing countries (especially the creation of environmental institutions), but its impact has also been negative (related to the erosion of national governments' authority). I explore the possibilities to overcome this drawback by constituting a 'countervailing power', focusing on the economic, political, and environmental realities faced by African countries. I conclude that civil society may play an important role to fill the environmental governance void in developing countries, but it will not provide a panacea.

INTRODUCTION

Differences in socio-economic (as well as ecological) conditions account for differences in the extent and impact of environmental policies. Thus, paying attention to environmental politics and policies in different macro-settings potentially makes sense, both rationally and empirically. I analysed environmental and resource-oriented policies in developing countries, and gauged their effectiveness. I viewed regional and global economic integration ('globalisation') as a crucial factor to take into account when seeking for strategies to make these policies more effective.

The structure of this chapter is as follows. The notions of globalisation and sustainable development are presented below. In the analysis, I consider international trade, foreign direct investment, and development cooperation programmes to be relevant aspects of globalisation. These affect economic development as well as the economic and political (macro-)environment in which states develop their policies, including those on sustainable development. This is discussed in general terms and at the meta-national level. The conceptual framework applied distinguishes as relevant categories of agents: (1) economic agents; (2) political agents at various levels; and (3) civil-society organisations. The relationships between these agents are addressed from a 'countervailing power' perspective.

Secondly, an analysis is made of environmental policies and agents in developing countries, and the forces that operate on both. Crucial questions are: what types and levels of intervention are envisaged in environmental policy regimes, what is the intensity with which they are implemented, and what are the prospects and constraints in this respect? An attempt is made to discern the dominant forces driving the dynamics of policy development in relation to the environment and natural resources. This may then lead to some suggestions as to what might be the most effective instruments or types of intervention in the specific context (or rather, *set* of contexts) of developing countries, given the overall politico-economic environment in which they must operate.

SUSTAINABLE DEVELOPMENT IN A GLOBALISING WORLD[1]

Environmental policies emerge in the face of environmental problems, and these are related to patterns and levels of economic activities and features of the natural environments in which they take place. Environmental policies also emanate from international exchange – whether in the form of negotiations and international fora such as those organised by the United Nations (UN), most recently the 2002 World Summit on Sustainable Development in Johannesburg – or in the form of pressures related to the evolving international economic and political systems.

---

[1] Much of this section is based on Opschoor, 2003a.

Environmental policies should not be analysed merely in terms of their relationships to the environmental problems they address, but must be seen as being the result of, and as operating in, a societal context in which a number of actors play decisive roles regarding the causes of environmental degradation and their remedies through policies and policy-based interventions. Moreover, they should be seen as also related to the tidal forces that operate on societies from higher-level systems, such as the economic order. As opposed to a merely 'technical' approach, I deal with environmental policies in an institutional framework. Environmental problems are taken to not result merely from pressures owing to pollution and resource depletion given a set of binding environmental constraints, but, given those constraints, from *institutional deficiencies* – that is, failures in the prevailing mechanisms regulating the orientation and magnitude of economic activities. These failures are not merely those of inefficient or ineffective administrations but also those resulting from inappropriate market signals, and worse, the market system per se. The latter system is one that has almost been imposed on developing and transitional countries through forces operative in the process of globalisation. The relevant concepts and processes are discussed briefly below.

Concepts

Sustainable development

Development is understood here as an expansion of the range of options open to human beings. It has long been assumed that economic growth, typically measured by the increase in income per capita, was the main vehicle for it. This is now being challenged, for instance, from a human development perspective. Environmental degradation is one specific manifestation of divergence between economic growth and (human) development.[2]

Concerns about this divergence have, through a series of international conferences, given rise to the concept of sustainable development.[3] These conferences were the UN Conference on the Human Environment (Stockholm 1972), the UN Conference on Environment and Development (UNCED, Rio de Janeiro 1992), and the UN World Summit on Sustainable Development (WSSD, Johannesburg 2002). The WCED's report (or 'Brundtland Report') triggered the UNCED and, thereby, the WSSD. It is interesting to see, through the evolution in the names of these conferences, the process by which the notion of sustainable development emerged. This process was the result of intensive debates between South and North about the content and the

2 Opschoor, 1996b.
3 WCED, 1987.

significance of the environmental agenda and the various elements thereof, going back to the Founex-Report of 1971.[4] Developing countries did not wish to discuss environmental issues in isolation from poverty-related issues. Moreover, they did not wish to discuss environmental issues specified by the North only. On both counts, they have managed to broaden the framing of the issues. They may be less successful in determining the institutional setting within which these issues are to be addressed.

The concept of sustainability cropped up for the first time in the World Conservation Strategy.[5] That strategy explicitly claimed that, for development to be sustainable, it must take account of social and ecological factors as well as economic ones. The next stage in the evolution of the concept was reached with the publication of the Brundtland Report, which defined sustainable development as 'a process of change in which the exploitation of resources, the direction of investments, the orientation of technological development, and institutional change are all in harmony, and enhance both current and future potential to meet human needs and aspirations'.[6] In practice, sustainable development became associated with patterns of economic development related to non-declining real income per capita, and with what this entailed in terms of maintaining and enhancing a capital base of which environmental capital (or natural resources) was only one element – one that, to a significant degree, was regarded as replaceable by other forms of capital (including knowledge and know-how). Sustainable development allows only for non-negative changes in resource endowments.

It is worth considering the emerging principles of sustainable development. The Rio Declaration established human beings as the main concern of sustainable development (Principle 1).[7] It reiterated that concerns about the environment and development are interrelated and demand an integrated approach. States have the right to exploit their own natural resources and the duty to do so without causing damage to the environment of other states (Principle 2). Poverty eradication is an essential precondition for sustainable development (Principle 5). The special needs of developing countries (Principle 6) and the different shares in responsibility for causing environmental degradation of developing and developed nations give rise to the notion of 'common but differentiated responsibilities' of these categories of states (Principle 7). States must end non-sustainable patterns of production and consumption, and facilitate appropriate demographic policies (Principle 8). A 'precautionary approach' is to be applied: in case of risk of serious or irreversible damage to the environment, cost-effective interventions will not

[4] UNEP, 1981.
[5] IUCN/UNEP/WWF, 1980.
[6] WCED, 1987:46. I prefer to substitute 'compatible' for 'in harmony'.
[7] UNCED, 1992.

be blocked by a lack of full scientific certainty (Principle 15). People's participation in developing environmental policies was laid down as the right to access to information; and effective recourse to judicial procedures must be ensured (Principle 10). The inclusion in prices of environmental costs according to the polluter-pays principle was recommended, together with the use of economic incentives to reduce the environmental impact of economic activities (Principle 16).

In developing countries, environmental policies were articulated to address the resource issues that plagued them. It is widely recognised that sustainable development, thus defined and operationalised, has become the most powerful stimulus for the emergence of environmental policies and policy agency in developing countries.

Globalisation

Development processes and development policies manifest themselves in a setting of internationalisation and globalisation. The process of growing international trade and division of labour goes back at least to the Middle Ages. Post-World War II efforts at stimulating economic development were, among other things, a particular phase in that process, with the underlying strategy of ensuring that as many of the newly independent nations as possible would keep their markets open and their resources available to the industrialised countries. After the Cold War, a type of world order emerged with a specific model of development defined by the basic tenets of the so-called 'Washington Consensus': a focus on liberalisation and privatisation combined with 'sound' macro-economic policies and monetary management. This was the model proposed, and almost literally imposed – via development programmes by the World Bank (WB) and the International Monetary Fund (IMF) – as the ultimate development strategy in the era of globalisation.

Globalisation has many features and aspects.[8] Below, I focus on economic aspects and their repercussions. First, it is important to note that the growing international cooperation in addressing issues of common concern may be regarded as one specific manifestation of globalisation at the political level. Thus, already in the mid-1940s, the foundations were laid for a political structure to address international concerns, especially about insecurity and conflict: the UN. An architecture was established to deal with financial issues related to post-war reconstruction, development, and financial stability: the WB and the IMF.[9] Recently, to focus the management of a series of social issues with development relevance, the UN organised global conferences on population, gender, social policy, habitat, and – notably – sustainability.

[8] For a broader discussion, see Zoeteman, 2002.

[9] The former General Agreement on Tariffs and Trade (GATT) and the current World Trade Organization (WTO) can be added to this.

Given what was observed above about the origins of environmental policy in developing countries in particular, it appears that globalisation has, at the political level, engendered in these countries environmental policy development and the creation of organisational structures to implement them: ministries, agencies, etc.

Economic globalisation can be defined as a process in which markets, technologies, and communication patterns become progressively more international over time. Much of today's economic globalisation is accelerated by pushes towards free trade and free mobility of capital. It implies the integration of cross-national dimensions into the identities and strategies of firms based on competitiveness. The result is a single integrated economy shaped by corporate networks and their financial relationships. Therefore, in my analysis, attention is paid to a series of economic linkages: through international trade, foreign direct investment flows, and other flows such as, particularly, development aid/cooperation.

Institutional failure

Economic activities claim resources and generate waste. The societal implications of this are not fully, if at all, reflected in the prices, costs, and benefits driving decisions in the economic domain by producers, investors, and consumers. These environmental impacts thus remain external to, or peripheral in, the decision-making process. The inability, under prevailing regimes, of markets to incorporate these externalities is one example of market failure, as has long been recognised in economic theory and the theory of economic policy. Checks and balances are needed to 'internalise' these externalities, and they can come only from outside the market system: from stakeholder pressure and/or via a societal response in terms of environmental regulation of economic activities and their impacts on environmental policies. However, such a response places environmental regulation in the domain of governance and administration by public authorities and that in itself poses problems. Governments, too, may fail to act according to the public interest (for example, owing to a lack of relevant policies, problems with the effectiveness of the policy instruments used, and a lack of enforcement of policies). There are various recognised types of government failure (or governance failure), such as inefficient delivery or implementation and lack of mandate.[10] Below, I describe some observations in the context of developing countries.

In a long-term perspective and in line with Habermas,[11] markets, corporations, and political structures may be seen as developments in

[10] See Opschoor (1996a) for a classification of both types of institutional failure, market and governance failure.

[11] Habermas, 1989.

societies in response to problems faced by people within their 'life worlds', initially on the basis of popular consent. Over time, these structures have gained an increasing degree of autonomy and have managed to occupy larger and larger parts of these life worlds. In Habermas' terminology, they have become systems colonising peoples' life worlds (and the biosphere generally, one might add) and have, to a large degree, become unresponsive to real public interests. Similar concepts can be found in Giddens' work on structuring features underlying (late) modernity: 'abstract systems' dislodging people from their life worlds, the 'colonisation of the future' to curb risks (including ecological ones as well as those related to global economic mechanisms), etc.[12] Habermas' position represents a deep pessimism in relation to the abilities of these systems to optimally act on social concerns, unless there is some adequate countervailing power (as Galbraith has called it) emanating from civil society. In a less pessimistic vein (such as perhaps Giddens'), these civil-society-based powers may be seen as operating on market forces through a democratic state, or, more generally, through political institutions at all necessary levels.

Institutional aspects are discussed below, in relation to agency and actors in the field of sustainable development.

Processes

Globalisation and development: The mainstream position

In the late 1990s, the OECD countries agreed on a combined development strategy based on the so-called Millennium Development Goals. In the context of the overall strategy of economic development compatible with the views of the WB and the OECD (see below), a set of targets was set concerning a wide range of aspects of development, including environmental features. To that extent, environmental policy was mainstreamed into development thinking. The set of objectives and targets was elaborated subsequently (e.g., in conferences such as the Monterrey Conference 2002 on development finance and the WSSD, also in 2002). The set of objectives and targets now acts as a compass in the international negotiation process relevant to (sustainable) development. Needless to say, such goals are essentially ambitions and they are as yet far from being realised. But it is clear that they do play their role as a compass in international negotiations and policy development – as witnessed, for example, by the WSSD and the 2003 negotiations in the context of the WTO.

The Millennium Development Goals and Targets typically relate to

[12] Giddens, 1991.

traditional issues such as poverty, education, and health.[13] The overall objective is to foster poverty reduction. The Millennium Development Goals also aim at ensuring environmental sustainability (Goal 7). It does so by setting three specific targets: to integrate the principles of sustainable development into country policies and programmes, and to reverse the loss of environmental resources (Target 9); to halve by 2015 the proportion of people without sustainable access to safe drinking water (Target 10); to have achieved by 2020 a significant improvement in the lives of at least 100 million slum dwellers (Target 11). Targets 10 and 11 were agreed at the WSSD. Target 9 is the original Millennium Development target; it is difficult to regard it as being at the same level of specificity as the other ones. Yet, as international programmes will be required for Targets 10 and 11, Target 9 is the most important one in terms of requiring policy formulation efforts to be made by developing countries themselves.

To achieve Millennium Development Goals, poor countries are asked by the UN Development Programme (UNDP) to mobilise and manage local resources effectively and equitably. These resources include finances for development generally, e.g., domestic savings. Rich countries are asked by UNDP to increase their aid to 0.7 per cent of gross domestic product (GDP) and to undertake debt relief in HIPCs (Heavily Indebted Poor Countries).[14] The latter mechanism is one in which the WB will play a key role and in which impacts on environmental policies may be expected (see below). Moreover, rich countries are required to provide market access and transfer technology.[15]

The current approach[16] requires that HIPCs, in order to qualify for debt relief, develop Poverty Reduction Strategy Papers: country-owned and -led strategy documents, developed in a participatory process and accepted by donors on the basis of conditionality on process (rather than substance), for instance, in terms of stakeholder involvement.

Globalisation and sustainability

Globalisation, where successful, is likely to increase international trade and economic growth. Open economies tend to grow faster, and per-capita income tends to rise with an increase in the degree of openness of economies.[17] However, globalisation – while raising average incomes in those parts of global society that are thus ‘connected’ – creates new threats to

---

13 UNDP, 2003.

14 UNDP, 2003.

15 The issue of market access was unsuccessfully discussed in the 2003 WTO negotiations in Cancún.

16 Maxwell, 2003.

17 Borghesi and Vercelli, 2003.

human security; the UNDP lists financial volatility, job and income insecurity, health and personal insecurity, political and community insecurity, and cultural and environmental insecurity.[18] Thus, globalisation is not always successful (at least, there are serious trade-offs); nor are all countries and regions of the world equally involved in the process.

Globalisation affects economic activity in several ways and, therefore, it is relevant from the point of view of environmental policy. Firstly, it enhances economic growth and thus leads to an associated increase in environmental pressure. Secondly, it potentially leads to an increase in average income – unless there is a matching growth in population. That may affect consumption patterns and lead to environmentally more damaging lifestyles. Thirdly, globalisation (through liberalisation) leads to more trade. Trade is a key accelerator of economic growth (and thus of the associated environmental repercussions already mentioned), but I must also point to the impact of trade on the international division of labour. Trade affects production patterns, both domestically and globally, and the changes therein also affect the environment – often in a negative way in developing countries.

In contrast to all of these potential boosters of environmental stress are the benign impacts of economic growth on the environment. Firstly, higher incomes may lead to higher environmental awareness and sensitivity. Secondly, higher growth levels and foreign direct investment (FDI) in domestic industrial development may result in environmentally more efficient technologies. Hence, as economies grow, the marginal and even average environmental impacts of economic activities may be reduced.

FDI may bring about significant environmental improvement if it is accompanied by the introduction to host countries of new, environmentally efficient technologies. Negative impacts may also be expected where regimes are lax or tolerant, and property rights are poorly defined. Many environmentalists expected this latter tendency to prevail, and industry to move to locations with little environmental protection, especially in developing countries (the 'pollution haven hypothesis'). Zarsky found evidence that both domestic and foreign firms in developing countries improve their environmental performance in response to national regulation and/or local community pressure.[19] Many new plants use advanced technologies. Yet, firms entering new countries may become lax in their environmental performance over time. Relatively pollution- and resource-intensive industries, such as the petrochemical and mineral-extraction industries, provide examples of FDI giving rise to pollution-haven-type phenomena. WWF conducted a study of the mining industries and their

---

[18] UNDP, 1999.
[19] Zarsky, 1999.

environmental regulation in the Pacific Region. Among its findings were the destruction of forests and rivers in Indonesia and the destruction of 50 per cent of the fish yields in several rivers in the Philippines.[20] Other resource-intensive sectors elsewhere have given rise to similar phenomena (e.g., oil spills in Ecuador and Nigeria, risks of environmental destruction in Chad, deforestation in many countries). Many countries compete for new investments by offering tax facilities and exemptions to environmental (and labour) regulation. Therefore, mixed evidence exists as to the pollution haven hypothesis. While the balance may be positive, certain sectors and locations show negative results.

In conclusion, it can be foreseen that, with the expected rates of economic and population growth, significant forms of environmental pressure at the global level will continue to grow far into the next century.[21] Thus, continued *economic* growth is not necessarily *environmentally* sustainable; nor will it automatically become sustainable in the Brundtland sense. The second conclusion is that markets by themselves cannot be relied on to deliver public goods at the desired levels and to incorporate the externalities they give rise to. Without far-reaching additional environmental policies and institutions to effectively address and curb inherent market failures, economic growth will push economies further on an unsustainable track. Policies at the national level to enhance reliance on market forces may, under such circumstances, have questionable social and environmental benefits: they may lead to more market failure. The institutional transformation inherent in the current mode of globalisation enhances inequalities, erodes social cohesion, and increases unsustainability.

Agency: Different types of actors

Various types of agency and different categories of actors (economic, political, and non-governmental) were mentioned above. It is important to distinguish between: (1) economic agents (national and transnational corporations and their organisations; consumers and consumer organisations; labour unions); (2) political agents (the state – environmental policy organisations and other organisations must be differentiated; supranational agents, such as the WTO, the WB, the IMF, multilateral environmental agreements (MEAs), and regional regimes); (3) civil-society organisations, including environmental and other non-governmental organisations (NGOs), academia, and media. Here, I consider only corporate agents (especially transnational corporations), both types of political agents (with emphasis on the national level), and NGOs.

---

[20] WWF, 1999.

[21] Opschoor, 2003b.

Globalisation affects the relationships between the various types of actors. States seem to be under pressure to proceed with liberalisation and deregulation. This initially appeared sensible in respect of states which were regarded with distrust for their inefficiency or for their allowing certain classes to appropriate a disproportionate share of the economy-generated income flows. Thus, in many circles, this was hailed as an opportunity for people to regain their claim to control over their livelihoods. However, it has become obvious that, in the process of globalisation, private corporate interests (especially those embodied in transnational corporations), rather than civil society, benefitted. States largely acted as competitors in the race to become the location of firms' financial headquarters and production units. Globalisation thus resulted in the erosion of vertical sovereignty (i.e., relationships between states and the private sector). Reinicke observed that, while globalisation integrates markets, it fragments politics.[22] The greater 'flexibility' inherent in economic globalisation has increased corporate power.[23]

The position of supranational political agents vis-à-vis globalisation has been varied. Some (like UN agencies) have shown deep concern about its impacts.[24] Others (such as the WB and the IMF) have been very much in favour of it, as manifest in the GATT/WTO's principal support for free trade. The Washington-based institutions had strong support from one of the industrialised world's leading think tanks: the OECD. Towards the end of the last century, the OECD was active in promoting the free flow of financial capital across national borders (through its aborted project on a multilateral investment agreement) and in furthering international trade-led development in what it called 'linkage-led development'.[25] It may be argued that tendencies towards internalisation and the globalisation of economic and other networks are inherent in economic evolution. It is equally correct to note that these processes have been enthusiastically boosted and stimulated in a particular institutional direction (i.e., that of the neo-liberal transformation of politico-economic relations) by that consortium of organisations. In fact, one could speak of a *project* to accelerate globalisation along neo-liberal lines undertaken by this consortium. Less powerful agents argue in favour of 'managed globalisation'.[26]

The relevant agents involved in development cooperation are organisations that are important in developing domestic policies (on environmental and other issues) in developing countries: the WB and bilateral donors. The

---

[22] Reinicke, 1997.
[23] Sideri, 1999.
[24] Ricupero, 1999.
[25] Opschoor, 2001.
[26] Ricupero, 1999.

former is dealt with below; first, some comments are in order on bilateral aid and its environmental impact. In the past, attempts were often made to operate in national development policies in support of specific sectors (such as health, education, rural development, gender, poverty, and the environment). Considerations related to the offering of aid were often cast in terms of other policies' impacts (e.g., human rights, environmental impacts) and programmes to be funded according to sectoral priorities set by the donor countries. This led to the feeling in receiving countries of being squeezed within a narrow set of 'conditionalities' and often resulted in a lack of domestic ownership of projects funded internationally. Many developing countries' governments have set up environmental agencies possibly more in line with what these conditionalities implied than they might otherwise have done, triggered by the massive international fora on sustainable development (see above) and hoping to augment the aid flows to their countries by complying with the wishes of donor countries and donor agencies.

World Bank Policies and Sustainable Development

The WB is one of the main actors in funding development processes in developing countries. It is the multilateral window for such funding (apart from windows for bilateral development financing from national sources in industrialised countries). The Bank is relevant to environmental policies in developing countries, and to economy-environment interactions in these countries, in different ways: as a funder of environmental projects and investments; as a stimulator of national and regional environmental planning and strategies; and as a stimulator of development in general.[27] Here, I discuss the last two features.

The WB adopts a position taken from Agenda 21: the call for 'country-driven' sustainable development strategies. It supports countries' environmental planning through national environmental action plans (NEAPs), country strategies, and economic studies. NEAPs describe countries' environmental problems, strategies, and investment plans. National governments are responsible for their development and implementation, but the WB provides advice and technical assistance. Moreover, the Bank sees NEAPs as a basis for dialogue with governments on their environmental strategies. The WB wants all relevant parties to be involved in the development of these NEAPs, especially those responsible for economic decision-making.

In the 1980s and 1990s, the WB was also involved in the development of economy-wide policies ('structural adjustment programmes'). In these, it

[27] World Bank, 1993.

emphasised macroeconomic stability and monetary policies and, at the micro level, it pushed for the removal of 'distortions' and the improvement of the efficiency of resource use. The Bank believed that programmes of liberalisation 'will generally be beneficial, both economically and environmentally'.[28] It felt that negative environmental effects were generally associated with old distortions and market failures. These adjustment programmes can be seen as part of the strategy to further globalisation along neo-liberal lines.[29] Analysis of the adjustment programmes entailed a consideration of their environmental implications, and in that context WB staff should 'review the environmental policies and practices in the country'.[30] Through such mechanisms, a streamlining of many aspects of economic planning and environmental policies took place, gearing developing countries to the process of trade-led economic growth. In the domain of environmental policies, this implied strategies aimed at 'getting prices right', terminating or reducing 'unsustainable' subsidies, relying on economic instruments as tools to achieve environmental objectives, and opening up markets (inside the developing world) to foreign trade and FDI.

Structural adjustment programmes also had an impact on environmental policies in an indirect way. Often, they entailed cuts in the budgets of spending ministries. This affected budgets in areas such as health and education, but also environmental ministries could be hurt. For instance, as part of the measures taken to limit the impact of the Asian financial crisis in the late 1990s, the budget for environmental infrastructure in Thailand was cut by one third.[31] This is not to say that the environmental effects of structural adjustment and other types of intervention by multilateral funding institutions is negative per se; rather, the WB's view of them may have been inappropriately positive.[32]

The WB is now replacing its structural adjustment lending with poverty-reduction support credits and will rely on budgetary support instead of project funding, in sector-wide approaches. Maxwell analysed the risks involved in this new approach.[33] One of these is that the preoccupation with poverty reduction will detract attention from other important components, including environmental concerns. Important trade-offs and conflicts of interest may be obscured and this, as we shall see, is of importance too when it comes to sustainable development. Maxwell argued that the Millennium Development Goals-based focus on selected targets and performance indicators may distort

---

[28] World Bank, 1993: 97.
[29] Opschoor 2003a, 2003b.
[30] World Bank, 1993: 159.
[31] Nicro and Apikul, 1999.
[32] For an empirically based assessment, see Munasinghe, 1999, and Opschoor and Jongma, 1996.
[33] Maxwell, 2003; see also AIV, 2003.

development efforts. In this context, the rather specific and somewhat vague environmental targets under Goal 7 should be borne in mind.

Globalisation and Countervailing Efforts

Markets increasingly predominate in processes of social change. Without countervailing power, this will exacerbate problems of inequality and unsustainability. Unsustainable development results from an inherent orientation within market forces and market actors towards short-term aspects (e.g., revenues and profits) as opposed to future needs, and from an inherent growth drive. Unless adequately checked, development in line with the neo-liberal globalisation project will also enlarge inequality, conflict, and insecurity; on inequality, the data generated by the UNDP is disconcertingly convincing already. But there *is* countervailing influence. Concern about the impact of neo-liberal development models promoted in the globalisation project is growing, as was manifest in a series of recent civil-society demonstrations (Response (1), below). Beyond these civil-society protests, three types of responses could be discerned: (a) adaptation of the globalisation project; (b) attempts to re-invigorate states and enhance interstate control over economic processes; and (c) the evolution of new forms of partnership.[34]

Response (1): Another globalisation

There is a rapid growth in global civil society. Examples of where these networks have been influential in the last decade include the international campaign to ban landmines, the contribution by NGOs to the abandonment of the OECD-based attempts to develop a multilateral agreement on investments, and the Seattle and post-Seattle actions against the neo-liberal globalisation project. These networks are also visible at the various UN conferences on social, economic, and environmental issues, such as the Monterrey conference on finance for development, the UNCED, and the WSSD. There are about 50,000 networks now, many accredited by the UN and thus having access to global conferences and their preparatory processes. To match the growing influence of the World Economic Forum and related manifestations of the de facto economic powers behind the globalisation project, a World Social Forum ('Porto Alegre') brings together many of the civil society, 'life world' voices and organisations.[35]

Global civil society and its networks are perfect examples of globalisation generally. NGOs reap the advantages of modern communication technology

[34] Opschoor, 2003a.
[35] Waterman, 2003.

and they have effectively sought access to the formal circuits of negotiation on issues such as trade and the environment.

Response (2a): Adaptation within globalisation
Enlightened elements in the neo-liberal consortium described above have identified a strategy aimed at curbing some of the adverse effects of globalisation.[36] It is recognised that, following structural adjustment, the provision of important public services (such as education and infrastructure) may fail and that more appropriate policies may be needed. Structural adjustment is still seen as being linked with liberalisation and with getting connected to the global economy, but specific action plans are now to be embedded in more integrated policies (see above).

Response (2b): Managing globalisation
The harnessing of countervailing power entails, in the view of many, a re-thinking and (re-) capacitating of the role of states and international institutions created by states. In this reformist approach, globalisation will have to be 'managed', which will require reform in order to enhance the capacities to implement environmental policies.[37] States will have to provide the legal and regulatory frameworks in which, ideally, market structures and market processes will be embedded. States should also regulate the access to resources (including natural resources) of users, today as well as in the future.

Response (2c): Public-private provision of global public goods
Others fear that countervailing power through political forces alone may not suffice to ensure equity and sustainability. States are no longer the loci of power. New organisational combinations or alliances emerge between public actors, on the one hand, and private and non-governmental ones, on the other hand:[38] civil-society organisations, corporations, (inter-) governmental agencies of various kinds and at various levels. This option of (public-private-NGO) partnerships was given a centre-stage position at the WSSD.

ENVIRONMENTAL POLICIES AND INSTITUTIONS IN DEVELOPING COUNTRIES

In this section, I address the following questions: what are the environmental policies of developing countries, where do they come from, how are they supported, and how (in)effective are they? Furthermore, what instruments are

[36] World Bank, 1998.
[37] Ricupero, 1999.
[38] Kaul et al., 2003.

available, and what barriers exist to these policies and instruments being more effective?

An important level at which to analyse environmental policy in developing countries is the national perspective. The second level is that of international environmental policy.[39] As far as the latter is concerned, there are two significant sub-levels: the regional and the global. At the regional level, reference should be made: (1) to the numerous agreements on regional cooperation and development with environmental repercussions, and the policies and agreements in relation to these; and (2) to the regional environmental agreements (e.g., on the management of common resources, such as shared river basins). As far as the former is concerned, the involvement of developing countries in international MEAs and regimes should be referred to, as should their involvement in other negotiations and conventions with potential environmental repercussions (such as the WTO negotiations). This second level is addressed below, but the focus is on the national level.

General Observations

Environmental issues in developing countries may differ from those in industrialised ones. They also differ between developing countries and even between regions in the non-OECD part of the world. In developing countries, land degradation is a pervasive issue, as are deforestation and water-related issues. There are differences in the degrees to which, at the regional level, issues related to biodiversity or conditions in human settlements are mentioned.[40] Environmental problems in developing countries arise from: patterns of production and consumption of the non-poor segments of populations, and poverty (some 900 million people live in absolute poverty in rural areas alone). Thus, reducing poverty can play a pivotal role in reaching environmental objectives. Ignoring environmental sustainability may lead to short-term financial gains but will undermine the sustainability of poverty reduction. Yet, many national environmental strategies fail to address the effects on other sectors and on poor people.

In the developing region, there is a sense of urgency in relation to the need to stimulate economic growth. Moreover, there is often resentment and distrust in relation to governments and the post-industrial countries. Governments typically show low levels of appreciation for environmental quality, and populations have low levels of environmental awareness.[41] Yet,

[39] See Chaytor and Gray (2003) for a recent assessment of that for one particular region (Africa) and from a regulation-oriented perspective.

[40] UNEP, as quoted in Steel et al., 2003: 159.

[41] Steel et al., 2003.

most developing countries have environmental legislation and a range of environmental institutions, and are signatories to most (if not all) international conventions and agreements. Most of the initiatives for this came from outside (often supported by domestic NGOs), and implementation or compliance is generally low.

It is interesting to see how differences in regional settings translate into differences in the views of citizens concerning environment-economy trade-offs, nature, etc.[42] At the regional level, these differences are relatively small, smaller than one might expect. For instance, in the (post-) industrial countries, over 58 per cent of a sample of some 10,000 citizens in 19 different countries chose 'the environment' over 'the economy', whereas, in developing countries 54 per cent of a sample of 25,000 people in 18 countries made this choice. Figures such as these show wide variation also within regions. Thus, in developing countries, this choice in favour of the environment ranged from 56 per cent (Philippines) to 83 per cent (China) – two countries almost equal in terms of popular support for higher taxes to prevent environmental damage (24 per cent and 41 per cent, respectively; support in other countries ranged from 52 per cent in Pakistan to 5 per cent in Brazil and Turkey).

National Environmental Policies

Below, I provide a review of national environmental policies on the basis of a number of case studies. The focus is on Africa. Other sub-regions are represented by a smaller number of studies. African countries are comparatively (and in general) amongst the least developed (in terms of GDP and GDP-growth per capita). This may account for some similarity in the environmental problems these countries perceive as the most pertinent, and for the shared conviction that there is a link between environmental degradation and poverty. Environmental issues raised in Africa include soil erosion, land degradation in general, desertification, deforestation, biodiversity loss, infectious and parasitic disease, air pollution in large cities, coastal ecosystem degradation, drinking-water problems, and lack of sanitation facilities. Whatever economic progress there is in many African countries is eroded by the forces of population growth, leading to growing pressure on the natural resource base.

Environmental resources are often perceived as land-related – in a way that reminds the modern reader of the analyses by Smith and Ricardo. In many contexts, this has politicised environmental issues: land, land distribution, and

---

42 Steel et al., 2003.

the colonial legacy.[43] The post-colonial era began with what has been dubbed 'authoritarian development', with land and other resources being opened up, often to transnational business. This had an impact on peoples in the form of displacement and/or curtailed access to resources; livelihood insecurity was typically the result of that. Besides this, there has been (and is) competition for limited resources. Salih described environmental concerns as arising often in a countervailing mode (i.e., as a process of liberation).[44] NGOs played a part as would-be or real catalysts in or champions of that process.

Africa's post-colonial environmental planning and policy formulation has typically been triggered by external forces, such as international conferences on environment and development. Thus, issues other than land-related ones were on the agenda as well. More recently (as described above), and under the influence of global actors such as the WB and the donor community generally, environmental policies evolved in the setting of the development of NEAPs. It was the 1980 World Conservation Strategy, with its call for national conservation strategies (NCSs),[45] that inspired the development of NEAPs, which actually emerged since the early 1990s. These have strongly influenced African environmental policies.

Salih and Tedla summarised an OSSREA-led review of NEAPs and NCSs in Eastern and Southern Africa.[46] They based their findings on case studies conducted in nine countries, the results of which were assembled in 1997. In his summary of this survey, Salih observed that environmental (and/or resource) management strategies in the continent were typically developed in a non-participatory way, without regard for the interests of surrounding communities.[47] This made such policies suspect from a popular perspective. All cases reviewed showed environmental policies in the spirit of the notion of sustainable development emanating from the Brundtland report and Agenda 21, as established at the UNCED. By the end of the last century, many plans and policies had been developed in Africa, especially in relation to sectoral issues. NEAPs and NCSs led to some progress, at least in terms of environmental awareness, sectoral approaches, and the constitutional anchoring of environmental concerns (as in Ghana and Ethiopia). Yet, Salih concluded that, in the evolution of environmental policies and strategies, African governments are still reactive rather than proactive. Moreover, they suffer from problems in implementation.

Explanations for the shortfalling implementation include the factor most often

[43] Salih, 1999a.
[44] Salih, 1999a.
[45] IUCN/UNEP/WWF, 1990.
[46] Salih and Tedla, 1999.
[47] Salih, 1999b.

mentioned: a lack of the technical, financial, and human resources needed to translate concerns about sustainability into practice. This factor is mentioned in almost all analyses of the environmental policies in developing countries, including Salih's.[48] It may be related to a more general lack of financial and human capital in developing countries, especially the poorer ones. But there are many more factors that merit thought and analysis.

One significant additional factor is the low intensity – if not lack – of political will to promote sustainable development. In international settings (such as the Rio and Johannesburg UN conferences and international negotiations on issues such as climate change and biodiversity), developing countries' governments often behave in accordance with the abstract Rio Principles. They sign agreements and framework conventions, especially when politically appropriate paragraphs are inserted about the links between poverty or shortfalling in economic development and environmental quality– usually couched in the language of 'common but differentiated responsibilities'. Nowadays, this even leads to integrated national strategies taking this point of departure in poverty reduction. But this is not a guarantee that implementation will occur or that, at the domestic level, environmental concerns will be given priority over the many other issues that face developing societies and their governments. Low levels of enforcement often reflect low levels of political will to, for instance, regulate pollution or polluters.[49]

Secondly, conflicts exist between the various sectoral institutions that may have overlapping responsibilities but are driven by totally different sets of priorities; these conflicts include competitive use of the environment and natural resources, and the use of different time frames in setting priorities and allocating scarce resources.

Thirdly, the phenomenon of an evaporating political will has been furthered by shortfalls in the behaviour of the OECD countries in living up to commitments made or suggested at large international gatherings such as the UNCED. Of the suggested substantial 'new and additional' financial means implicit in Agenda 21 at the level of approximately 50 billion US dollars, only a fraction was effectively made available through the Global Environment Facility (GEF), and development aid halved since 1992.

Related to the point of political will raised above in relation to government policies is the question of support for environmental policy statements, such as NCSs and NEAPs. Many were developed without (real) consultation with the communities that have a stake in them, or without consultation with the NGOs which have an interest in the issues involved. Many NEAPS remained

---

[48] Salih, 1999b.

[49] See, for example, the case of Nigeria, described in Ahmad, 2003.

only documents, seen as coming from the capitals, and – even worse – inspired (if not written) by external forces or their representatives.[50] An example is the 1994 Tanzanian NEAP.[51] It was intended to ensure the sustainable and equitable use of resources, and to prevent and control the degradation of land, water, and air. A local NGO was contracted to prepare the NEAP under the guidance of the environmental authorities. The WB provided the assistance of an expert, and many ministries, NGOs, and academics were formally part of the structure and/or the process. Awareness was low except amongst the ministries most directly concerned, and public participation was virtually absent. The NEAP remained an elitist document.

Fifthly, the problems of providing the institutions with resources may affect the quality of the information on the basis of which decisions on policy and implementation are made, hence affecting their effectiveness. This is part of a larger set of institutional failures or deficiencies in effective governance, which also includes factors such as logistical problems in large countries, differences in strength between the various layers of administration and management, and – not to be overlooked but too often politely ignored – corruption. Several studies of African environmental policy experiments mention the latter factor.[52]

Finally, when comparisons are made between environmental policies in OECD states and developing or emerging countries, this is often on the basis of the sets of instruments that are applied by the environmental institutions in these countries (see above). And there has been a heavy and constant pressure on administrations in developing countries to use certain instruments, which may be more effective in some settings than in others. Almost invariably, economic instruments have been designed and applied firstly in the setting of OECD countries. Even there, their success is generally far from convincing.[53] In the settings of societies and economies in developing countries, there may be more institutional failures affecting the effectiveness and efficiency of environmental policy instruments, including the obvious types of market failure and deficiencies at the policy and administrative levels. Salih observed that most African countries have (at least rudimentary) systems of environmental institutions, but they are not as effective as they are supposed to be.[54] The implementation of policies and of strategies is lagging behind policy formulation, and the missions of these institutions are not translated into adequate action. Among the elements Salih identified are: the institutions

---

[50] The WB is accused of this perhaps more than any other organisation (Nicro and Apikul, 1999).

[51] Mwalyosi and Sosovele, 1999.

[52] Adel Ate and Awad, 1999; Ahmad, 2003.

[53] Turner and Opschoor, 1994.

[54] Salih, 1999b.

are relatively weak, operate on a sectoral basis with inadequate coordination, and suffer from inadequate knowledge; underdevelopment and associated market failures show up in inadequate pricing and institutional constraints, and slow down the effective introduction of economic incentives; regulatory instruments are in use but they often date back to colonial days and are regarded as such, and thus suffer from a lack of support; the regulatory framework is dysfunctional and personnel are inadequately trained.

Similar observations result from case studies in Asia.[55] In a case study of Tanzania, Mbelle listed, among other things, the following additional obstacles: cultural rigidities (e.g., resistance to the destocking of rangelands), making almost all policies and many instruments ineffective; laxity in enforcement; limited options for (alternative) economic activities; limited transportation and communicative facilities; a narrow base for environmental taxation (in terms of taxable money flows and in terms of possibilities for effective tax collection).[56]

In several African countries (for example, Ethiopia, Ghana), the right of people to be consulted on projects and policies relevant to them has been laid down constitutionally. This is an important provision on which to base a more interactive, if not bottom-up, approach to the development of environmental strategies and policies. It could help bridge the distance felt by communities to exist between their interests and policy articulation. A new approach is seen as emerging, including elements such as participation, co-management, and benefit sharing.[57] Pleas for a similar approach have ensued from other case studies, and the point has been made that such participatory practices would enhance the chances of effective implementation.

The conclusions drawn with regard to the experiences and assessments of policies at the national level may, to a large degree, be taken to mirror those that would hold at the international level. Chaytor and Gray, for instance, echoed a finding on the impact of the Desertification Convention in Africa, where – as they noted – implementation of the Convention is a matter of survival. Yet, the degree of implementation appears to depend heavily on the levels of financial support and technology transfer: 'the lofty objectives of the Desertification Convention will only be met by increased development assistance in addition to the political will by African countries'.[58]

In looking at the powers of the various actors, Steel et al. observed that developing countries tend to have 'soft states', a notion they defined as 'weak, inefficient states, lacking in legitimacy and suffused with

[55] Nicro and Apikul, 1999.
[56] Mbelle, 1994.
[57] Tedla and Lemma, 1999.
[58] Chaytor and Gray, 2003: xii.

corruption'.[59] They also noted the variation in this qualification. For example, China is presented as a strong state. There may be ample reason to qualify and differentiate beyond this generalisation. For instance, corruption may be less prevalent, if not absent, in some developing countries, and should not be ignored in a number of post-industrial countries. Yet, as a generalisation, this statement has validity. Coupled with the above, Steel et al. argued that developing countries are easily dominated or manipulated by wealthy domestic elites and powerful foreign interests (especially transnational corporations). Elite interests are often linked with those of corporations. Here, too, one must be careful not to generalise. Ahmad stated that, in terms of their contributions both to environmental problems and to the types of interventions necessary to curb them, one should distinguish between domestic (private) corporations, multinational corporations, and domestic public (municipal, regional, national) corporations.[60] Different prospects, options, and mechanisms for rent appropriation are involved in each of these cases.

In terms of what can or should be done to alter this situation, Steel et al. advocated enhanced governance in the countries concerned, and urged post-industrial countries to promote that, via appropriate incentives, penalties, and rewards.[61] Much of what is disclosed by the studies reviewed here supports their emphasis on governance. But it also casts doubts on the adequacy of their 'sticks and carrots'-based approach. That approach may be conducive to, and perhaps in the short run even a necessary condition for, the development of a more effective environmental policy, but it will be far from sufficient. The *content* of the policy needs to be related to the relevant set of issues and to the possibilities of countries and their governments to effectively intervene in economic processes. This requires a contextualised, setting-specific, *institutional* approach. An obvious but time-consuming way to address the issues put forward by Steel et al. would be to enhance the quality and transparency of governance, administration, and development management in general. This is at the core of the concerns of donors (bilateral as well as multilateral ones) and civil-society organisations in developing countries, as witnessed by the growing number of programmes aiming at improved governance and democracy. This road is one that will take time. Ultimately, there must be an endogenous, embedded political will, as opposed to an exogenously manipulated one, to address these issues. And that brings us back to another set of actors: direct stakeholders and the agencies acting on their behalf: civil-society forces.

---

[59] Steel et al., 2003: 163.
[60] Ahmad, 2003.
[61] Steel et al., 2003.

Civil-Society Participation

The participation of civil society has been recognised as a crucial element in environmental policy-making and it is beginning to occur. One of the Rio Principles discussed above deals explicitly with the issue of participation and the conditions necessary for it to occur (Principle 10). A number of rationales have been put forward in support of citizen participation in environmental decision-making. Kufuor mentioned participation as a vehicle for achieving social justice.[62] He also regarded it as a check on the overriding influence of economic powers on decision-making and the associated risk of myopic environmental policies. Thirdly, he pointed out that it reduces conflicts between local communities and industry. Lastly, he suggested that participation might reduce transaction costs. Above, I have suggested that such participation might also enhance acceptance of – or commitment to – policies as well as the possibility of their implementation. Principle 10 of the Rio Declaration refers to: (1) the appropriateness of involving civil-society stakeholders in environmental decision-making; (2) their rights of access to environmental information available to government; (3) the obligation to inform and educate the public; and (4) access to legal recourse. It effectively prepared the ground for the 1998 UN Aarhus Convention with its three pillars of environmental governance: access to information, public participation in decision-making, and access to judicial and administrative redress. The extension of these ideas or similar ones to other regions (including Latin America, Asia, and Africa) is an ongoing process.[63]

At a more concrete level, questions arise as to how to conceive of public participation and how to achieve it. Options include the individual stakeholders in the public at large, community groups emerging through group formation processes, and (environmental) NGOs as crystallised or solidified representatives of certain interests. This cannot be elaborated upon in this context but, obviously, positive as well as negative aspects are associated with each of these options. Participation by stakeholders is a requirement in many national policies on environmental impact assessment, in the development of new poverty-reduction policy documents, etc. It is clear that there is still a gap between policy rhetoric and real involvement, but the beginnings of a process of increasing public participation exist.

Nicro and Apikul discussed the case of Thailand.[64] They claimed that civil society was the main engine in putting the environment on Thailand's development agenda. The environment became a topic that enabled the expression of dissatisfaction with governmental policies more generally. For a

---

[62] Kufuor, 2003.
[63] Bruch, 2003.
[64] Nicro and Apikul, 1999.

variety of reasons, the government in Thailand incorporated notions such as human rights and democracy. Internationally, too, 'people' were put increasingly, at least theoretically, at the centre of the development effort. This facilitated the recognition of the need for a participatory approach. In Thailand, NGOs were catalysts for local and public community awareness and grassroots action. Nationally, NGOs were capable of involving themselves in the processes of policy development and implementation. What may have helped much in all of this is the cooperation of the media with NGOs in drawing public attention to the issues raised by the latter.

One aspect related to participation discussed by Nicro and Apikul is the fact that regional programmes to address environmental issues are often top-down and show little involvement of local governments or civil-society representatives. This may be true in a general sense. Similar critiques have been voiced in relation to NEPAD – the new strategy for development in Africa, formulated by governments and informed by experts close to the large multinational development funders. Such observations may imply that efforts to develop regional approaches to environmental issues may lack support even more than do national strategies.

Civil-Society Involvement: No Panacea

Many social scientists have regarded the emergence of civil-society organisations (CSOs) as intrinsically beneficial: more CSOs implies less state power; more CSO influence is tantamount to more democracy, etc. Globalisation was seen as providing CSOs with a new role: globalisation 'from below' to counter globalisation 'from above', leading to social transformations that political agency had failed to bring about. Many assumed that CSOs would almost spontaneously work together as they had the same, or parallel, interests, norms, and values, and thus create effective and hopefully adequate countervailing power. Such views are now being recognised as naïve. Theories on agency in (global or national) policy development need to be refined to take such lessons into account, for example, by incorporating more realistic perspectives on the roles of the private sector and the state. I fear that the warnings of Habermas are still more pertinent than we may like, and that the notions of countervailing power should not be based on the presumption that anything that comes from civil society is good per se.

Dwindling states do not in themselves make for more civic autonomy, as market forces fill the voids more effectively. Moreover, CSOs do not always have parallel interests, and they are not necessarily democratic. A more realistic approach (analytically, politically) is emerging, recognising a number of ambiguities and contradictions in relation to CSOs and their impacts.

These include the limited extent to which environmental issues can be presented with adequate strength in contexts characterised by large-scale poverty and deepening inequality, or in contexts in which the freedoms of speech and organisation necessary for effective NGOs are absent. These limits have to do with the relative lack of financial resources in NGOs in poor countries as well as restricted accountability (or representativeness) of and within NGOs.

A research project in the field of environmental issues and actions in Mexico and the North American Free Trade Area (NAFTA) illustrated these problems.[65] Moreover, it shed light on the transborder interactions of environmental NGOs in that region in dealing with the setting up of the NAFTA, and thus provided a micro-model of what may be at stake globally. The findings of the project indicated that transnational cooperation between CSOs can be effective, but they also showed that there were differences in priorities, views, and analyses that had to do with contextual differences. There was much discrepancy in terms of constituencies, financial power, access to information, etc. There were asymmetries, notably between (relatively rich) US-based and (relatively poor and unconnected) Mexican environmental organisations to the effect that, at the end of the period of research, the former's priorities were apparently more effectively pushed than the latter's. There were, of course, also differences which were unrelated to the geographical context, such as those between radical and moderate organisations. All of these elements proved to be important and qualified the performance of the various actions (concerted or not) of the CSO networks concerned. There is a justified debate about the degree of democracy and accountability in NGOs. Often, also in developing countries, they reflect elitist positions. The NAFTA case showed that such regional development plans/strategies can indeed bring NGOs together in a transboundary network. They may also lead to enhanced environmental policies and/or implementation programmes, as was clear in Mexico. To some extent, the support these networks provided to grassroots environmental organisations may enhance democracy in the form of effective participation. But the study also showed how easily that is eroded by state agency and elitist NGOs dominating the effective developments.

Furthermore, the study illuminated another point made about the development of environmental policy in the context of globalisation. Much of the initial advance in Mexican environmental policy implementation in the wake of the NAFTA evaporated as a result of the impact of the Mexican

---

65 Hogenboom et al., 2003.

financial crises and their management. The 'Peso-crisis' led to a disproportionate cut in environmental budgets.[66]

Much of the idealism and wishful thinking of the early days of NGOs is now being replaced with empirically based realism. CSO involvement is no panacea for important national, regional, or global problems. Nevertheless, their involvement and empowerment – if accountability and representational issues can be resolved – are important, not least in environmental policy matters.

Improving Governance Processes

If it is true, as is argued in the first part of this chapter, that differences in (social, ecological, and cultural) contexts play a significant role and that differences in economic development level, knowledge, etc., are also important elements, then education plays a crucial role in transforming the ideological infrastructure for setting relative priorities in policies. This is a long-term dynamic element that should be given emphasis, but I cannot dwell on it here. More is at stake than priorities, values, and attitudes, and it is on these other aspects that I focus here.

In terms of policy instruments, the common practice in developing countries, as in OECD countries, has been to put in place a set of regulatory instruments, occasionally enhanced with (smallish) fines. Globalisation and the free-market economics it entails widens the set of options to include economic instruments, giving decision-makers incentives to enhance their environmental performance. It is mainstream wisdom that this is a better way of achieving sustainable development. It is proposed as an effective and efficient way forward in the Rio Declaration. Legal experts point out that the use of economic instruments may have advantages as the implementation of command-and-control instruments has been weak in developing countries as a consequence of institutional failure, lack of enforcement, duplication, and/or fragmentation of authority and other administrative failures.[67] Economists presume – even based on *a priori* arguments alone – that incentive-based approaches will be more efficient in achieving objectives (including environmental policy goals) if properly designed. In reality, globalisation means pressure for, and perhaps a governmental interest in, using instruments of environmental policy that appear to be compatible with the basic tenets of the globalisation project (i.e., economic instruments or incentives rather than regulatory approaches).

We must be careful not to generalise too easily here, and to avoid

---

[66] Hogenboom et al., 2003.
[67] Chaytor and Gray, 2003.

dogmatism or zealousness. Analysts of environmental policy instruments as they are used in real economic systems have found that these instruments do not come in pure form, but in mixed forms with, typically, a regulatory as well as an incentive component.[68] They have also found that, in order to assess instruments' effectiveness and appropriateness (including their cultural compatibility), one should look at the domestic and regional contexts in which these instruments are to operate. On these grounds, I conclude that a tailor-made approach is required in environmental policy instrument design rather than a generalised approach based on deductive arguments alone. Nnadozie, discussing the use of economic incentives in Nigeria, observed, 'Certainly, more sophisticated, market based environmental policies will definitely need more understanding, fiscal discipline, political will and an effective state machinery than exist in this country at the moment'.[69] I believe this illustrates the point I wish to make.

The above should not be taken as a plea against the use of economic instruments. Relying on incentive-based approaches can help much in enhancing effectiveness. In some cases (for instance, in South-East Asia), economic instruments have been very successful. Still, Nicro and Apikul, along with others, pointed to the disadvantages of using instruments exported to settings where they may be inappropriate given the differences in economic, political, social, and cultural contexts.[70] In so far as successes are real, one should be aware of the fact that, in South-East Asia, there has been a tradition of open economic and trade policies. Economic instruments are as likely as any others to have adverse or perverse effects. Tansini, describing energy and forestry policies in Uruguay, showed how a tax on fossil fuel initially led to accelerated deforestation, until accompanying policies stimulating hydropower and subsidies for reforestation curbed that effectively.[71] Subtle and balanced approaches are needed.

Policy instruments in any setting should be compatible with the relevant institutional capacity, stakeholder acceptance, and the political will to enforce them. A decade ago, Sterner formulated some general criteria for the successful development of environmental policy in developing countries involving economic instruments.[72] First and foremost is the availability of adequate knowledge of the environmental and economic aspects involved. Second is that the societies concerned must, to a reasonable extent, be characterisable as having functioning competitive markets. Third, there must be appropriate legal structures and sets of institutions defining property rights

---

68 O'Connor, 1999; Opschoor and Turner, 1994; Seroa da Motta et al., 1999.
69 Nnadozie, 2003: 126.
70 Nicro and Apikul, 1999; Parikh et al., 1999.
71 Tansini, 1994.
72 Sterner, 1994.

and establishing frameworks within which environmental authorities must work, with governments that have both the backing and the means to implement the policies. The general lesson he drew from a comparison of experiences in both developing and transitional economies is that it is best to envisage a phased process, beginning by pricing policies related to natural resources and in a later stage developing economic instruments affecting the environmental performance (e.g., pollution) through taxes and levies.

CONCLUSIONS AND RECOMMENDATIONS

On the list of priorities of developing countries, environmental policies hold third position at best, after those directed towards economic growth and poverty alleviation, and sectoral concerns in fields such as education and health. This reflects, of course, the political will to address environmental matters. In contexts characterised by a general lack of financial resources and human capital, the impact of this relatively low priority is that the effectiveness of environmental policies leaves much to be desired. Environmental policy agencies cannot be expected to effectively ensure sustainable development. This is even more true where these policies are effectuated in contexts of weak and weakening states, strong elites, and powerful links between these elites and corporate interests in the private sector; this is a second, important point. A third feature to bear in mind is that, within the national environmental policies, there has been a traditional leaning on inherited approaches now dubbed 'colonial' – where that phrase applies – and, rightly or wrongly, discarded as such. In past decades, the same feature arose in different guises: (a) the influence of Northern priorities on environmental policy agendas in the South, and (b) the choice of policy instruments. There has been an indirect influence, too: (c) development funding in situations of financial crisis has in many cases led to budget cuts affecting policy-driven approaches to dealing with social and environmental issues.

The development of environmental policies in developing countries has been influenced heavily by globalisation, both positively and negatively. Global discussions and negotiations about the environment and development as well as about poverty eradication and future needs for natural resources have provided an impetus to the development of environmental policies in developing countries. Globalisation has, at the political level, engendered environmental policy development and the creation of organisational structures to implement these policies: ministries, agencies, etc. However, the ideological perspective prevailing at present in the globalisation project places pressure on environmental policy in these countries to resort to market-compatible systems of economic incentives to achieve environmental objectives, and also pressures these countries to look at a particular set of reform measures to boost development or deal with crisis: institutional reform in neo-liberal style.

Turning to the real setting of economic activity, globalisation has impacts on economic activity levels and on activity patterns. It generally comes with an increase in trade. That in itself accelerates economic growth and changes the production patterns or the international division of labour. As market forces overtake political ones, the effectiveness of the checks and balances applied to economic agents by the political sphere diminishes. In such a setting, continued *economic* growth is not necessarily *environmentally* sustainable, nor does it automatically become sustainable in the Brundtland sense. The institutional transformation inherent in the current mode of globalisation increases inequalities, crowds out care, erodes social cohesion, and increases ecological unsustainability.

Without countervailing power exercised from outside the market system, this will exacerbate the problems of inequality and unsustainability, as well as other forms of human insecurity. CSOs may provide part of a corrective push to this biased dominance of short-term economic considerations institutionally embedded in the market system. In addition, there is a need for a re-thinking of the role of states and political agents (and checks on them!) in taking responsibility for managing the process of globalisation in relation to reaching human development goals, including the Millennium Development Goals. This sounds utopian – and it is, to some degree. But trends may be discerned which suggest an emerging re-orientation in this direction. The change in policy focus within the WB is one indication of this. Attempts at defining approaches to globalisation labelled 'managed globalisation' in UNCTAD are another indication. The strong call (and the response to it!) for new partnerships between private-sector actors, NGOs, and national and regional governments at the WSSD is a third indication.

In this context, there is a need to enhance the effectiveness of environmental policy in less-developed countries. Effective international policy development and institutionalisation to deal with global environmental problems and with the problems of globalisation are also required. At the national level, I have argued that the lagging effectiveness of the current environmental policies should not be seen merely as a sign of the need to provide more resources and raise budgets. Various ideological and institutional elements play a role as well, and must be explicitly taken into account in the tailor-made renovation exercises that are needed. Political systems and governance need to be strengthened, and the position of priority within the political culture of responsibility for future generations and biodiversity ensured. States should be strengthened in these ways. Improving institutions and governance includes establishing property and user rights as well as a culture of compliance with standards and of community involvement in resource management. Environmentally effective institutional reform is also needed to ensure that environmental protection and management issues are addressed in all relevant sectoral policies and development strategies, and not left to isolated agencies.

There is no single recipe for making environmental policy more effective in all developing countries. These countries differ in the environmental challenges, the available economic, human, and social capital, and cultural characteristics. They differ very much in terms of the economic mechanisms that have prevailed in their societies. These countries will, by definition, have to develop their specific policy agendas, knowledge centres, and institutional responses to the challenges they face. Given that variety, the usual approach of flying in with models which may work – or, to be more precise, may be regarded as working – in the wealthier parts of the world is bound to lead to frustration. Thus, the overemphasis on economic approaches to environmental degradation should be recognised and curbed.

In developing countries, specific issues will affect political systems and their effectiveness in achieving social welfare. There may, even more than in the North, be a role for CSOs, to represent, inform, and train people in terms of environment-friendly behaviour, and to influence the operations of other institutions. Education and related systems of value and attitude formation should be relied on to strengthen the robustness of the support base for environmental policies and sustainable development. Ownership and participation are important concepts in this regard.

In terms of the different categories of actors discerned here, globalisation enables action by CSOs. These organisations are forming global networks and developing concerted approaches to the systemic developments that prevail. Global civil society has shared values to a large degree, but the divergences of interests within these networks should not be overlooked. It should not be expected that any manifestation of civil-society agency will lead to a more democratic management of environmental resources. In focusing on occupying the void left behind by a weakening state-based political system, global civil society should not overlook the fact that the market powers may in fact be much more effective in gaining control over people's lives. Effective countervailing power vis-à-vis market forces may require smarter and stronger states, and a publicly managed approach to globalisation.

New partnerships between agencies of strengthened – but still fallible – states, progressive and responsible parts of the private corporate sector, and civil society may have to be mobilised in order to achieve sustainability on the ground. The North should help developing countries in overcoming some of the practical hurdles between where they are now and where they should be in terms of effective environmental policy. I see roles for cooperation in making resources available, but more so in assisting in domestic capacity development to deal with both national and international environmental concerns.

REFERENCES

Adel Ate, H. and N. Awad (1999), 'Effectiveness of environmental planning in Sudan', in M. Salih and S. Tedla (eds), *Environmental Planning, Policies and Politics in Eastern and Southern Africa*, London: Macmillan Press, and New York: St Martin's Press.

Ahmad, A. (2003), 'Policing industrial pollution in Nigeria', in B. Chaytor and K. Gray (eds), *International Environmental Law and Policy in Africa*, Dordrecht: Kluwer Academic Publishers.

AIV (2003), *Pro-Poor Growth in de Bilaterale Partnerlanden in Sub-Sahara Afrika (Pro-poor Growth in the Bilateral Partner Countries in Sub-Saharan Africa)*, AIV, 29, The Hague: Advisory Council on International Affairs.

Borghesi, S. and A. Vercelli (2003), 'Sustainable globalisation', *Ecological Economics*, **44**, 77-89.

Bruch, C. (2003), 'African environmental governance: Opportunities at the regional, subregional and national levels', in B. Chaytor and K. Gray (eds), *International Environmental Law and Policy in Africa*, Dordrecht: Kluwer Academic Publishers.

Chaytor B. and K. Gray (eds) (2003), *International Environmental Law and Policy in Africa*, Dordrecht: Kluwer Academic Publishers.

Giddens, A. (1991), *Modernity and Self-Identity: Self and Society in the Late Modern Age*, Cambridge: Polity Press.

Habermas, J. (1989), *The Theory of Communicative Action, Vol. 2: Lifeworld and System*, Boston: Beacon Press.

Hogenboom, B., M. Cohen, and E. Antal (2003), *Cross-border Activism and its Limits: Mexican Environmental Organisations and the United States*, Amsterdam: Cuadernos del CEDLA 13.

IUCN/UNEP/WWF (1980), *World Conservation Strategy: Living Resource Conservation for Sustainable Development*, Gland: IUCN.

IUCN/UNEP/WWF (1990), *Caring for the World; A Strategy for Sustainability*, 2nd draft, Gland: WWF.

Kaul, I., P. Conceição, K. le Goulven, and R. Mendoza (eds) (2003), *Providing Global Public Goods; Managing Globalization*, UNDP, New York: Oxford University Press.

Kufuor, K. (2003), 'The evolution and structure of popular participation in environmental decision making: The case of Ghana', in B. Chaytor and K. Gray (eds), *International Environmental Law and Policy in Africa*, Dordrecht: Kluwer Academic Publishers.

Maxwell, S. (2003), 'Heaven or hubris: Reflections on the new 'New Poverty Agenda', *Development Policy Review,* **21**(1), 5-25.

Mbelle, A. (1994), 'Environmental policies in Tanzania', in T. Sterner (ed.), *Economic Policies for Sustainable Development*, Boston: Kluwer Academic Publishers.

Munasinghe, M. (1999), 'Structural adjustment policies and the environment: Introduction', *Environment and Development Economics*, **4**, 9-18.

Mwalyosi, R. and H. Sosovele (1999), 'National environmental policies in Tanzania: Processes and politics', in M. Salih and S. Tedla (eds), *Environmental Planning, Policies and Politics in Eastern and Southern Africa*, London: Macmillan Press, and New York: St Martin's Press.

Nicro, S. and C. Apikul (1999), 'Environmental governance in Thailand', in Y. Harashima (ed.), *Environmental Governance in Four Asian Countries*, Hayama: Institute for Global Environmental Strategies (IGES).

Nnadozie, K. (2003), 'Environmental regulation of the oil and gas industry in Nigeria', in B. Chaytor and K. Gray (eds), *International Environmental Law and Policy in Africa*, Dordrecht: Kluwer Academic Publishers.

O'Connor, D. (1999), 'Applying economic instruments in developing countries: From theory to implementation', *Environment and Development Economics,* **4**(1), 91-110.

Opschoor, J. (1996a), 'Institutional change and development towards sustainability', in R. Costanza, O. Segura, and J. Martinez-Alier (eds), *Getting Down to Earth: Practical Applications of Ecological Economics*, Washington, DC: Island Press.

Opschoor, J. (1996b), *Sustainability, Economic Restructuring and Social Change*, Inaugural address, The Hague: Institute of Social Studies, 23 May.

Opschoor, J. (2001), 'Economic development in a neoliberal world: Unsustainable globalization?', in M. Munasinghe, O. Sunkel, and C. de Miguel (eds), *The Sustainability of Long-term Growth: Socioeconomic and Ecological Perspectives*, Cheltenham: Edward Elgar.

Opschoor, J. (2003a), 'Sustainability: A robust bridge over intertemporal societal divides?', in P. van Seters, B. de Gaay Fortman, and A. de Ruijter (eds), *Globalization and its New Divides; Malconditions, Recipes, and Reform*, Amsterdam: Dutch University Press, pp. 79-100.

Opschoor, J. (2003b), 'Sustainable human development and the North-South dialogue', in M. Darkoh and A. Rwomire (eds) (2003), *Human Impact on Environment and Sustainable Development in Africa*, Burlington: Ashgate.

Opschoor, J. and S. Jongma (1996), 'Structural adjustment policies and sustainability', *Environment and Development Economics*, **1**, 183-202.

Opschoor, J. and K. Turner (1994), 'Environmental economics and environmental policy incentives', in K. Turner and J. Opschoor (eds), *Economic Incentives and Environmental Policies: Principles and Practice*, Dordrecht: Kluwer Academic Publishers.

Parikh, J., T. Raghu Ram, and K. Parikh (1999), 'Environmental governance in India, with special reference to freshwater demand and quality strategies', in Y. Harashima (ed.) (1999), *Environmental Governance in Four Asian Countries*, Hayama: Institute for Global Environmental Strategies (IGES).

Reinicke, W. (1997), 'Global public policy', *Foreign Affairs*, **76**(6), 127-139.

Ricupero, R. (1999), *The Challenge of Development in the Knowledge Era*, Address at the Institute of Social Studies, The Hague, 15 July.

Salih, M. (1999a), *Environmental Politics and Liberation in Contemporary Africa*, Dordrecht: Kluwer Academic Publishers.

Salih, M. (1999b), 'Environmental policies and politics in Eastern and Southern Africa', in M. Salih and S. Tedla (eds), *Environmental Planning, Policies and*

Politics in Eastern and Southern Africa, London: Macmillan Press, and New York: St Martin's Press.

Salih, M. and S. Tedla (eds) (1999), *Environmental Planning, Policies and Politics in Eastern and Southern Africa*, London: Macmillan Press, and New York: St Martin's Press.

Seroa da Motta, R., R. Huber, and H. Ruitenbeek (1999), 'Market-based instruments for environmental policymaking in Latin America and the Caribbean: Lessons from eleven countries', *Environment and Development Economics*, **4**(2), 177-203.

Sideri, S. (1999), *Globalization, the Role of the Sate, and Human Rights*, Valedictory address, The Hague: Institute of Social Studies.

Steel B., R. Clinton, and N. Lovrich (2003), *Environmental Politics and Policy: A Comparative Approach*, New York: McGraw Hill.

Sterner, T. (ed.) (1994), *Economic Policies for Sustainable Development*, Boston: Kluwer Academic Publishers.

Tansini, R. (1994), 'Energy, forestry and environment in Uruguay', in T. Sterner (ed.), *Economic Policies for Sustainable Development*, Boston: Kluwer Academic Publishers.

Tedla, S. and K. Lemma (1999), 'National environmental management in Ethiopia: In search of people's space', in M. Salih and S. Tedla (eds), *Environmental Planning, Policies and Politics in Eastern and Southern Africa*, London: Macmillan Press, and New York: St Martin's Press.

Turner, K. and J. Opschoor (eds) (1994), *Economic Incentives and Environmental Policies: Principles and Practice*, Dordrecht: Kluwer Academic Publishers.

UNCED (1992), *Verklaring van Rio en Agenda 21 (Declaration of Rio and Agenda 21)*, The Hague: Distributiecentrum Ministerie VROM.

UNDP (1999), *Globalization with a Human Face*, New York: Oxford University Press.

UNDP (2003), *Human Development Report 2003: Millennium Development Goals; A Compact among Nations to End Human Poverty*, New York: Oxford University Press.

UNEP (1981), *In Defence of the Earth*, Nairobi: United Nations Environment Programme.

Waterman, P. (2003), *Place, Space and the Reinvention of Social Emancipation on a Global Scale: Second Thoughts on the Third World Social Forum*, working paper 378, The Hague: Institute of Social Studies.

WCED (1987), *Our Common Future*, Oxford: Oxford University Press.

World Bank (1993), *The World Bank and the Environment*, Washington, DC: World Bank.

World Bank (1998), *World Development Report*, Washington, DC: World Bank.

WWF International (1999), *Foreign Investment in the Asia-Pacific Mining Sector*, Gland: WWF.

Zarsky, L. (1999). *Havens, Halos and Spaghetti: Untangling the Evidence about Foreign Direct Investment and the Environment*, paper for the Conference on Foreign Direct Investment and the Environment, OECD, The Hague, 28-29 January.

Zoeteman, K. (2002), *Globalisation and Sustainability: On Governance and the Power(lessness) of the Nation State*, inaugural address, 24 May, Tilburg University.

PART II

Societal Perspectives

11. Drivers of Business Behaviour in the Realm of Sustainable Development: The Role and Influence of the WBCSD, a Global Business Network

Björn Stigson and Britta Rendlen

SUMMARY

The topic of this chapter is how and when business can best contribute to sustainable development, with an emphasis on the role of the World Business Council for Sustainable Development (WBCSD). The origin and evolution of the WBCSD are described and good reasons for conducting sustainable business are highlighted, including securing human and material resources, enhancing operational efficiency and effectiveness, and creating market value and business opportunities. Then, the work programme of the WBCSD is explained, including its emphasis on energy and climate, accountability and reporting, and sustainable livelihoods. The WBCSD's activities aim at changing mindsets, developing policies, building capacity, and reducing the environmental impact of companies. However, there are roadblocks on the sustainable-development path, in particular, profitability imperatives, the lack of adequate regulation, market imperfections, and capacity problems of governments. To meet the future sustainability challenge, government should eliminate regulative flaws, while business should learn to change, engage in partnerships, inform consumers, innovate, and make markets work for all.

INTRODUCTION

Imagine the following situation: a large multinational corporation is in the process of building operations in a developing country with low environmental standards, and needs to decide to what extent it will invest in measures that reduce its impact on the local environment. Any investment going above and beyond compliance with local legislation would be, from a legal point of view, unnecessary. Would a company deciding to go beyond legislative requirements really be wasting capital and destroying shareholder value? What drives a company to make such a decision? What additional factors does a firm need to take into account?

The WBCSD aims to demonstrate that there are many factors other than the rule of law, or short-term returns, for which companies should account when making strategic as well as operational decisions. Its mission is to provide business leadership as a catalyst for change towards sustainable development, and to promote the role of eco-efficiency, innovation, and corporate social responsibility. To the WBCSD, sustainable development is about resource efficiency, environmental balance, social progress, justice between generations, and the economic development of poor countries. The WBCSD is a child of the globalisation process, which is the main driver of its work and the reason that sustainable development is a key topic on today's political and business agendas.

It is difficult to say how much credit industry networks can take for the individual actions of companies, as organisations such as the WBCSD are certainly not the only influence on corporate decision-making processes. However, we strongly believe that the WBCSD does have some influence on member companies, and potentially on the business community at large. This firm belief in the value of the WBCSD's work is grounded in the many tangible results, verifiable through member case studies as well as member-led sector projects.

In this chapter, we first give a brief overview of the history and mission of the WBCSD. We then present the sustainable development agenda that drives current trends and actions, focusing on the role of the business community. Building on the business case for sustainable development, the fourth section focuses on the WBCSD's approach to today's challenges and highlights concrete results of our work. Finally, we take a look at the future trends and framework conditions necessary to achieve a sustainable society.

SUSTAINABLE DEVELOPMENT AND THE WBCSD

The history of sustainable development is a relatively modern one. It started in 1972, when the United Nations (UN) Conference on the Human Environment took place in Stockholm, focusing on global environmental challenges. Eighteen years later, at a UN conference in Bergen (Norway), business representatives discussed how they could contribute to the making of a sustainable society. Stephan Schmidheiny, a Swiss industrialist, argued at this meeting that the protection of the environment and business interests do not exclude each other but can be mutually beneficial. Schmidheiny later led a group of 48 other Chief Executive Officers (CEOs) of major companies in formulating the business view on sustainable development. This group was named the Business Council for Sustainable Development (BCSD). It was charged by Secretary-General Maurice Strong to represent business interests at the 1992 Rio Earth Summit, which focused on both environmental and economic development, recognising the close link between these two issues. The group published a book for the Earth Summit[1] and coined the term 'eco-efficiency' to represent a solid way for business to benefit from actions that protect the environment. In 1995, the World Industry Council for the Environment (WICE), founded by the International Chamber of Commerce (ICC), merged with the BCSD to become the WBCSD. The goals were to send out a concerted business message and to avoid splitting resources and efforts. Finally, at the 2002 World Summit on Sustainable Development (WSSD) in Johannesburg, the WBCSD, together with the ICC, mobilised the business community under the Business Action for Sustainable Development (BASD). As the Summit looked at the three pillars of sustainable development – social responsibility, environmental progress, and economic viability – in an integrated way, there was a great incentive for business to actively participate and contribute. And indeed, there were more CEOs than heads of governments present at this meeting. Today, the WBCSD is a coalition of 165 international companies united by a shared commitment to sustainable development. Our members are drawn from 35 countries and more than 20 major industrial sectors, and are represented in the Council by their CEOs or equivalents. We also benefit from a Regional Network of 44 national and regional BCSDs and partner organisations, involving some 1,000 business leaders globally. This Network represents the views of a wide coalition of business interests drawn from many different industrial sectors and geographical regions, mostly in the developing world.

Our objective is to participate in policy development to create a framework that allows business to contribute effectively to sustainable development. We work to demonstrate business progress in environmental and resource management and corporate social responsibility, and to share leading-edge practices among our members. In this context, we recognise the importance of global outreach to contribute to a sustainable future for developing countries and nations in transition.

Aggregated member company turnover of the WBCSD amounts to some 4,000 billion US dollars (USD). The full market capitalisation of the

---

[1] Schmidheiny, 1992.

WBCSD's members is close to 3,400 billion USD, and they employ about 11 million people. All of our member companies combined reach about 2 billion customers daily – a third of the world's population – through either the sale of products or the provision of services.

One could argue that we are a superfluous organisation that is 'preaching to the converted'. Should we not reach out to those companies that do not recognise their role in correcting environmental and social problems? There is much to do in creating greater awareness about what the corporate sector should contribute to sustainable development. However, the WBCSD has set its priorities elsewhere.

Change always requires leadership of the enlightened, and the WBCSD plays an important role in facilitating this leadership. Our member companies, proactive and progressive as they may be, do not know all the answers. Even though these companies recognise their responsibilities and the need to act, they do not necessarily know how this can be done in the most effective and efficient way. The idea of the triple bottom line is fairly recent, and still makes business strategies and operations complex tasks. Many uncertainties exist, both within the business community and outside, as demonstrated by the business and public policy agendas, which we discuss later.

The business community feels the need for a safe environment in which to explore ways of addressing the sustainability challenge. The WBCSD provides for this need by giving members a unique platform for exchanging, learning, and experimenting. Our members strive to become leaders in sustainability matters, and trust in the nature of the market to influence other companies to follow their example. The WBCSD is a member-led, member-driven organisation. Our work programme and policy statements are entirely influenced by the needs and realities of our members, who pledge their support and contribute by making available their knowledge, experience, and appropriate human resources. Membership of the WBCSD is by invitation only, and requires the personal commitment of the CEOs, who act as Council Members. They are supported by Liaison Delegates, senior managers who provide feedback to their companies and support their CEOs in the WBCSD's work.

DRIVING FORCES OF SUSTAINABLE BUSINESS

Globalisation as the Overarching Driver

Globalisation is facing a crisis of legitimacy. The past few years have seen a groundswell of anti-globalisation feelings, exemplified by demonstrations in Seattle, Davos, Genoa, and Geneva. Global markets and their possible negative consequences are typically the principal cause of these anti-globalisation protests. However, globalisation is a much larger concept than that. It is a direct result of human nature: our need to socialise and communicate, our will to learn and explore, and our desire to take advantage of opportunities when they arise. Globalisation can be described as follows: 'the rapid acceleration in the global exchange of information, ideas, goods, services, technology, values, people, culture, microbes, weapons, and capital. It implies a broad-based interdependence among nations in the three areas of sustainable development: economic, social and environmental.'[2]

International economic activity cannot be viewed in isolation from the ecological or social impact it may have. Neither can business ignore the protests and resistance of non-governmental organisations (NGOs) and other civil-society groups. We believe that business can benefit substantially from addressing the different dimensions of globalisation in an integrated way, as we show below.

Globalisation and its ecological and social impacts are complex. The issue becomes even more confusing when one tries to determine how sustainable development fits into the concept, and what the role of business should be. The need to establish clarity about facts as well as risks and opportunities, and to address the present global challenges in an effective way was a major impetus to the creation of the WBCSD and is still at the core of our mission. Our creation resulted from the desire to mitigate the negative consequences of globalisation, as well as the opposition to it caused by fear and anxiety.

Good Business Sense as a Driver

We see three kinds of companies in the corporate world: those that see business opportunities in sustainable development; those that reduce sustainable development to a public-relations issue; and those that ignore sustainable development. In the WBCSD's membership, we see predominantly the first sort, which understands that sustainable development makes good business sense. Clearly, this group makes up a minority of the business community at large. Yet, considering that twenty years ago the idea of integrating environmental and social aspects into a business plan was generally unheard of, it is encouraging to see even a few (and the number is getting larger!) of the world's largest corporations commit themselves to sustainable development. This change of mindset has happened quickly and is exemplified in the drastic difference of corporate representation at the Rio and Johannesburg Summits.

---

2 Gladwin, 1998: 29-30.

Image-building may be a consideration for companies to join the WBCSD, but it is, in our view, not a primary reason. Our organisation is often viewed as a 'window to the world', because we keep companies informed about new thinking and breaking developments that pertain to sustainable development. Member companies also benefit greatly from the learning-by-sharing-ideas on how other companies or industries grapple with their sustainable development challenges.

The Public Policy Agenda

Looking to the future, two specific agendas drive the business community towards sustainable development. The first is a public policy agenda driven by forces outside the corporate world. This agenda deals with the framework conditions and policies that society sets for business. It is based on the UN Millennium Development Goals (MDGs), Agenda 21 and the Plan of Implementation from the Rio and Johannesburg Summits, the Doha Declaration on international trade and socio-environmental concerns, the Monterrey Consensus on financing for development, and the different Multilateral Environmental Agreements (MEAs) entered into during the last ten years. The key issues of the public policy agenda include global governance, poverty eradication, sustainable production and consumption, the health of the ecosystems, energy and climate change, the role of innovation and technology, accountability and reporting, and risk. It should be stressed that the public policy agenda is not something companies can choose to deal with. It will have an impact on business, whether we like it or not.

The Business Agenda

The involvement of the business community in sustainable development is ultimately dependent on a strong business case for sustainability, which sets out why addressing sustainable development makes good business sense. Elsewhere, we outlined the business case from a societal perspective.[3] Here, we address the business perspective. The basic notion of the business case is that, in a world in which environmental and social issues come increasingly into focus, the long-term success of companies is dependent on how compatible their business models are with the principles of sustainable development. Below, we examine the significant opportunities and the existing risks in corporate sustainable behaviour.

---

[3] Holliday et al., 2002.

Enhancement of Brand Value and Generation of Goodwill

Effect on market valuation

Historically, mainly physical assets like land and access to raw materials defined the value of companies. At present, however, available data indicates that intangible assets may represent over 75 per cent of the market capitalisation of public companies. These intangible assets include reputation, brand, ability to participate in constructive dialogues and partnerships with stakeholders, ability to adapt and change, and the public risk perception. The value of these assets is highly influenced by the ability of companies to manage the public policy agenda. The growing number of sustainability indices and other evaluations indicate that there is a positive correlation between leadership in sustainable development and superior performance on the stock exchange.

Effect on licence to operate and innovate

Business does not function in a vacuum. On the contrary, it is intricately woven into the social and environmental fabric. Corporate responses to social and environmental challenges will affect how companies are perceived by society. This, in turn, influences companies in various ways: with regard to operations – boycotts, strikes, and other activities aimed at hindering the smooth operations of a company are less likely to happen when companies have built a reputation of integrity, responsiveness, and responsibility; with regard to investments – it is becoming standard practice for investment approvals in developing countries to report on the environmental and social past and potential future performance of companies in addition to the financial terms; and with regard to innovations – there is an increasing public distrust of science, specifically of corporate scientists, so goodwill is crucial for companies to achieve public acceptance of their innovations.

Operational Efficiency and Effectiveness

A number of business practices increase operational efficiency and profitability while improving the environmental and social performance of companies, and hence drive society towards higher resource efficiency: eco-efficiency, safety, and local community interaction. These practices can also stimulate ongoing creative processes and, therefore, heighten the level of innovation and product development, leading to a competitive edge.

Eco-efficiency

The combination of environmental and economic gains results from: optimised processes – moving from costly end-of-pipe solutions to approaches that prevent pollution in the first place; waste recycling – using the by-products and wastes of certain industries as raw materials and resources for others, thus creating zero waste; eco-innovation – using new knowledge to make old products more resource-efficient to produce and use; new services – for instance, leasing products rather than selling them, thus spurring a shift to product durability and recycling; networks and virtual organisations – shared resources increase the effective use of physical assets.

Safety

Safety is another area that has proven to bring value to companies. A safe working environment is by nature a more efficient environment. If machinery is well maintained, staff adequately trained, and accidents reduced to a minimum, companies will reap rich benefits in efficiency savings, as well as eliminating the danger of costly and reputation-damaging law suits.

Local community interaction

Being a good neighbour creates a more stable and efficient business environment for plant operations and facilitates approvals for new investments and government regulations.

Recruitment and Retention of Talent

Sustainable development gives companies a competitive advantage in the fight for talent. The best and brightest increasingly review the values and principles of employers, and are hesitant to work for firms with a questionable reputation.

Risk Reduction

Perceived and real weaknesses in environmental, social, or health-related performance can present great risks for companies. For example, asbestos litigations have driven a number of companies to filing for bankruptcy, or close to it. Also, pollution spills and clean-up can be costly and substantially reduce the value of land and buildings. Coherent and thorough sustainable development policies and procedures allow companies to stay ahead of the game, as they anticipate the potential negative social, environmental, or health-related effects of products, services, or manufacturing processes.

In addition, the financial industry increasingly demands information about the environmental and social exposure to risk of companies in order to assess both the availability and cost of funds and insurance cover. Companies with a lower exposure to risk enjoy better borrowing or insurance conditions.

Market Opportunities

The focus on sustainable development leads to both the emergence of new markets and increased value in enhancing products, services, and brands with positive environmental and social properties. Leading consumer goods companies increase their efforts to build environmental and social values into their brands, as they are aware of growing customer sensitivity to these matters. We note that sustainability is referred to more clearly in company advertisements, indicating that companies seem to derive a positive value from it.

Another new market opportunity lies in the developing world. The growth in future consumer and labour markets will, to a high degree, come from the emerging economies. To stimulate these huge markets, dominated by young populations, it is in the interest of business to support a supply of products and services that are affordable and increase the quality of life for the poor. This includes the fulfilment of basic needs such as sanitation and electricity, which stimulate the demand side by raising the purchasing power of these populations, as well as the provision of other affordable consumer goods. The social and economic well-being of these new markets is necessary if we want to maintain open and stable global markets.

Finally, another type of new market is represented by the growing use of market instruments for environmental protection. Sulphur emissions trading has existed for a long time in the United States (US). We now see emerging carbon dioxide emissions trading in the European Union as well as in the US via the Chicago Climate Exchange (CCX).

Protecting the Resource Base of Raw Materials

We currently consume many natural resources at a faster rate than they can be replenished. Major industry sectors depend on the sustainability of the natural resource base for their existence and success. From a plain utilitarian perspective, business needs to preserve these resources. Otherwise, it will cease to exist.

This is especially true for industries built on the use of renewable resources, such as forest products, food, and pharmaceuticals. The proper functioning of ecosystems such as climate, rainfall and waterways, fisheries, top-soil, and forests is a prerequisite for these companies to stay in business. For non-renewable resource-based industries such as mining and the fossil-fuel-based energy sector, the conditions under which we extract and use the natural resources will determine the viability of these industries, both in the short and in the long term.

Preserving for the Next Generation

In most religions and cultures, there is a strong sense of responsibility towards future generations. Business, being an integrated part of society, needs to share the responsibility for maintaining the world in a state that allows the next generations to meet their needs. By helping to alleviate poverty and build sustainable livelihoods, business can secure the natural resource base that the poor would otherwise have to consume to survive. According to Nobel Prize winner Murray Gell-Mann, sustainability means living on nature's income rather than its capital.[4]

THE WBCSD'S WORK

WBCSD Work Programme

The WBCSD's work programme for 2003 focuses on three crucial aspects for achieving sustainable development in which business plays a prime role: energy and climate, accountability and reporting, and sustainable livelihoods. These are all key topics in the post-Johannesburg world. The climate and energy debate now focuses on the future energy infrastructure rather than on climate change, which dominated the agenda since Kyoto in 1997. Since the 2001 Marrakesh agreement, our energy and climate working group has been concentrating on creating a common business position on the different pathways to the use of sustainable energy in the future. Our work on sustainable development reporting culminated in a report in January 2003 that is both a guide to help companies produce reports and a document offering insights into the reporting standardisation debate. The project is now pursued as a new and broader accountability and reporting working group. Under the rubric of sustainable livelihoods, we investigate how companies can improve the quality of life of the world's poor, by exploring synergies between private investment and development assistance, and by identifying new business models. We also bring our members into contact with local agencies that address poverty on the ground in the developing world.

A unique feature of the WBCSD's work is our so-called sector projects: voluntary, industry-specific initiatives. These projects demonstrate the power of partnerships and the commitment of companies to addressing some of the most difficult dilemmas in their industries along with the whole value chain. These sector projects adopt a participatory approach to developing solutions broadly supported by stakeholders. Their involvement is critical for the

---

[4] Gell-Mann, 1990.

credibility of our projects. The WBCSD is involved in research in four specific sectors: forestry, mining and minerals, cement, and mobility.

Overarching issues in the WBCSD's work programme are addressed under the umbrellas of cross-cutting themes and capacity building. Cross-cutting themes are a way to keep a strategic focus on important topics for which we currently do not have working groups. In this way, we ensure that we both track progress on past WBCSD initiatives and respond to emerging issues and breaking events. Cross-cutting themes include eco-efficiency, innovation and technology, corporate social responsibility, ecosystems, sustainability and markets, and risk. The WBCSD promotes capacity building through various education and training programmes, stakeholder dialogues, a vast case studies collection, learning by sharing initiatives, and the young managers team. Given the post-Johannesburg focus on implementation, it is important to build capacity so that business people can master sustainable development issues and learn how to address pressing challenges before they turn into crises.

Delivering Results

Our work spans a wide array of activities, ranging from awareness creation to policy development and implementation of sustainability principles, which lead to more or less measurable results. In this section, we give some examples of the more tangible achievements. It is important to note that the impact of the many dialogues we have been engaged in with NGOs, civil society, and government and business representatives often mark the beginning of a long-term process of change, the results of which are hard to measure.

Changing mindsets

The main achievement of the WBCSD has been to raise the profile and the acceptance of business as a contributor to the debate on sustainable development, and to champion its role as a solution provider. Overcoming widespread distrust of business motives at the Johannesburg WSSD is a good example. Our joint effort with the ICC to set up the BASD succeeded in presenting business with a human face and as a constructive, proactive force.

Over the past decade, we have developed eco-efficiency from an idea to a general efficiency concept within our member companies and beyond. We have also spawned national BCSDs around the globe as part of our Regional Network, and have put together a large number of case studies documenting numerous projects that advance the concept of eco-efficiency.

The WBCSD has a long history of dealing with forestry and forest products sustainability issues. In 1996, we were the first organisation to conduct a sustainability study on the entire paper cycle, including forestry, production, use, recycling, energy recovery, and final disposal. The WBCSD sustainable-forest-product working group provided the first forum for a number of large forest products companies with an international focus to work closely together to build knowledge and cultural understanding, as well as to forge a common business agenda based on sustainable development for the sector. Before we started our project, there were no international business networks and stakeholder dialogues focusing on forest issues. Our work has resulted in the creation of several such networks or partnerships: the Forests Dialogue (TFD), which includes a wide range of stakeholders; the Global Forest Industry CEO Forum; a collaborative framework agreement between the WBCSD and WWF International to undertake joint efforts to improve sustainable forest management outcomes; a quarterly dialogue forum with the World Conservation Union (IUCN) to strengthen cooperation in sustainable forestry and biodiversity conservation. The WBCSD has been instrumental in demonstrating that sustainably managed forest plantations represent a way of providing timber, crucial ecosystem services such as carbon sequestration, watersheds, and biodiversity, and benefits to the local community.

Policy development

The International Emissions Trading Association (IETA) was launched in June 1999 to help meet the objectives of the Kyoto Protocol by establishing effective systems for market-based approaches. The WBCSD was instrumental in forming the IETA, a non-profit member organisation made up of leading international companies representing a wide variety of private sector interests. The success of trading schemes will depend on companies providing transparent, reliable data about their emissions. To that end, in 2001, the WBCSD and the World Resources Institute (WRI) released the 'Greenhouse Gas Protocol', a first step towards internationally accepted greenhouse gas accounting and reporting practices.[5] The design process involved businesses, NGOs, governments, and intergovernmental organisations over a period of three years. The corporate accounting and reporting standard was tested by over 30 companies in nine countries. The European Commission has adopted this protocol as a basis for developing emissions trading policies.

Capacity building

As part of our capacity building activities, the WBCSD has joined with Cambridge University's Programme for Industry to produce a web-based tutorial called Chronos, which enables companies to explore the business case for sustainable development. The WBCSD has developed a training module

---

5 Rangathan et al., 2001.

for corporate employees on the topic of energy and climate. Five further modules are designed on corporate social responsibility, sustainable livelihoods, eco-efficiency, and the business case for sustainable development.

Our Regional Network plays a key role in reaching out to the numerous small and medium-sized enterprises (SMEs), which tend to have the greatest difficulty in improving their environmental and social standards. Several BCSDs have conducted training programmes and discussion panels, or established information/training centres to reach out to SMEs.

Implementation

In December 2001, the Brazilian BCSD, together with BP, the WBCSD, and the UN Development Programme (UNDP), launched a pilot project in accordance with the Clean Development Mechanism (CDM) model of the Kyoto Protocol. BP Solar installed solar energy in 1,852 schools in 11 states in Brazil. The project, valued at 10 million USD, installed over 1.3 megawatts of solar power in 9 months. The purpose was to 'road-test' the CDM and communicate the learning outcomes to other companies that want to participate in this part of the Kyoto Protocol.

Under the umbrella of the WBCSD's Cement Sustainability Initiative (CSI), 10 cement companies have developed and agreed upon a method for monitoring and reporting carbon dioxide emissions from cement manufacturing: the Cement CO_2 Protocol. The protocol aims at harmonising the methods for calculating the carbon dioxide emissions from cement production, with a view to reporting these emissions for various purposes. The Protocol is intended as a tool for cement companies worldwide. It allows the monitoring and reporting of all direct and indirect carbon dioxide emissions from the cement manufacturing process, both in absolute and relative terms. The CSI also signed an agenda for action, which requires companies, among other things, to respond to the recommendations of a health and safety task force, develop a climate-change-mitigation strategy and publish targets and progress by 2006, apply the above-named protocol to measure and monitor emissions and to hence achieve the emission targets, publish a statement on business ethics by 2006, and develop documented and auditable environmental management systems at all plants.

Roadblocks

WBCSD members put significant resources behind their efforts of improving their social and environmental performance. However, difficulties will always be encountered.

Profitability of businesses

One of the most obvious roadblocks to achieving sustainability is that a business has to be profitable in the long run, otherwise it will cease to exist. In essence, this means that while the term 'triple bottom line' is appealing, the reality is that there is only one bottom line. The task of the WBCSD is to show that good results on the environmental and social bottom lines will enhance the financial bottom line of companies. However, this is not a default effect: not all environmental or social improvements lead directly to a better financial performance. This is why it may be problematic for companies to weigh the costs against the benefits of environmentally or socially benign measures.

Lack of legislative frameworks

The WBCSD views the inaction of governments with regard to issues related to sustainable development as another impediment. Corporate voluntary emissions-trading schemes, pollution-reduction targets, and energy-efficiency programmes have long demonstrated that business has understood the importance of sustainable development. But business can only go so far without the support of sensible regulation.

Two legislative measures are of particular importance. The first is the introduction of full-cost pricing. This means requiring producers and consumers to account for the neglected social costs of environmental damage that is caused by the production or use of a good. Full-cost pricing can be realised through the application of pollution charges, tradable pollution permits, and fees for natural-resource use, as well as by establishing property rights over natural assets. These practices are the most fundamental and cost-effective ways in which markets can encourage, rather than discourage, sustainable development. An important effect of full-cost pricing is that it can – but does not necessarily have to – make costly measures to improve the environmental performance of, for example, a production process financially viable. Indeed, the cost of implementing these measures can be lower than the cost of the pollution that is avoided. We recognise that it is difficult to value particular externalities, but that should not stop us from trying. Full-cost pricing would also improve macro-economic policy-making by integrating social and environmental values into decision-making, and by allowing governments to benchmark progress towards their sustainable development objectives.

The second necessary legislative measure is the provision of a level playing field that ensures that companies investing in environmental and social improvements are not at a competitive disadvantage to free riders. Companies which do not self-impose restrictions, or manage to avoid them owing to a lack of effective monitoring or penalties for non-compliance, can incur lower costs. As with full-cost pricing, the goal is to encourage whole market segments to change so that supportive companies are not doomed owing to unfair competitive relations.

One way of achieving a level playing field is by introducing pollution-reduction targets. While these targets need to be set by governments, it should generally be up to the individual companies to determine how to reach them. An effective complement to pollution-reduction targets is market-based mechanisms. These reward sustainability-oriented behaviour and discourage pollution by making the polluters pay. Such instruments have a number of advantages: they are cost-effective, display a lower administrative burden, allow greater flexibility in the choice of means, and provide stronger incentives for improvement and innovation. A prototypical case is that of the sulphur dioxide emissions-trading programme created by the US Clean Air Act of 1990. In this programme, the companies were given the freedom to choose the most efficient and inexpensive way to control their emissions. Owing in large part to this flexibility, the sulphur dioxide trading programme has become a success: reduction targets have been exceeded by 39 per cent.

Command-and-control approaches tend to be less flexible and, therefore, can stifle innovation, removing the incentive for continuous improvement. They can also result in a technological lock-in (i.e., commitment to existing technologies). To be fully effective, any policy mix should be supported by horizontal mechanisms such as education and research. Voluntary initiatives can provide an effective alternative to legislation, but unfortunately also leave room for laggards.

Imperfect market frameworks

The WBCSD believes that sustainable human progress is most likely to be achieved through open, competitive, international markets that encourage efficiency and innovation. Today, we witness a virulent debate between those opposed to the global market and those in favour of it. Yet, there is no true global market, only the flawed, shaky beginnings of one. A number of conditions, such as imperfect local market conditions, distorting subsidies, political opportunism, and a lack of full-cost pricing, make it difficult for markets to support sustainable development.

Competition. Without competition, innovation is stifled, quality suffers, and monopolists, whose customers have little negotiating power, control prices. Competition also plays a vital role in driving business towards resource-efficient provision of goods and services, an essential aspect of sustainability.

Subsidies and trade barriers. One major barrier to efficient markets is subsidies that encourage damage to natural resources. Elimination of such distorting subsidies would free up tax revenue that could be better spent elsewhere. The amount spent on subsidies that undermine sustainable development dwarfs the aid money aimed at encouraging sustainability in developing countries. Globally, almost 1.5 trillion USD is spent each year to subsidise activities that cause significant resource damage or overuse, representing roughly 30 times the total amount spent on official development assistance. Eliminating even a portion of this amount could create significantly different cost and price structures.

Tariffs constitute another barrier that the industrial countries use to protect their domestic markets. Tariffs make it harder for developing countries to export their natural-resource-based and agricultural products, which are generally produced at lower costs. The inefficiency is hence twofold: first, consumers pay prices which are higher than the market values; and second, developing countries cannot benefit from their comparative cost advantage.

Pricing natural capital. Many of the Earth's raw materials, products, and services do not yet have a monetary value. Establishing such prices, in ways that do not cut the poor off from crucial resources, could reduce inefficiency of resource use and environmental pollution, and would, therefore, be a great step towards sustainability.

Quality and capacity of governments

Optimum government frameworks have as their core objectives the generation of economic value, the regeneration of the environmental resource base, the reduction of poverty and inequality, and the overall improvement of standards of living – all within an open, accountable system of governance. Frameworks that encourage business enterprise tend to reduce poverty, while those that make it hard for business keep people poor. Basic prerequisites include the rule of law and effective property rights so that risk-taking in the context of innovation can be rewarded, and the predictability of government intervention. Corruption is a serious impediment to enabling sustainability. Tens of billions of USD annually go towards bribes and kickbacks. This results in goods and services becoming more costly, and unaware consumers receiving poor and dangerous products.

FUTURE DEVELOPMENTS

It is clear that sustainable-development problems are too large and too complex to be solved by one sector only. Certain steps need to be taken by government and business.

Government Action

Governments, in consultation with business and civil society, need to correct the legislative and market-related flaws mentioned above. They also need to put conductive regulations and multilateral agreements in place, and lead societies towards sustainability, even if this means that unpopular decisions have to be made.

Business Action

Learning to change
Those who have already begun transforming their organisations tend to agree that the sustainability learning curve is mainly about vision, leadership, employee empowerment, stakeholder engagement, and the creation of new performance indicators. The required changes are not all sweeping and radical: an implementation process tends to use management tools and systems already in place. But leaders must set an example and 'walk the talk', serving as models and providing enough resources, education, direct challenges, encouragement, and follow-up assistance to their employees. This leadership needs to be complemented by training programmes on sustainable development, and application of this in businesses.

Engaging in partnerships
A particular element of the Business Agenda is the need for partnerships to address the sustainable-development challenges, which are 'too big to handle on your own'. The following kinds of partnerships are required:

Partnerships within business. Business shows a growing interest in looking at the sustainability performance and challenges for the whole value chain of industry sectors. Companies want to better understand future challenges in order to create more stable platforms for long-term investments, and to strengthen their licence to operate. A typical example, again, is the WBCSD sector projects. Companies can also influence the performance of their peers via voluntary, industry-specific standards, such as the chemical industry's Responsible Care programme.

Partnerships with government. Sustainable development is a relatively new concept for which the proper framework conditions have yet to be established by governments. Business has to remain an active player in designing these policies and, therefore, should engage in open dialogues and partnerships with governments.

Partnerships with civil society. The involvement of civil society (NGOs or other interest groups) in corporate efforts to achieve sustainable development is a way to gain credibility and the trust of the public. In addition, specialised NGOs provide expertise and new perspectives, which can help companies improve their sustainable-development performance. However, we see a substantial problem in relation to a lack of capacity and funding within the NGO community.

Informing and providing consumer choice
To achieve sustainability through market mechanisms, consumers must be well informed about the environmental and social impacts of products. This task needs to be completed by business in cooperation with NGOs and governments. But even when consumers are fully enlightened, it is debatable whether they are willing to pay a premium for environmentally or socially 'sound' products. Either way, they can be quick to turn away from companies that are seen to be ignoring environmental concerns.

Innovating
Even though there is no guarantee that technological innovation is sufficient to solve the problem of a growing population, increasing living standards, and the physical limitations of the planet, it is an important tool to address sustainable development. As the main source and user of technology, business clearly has a prominent role to play in meeting this challenge by investing in research and development.

Making markets work for all
Business cannot succeed in societies that fail. Making markets work for everyone involves two basic measures: enabling access to effective markets and spreading consumer purchasing power. These two measures, representative of supply and demand, go hand in hand. The first measure provides access to choice by developing and implementing innovative business models. The second measure ensures the freedom to choose by fostering income creation by the poor (via, for example, the provision of solid job opportunities and micro-loans). There will be rewards for companies that creatively meet this challenge.

CONCLUSION

Sustainable development is an ongoing process that requires the continuous evolvement of thinking and best practice. Despite this ever-changing environment and the many inherent uncertainties, it is clear that business plays a crucial role in achieving sustainability. While this represents vast opportunities for the business community, it also puts on it much pressure and responsibility. Business networks such as the WBCSD are crucial to provide guidance and support to successfully steer business operations in the right direction.

We stress again, however, that business cannot reach the goals by itself. It is only together that we can move from the current overuse of the planet's resources towards sustainable development. Government, civil society, and business need to work together and channel their efforts towards the same objective. While much has been achieved since Rio, great challenges remain ahead. In this sense, we agree with Henry Ford, who argued that you cannot build your reputation on what you are going to do.

REFERENCES

Gell-Mann, M. (1990), *Visions of a Sustainable World*, speech held in Santa Fe, New Mexico, May.

Gladwin, T. (1998), 'Economic globalization and ecological sustainabilty: Searching for truth and reconciliation', in N. Roome (ed.), *Sustainability Strategies for Industry: The Future of Corporate Practice*, Washington, DC: Island Press.

Holliday, C., S. Schmidheiny, and P. Watta (2002), *Walking the Talk: The Business Case for Sustainable Development*, Sheffield: Greenleaf Publishing.

Rangathan, J., D. Moorcroft, J. Koch, and B. Pankoy (2001), *The Greenhouse Gas Protocol: A Corporate Accounting and Reporting Standard,* Washington, DC/ Geneva: World Resources Institute/World Business Council for Sustainable Development.

Schmidheiny, S. (1992), *Changing Course: A Global Perspective on Development and the Environment,* Cambridge: MIT Press.

12. Globalisation, Public Utility Suppliers, and the Environmental Agenda

Jan Hol

SUMMARY

The relationship between trade liberalisation and environmental concerns in the European energy market is the central topic of this chapter. Nuon is one of the major utility suppliers in the Netherlands, aiming at sustainable business – which includes a high share of renewable energy. While Nuon's efforts have been fruitful so far, the ongoing liberalisation of the European energy market threatens the sustainability ambition. Competition increasingly takes place at the European level, where much larger companies operate. Besides, consumers are very sensitive to prices. The outlook for renewable energy in a liberalised market is dim because of the enhanced need for efficiency and the disappearance of a favourable fiscal regime. Therefore, the government should (continue to) stimulate investments in renewable energy. Nuon's response to liberalisation is international expansion, scale enlargement, and vertical integration. It believes that a future balance between environmental and business imperatives can and should be found.

THE ENERGY MARKET AND THE ENVIRONMENT

At first sight, the environmental performance of the Dutch energy market as a whole, and of energy supplier Nuon in particular, looks like a success story. The statistical evidence is impressive, especially in comparison with that of most other European countries. One headline figure summarises the difference: about 20 per cent of all electricity supplied to households and businesses in the Netherlands is generated from renewable sources, compared to an average of between 2 and 12 per cent in other countries of the European

Union (EU). Nuon has been an acknowledged pace-setter in stimulating the use of renewable (or 'green') energy by consumers and business customers throughout the past decade.

In its 2002 annual report, Nuon reported figures that suggest we are among the top-performing companies in Europe for the supply of renewable energy.[1] The following key figures apply to Nuon:

- The investment of more than 100 million euro (EUR) in renewable-energy developments during 2002.
- Over 600,000 customers taking green energy, which amounts to approximately 25 per cent of the total customer base.
- Two different green-energy products, one based on wind and water power, the other on biomass and other renewable sources.
- Leading company for the supply of solar photovoltaic energy in the Netherlands.
- Active in the supply of green energy, not just in the Netherlands but also in the United States (US), Germany, and Belgium.

Good though these figures are, it would be foolish of us to feel complacency concerning the success of the environmental agenda in the energy industry. The steadily growing interest in renewable energy in the Netherlands is a success story, but its long-term future is by no means secure. Gains have been made, not simply because individual consumers have committed themselves to green energy, but also because of specific pricing and taxation policies that were designed to foster increased use of renewable energy.

Some of these policies are either being reversed or may be negated by the larger forces at work in the marketplace. The European energy market is being transformed and everyone with an interest in promoting a responsible attitude to the environment needs to be aware of the impact this process of change could have on policies towards sustainability. In this chapter, we look at the challenges that the coming years will bring for environmentally conscious energy providers, and at how at least one of them – Nuon – is likely to respond.

Forces for Change

The world's energy markets have been going through a period of rapid and far-reaching change for some years now. The pace of change is, if anything, increasing and the long-term effects on key stakeholders remain difficult to predict. A number of important forces are interacting to reshape the

[1] Nuon, 2002a.

marketplace, and all of them play a significant part in affecting future strategic planning for business leaders and politicians. They also influence the economic well-being and quality of life of ordinary citizens around the world. The Nuon experience provides a useful and interesting study of how a relatively balanced approach to business and environmental issues can positively support the interests of all stakeholders in the marketplace. It also offers a relevant example of how a new and different kind of balance needs to be found in order to support improved environmental performance in what might be a very different future marketplace.

Rising Demand

The Boston Consulting Group (BCG) recently analysed the trends within the European marketplace in terms of likely future demand, the effects that environmental policy will have on fuel mix, and the financial implications of sourcing appropriate materials to supply that demand.[2] Some of the outcomes make for uncomfortable reading. It is likely that an extra 65 gigawatts of production capacity will be required by 2012 in the EU and IEA (International Energy Agency) countries. Virtually all of this capacity would come from clean-burning natural gas. If stipulations of the Kyoto Protocol – aimed at reducing the human impact on climate change – are fully implemented by all countries concerned, however, the requirement for new capacity will rise to an extraordinary 165 gigawatts, as generation from coal, orimulsion, and other sources that produce high levels of pollution is phased out. Almost all of this would need to be imported either as Liquefied Natural Gas from the Middle East or Africa, or through pipelines from Russia and other former Soviet states. The result is likely to be a substantial increase in energy prices across Europe, with some estimates suggesting that rises of 30 per cent could occur. There is good reason to believe that European consumers may find this difficult to accept.

The Risk of Boom and Bust

This is only the starting point for some difficult decision-making for energy-company executives and governments. The long-term need to satisfy growing demand still makes it possible for short-term oversupply to take place, leading to downturns in wholesale prices and commercial difficulties for many companies. In the United Kingdom (UK), for example, the so-called 'rush to gas' led to the building of excessive capacity, with several plants being left idle while general price cuts have crippled the nuclear electricity

---

2 Jansen et al., 2003.

industry that once supplied over 20 per cent of the UK's power. It is beyond doubt that extra capacity will be needed, but investing in the wrong way at the wrong time could lead to financial disaster. Yet, the failure to invest on time, as we have seen in California during recent years, can lead to chronic shortfalls and major economic damage.

The classic potential problem here is the 'boom-and-bust' cycle seen in both the UK and California, where perceived shortages lead to massive investment very quickly at a time of rising prices (boom), only for too much new capacity to come onto the market at the same time, leading to price slumps and the economic collapse of major players (bust). The BCG analysis is that exactly such a cycle (or probably several such cycles) will probably happen across Europe in the next decade.

Economic Globalisation

Globalisation has been a common topic of conservation for many years, but it is only since the creation of the World Trade Organization (WTO) that potentially globalised markets have started to become reality. For energy markets, globalisation remains a potentially important development but one that still has a long way to go. In the meantime, Europe has remained in a halfway position: moving towards a not fully clear liberalised future while continuing to use most of the mechanisms of the past. The move from local to national to large regional markets for electricity has changed the rules of the game dramatically. These steps on the path towards globalisation have opened up the possibility of a new breed of international electricity companies. Rapid consolidation is taking place at many levels and in many places. Cross-border ownership, once unthinkable in national electricity markets, is becoming commonplace.

Market Liberalisation

Many of the changes associated with globalisation are also intimately linked to the concept of liberalisation. It should be noted that, in most developed countries, there has been a growing movement away from publicly owned authorities that generate and distribute electricity over the past two decades. The speed of change has been uneven and this has caused problems in some European markets, where the national champions of countries slow to liberalise have had a temporary advantage over those that moved more quickly.

In general, however, liberalisation has introduced some new characteristics to the marketplace. Competition between different suppliers is growing rapidly, giving businesses and domestic consumers the ability to select their own suppliers based on service and price criteria. This growing emphasis on price as one of the key factors in consumer choice has a serious impact on the development of environmental policy. Liberalisation has also added a strong impetus to the drive for internationalisation, with national markets becoming more widely opened up to foreign entrants. Finally, as the electricity market becomes increasingly dynamic, it becomes a greater-than-ever subject for speculation. That has fuelled the growth of electricity trading as a core line of business for major electricity companies.

Despite some well-publicised problems in 'early adopter' markets, there seems no doubt that a more liberal model will eventually predominate within developed countries in the near future. In all such cases, liberalisation is changing the nature of the marketplace, replacing the ethos of publicly owned utilities with the commercial reality of heavily branded companies competing for customers on the retail model.

New Regulatory Frameworks

These changes in the market have led to the development of new approaches to regulation. Within European national markets, in particular, there is a general trend for regulation to be taken out of direct political control, exercised by Energy Ministries, and placed in the hands of an independent, specialist regulator. As the countries of the EU move further along the road to a single European energy market, these national authorities are likely to become subordinate to a pan-European regulatory regime. There has been little resistance to the concept of regulation vested in an objectively managed body rather than direct political control, but the actions of individual regulators have caused – and continue to cause – controversy.

The Environmental Agenda

The drive for open markets may directly conflict with an environmental agenda that has been an increasingly important factor in both uniting and profoundly dividing politicians, societies, and large enterprises. This subject is of particular importance to the electricity industry, because it has been targeted as a key contributor to the process of global climate change. The electricity industry accepts that 80 per cent of its power-generation activity leads directly to the emission of carbon dioxide, the main heat-retention (or 'greenhouse') gas. This is in itself a major incentive for change and, in addition, we can now put a definite time limit on the availability of fossil fuels. Oil and gas production are both predicted to peak between 2010 and 2025. Thereafter, the long-term trend will be downward.

The Kyoto Protocol of 1997 led to wide-ranging agreements on emission controls, reduction in greenhouse-gas generation, and a related tightening of regulations covering electricity production, distribution, and use. Despite the well-publicised controversies over ratification of Kyoto, many aspects of the agreement are already being implemented in Europe. Some of these have already led to significant changes in strategic priorities for many energy companies. These include a growing emphasis on renewable energy sources, based on ambitious targets set by a growing number of national governments. According to the BCG, full implementation of the Kyoto Protocol will require a further 70 gigawatts of renewable energy production by 2012, and the trend in this regard remains upward.[3] As a result of this, a lively and fast-growing market in emission trading has started to develop, enabling companies to offset their own energy emissions through credits (such as 'green certificates'), traded legally on international markets. Finally, an important issue for consumer-focused companies such as Nuon is the tendency of greater emphasis on environmentalism in branding and new product development.

These major forces affect the entire world energy system: no country and no company is immune to their influence. Nevertheless, strategic responses vary greatly from place to place and, in the Netherlands, there is a highly distinctive approach to market development that reflects a specifically Dutch tradition and sensibility.

THE DUTCH EXPERIENCE

The Dutch market for energy has much in common with other European marketplaces, but there are also some significant differences. We have already seen that the green electricity market is relatively well developed in the Netherlands; possibly because Dutch citizens are more environmentally aware than many of their European counterparts. Yet, the higher uptake of renewable energy in the Netherlands has also happened because this policy has been strongly fostered by successive governments, and supported through a targeted tax regime.

Nuon serves as a representative example of what has happened in a national market that changed steadily throughout the 20th century. Consolidation has taken place within the Dutch market since at least 1920. In the intervening years, the number of utilities has been reduced from over 300 to around 20, with the number of generators dropping from 16 to 4. The creation of Nuon is a classic example of this form of national consolidation,

---

[3] Jansen et al., 2003.

in which many local utilities – publicly owned and focused on small geographical areas – formed a single, much larger, and fully national body by a process of long drawn-out accretion. The ownership issue remains as it was, with the provincial authorities that owned the original utilities remaining the sole shareholders today.

The formation of the present organisation occurred over a five-year period between 1994 and 1999. The former Nuon was created in 1994 through the merger of three provincial utilities in electricity, gas, and water. These were largely concentrated in the north and east of the Netherlands. One more company was taken over in 1995, and two other utilities merged with Nuon in 1998. Meanwhile, a similar process of consolidation was happening in other parts of the country. By 1996, a further eight utilities from the north-west had come together to create Energie NoordWest. Finally, in 1999, the two larger groups formed from this consolidation of fourteen utilities merged to become the Nuon that exists today. The group remains owned by the provinces served by the original 14 local utility companies.

In the early 1990s, the profile, performance, and business mix of the companies that would later form Nuon was not in any major way different from the position of its major competitors. Since then, however, environmental performance has improved considerably, with heavy investment in sustainability. This strategy was adopted before the first Earth Summit – at Rio de Janeiro in 1992 – and represents a long-term commitment to the vision of sustainability, carried out not by a single management team but by successive groups of managers over a period of great change for the businesses concerned. In the past decade, we have witnessed: consolidation, with four companies becoming one; expansion, including substantial international investments; movement towards a new business structure; and tremendous upheavals in the marketplace, with liberalisation and a new regulatory regime. Throughout all of these changes, Nuon's commitment to sustainability has remained constant.

The reasons why Nuon started to make such a high-profile commitment to sustainability are complex and related to a number of different factors. Three of these stand out from the other and are as follows.

Core Values

Nuon's background is not significantly different from that of other Dutch utility companies, but is quite special when compared with the outlook and underlying principles of most European energy companies. Based on a distinctively Dutch sensibility, in which care for the environment and respect for the community are vital concepts for managers, customers, and employees, there is a stronger innate drive for sustainable working methods than elsewhere. As the figures show, Dutch utilities in general invest more heavily in sustainability than their European counterparts; not just in terms of fuel mix but also in other aspects of their business lives.[4] It appears easier to find general support for environmental ideas in the Netherlands than in many other European countries.

Differentiation

We should not disguise or attempt to deny the role that commercial considerations have always played in the development of a sustainability strategy. As a new company with ambitious growth targets, Nuon was determined to stand out from its competitors, in the Netherlands and abroad. Its brand policy was consciously developed to be different, and the environmental message has become an increasingly important component in this effort.

Financial Issues

Dutch government policy was, until recently, a major factor in support for sustainability. Thanks to its tax concessions in favour of renewable-energy consumption, it made good commercial sense for suppliers and users to prioritise renewable energy.

STRATEGIC FOCUS AT NUON

The background to Nuon's general strategy is the concept of balance. This means understanding that there are four key groups of stakeholders in the business: customers, employees, shareholders, and society (including the environment). Though it remains a strongly commercial organisation, Nuon's official goal is to achieve a productive balance between these four key interests, with the aim of ensuring high-quality performance for all of them. It seems increasingly likely that a balanced approach on similar lines may need to be followed by more and more energy companies in the future. In the face of stronger and possibly more restrictive regulatory regimes, much greater consumer pressure, and increased environmental concern, it is clearly essential for energy companies to learn how to deal successfully with this very different trading environment. Striking the right balance will become a key strategic requirement.

---

4 Nuon, 2003a.

To give one example, public opinion increasingly holds energy companies accountable for the entire production chain. This means that an electricity-generating company in Western Europe may risk severe damage to its reputation because of problems in waste processing related to the supply of raw materials for power generation – even when this processing is carried out abroad. In earlier times, municipal authorities processing coal into gas for use as a power source did not have to consider the likely future cost of decontaminating the land used for their operations. This form of cost *must* now be factored into their long-term planning and, of course, financial reporting. It is in basic, practical ways like this that the need to balance different stakeholders becomes embedded within strategic planning. Nuon believes that the best and ultimately most profitable way to deal with such issues is through fundamental business planning, driven by a clear strategic vision. This, rather than reluctant, ad hoc responses to external pressure, is the key to a successful future.

The starting point for review of Nuon's strategic response to the major forces shaping the European energy markets is the basic reality that, as a large player in a small country, Nuon does not have the critical mass possessed by the national champions of larger countries such as France and Germany. As the drive for consolidation gathers pace within a liberalising, globalising market, Nuon's options are necessarily limited and priorities have to be chosen with care. In the following sections, we look at the headlines that explain the broad shape of Nuon's strategic development.

International Presence

When Nuon was formed in 1999, it included not just a large capability to generate and distribute energy but also one of the largest water companies in the Netherlands. As a result of a clear direction from the Dutch government that water resources would not be included in any future privatisation schemes, Nuon decided to externalise its water assets within a newly formed company called Vitens – in which Nuon is the largest single shareholder. Within the core business (energy generation, distribution, and sale), Nuon's strategy reflects the need to prepare itself for the single European energy market that is prefigured by developments in the EU regulatory framework. In practice, the current position does not offer a level playing field for all entrants, owing to the different historical trends within individual European countries. Companies with large home markets (such as EdF in France, E.ON and RWE in Germany, and ENEL in Italy) have a greater critical mass than smaller companies (such as Nuon and Essent in the Netherlands and Vattenfall in Sweden). More important still is the fact that some of these emerging international players have maintained a near-monopoly status in their home markets. In France, for example, the top-three players in a market that nominally includes 150 competitors still account for 97 per cent of total energy output. In a pan-European market, this gives them a potentially decisive built-in advantage.

Building on this strength, a number of companies have already developed strong international holdings. Medium-sized businesses such as Vattenfall and Scottish Power, with comparatively small home markets, each earn around half of their income outside their own countries of origin. E.ON, RWE, EdF, and Endesa (Spain) have built on their strength by each achieving between 34 and 43 per cent of their income from business abroad.

A Wider 'Home Market'

In response to these unarguable facts, Nuon has adopted a commercial strategy that is designed to increase the scale of its 'home' market while also establishing differentiators that enable it to add value to its offerings in a market that is becoming both more competitive and more commoditised. From mid-2002, Nuon has started to see not just the Netherlands but also a much larger territory that includes both of its nearest neighbours (Germany and Belgium) as the true home market. This approach reflects not simply a need for scale but a true, continuing convergence that is taking place between these three countries in parallel with the slower, broader movement to a single market within the EU as a whole.

Vertical Integration

This highly pragmatic approach to international activities in energy distribution, wholesale, and retail sales is conditioned by the pace of European liberalisation. It is clear that, in the future single market, companies possessing some degree of vertical integration, combined with a large home market, will have a decisive competitive advantage. There are clearly some regulatory obstacles in the way of vertical integration. In the Netherlands, for example, the trend over the past decade has been strongly towards unbundling assets: separating generation from distribution. Nuon played its part in that process by selling off many of its power-generation assets.

It seems that this strategy now needs to be reversed quickly. Given the unpredictable nature of the forces shaping the single energy market in Europe, it is becoming increasingly clear that only companies with a strong presence across the entire value chain can show the resilience needed to cope with unexpected events. Nuon's need is to extend its home base and ensure the right balance between production and distribution in order to survive and prosper in the liberalised marketplace. Other strategic goals have to take second place to this, which explains why Nuon has recently made significant moves to acquire additional generating capacity, thus reversing the process of unbundling that seemed logical in a national market but is clearly untenable in a single European market.

Renewable Energy Sources

Sustainability is a long-term differentiator for Nuon and recent strategic developments have focused on ways of using leadership in renewable energy not just to strengthen Nuon's position in its home market, but also to establish value propositions that can be extended to other areas within a single European marketplace. In this context, the drive to develop renewable energy sources has led to business activities that extend well beyond the north-west European heartland of Nuon's retail and distribution business.

Nuon has interests in most of the currently achievable forms of renewable energy production: biomass, geothermal, solar, hydro, and wind energy. Several of these sources have little relationship to international business. Biomass is based on sources such as landfill waste gas, wood residue, and organic by-products from the paper industry. As with fossil fuel, using biomass to generate electricity causes some pollution, but the source material is renewable. Geothermal energy production involves the use of heat pumps to extract warmth from groundwater, using this to heat domestic water before returning the groundwater to the earth. This normally provides power to individual homes or small developments. Nuon is a significant innovator in solar power, and is involved in around half of all solar-power projects in the Netherlands. In hydropower, Nuon's five Dutch developments are complemented by two additional schemes in France. In wind power, Nuon is rapidly becoming one of the largest players in an increasingly significant European marketplace. By the end of 2002, the company managed a total of 39 large wind farms in Europe. Of these, 23 are located in the Netherlands and 16 are based in four other European countries. Wind energy is an increasingly international business, with joint ventures to develop large schemes in technically challenging areas likely to become more common in the years ahead. This is an area in which demand, fuelled by EU and national government targets, currently outstrips both available supply and the specialised skills needed to develop the necessary capacity. In this area, building a strong track record, as Nuon is doing, could provide a much-needed point of differentiation within the single European energy market of the near future.

Renewable Energy Products

As indicated above, the Dutch energy market is perhaps the greenest in Europe. By mid 2003, no less than 23 per cent of Nuon's customers had opted for energy products based on renewable sources, numbering over 500,000 households altogether. There are several key factors in this continued, rapid rise in the uptake of renewable energy, including product design, credible measurement procedures, and tax regime.

For the consumer market, Nuon has developed two distinctly positioned branded energy products in the renewable energy space. Nuon 'Natural power' is guaranteed to have been generated only from non-polluting, renewable sources. This means wind, solar, and hydropower energy. Nuon 'Green power' is also generated from renewable sources, but may include biomass – which makes the product less clean than the former type. Total sales for these two products in 2002 reached more than 1.1 billion KWh.

Following a long period of consultation and development carried out under the auspices of the EU, a Europe-wide scheme for certifying renewable energy production was introduced in 2003. This scheme, known as RECS (Renewable Energy Certificate System), will effectively 'label' renewable energy in packages of 1 KWh. The scheme should provide European citizens with the confidence they need concerning the real origins of the energy they use. This is a much-needed form of reassurance to a public that, even in the Netherlands, is not wholly convinced about the claims made for green energy.

Financial Incentives for Green Energy Use

The growth in use of renewable energy within the Netherlands has been greatly encouraged by the favourable tax regime followed by successive Dutch governments. This had the effect of imposing a differentiated tax burden on energy supplies, with by far the lowest level of tax being charged on renewable energy. With effect from early 2003, this policy has started to change, and energy tax is now levied on renewable energy as well as on all other kinds, though at a lower level for renewable types.

The general direction of Dutch government policy is towards continued reduction of tax differentials, to be replaced by direct grants to producers. Even this is limited to renewable-energy production within the Netherlands: an especially unfortunate move, given the increasingly international nature of the wind-power market in particular, where part of the renewable energy generated by Nuon in other countries is imported for use in the Netherlands. This somewhat incoherent policy runs directly counter to the stated goals, both of the EU and the Dutch government, to reduce dependence on conventional (or 'grey') energy and increase the proportion of renewable energy. The long-term influence of this and other equally contradictory policy developments is addressed below.

The Wider Meaning of Sustainability

Nuon defines environmental concerns, expressed by the notion of 'sustainability' in the broadest possible way. To us, it does not simply mean finding clean and renewable energy sources, but is rather a philosophy that permeates every aspect of the business. A company that takes sustainability seriously takes strong measures to cut down on all activities that tend to create pollution or unnecessary wastage. For Nuon, this has led to a wide range of initiatives, detailed in the 2002 sustainability report.[5] Some of the most notable include: specialising in street lighting (through a subsidiary company) that uses less energy than normal and requires less frequent maintenance; reducing the use of lubricants in plants, cutting down on leaks, and more effectively recycling waste oils; and restricting travel through car-sharing schemes and similar measures. Large or small, these initiatives and many others like them make a measurable difference, but also reflect a specific vision and management style.

With the recent move towards building a strong position within an extended 'home market', we can see evidence of territorial consolidation linked to a push for growth within an added-value segment of the market. Sustainability has been a core component of the commercial strategy for many years but the question is how well such an approach will survive the reality of the forthcoming single energy market.

MARKET LIBERALISATION AND ENVIRONMENTALISM

In European countries, the progress of liberalisation is happening in an uneven way, but it is now an inescapable fact of life for all. At the moment, several national markets are already completely open to competition to external companies. These include the UK, Germany, and some of the Nordic countries. By the end of 2004, the Netherlands, Spain, Belgium, and Denmark will also be fully liberalised. France and Italy are moving at a significantly slower pace than their neighbours, but even they are finally on track for full liberalisation by around 2007.

---

[5] Nuon, 2002b.

Multi-Speed Liberalisation

Under the general heading of liberalisation, there are varying levels of commitment, and these can be objectively measured according to a number of key criteria. First of all, several European countries contain dominant players with sufficient critical mass to gain a significant advantage in a single European market. This status is due largely to historical reasons, based partly on the sheer size of the countries concerned but especially on the centralised regulatory approach followed in these countries. In France, therefore, dominant players account for 92 per cent of all electricity sold. In Germany, the figure is similar: 89 per cent. Some other national markets remain far more fragmented, with fewer dominant players, resulting in smaller-scale companies less able to compete in a cross-border marketplace.

Tax regimes also have a large influence on the freedom of companies to operate in their national markets. Generally speaking, in countries where tax forms a smaller percentage of total energy cost to consumers, major players have greater freedom to manoeuvre in price policy. High tax regimes correspondingly reduce options in this respect. Tax percentages in Europe range from a low of 6 per cent in Norway, through a median band of 11.5 per cent in the UK, to a much higher level. The Dutch tax rate is 19.2 per cent, among the highest in Europe.

Consumer Power

As markets become freer, it becomes easier for consumers and large business customers to switch suppliers, and we can already see huge differentials in this regard. In the Netherlands, where commercial and renewable energy sectors are now fully open to competition, churn (i.e., shift to another supplier) in the first year of open trading reached 15 per cent. In the UK, where the market has been fully open for about ten years, churn has stabilised at around 10 per cent. In Germany, however, where liberalisation has yet to arrive, the rate is still no higher than 1 per cent. Even at a purely technological level, there are still major differences in true openness. Where import and export are concerned, for example, the Netherlands is again at the top of the list, with easy access to imports from Germany and Belgium, potentially making this country an attractive target for would-be exports from neighbouring countries.

Regulatory Differences

Finally, large discrepancies still remain in many aspects of the regulatory regimes in force today, ranging from the whole-hearted early adopters, such as the UK and the Netherlands, to the much more reluctant latecomers, of which France and Germany are the most obvious examples. In Denmark, the UK, and the Netherlands, for example, the legal obligation to unbundle assets has been much more strictly enforced than in France and Germany. This makes consolidation in the more open markets significantly more difficult than in the less free national markets.

The same differences can also be seen in some aspects of basic market regulation. Germany is, for example, nominally a fully open trading environment, but there is as yet little available information about the regulatory regime that will ensure adequate access to the power grid. The EU target is for independent regulation via a fully empowered sectoral authority, in place from the outset of market liberalisation. In Germany, access is to be managed by third-party agreement on a case-by-case basis. In the Nordic countries, as well, regulatory authorities will have less influence than in the rest of Europe. In France, not surprisingly, the ultimate accountability will rest with a minister rather than with a regulator.

The Liberalisation 'League Table'

It is possible, therefore, to grade the countries taking part in the open energy market according to the completeness of their liberalisation agenda. When the process is complete, the UK, the Netherlands, Sweden, and Belgium will be the most thoroughly liberalised, truly open national markets. France, Norway, Denmark, Finland, and Austria will form a second, slightly less liberalised tier, and Germany, because of the lack of clarity about its intentions, is currently seen as the least liberal national market of all.

How Liberalisation Affects Business Priorities

The rationale for wholesale liberalisation is based on criteria that are partly commercial and partly social, and also have major political significance. In commercial terms, the market is undergoing a process of consolidation, in exactly the same way as many other service and manufacturing industries throughout the world. Major producers and retailers are looking for ways to achieve economies of scale, to satisfy their growth ambitions, and to open new opportunities and new markets for exploitation. In social terms, national and supranational authorities (in this case, the EU) are seeking for ways to improve quality standards and reduce prices, partly to give their citizens a better deal and partly to raise the efficiency levels of other economic activities that depend on steady, reliable, and competitively priced energy sources. In political terms, the EU has targeted the energy industry as one of many areas in which European levels of efficiency were seen as falling below best world standards (notably those in the US), to the detriment of Europe's general economic performance. In this set of calculations, which date back many years, environmental concerns have always been present, but not as a driving force for major structural change of the kind occurring now. On the contrary, the main lines of liberalisation as a strategy were set in place long before the Kyoto Protocol placed environmentalism higher up the policy agenda than before.

In practice, therefore, it seems as if liberalisation and environmentalism are not properly connected. The key question is whether they are even compatible – in which case a creative tension exists – or are mutually exclusive (i.e., constitute a contradiction in terms).

The Pricing Mechanism

Experience in previously liberalised markets, the best example of which is probably the UK, suggests that once the initial start-up phase is past, the price issue becomes paramount. It is true that British consumers have been given a relatively good deal in price terms by the new regulatory framework and the increase in competition within the marketplace. Yet, it is also clear that a single-minded focus on low pricing within the framework of greater consumer choice may run counter to the needs of a strong environmental policy. Recent developments in the Netherlands have seen some of the policies that encouraged a greater uptake of green energy over the past decade reversed at the very moment when European governments committed themselves to more stringent environmental targets. This apparently illogical action has nothing to do with environmental concern, it is true, but it has everything to do with consumer-focused market competition. At the very moment when concerns about environmental change are reaching new levels, the commercial and regulatory changes taking place in Europe tend to conflict directly with the environmental agenda. This and similar developments in other national markets also reinforce the point that, even in a liberalised environment, political issues can be used to manipulate the performance of the marketplace as a whole. Ground rules will be set by regulatory bodies, and it will be a long time before a true level playing field exists from one end of Europe to the other. In the meantime, changes made by independent regulators exert a powerful influence on business relationships and even on the viability of entire market sectors.

National governments still exercise a decisive influence on the development of the more liberal markets now emerging and, therefore, also on both the service that consumers receive and the commercial prospects of electricity suppliers. Nuon believes that some national governments have been naive in their attitudes, imposing somewhat doctrinaire policies without due regard for what is happening in neighbouring countries. The very existence of a multi-speed European marketplace, with some countries moving much faster towards liberalisation than others, is clear evidence of this. Governments do not help their own economies, the convergence of European markets, or the cause of environmental responsibility by moving too fast without insisting on corresponding moves from other European governments. In the end, Europe's best interests will be served by creating a level playing field within a consistent framework that gives due priority to the need for sustainability. National governments would do posterity a favour if they paid more attention to negotiating this kind of long-term settlement and less attention to being the first movers at any cost.

The Path to Privatisation

One final point needs to be taken into account when considering the impact of liberalisation: the relentless drive towards privatisation in the energy market. There has never been a standard format, size, or shape for energy companies, but in most European countries, they tend to have been publicly owned from the late 19th century onwards. Some were extremely local in scope, while others became extremely large, centrally controlled, nationalised industries. Whatever their backgrounds, however, they are almost without exception aiming for full privatisation in the near future.

This will be enough in itself to cause profound changes in culture, in some aspects of behaviour, and, surely, in attitudes to commercial priorities (including environmental policy). Concerning this latter point, it is important to make some predictions about likely outcomes. First of all, privatised companies operating in a price-sensitive, competitive market will inevitably find themselves operating under similar rules to those that prevail in the retail market. This should have the effect of making the companies within that market become automatically more customer focused and service oriented than before. Customers can change suppliers relatively easily in free markets, and service failures are likely to be punished as long as suitable, credible alternatives are available. In this respect, they follow the same path as other industries that were once monolithic, monopolistic, and sometimes disdainful of consumer issues. These include industries such as telecommunications, financial services, and transport. We can assume that the energy market will show some similar characteristics as it moves further down this path.

We can also expect to see some other companies expanding into what can certainly be seen as inappropriate sectors for utilities, but may be quite normal for retailers. The former state-owned British Gas was divided in two on privatisation. One of the resulting components, Centrica, is now a classic example of a pure service conglomerate, with energy distribution, electricity trading, insurance, motoring services, and other financial-service interests all under the same umbrella. The long-term cffects of similar developments throughout Europe cannot be accurately predicted.

In summary, market liberalisation changes the rules of the game for two reasons:

- Free markets strengthen the price mechanism to an unprecedented level. Instant consumer satisfaction is the key requirement and policies that may stand in the way of this classic retail behaviour is instantly put at risk.
- Privatised companies are subject to financial market pressures in ways that utilities can only guess at. Stock prices, shareholder satisfaction, market sentiment: all of these factors have an impact on executive decision-making and strategic developments.

It is hard to see how any of these changes will positively effect the environmental performance of energy companies, unless regulation is relentlessly (and equitably) focused on this area, or unless companies can find intelligent ways to make sound environmental policy also become sound business sense.

RIGHT BALANCE OF ROLES

There is no clear right or wrong way to operate successfully in a liberalised environment. Nor can we say with any certainty how best to reconcile the logic of the open market with the growing need to improve environmental performance throughout the energy industry. It is possible, however, that Nuon's own strategy offers useful experience that will become increasingly relevant as the process of liberalisation progresses.

The Importance of Balance

Nuon's strategy is based on three key principles:

- To be a sustainable entrepreneur, an organisation that aims to create value in three dimensions: profit, people, and planet. The aim is thus to ensure that commercial success is linked closely to successful personnel management and environmental policies.
- To achieve a sense of balance between all of its stakeholders. This means excellent customer service, good working conditions, an above-average return on investment, and caring for the environment.

- To extend the concept of sustainability by embedding sustainability in all aspects of the business.

Can Sustainability also be Good Business?

This strategic approach is based on the belief that sustainability is good for business at every level and in every area. It has led to the implementation of policies that go far beyond regulatory requirements. It aims for an ideal blend of commercialism, delivering real competitive edge in an increasingly competitive marketplace, and social conscience, based on care for employees and the environment. The sustainable approach has been made part of Nuon's business routines. The evidence suggests that showing concern for the individual in this way significantly raises morale, improves operational efficiency, and has a positive effect on the serviced standards the company delivers. In this dimension, at least, it seems that the commercial and the socially responsible objectives can co-exist quite happily. Over the past decade, the figures suggest that Nuon has succeeded in making its position on sustainability become a commercial advantage rather than a financial liability. Yet, the fact that a policy has generally worked satisfactorily over the past few years is no guarantee that it will continue to do so in the future. The way ahead will probably not be easy, and some difficult choices will need to be made – by companies, consumers, governments, and regulators – if we are to have the energy market that we need and want. In this changing market, Nuon will continue to support the strategy that has served well in the past decade: an approach that combines a responsible attitude to the environment, and a determination to deal fairly with all stakeholders while still being able to compete strongly in an increasingly liberal marketplace.

The Consumer Priority

The essence of liberalisation, and perhaps its main justification in the eyes of its apologists, is that a liberalised marketplace is, by nature, more strongly focused on the interests of consumers. Nuon recently commissioned research in the Dutch marketplace concerning consumer attitudes to renewable energy in order to test real attitudes to sustainability, pricing, and the all-important relationship between the two.[6] The results were interesting but not encouraging in some ways.

Awareness of green energy is high, close to 100 per cent, and most people were able to say how green energy is produced – with discrepancies based on age and educational attainment. Up to 40 per cent of these respondents

---

[6] Nuon, 2003b.

already use green energy, though it should be noted that use by the lowest age group, 22-35 years old, is half of that by the group of people aged 35-50 years. People's reasons for using and for not using green energy were also clearly explained. Around 80 per cent of all those who currently use green energy, or who plan to, stated that environmental reasons drove their choice. The real problem emerges when we look at why people do not (plan to) use green energy. Around 25 per cent said that they simply did not trust their suppliers to give them only green energy through their power sockets. There is no special green outlet for energy and no clear proof that energy is sustainably produced, and a significant proportion of consumers do not trust the industry to deliver as promised. A higher proportion, around 35 per cent, think the price is either too high or is likely to move upwards too fast in the future. Others, around 10 per cent, simply felt that care for the environment is a matter for the government and not the duty of consumers.

Price sensitivity turned out to be another difficult issue. When asked whether they would change back from green to grey energy should the price increase significantly in the future, a large proportion said they would. The difference among the age groups is striking. Of older people, 70 per cent said that a substantial price increase would make them cancel green energy use. When asked how much more per month they would pay for green energy compared to grey, more than 50 per cent said to be willing to pay an extra 5 euros. But as the price goes up, loyalty goes down. Once the price of green energy reaches a level of more than 20 euros per month more than grey energy, virtually nobody would be prepared to keep buying it.

Even in an environmentally conscious country like the Netherlands, there is thus a clear limit to what people will pay. A small price premium is acceptable, but that is as far as it goes. Added to the concerns felt by consumers about the true provenance of green energy and their belief that sustainability is a matter for the government rather than for individual consumers, we can see where the political battlelines are likely to be drawn in the years ahead.

Strategic Imperatives for Nuon

From the Nuon viewpoint, the following conclusions can be drawn. First, size is crucial. In an emerging European market, it is more important than ever for the continuity of companies to gain critical mass. Dutch energy companies are, at best, middle rankers in the European order of merit. The drive for scale is irresistible, and Nuon is no more immune to this than any other player in the new reality. Medium-sized national markets do not have the luxury of enforcing strict competition criteria within a national context. The energy market is no longer national but continent-wide. The emergence of national champions seeking to compete well beyond their traditional borders seems almost inevitable. There will always be a role for niche players that remain small, but they will need to become increasingly specialised in their activities in order to maintain a valid role for themselves. They will also be subject to predation from the industry giants, eager to supplement their own activities with additional sources of higher-value business.

Consolidation can lead to problems such as integration, inertia, and rigidity – which may eventually lead to break-up and divestment. Yet, we strongly suspect that the process of consolidation within the energy market will be a key factor for a long time to come. The advantages, which include a larger customer base generating more revenues, a stronger negotiating position over energy sources, more financial muscle, and a greater capacity to deal with risk and unexpected market developments, all greatly outweigh the potential disadvantages. The electricity marketplace is, after all, essentially about supplying a key commodity (perhaps *the* key commodity) to as many people as possible. In this kind of market, more than in any other, size really counts.

Second, vertical integration is important. This may be a controversial point to make in reference to a national market in which unbundling and separation of distribution from generation have been so aggressively promoted. The experiences of countries such as the UK make it clear that, because the energy market is likely to remain extremely volatile in the medium term, companies that control their entire value chains are much better placed to cope with unpredictable changes and manage the inevitable ups and downs in the market. Further vertical consolidation in this regard seems inevitable.

As we move closer to a single European energy market, the need for a consistent regulatory framework becomes all the more urgent. At present, those companies permitted by their own national regulators to achieve more complete levels of vertical integration have a strong built-in advantage over those operating under more restrictive regimes. Operating in the European marketplace will involve risk on a larger scale than could ever exist in a medium-sized national market. The ability to mitigate risk through vertical integration is a basic requirement for implementing the next stages of a common European energy strategy.

Third, pricing is a key issue. In a liberalised marketplace, retail rules prevail. This means that customers will buy at the lowest price as long as quality remains acceptable. Nuon's own green-energy-attitude survey reinforces the point that no principle, however respected and sincerely held, will – in the long run – survive the harsh impact of competitive pricing. As we have seen, the implementation of environmentally responsible policies comes at a high price and consumers do not see why they should bear the full cost of such a politically charged strategic change.

This is likely to have a negative impact on the mix of fuel sources used for electricity generation. We have no choice but to accept that, even as a growing trend towards the use of renewable energy sources develops, most of the electricity sold in European countries will be generated from grey sources. In a more liberal, European market, it may – from a legal viewpoint – become simpler to import the cheapest energy stock from a wider range of suppliers than before. In reality, however, a low-cost revolution in energy sourcing is unlikely, because restrictions based on capacity and capital investment costs will not be removed as a result of liberalisation.

Neither Nuon nor its competitors will change their policy towards energy generation as a result of market changes. The trend towards higher standards in emission control, lower levels of pollution, and higher standards of corporate responsibility towards electricity generation will remain in force. Gray energy will remain dominant for years to come, but it will not become more gray as a result of liberalisation.

Resolving the Contradictions

A point we need to make is that the current political and regulatory climate contains a number of extraordinary contradictions, and these must be resolved before there is any real prospect of an effective, long-term renewable-energy policy across Europe. Policy-makers have in the recent past expressed many ambitions: a more competitive marketplace, driving down prices for Europe's consumers; price stability, linked to reliable supplies of fuel; higher-quality service, with greater customer satisfaction; lower dependence on imports, making Europe truly energy independent; sustained and continuous reduction in greenhouse-gas emissions; and reduction in the use of nuclear power. It is simply not possible to achieve all of these goals at the same time. Priorities need to be chosen, even if the choices turn out to be extremely difficult.

Finally, the question may be raised if green energy has any future in a world in which price mechanisms, consumer power, and relentless competition will be the norm. Nuon believes that a balance between competing interests can and must be struck eventually, though there may be an uncomfortable period of adjustment until the right modus vivendi is achieved. In practice, the move towards greater use of renewable-energy sources is gathering momentum at the exact time at which market liberalisation is also moving ahead. Governments across Europe encourage large-scale investments in wind and water power because of their concerns about the long-term prospects of other energy sources. There is a dichotomy between the strategic need for greater sustainability and the urge to harvest the perceived benefits of less prescriptive regulatory regimes and cost reductions arising from competitive pressures.

It is probably true to say that we, as societies, cannot have everything that we want: energy that is renewable and non-polluting, combined with relentlessly falling prices. A balance has to be struck in the energy market. We are confident that a way will be found to ensure that sustainable energy is enabled to compete successfully even in the most liberal of markets. Nuon will continue to invest in green energy and to keep it as a key focus of attention. We are sure that governments, regulators, and consumers will increasingly see for themselves that there is no viable alternative to giving electricity from renewable sources a high and growing priority in our long-term energy mix.

Nuon remains as committed to its long-term strategy as ever. The ability to balance the interests of different stakeholders is of fundamental importance; combining improved customer service with respect for employees, the environment, and society remains our ambition. Yet, we foresee considerable challenges in the coming years as we try to make sense of the need to grow in scale, and the need to follow a sustainable agenda in a political climate that, despite all the rhetoric, is less favourable to environmental concerns than before. The coming years will bring an increased pace of change. Nuon will proceed proactively where possible within this changing market, working to keep our underlying vision a living reality.

REFERENCES

Jansen, Y., C. Brognaux, and J. Whitehead (2003), 'Keeping the lights on: Navigating choices in European power generation', Boston Consulting Group, *http://www.bcg.com/publications/files/KeepingLights%20On_rpt_May03.pdf.*

Nuon (2002a), 'Jaarverslag 2002', *http://corporate.nuon.com/nl/Images/81_9379.pdf.*

Nuon (2002b), 'Duurzaamheidsverslag 2002', *http://corporate.nuon.com/nl/Images/81_9406.pdf.*

Nuon (2003a), 'Jaarverslag 2003', *http://corporate.nuon.com/nl/Images/81_9390.pdf.*

Nuon (2003b), *http://corporate.nuon.com/nl/content.jsp?page=/nl/duurzaamheid/.*

13. Unilever and Sustainable Development

Chris Dutilh

SUMMARY

Unilever is a multinational company involved in food as well as home and personal care. In order to meet its ambition of sustainable business operations, Unilever has taken environmental and social initiatives in the entire product chains. These include programmes in sustainable agriculture, sustainable fisheries, and clean water. The interaction among four types of societal actors (authorities, companies, consumers, and citizens) are analysed in order to understand Unilever's relationships and motivations for action. Consumers are different from citizens: the former offer companies a licence to sell, while the latter provide business with a licence to operate. Governments have two major roles: the technocratic role of developing and implementing legislation and the educational role of informing citizens how to become more balanced people with a higher degree of responsibility for sustainability in society.

INTRODUCTION

From the viewpoint of industry, environmental policy development should always be seen as part of a much broader approach, which is generally referred to as sustainable development. In that approach, environmental issues are only part of a wider range of issues. Multinational companies such as Unilever operate on a global scale, so globalisation is not new to them. When Unilever was formed in 1929 following the merger of the Dutch Margarine Union and the British Lever Brothers, the company was already actively involved in business in 35 countries all over the world. Indeed, one of the main reasons for the merger was the common interest in both raw-material supply chains and consumer markets. Since the merger, Unilever has

continued to develop its international presence, but always with an emphasis on local needs, rather than on imposing a centralistic policy. Hence, the company usually calls itself a truly multilocal multinational.

Local needs are composed of consumers' needs as well as social and environmental needs. Unilever's corporate mission states the company will 'meet the everyday needs of people everywhere with branded goods and services which contribute to the quality of life'. Everyday consumers' needs may vary substantially from country to country, as do Unilever's products. For economic and quality-assurance reasons, however, the supply chain has been considerably simplified over the last decade. The number of suppliers has been drastically reduced, as has the number of brands. General standards for quality control and supplier selection have been formulated centrally. Operating companies are responsible for executing those standards, while both companies and third-party suppliers are audited regularly in order to ensure the correct implementation of those standards.

As a global company, Unilever aims to play its part in addressing global social and environmental concerns, such as micro-nutrient deficiency, health and hygiene, water quality, and sustainable agriculture. It does not believe, however, that it is practical to address these issues at a global level only, or that companies such as Unilever can make a difference without working in partnership with others. This is why Unilever addresses global concerns with local actions and works in partnership with local agencies, governments, and non-governmental organisations (NGOs).

In this chapter, a brief introduction to Unilever is provided, explaining what it is and how it operates, particularly in the field of sustainable development. Unilever's experiences are put in the wider context of corporate social responsibility.

UNILEVER

In 2003, Unilever's turnover was about 43 billion euro (EUR). It had 250 thousand employees, 5,500 of whom were based in the Netherlands. Unilever's core business is to sell branded products in the fast-moving consumer-goods market. The company sells products in about 150 countries around the world. The battle on the shop floor has to be won 150 million times a day. Shoppers buy Unilever products on their own initiative, so they must have something in the back of their minds which tells them that those products are more attractive than all others available. To do so, the company proceeds as follows.

Unilever is composed of two divisions: a foods division (Unilever Bestfoods), with about 55 per cent of the turnover, and a home- and personal care division (Lever Fabergé), with about 45 per cent of the turnover. Both divisions are organised in business groups with regional responsibilities. Products are made close to the points of sale, while the management is, in principle, composed of local people. Unilever maintains a substantial exchange of senior staff globally, in order to support an effective network and to make sure that experiences around the world are optimally shared. Technical support for the factories and product development are organised by the divisions. National legislation always forms the minimum requirement; in many countries, Unilever standards are substantially higher.

Unilever is a leading supplier in a number of product categories, including culinary products, spreads, tea, and ice-cream in the food area, as well as personal hygiene products, deodorants, and detergents in the home- and personal care area. Major food brands are Becel, Bertolli, Knorr, Hellmanns, and Iglo. Home- and personal care products include Axe, Cif, Lux, Omo, and Sun.

Sustainable Development

Sustainable development was defined by the Brundtland Committee as a way of life in which 'the present [generation] meets its needs without compromising the ability of future generations to meet their own needs'.[1] This definition is clear and generally accepted, but when we ask ourselves what is meant by current needs, we see that there are as many different answers as there are people around the world. Although the definition is clear, it does not specify what sustainable development should involve, because people have different needs.

Unilever has formulated a corporate mission which states that 'we believe that to succeed requires the highest standards of corporate behaviour towards our employees, consumers and the societies and world in which we live'. This statement is based on well over one hundred years of experience, and it has been translated into various policy statements. For example, the environmental policy reads: 'Unilever is committed to meeting the needs of customers (e.g., professional kitchens) and consumers in an environmentally sound and sustainable manner, through continuous improvement in environmental performance in all our activities.' Two elements in this policy require specific attention:

- The company does not believe in a stepwise change, with a vision of a well-defined sustainable future in mind. Instead, it prefers the concept of continuous improvement, which implies that it can do better tomorrow

[1] WCED, 1987: 8.

than it did today. In order to improve, Unilever carefully monitors its current performance.

- Improvements are directed in a sustainable manner, which indicates that, in addition to the environment, other aspects are taken into account, such as the social conditions in which Unilever operates and the impact of its operation on local economies. A real contribution to sustainable development can only be achieved if these three factors are looked at in combination.

The Unilever Approach

Unilever starts its improvement processes by analysing current situations. Opportunities for improvement are identified, progress is reported regularly, and evaluations are made. Environmental management systems have been installed in all its factories around the world. These are used to measure the impact of Unilever's activities (for example, on the environment) and to formulate improvement targets. Following this, improvements are implemented. Each factory is audited every three years in order to see whether the systems are in place and the processes are proceeding according to expectations. Only part of the total environmental impact of Unilever's products is generated in its own operations (see Figure 13.1). A large share of the environmental effects take place upstream (in the production of raw materials) or downstream (in using the products). Most Unilever factories represent 10-20 per cent of the overall impact of their final products.

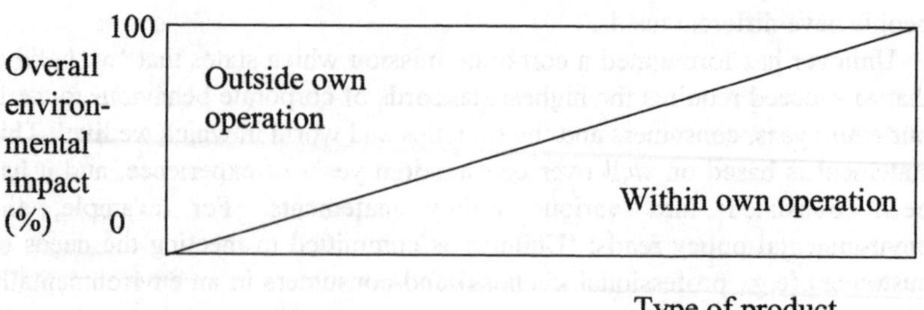

Figure 13.1 Relative environmental impact of Unilever's activities

In the making of margarine, for instance, raw materials are put together, packaged, and sent to a distribution centre. Energy is used, and waste water is generated. But the raw materials were produced elsewhere, and the products go to households, which have their own environmental effects (such as the use of energy by refrigerators, the contamination of waste water by detergents used during washing, and the disposal of packaging waste). It is clear that only a small part of the overall impact of a product is within Unilever's direct control; the rest occurs elsewhere in the product chain. Most companies focus on the impact of their own activities – at least in the beginning – because, in the lower triangle of Figure 13.1, everything companies do generates money or prevents the waste of money, and, therefore, pollution prevention pays. But in the upper triangle, companies also have responsibilities. Firms can significantly influence the overall impact of their products by selecting the suppliers of raw materials and by designing the products in such a way that they do or do not have impacts elsewhere. In order to be able to judge these effects, Unilever requires an overall environmental evaluation to be made for all its development processes. Suppliers have to meet the same requirements as met by Unilever operations.

A Life-Cycle Approach

Companies using a life-cycle approach must consider not only the performance in their own operations, but also activities up- and downstream: only by understanding the entire cycle can companies see whether local improvements lead to overall improvements. Figure 13.2 exemplifies such interactions for a food product chain.

Unilever started to conduct life-cycle studies of its products in the late 1980s. It had to identify what function it fulfilled for its customers. Unilever discovered that, in order to identify opportunities for improvement, it is not enough to look at technical functions in isolation: each product also has an emotional function. Only the combination of the two provides the overall function, as is shown schematically in Figure 13.3 (where F_t represents the technical dimension, F_e the emotional aspects, and F_o the overall effect). This holds for most functions in society: they have a technical as well as an emotional dimension. Environmental policies, for instance, aim at the reduction of waste and energy consumption. But they also have emotional elements, for example, regarding animal welfare or nuisance in a neighbourhood. In the technical domain, the issues have to do with limited resources (money, minerals, clean water, arable land, etc.), while the purpose is to maximise yield. Surpluses can be sold or stored for later use.

In the emotional domain, the issues have to do with resources which can be generated as needed by human beings (such as care and attention), with the objective of providing just enough. In this domain, it is not possible to store surpluses. Human needs can also be split into these two dimensions: people need food and clean air, but they equally need affection. It is relevant to note that both needs can be influenced by human interaction. These two domains can be related to the masculine and feminine nature of human beings.[2]

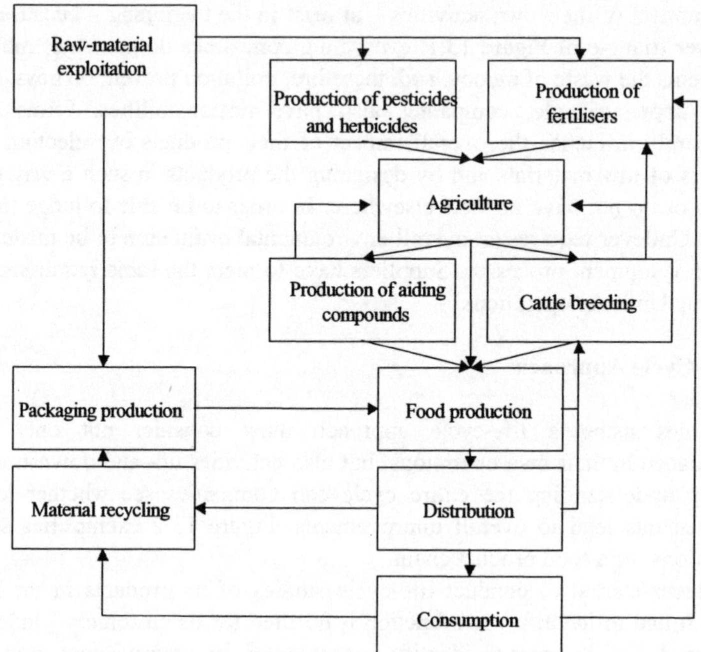

Figure 13.2 Life-cycle links in a food product chain

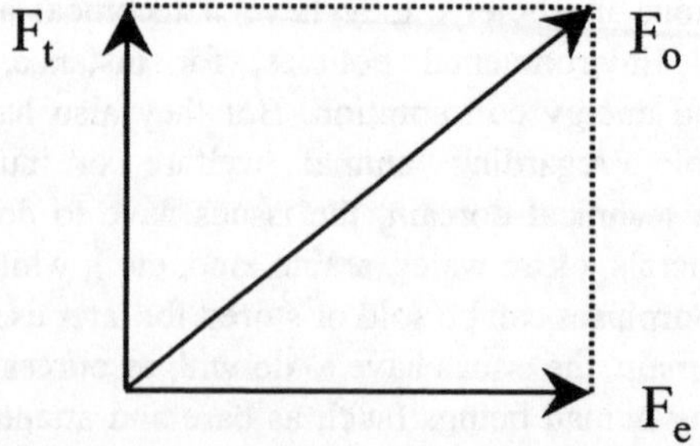

Figure 13.3 Dimensions of products

[2] Casimir and Dutilh, 2003.

Sustainable Development Initiatives

On the basis of its life-cycle studies, Unilever conducted a so-called 'overall business impact analysis', in which it looked at the total impact of the raw materials it purchases, the impact in its factories, and the impact of its products in households. On the basis of these studies, Unilever formulated three key sustainability programmes. Two of these are related to the raw-material inflow (viz., vegetables and fish). The third concerns clean water, which is important in every stage of the production and use of Unilever's products (viz., in agriculture, in factory operations, and in the home).[3] Clean water is required to make a cup of tea, but also to have a shower or to do the washing. Unilever products affect the water quality as well; in particular, detergents end up in the sewage system. Unilever has undertaken three projects on which it focuses in relation to sustainable development: sustainable agriculture, sustainable fishing, and clean-water stewardship. Significant programmes have been started in all three areas.[4]

Sustainable agriculture programme

For the sustainable agriculture programme, Unilever identified the main crops it uses. Some are annual crops (such as tomatoes, spinach, and green peas); others are perennial crops (such as tea and palm oil). By developing improvement programmes, Unilever learned what sustainable development would look like. It invited farmers, together with representatives from local authorities, scientists, and NGOs, to identify potential opportunities for improvement. One of the outcomes of this process was a 'best-practice' guideline for each crop in the various regions involved, with concrete advice on how to improve production conditions. Those guidelines are available in the local language to everyone interested.[5] Unilever's buyers and quality control departments use these guidelines to formulate specifications.

Participants in the projects operate in various places. For instance, the tomato programme is conducted in Australia, Brazil, and California. In all three countries, improvement programmes have been defined, and farmers as well as other people involved are optimistic about the outcomes. However, different targets for improvement were formulated in the different countries. Unilever does not consider this a problem, because its priority is to start improvement programmes. Only for reporting purposes are such differences important, as it is sometimes difficult to aggregate results from different regions. Such complications are important to those who focus on reporting standards. They request harmonised programmes, because otherwise the

---

[3] Unilever, 1998.
[4] Unilever, 2004.
[5] Unilever Sustainable Agriculture Programme, 2004.

results cannot be aggregated, and only standardised figures can be controlled. This example shows why there is sometimes tension between the people who try to implement sustainable development in practice, and those who watch and monitor such endeavours from behind a desk.

Sustainable fisheries

In the domain of sustainable fish supply, Unilever has also initiated a programme. In 1996, together with WWF, Unilever was involved in the foundation of an independent body, the Marine Stewardship Council (MSC). This body was asked to formulate principles for sustainable fishing and to come up with standards according to which fishermen must operate in order to get certified. The three standards adopted relate to the depletion of the fishstocks, the impact on the eco-structure of the fishing grounds, and the management system of the fishery. The standards may not be the same for all areas where a certificate has been granted. Practical measures for fishermen in Australia, fishing for lobster, can be quite different from those for fishermen in the United Kingdom (UK), fishing for herring, because the maintenance of the fishstocks may require completely different measures in both regions. Again, it is more the process than the concrete actions to be taken that are standardised.

At the moment, over 30 institutions and foundations support the MSC, Unilever being just one of them. Eight fisheries have been formally certified so far, and over 160 MSC-labelled products are on sale around the world, particularly in North-Western Europe and North America.[6]

When the MSC was founded, the Unilever chairman Anthony Burgmans said that by 2005 Unilever would use only sustainably caught fish, provided that sufficient raw material would be available by that time. This statement helped the initiation of the MSC greatly, because it gave fishermen a clear incentive to participate. There were also problems, however, because consumers did not show any willingness to pay more for labelled products. In 2003, Unilever obtained about half of its fish from sustainable sources, so the initial target will be difficult to meet. Nevertheless, the company considers that result to be a major achievement.

Clean-water stewardship

Almost all of Unilever's products consume water: in growth, in manufacture, and in use by consumers. On the one hand, Unilever makes efforts to reduce water use: on its factory sites as well as by, for instance, reformulating detergent products to allow washing procedures to be followed which require much less water. In that context, Unilever works with consumers to change

6 Marine Stewardship Council, 2004.

product-use patterns and foster the responsible use of water. One such project is the Lifebuoy handwash campaign, which is conducted by Hindustan Lever with the objective of changing the way consumers in India use water and soap products in order to promote disease prevention.

On the other hand, Unilever is involved in various projects which aim at the improvement of surface water around the world. With the assistance of the UK sustainability organisation Forum for the Future, Unilever has developed a set of principles which incorporate a practical approach to ensure that the community water partnerships it engages in are effective and successful. One such initiative is the Clean River Brantas project, in which Unilever Indonesia has adopted four villages along the Brantas River. In partnership with these communities, a local university, NGOs, and government agencies, the local Unilever company works to improve environmental awareness, sanitation systems, waste management and recycling, tree planting, and housing developments along the riverbank. As a result of these initiatives, the river now generates income for the villages through small-scale fish farming and the cultivation of Java Noni fruit crops for export.[7]

Sustainable development in context

In order to understand why a company such as Unilever would become involved in this process, the mechanism behind all these actions must be understood. In this process, represented in Figure 13.4, four societal parties fulfil key roles.[8] Together, they control the whole process.

The authorities set the rules and monitor implementation; industry takes action; and between those two is a regulatory/control mechanism. Companies do not operate well if they cannot sell products, so they have – directly or indirectly – commercial interactions with customers; the people who buy products are of vital importance. The fourth group consists of citizens. Some people consider the distinction between consumers and citizens artificial, but in my view it is not. For instance, Unilever introduced liquid margarine in a bottle to the Dutch market in 1997. In that year, the company won two prizes with that product. The first prize was from Friends of the Earth, which had invited people to tell what the most ugly product in the shops was. A special prize was presented in a popular television programme. Clearly, a significant part of the Dutch population did not like the product. However, in the same year and for the same product, Unilever won a prize from the retail association for the innovation which had generated the most new profit. This shows that there are two groups in society, one for which the product had

7 For more details, see Unilever, 2004.

8 It is postulated that each party has a direct influence on only two other actors.

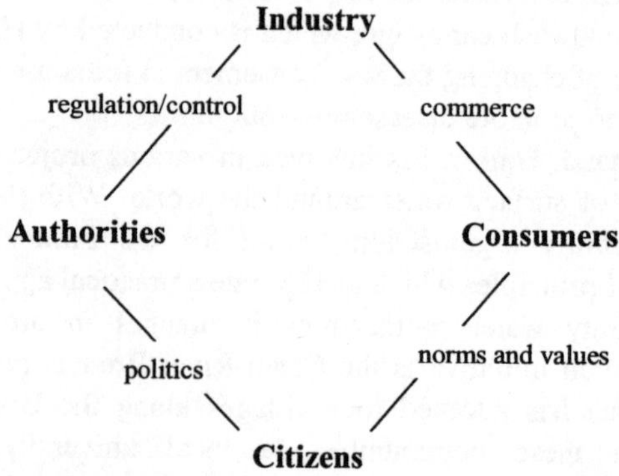

Figure 13.4 Parties determining the sustainable development process

been designed, which was pleased with it, and which bought even more of the product than was expected. This group is referred to as consumers. The other significant group, which did not buy the product, objected to the product and expressed its dissatisfaction. This matters to Unilever, because dissatisfied persons may dissuade others from buying the product. That interaction between individuals in society is influenced by the notion of what sustainability means regarding food products.

For that reason, Unilever conducted an investigation which led to two important findings.[9] First, people believe that sustainable food products should be healthy and safe. If products are not healthy and safe, they are not regarded as sustainable. The other finding resulted from an analysis of publications on sustainable products.[10] The results of that study are shown in Figure 13.5, which shows that public attention develops in peaks.

Around 1990, public attention was focused on phosphates in detergents; following this, attention was paid to waste, and packaging waste in particular. Around 1997, there was a break-out of swine fever in the Netherlands, and that brought animal welfare to the top of the public agenda. Around 2000, much attention was given to organic food, but that interest has waned. Therefore, there are peaks of attention related to subjects which are of interest to society. A similar investigation in Brazil or in Australia would probably have revealed the same pattern, though with different issues for each of the peaks. This clearly demonstrates that sustainable development changes over time and differs from place to place.

---

9 Dutilh and Mostard, 2001.

10 All publications on sustainable products in 12 volumes of the magazine AllerHande were counted. This free magazine is issued by Albert Heijn, the largest Dutch retailer.

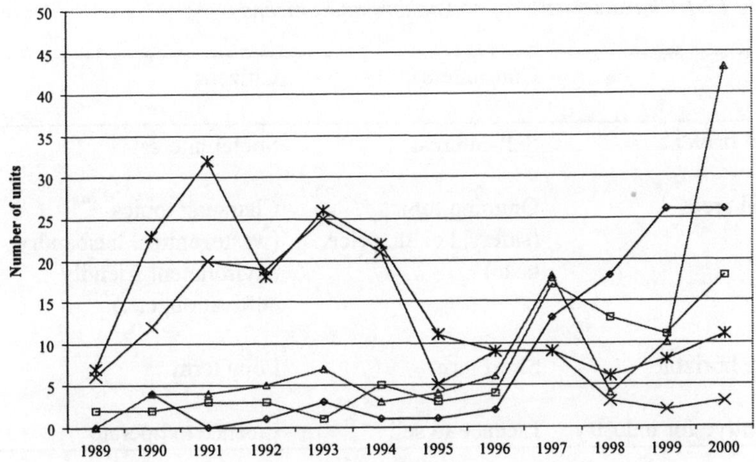

Figure 13.5 Evolution of public attention to sustainability issues[11]

CONCLUDING REMARKS

Table 13.1 shows that, on the one hand, there are subjects related to short-term self-interest, which are particularly relevant to consumers. Those subjects are invariant, and include safety and health; they determine whether persons will buy specific products. On the other hand, there are subjects of social interest which are related to the desires of citizens. These subjects have a much longer-term impact. They include waste, animal welfare, and child labour. Figure 13.6 suggests that industry looks to the authorities to provide a licence to produce. Such interactions are predictable and reversible; they concern permits, legislation, etc., with which companies must comply in order to continue their activities. The interactions with consumers concern the licence to sell. If consumers do not like specific products, they simply do not buy them. In this domain, we find competition: companies compete for the same consumers. Developments are also largely predictable and reversible.

The third domain, which is shown with a dotted line, concerns the licence to operate. In that domain there is no competition, and the processes are hardly predictable and virtually irreversible: if something goes wrong, it is hardly possible to return to the previous situation, as in the other two domains. Therefore, companies active in that field usually look for partners, and get worried about free riders who do not participate. Companies get

[11] Subjects referred to are related to non-food (double crosses), packaging issues (crosses), organic issues (squares), animal welfare (diamonds) and other food issues (triangles).

Table 13.1 Characteristics of consumers and citizens

	Consumers	Citizens
Main driver	Self-interest	Social interest
Focal areas	Ongoing topics (safety, health, price, taste)	Changing topics (waste, animal husbandry, environment-friendly cultivation, etc.)
Time horizon	Short term	Long term
Incentive for industry	Licence to sell	Licence to operate

involved in this area because people do not like to work in companies that do not work in this domain. Referring to Figure 13.3, we see that companies which provide wages, money, cars, and other material forms of remuneration but which do not pay attention to the emotional interests of their employees will lose their staff. Therefore, this is also a crucial domain.

The interaction between citizens and industry is sometimes referred to as 'corporate social responsibility'. At the end of the 19th century, the role of the authorities was rather restricted. Enlightened entrepreneurs (such as Lord Lever, one of the founders of Unilever), built houses, medical care centres,

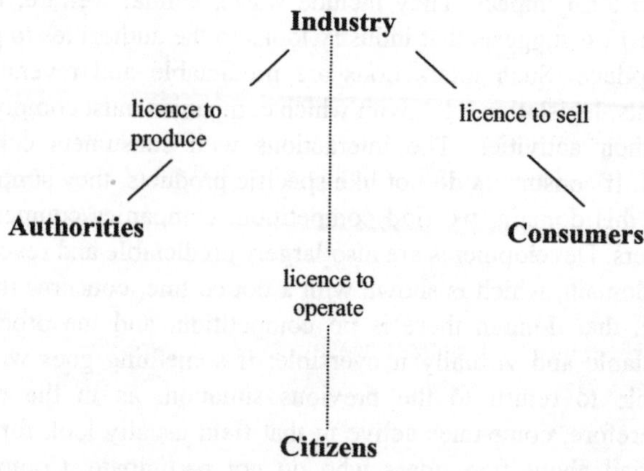

Figure 13.6 Interaction between industry and other parties

and schools for employees and their families. At some stage, the notion developed that too many people and companies did not contribute to the social well-being, and hence the authorities were empowered to introduce a tax system and develop school and medical-care systems. Those systems worked satisfactorily until, about two decades ago, various parties in society started to feel uneasy about overregulation. At that time, the direct interaction between citizens (or NGOs as their representatives) and business parties had taken on significant proportions. As Figure 13.6 shows, this is merely a matter of the public domain, and hence there is no role for the government here. The authorities do have two essential roles to play, however: on the one hand, to educate citizens to behave properly and, on the other hand, to formulate and uphold a legislative structure which composes the bottom-line for all members of society.

In practice, most attention is given to the second aspect, while the first is ignored in most policy development plans. Governments use only technical measures to solve the problems we are facing: the formulation of new legislation, the introduction of fees or levies, and the development of new technology. The authorities rarely consider how to educate people in society to become more responsible people or to reproach others for poor behaviour. Environmental policy planning is not just a technical problem, but has also many emotional aspects. It has to do with the ways in which people interact with each other and with how governments help people to become more responsible and more valuable citizens.

REFERENCES

Casimir, G. and C. Dutilh (2003), 'Sustainability: A gender studies perspective', *International Journal of Consumer Studies*, **27**(4), 316-325.

Dutilh, C. and L. Mostard (2001), 'Nederlanders over duurzaam geproduceerde voeding' (Dutchmen about sustainably produced food), *Voeding Nu*, **3**(4), 29-31.

Marine Stewardship Council (2004), *http://www.msc.org*.

Unilever (1998), *Environment Report 1998*, Rotterdam: Unilever.

Unilever (2004), *http://www.unilever.com*.

Unilever Sustainable Agriculture Programme (2004), *http://www.growingforthe future.com*.

WCED (1987), *Our Common Future*, World Commission on Environment and Development, Oxford: Oxford University Press.

14. Globalisation and National Environmental Policy: The Influence of WWF, an International Non-Governmental Organisation

Claude Martin

SUMMARY

This chapter addresses WWF's role of protecting the environment in an era of globalisation. While economic globalisation has brought significant benefits, it has also engendered important environmental and social problems, because free-market capitalism is based on flawed assumptions and erodes the protective power of national governments. Solving these problems requires greater clarity of trade rules, cooperation among supranational environmental institutions, a reformation of the World Trade Organization, and new tools for national governments. WWF has pursued its conservation aim by focusing on six global environmental issues and 200 ecologically valuable regions, and with the help of modern communication technologies. WWF speaks out for millions of citizens; surveys have shown the legitimacy of non-governmental organisations. WWF has engaged in partnerships with business and supports environmentally sound market mechanisms like eco-labelling. Cooperation among enlightened governments, companies, and non-governmental organisations is required to meet the future environmental challenges we face.

INTRODUCTION

Globalisation and sound environmental policy are often perceived as being incompatible or at opposite ends of the spectrum. One purports to represent trade liberalisation and a free-wheeling mentality, while the other aims to create safeguards and controls. In reality, the two are intertwined, because any long-term economic activity depends on a healthy environment and any effective environmental policy takes into account the true economic value of resources. The problem is not globalisation per se, but the fact that market liberalisation, as its main driver, has been portrayed as a panacea. It has been sold as a tide that lifts all boats, but some boats are not seaworthy.

The increasing global integration that we have experienced over the past few decades has brought enormous changes that influence every aspect of our lives – economic, environmental, and social. With this revolution have come not only numerous opportunities, but also daunting challenges, which require urgent attention.

All indications suggest that the ecological footprint of mankind is growing beyond sustainable levels. If current trends continue, humanity's ecological footprint will increase to twice the Earth's regenerative capacity within the next 50 years.[1] It is all too evident that it will be the poor of the world who bear the brunt of the resulting resource degradation. Who would have predicted when the Climate Change Convention was signed at the 1992 Earth Summit in Rio that, within 10 years, we would witness such severe effects on so many ecosystems, with melting ice masses in Arctic regions threatening the lives of many indigenous peoples, coral bleaching and tropical storms wrecking havoc on the economies of so many coastal populations and small island states, and drought and severe flooding causing misery and chaos, particularly in deforested, poverty-stricken areas? What emphasises more strongly the need for a multilateral system to address such global issues?

While some progressive governments have shown sound leadership and continue to push for environmental reform in an attempt to keep pace with the new demands facing our environment and global society, many non-governmental organisations (NGOs) – including WWF – find governments' responses to be slow, inadequate, and often opaque, resulting in too little being done too late. Because adequate steps by government are often politically unaffordable, the smaller steps usually taken do not go far enough to solve environmental problems, much less address their root causes. However, some progressive leaders and communities do support alternative initiatives, create new incentives, and demonstrate the benefits of social responsibility and sustainable development. Such efforts, though, are not a substitute for sound legal and policy frameworks, or committed government oversight. This is particularly the case with regard to trade at the national and international levels. The world needs willing decision-makers with foresight and fortitude to manage the impacts of globalisation on our environment.

Not long ago, environmental NGOs were accused of inhibiting economic growth and wanting humanity to go back into the caves. Ironically, governments are nowadays withdrawing into their own caves, influenced by

---

[1] WWF, 2002.

short-term economic interests, rather than showing the leadership needed to address the obstacles that stand in the way of sustainable development and poverty eradication. A world of free trade without authoritative intergovernmental regulatory institutions will never be able to reach the global goals set by the Rio Earth Summit and the 2002 World Summit on Sustainable Development (WSSD) in Johannesburg. Growing coalitions of concerned citizens, representing a variety of interests, see the importance of international and national regulations; and they want more government action – not less – to halt unfair and environmentally harmful trade practices around the globe.

While WWF understands their concerns, it is engaging in a wider variety of activities, which it hopes will influence the trade debate at many different levels and also ensure that sustainable-development principles become an integral component of all future trade agreements. WWF accepts that there is no turning back the clock. Globalisation is here to stay. However, WWF also feels that safeguards are urgently needed if those adversely affected by such activities, particularly in developing countries, are to be protected in ways that also secure the sustainability of the fragile, and often finite, natural resources on which their livelihoods depend. All of this underscores the need for greater commitment on the part of governments to work together and implement effective environmental and trade policies; a need that is more pressing now than ever before.

We live on a bountiful planet – but not a limitless one. Environmental policy-makers play a critical role in managing the balance between meeting human and environmental needs, and in ensuring that the long-term benefits last for future generations. The term 'sustainable development' is widely used today, as much by governments as environmentalists. But to WWF, it means the same thing as it did more than 10 years ago, when WWF collaborated with the IUCN and the UN Environment Programme (UNEP) to carry out a broad consultation process as part of the development of a new global conservation strategy,[2] which built on the earlier 'World Conservation Strategy' (1980). In that report, WWF defined sustainable development as 'improving the quality of human life while living within the carrying capacity of supporting ecosystems'.[3] If this is our end goal, then surely securing the access to basic resources and improving the health and livelihoods of the world's poorest people cannot be addressed separately from caring for the planet. Given the escalating threats of climate change, deforestation, overfishing, freshwater shortages, species depletion, and toxic pollution, policy-makers and their governments must adopt a more holistic approach

---

[2] IUCN/UNEP,WWF, 1991.

[3] IUCN/UNEP/WWF, 1991: 10.

that takes into consideration the impacts of globalisation on the environment.

This chapter provides an overview of the environmental problems exacerbated by growing globalisation and emphasises the need for more political will to solve them. The flaws of the current economic policies being pursued are specifically addressed and their environmental and social consequences are highlighted. Some suggestions are put forward as to what governments, businesses, and industry might do to better manage these issues. During the past decade, WWF has also undergone a transformation to meet the challenges posed by globalisation more effectively. It is explained how WWF's international network has established new conservation principles and priorities in order to influence the current political debate on critical environmental issues. It is also illustrated how the organisation works at many different levels of government to mitigate the negative impacts of globalisation. As efforts to address pressure on natural resources continue to increase along with the economic inequities, WWF is hopeful that it can foster greater participation and commitment at all levels of society to help build a sustainable future.

THE IMPACTS OF GLOBALISATION

Over the past two decades, global economic integration has brought significant economic benefits to some countries but shattered the hopes of many more. The opening of markets to international investors and the opening of borders to the freer flow of goods, capital, and services has provided opportunities for growth, employment, and economic diversification to a number of dynamic middle-income and industrialised countries. At the same time, however, economic reforms and liberalised trade policies have brought declining commodity prices, financial volatility, and repeated economic shocks to scores of developing and vulnerable middle-income countries.

The unequal distribution of costs and benefits threatens the viability of a number of developing countries, which for years have tried to restructure their economies, often at the instigation of external development partners. The result has often been increased poverty, heightened vulnerability, and greater economic uncertainty. Much of the growth experienced in poor developing countries comes from export revenues derived from primary products. This has generated comparatively limited economic diversification and alarming rates of environmental damage. For example, unsustainable timber extraction in Indonesia will provide only a short-term economic boost to the economy if the government there is not able to take the drastic measures needed to control the largely illegal extraction of this limited resource. As we now know all too well, the end result of such rampant deforestation is increased poverty and environmental degradation, dwindling water supplies and erosion, and little hope of future development for the people who depended on this resource for their livelihoods.

Flawed Assumptions

During the past 20 years, the pre-eminent feature of the current global economic policy, called neo-liberalism, has been the promotion of a set of economic principles that all countries, large and small, North and South, have been pressured to adopt. The essence of this economic prescription has been the opening of national borders to the flow of international capital, the privatisation and liberalisation of domestic markets, and the diminution of the state as an economic agent and guardian of the public welfare. Private financial flows and international development assistance have been measured according to the degree by which national development policy has conformed to this set of economic tenets, instead of allowing countries to determine national policies that are most appropriate to their level of development. Therefore, while neo-liberal economic policies have generated wealth for some groups and individuals, they have left many others in the margins because of the flawed assumptions on which the policy was constructed. As a result, such policies have not been the panacea that their supporters imagined.

First, neo-liberalism was predicated on the grounds that the freer flow of goods and capital would provide growth opportunities for all countries as each sought to specialise in areas in which it enjoyed comparative advantage. In reality, however, international markets are greatly distorted by the world's largest economies, notably the United States (US) and the European Union (EU), to the point that smaller, more vulnerable economies cannot compete effectively and have little chance to benefit in this skewed power relationship.

Second, trade and investment liberalisation promised to unleash market forces from which sustained growth in all economies would flow and from which more resources for poverty alleviation and environmental protection would become available. But long-lasting economic difficulties, afflicting more than 120 vulnerable economies, belie this claim, as does the rising incidence of poverty and environmental degradation in countries around the world where resources for combating poverty and improving the environment have actually declined.

Third, removing the state as an economic agent was expected to free up domestic markets, stimulate growth, and improve the welfare of all citizens. There is no doubt that the diminution of the economic role of the state was long overdue, and it has opened opportunities to entrepreneurs, as well as increased opportunities for accountability. However, most developing countries remain unable to provide many regulatory functions, previously overseen by the state, to ensure that private economic activity enhances public welfare. These governments also remain unable to provide many economic services needed to facilitate national economic diversification.

Some proponents of free trade suggest that developing countries, no matter how poor, will eventually reap enough economic rewards to be able to invest in strengthening environmental legislation and greater resource-efficient technology. However, experience and case studies to date show that there is no automatic (positive) link between trade liberalisation on the one hand, and environmental and social improvement on the other hand. In many cases, the economic gains arising from trade liberalisation are actually eroded by a loss of the natural capital upon which many people depend, especially the rural poor.

Proponents of unregulated growth draw on the assumption that, as the per-capita income in developing and vulnerable countries begins to rise, environmental degradation will also peak and reach its threshold, by which time these countries will be able to pay to clean up their pollution and restore the environment. This theory has been represented by the environmental Kuznets curve (see Figure 14.1). It is illustrated in an inverted U pattern, and suggests that environmental regulation will eventually reverse any negative impact once countries reach a certain economic status. Thus, it promotes the flawed assumption that economic growth is itself a sustainable development strategy.

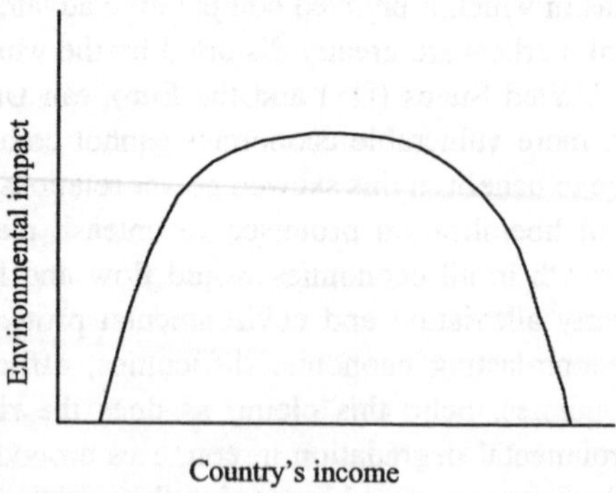

Figure 14.1 The environmental Kuznets curve

The paradigm represented by the environmental Kuznets curve leads to a pollute-now-and-clean-up-later mentality, which raises more concerns than it addresses. It does not account for the irreversible losses in biodiversity – which can often not be replaced, the relocation of polluting industries – which in effect just moves the problem elsewhere,[4] and the impact of resource extraction and the environmental footprint outside country borders.[5] In contrast to the dominant economic perspective that establishes compliance with economic orthodoxy as the prime criterion of trade and investment regimes, WWF believes that the objectives of sustainable development must shape trade and investment policies in each and every country. This means that promoting social and economic equity, and environmental protection, should be the prime determinants of governments engaging in trade and investment policies.

Market-Driven Decision-Making

Unfortunately, governments appear to be playing a diminishing role when it comes to protecting the environment from the threats posed by globalisation. Decisions made by industry, international corporations, and investment firms increasingly determine our fate. The late 1980s saw more than a hundred governments negotiate a phasing-out of chlorofluorocarbons (CFCs) under the Montreal Protocol on Substances that Deplete the Ozone Layer. This is now regarded as a major victory for international environmental policy. But behind this problem were a mere 12 companies manufacturing ozone-destructive substances. More recently, the US government blatantly refused to support the Kyoto climate treaty. Despite being the world's largest emitter of greenhouse gases, the US refuses to ratify the global treaty, largely out of self-interest and pressure from its oil, coal, and car lobbies.

Further evidence of the influence of market forces is provided by the impact of foreign direct investment (FDI) in developing countries. Hundreds of billions of US dollars (USD) are spent each year to underwrite manufacturing plants, mining development, power stations, port development, and the like. While these projects clearly have direct impacts on the environment, they may also contribute to the creation of 'pollution havens' – areas where companies relocate to in order to take advantage of less stringent environmental regulations. The pollution-havens debate must not be conveniently dismissed, as there is clear empirical evidence that certain resource- and pollution-intensive industries have a locational preference for areas of low environmental standards. Though environmental regulations may not be the primary influence on firms' investment decisions, they are important for pollution- or resource-intensive firms when choosing between

[4] For example, during the 1980s and 1990s, Japan moved many of its pollution-intensive industries to South-East Asia, thus nullifying any overall positive steps towards reducing pollution outputs.

[5] For example, the consumption of unsustainably produced timber has a huge impact on producer states.

countries in the same trading region, and between different locations in the same country. There is also evidence that host countries have neglected, or not enforced, domestic standards in order to attract investors.

The 'pollution halos' argument is that foreign firms, which are subject to more stringent regulations at home, use newer, cleaner technologies and environmental management systems and standards that they diffuse to the host countries. Empirical evidence shows that in some sectors, particularly those which are energy-intensive or require high technology, there is support for the 'pollution halos' hypothesis.

A recent report by WWF demonstrated that some resource- and pollution-intensive industries have a preference for, and an influence in, creating areas of relatively low environmental standards.[6] Furthermore, there is ample case-study evidence that the environmental performance of many transnational corporations is worse in developing than in developed countries, though sometimes still better than local practices.

One example of where foreign investment has contributed to destructive environmental practices is Indonesia's palm oil industry. Palm oil plantations grew from some 600,000 hectares in 1985 to 3 million hectares in 2000, largely at the expense of primary forests; and the industry is expected to double its current output by 2020. Investors, traders, and retailers in Europe, North America, and East Asia could save the remaining forests by developing and promoting sound management practices. However, few investors to date have been willing to adopt a pioneering and socially responsible policy that promotes sustainable development.

Some doomsayers warn that we have achieved all that we possibly can with regulations. As suggested by Zarsky, globalisation creates forces that lead to a domestic 'environmental paralysis', and puts the market in the driver's seat with regard to environmental policy and performance.[7] However, while globalisation may impair individual governments' abilities to regulate the environment, it also offers opportunities for governments to coordinate their domestic environmental policies and practices, and manage their limited natural resources more effectively for the common good.

Poverty, resulting from inequitable international trade by increasing direct and unmitigated dependence on environmental goods and services, can be thought of as an indirect effect of international trade on the environment. As poverty is also one of the root causes of biodiversity loss, WWF believes it is crucial for the world's governments – particularly those responsible for driving global trade – to lever market forces in support of sustainable

---

6 Mabey and McNally, 1999.
7 Zarsky, 1999.

development. The world's fragile ecosystems cannot afford to be further pillaged and ignored.

The Environmental Consequences

As long as governments continue to accommodate reduced barriers to trade and investment, the world will continue to witness unprecedented threats to human health and the health of our planet. Whether resulting in climate change, toxic chemical contamination, deforestation, species extinction, overfishing, or freshwater depletion, the past few decades have brought about unprecedented economic activity, which governments have failed to match with adequate regulations and political capital.

In the past two decades, it has become increasingly clear that climate change and global warming are serious threats to the world's biodiversity, as well as to the well-being of many people around the globe. Human-induced carbon dioxide emissions contribute to unpredictable weather patterns, melting icecaps, and rising seas, endangering small island states and their populations, as well as biologically important ecosystems, such as the world's coral reefs. After years of investigating this phenomenon, governments have finally agreed to the Kyoto climate treaty, which sets modest but realistic targets to control greenhouse gas emissions. Its entry into force, even without the US, sends a signal to the global market and policy-makers that binding action is required.

Contamination from synthetic chemicals, which are now known to be pervasive and global in nature, has spread as economic activity and trade have increased. Ironically, the same chemicals that were created to control disease, increase food production, and provide extra convenience in our daily lives are now recognised as being responsible for serious health effects on people and wildlife owing to their toxicity and persistence. These chemicals include lindane and endosulfan, used in pesticides, and bisphenol A, used in the lining of tin cans. Their impact now threatens every corner of the globe. While some positive steps have been achieved in the past few years, four international treaties that could help ban the most dangerous chemicals and promote ecologically safe alternatives are still pending government ratification.

International trade in forest products is also on the rise, as is – not surprisingly – deforestation, with commercial logging now affecting 70 per cent of the old-growth forests. This trade clearly has an impact on biodiversity, accelerating erosion, impeding access to important genetic resources, and threatening indigenous people who depend on this resource for their livelihood. Yet, efforts by governments to halt the illegal trade – estimated to be as much as 30 per cent of the overall trade in some countries

– and adopt sustainable forest practices have been painfully slow at the best of times. Recently, Brazil, guardian of the world's largest remaining rainforest, discovered that it had lost an area the size of Belgium between 2001 and 2002, which is probably its largest loss in a decade.

Globalisation has also had an impact on the trade in wild plants and animals. Over the years, WWF has learned that threatened species and habitats cannot be protected in isolation from commercial activity. Under the Convention on International Trade in Endangered Species (CITES), it has worked with governments to monitor the legal trade and stem the illegal trafficking in endangered animals and plants. But with increased access to remote areas, pressure on local communities to compete for hard-earned currency, and ineffective law enforcement, especially in developing countries, ensuring the effective implementation of this important treaty is extremely difficult.

Despite warnings from the Food and Agriculture Organization (FAO) nearly 10 years ago that about 70 per cent of the world's commercial fisheries were either overfished or fished to the limit, the fishing industry still expands. Buoyed up by perverse government subsidies and weak regulatory oversight, the industry stays afloat by fishing further from home and further down the food chain. To avoid complete depletion of their own fisheries, large fishing nations (such as the EU, Japan, and Russia) increasingly opt to fish in the rich waters of developing countries in Latin America, West Africa, and elsewhere. Their unregulated industrial fishing takes its toll on these marine environments, as well as affects local fishermen.

Currently, 1.5 billion people lack ready access to clean drinking water, and, if consumption patterns continue, at least 3.5 billion people, or 48 per cent of the population, will live in water-stressed areas 25 years from now. This is partly due to ill-conceived dams and river-training works, which were once regarded as the solution to the ailing economic needs of countries. Developers and governments alike now recognise that dams can kill river ecosystems, leaving less water for irrigation and other local activities.

There have been many warnings that the Earth's fragile ecosystems are in danger. Nevertheless, despite all of the public-awareness efforts, national debates, and international negotiations, governments appear to lack the political will not only to address these environmental problems, but also to change the economic policies that clearly contribute to them.

SOLUTIONS FOR A SUSTAINABLE FUTURE

Unless governments and environmental policy-makers are willing to take the hard decisions required to obtain long-term economic and environmental security, the world will continue to experience environmental degradation, and this will undoubtedly lead to further human suffering. Collective global action is essential to remedy the Earth's ailing ecosystems.

Greater Clarity and Cooperation

The World Trade Organization (WTO) has a mandate to establish the rules for trade between countries. It also has an obligation to do so in cooperation with other relevant intergovernmental agencies that are concerned with sustainable development, such as the United Nations Environment Programme (UNEP), the UN Committee on Trade and Development (UNCTAD), and the Commission on Sustainable Development (CSD), as well as with the many multilateral environmental agreements (MEAs). While at least 20 MEAs include some sort of trade component, the WTO trading system should neither be a threat to, nor consider itself superior to, important environmental treaties such as the Kyoto Protocol and the Convention on Biological Diversity.

Success in clarifying the relationship between the rules that govern trade and investment on the one hand, and the environmental regime on the other hand, is extremely important and resides largely in enhanced cooperation between different government departments and the effective participation of all governments. Governments should, for example, recognise the WTO and MEAs as equal bodies of law that need to support the environmental, social, and economic pillars of sustainable development. Building capacity and increased cooperation between the trade and environment policy-makers will help strengthen global environmental governance.

Urgent Reforms Needed

Reforms are urgently needed within the WTO if it is to become a fair, environment-friendly, and credible organisation, which contributes positively to the ultimate objective of sustainable development. These reforms include:

- Changing the WTO's negotiating, dispute-settlement, and other decision-making processes to make them more open and responsive to environmental and sustainable-development concerns.
- Ensuring that legitimate environmental regulations are protected from challenge by the WTO.
- Promoting environmentally and socially beneficial trade through the use of new rules and voluntary measures, such as fishing-subsidy disciplines and voluntary eco-labelling.

- Increasing protection and benefits to the poorest countries and communities, including increased market access, the transfer of financial and technical resources, and the use of adequate safeguards to avoid protectionist abuse.
- Preventing the expansion of the WTO into areas where it currently has no competence, such as foreign investment and government procurement.

There is much NGO scepticism vis-à-vis the WTO. Yet, there are clear opportunities to create 'win-win' scenarios that could benefit both sustainable economic trade and the environment, for example, the elimination of harmful, trade-distorting subsidies in the agricultural and fisheries sectors. It is important, however, that these negotiations facilitate meaningful participation by all affected stakeholders, from the responsible governments and recipients of the subsidies to the governmental agencies monitoring their impact and civil society. Such inclusivity is particularly important if more than one institution is involved in implementing the ensuing rules.

New Tools for National Governments

While there is much to be gained from acting together to address these common concerns, national governments should respect the sovereign rights of countries to set their own environmental and health standards, so long as this is done in a non-protectionist manner and they are in line with international trade and investment rules. There is an urgent need for more national leadership and innovative solutions to meet the challenges posed by globalisation.

National sustainability assessments represent an opportunity for governments to show such leadership by examining the true economic, environmental, and social effects of trade and investment. Such evaluations provide a critical tool for gathering much-needed information on the effects of trade and investment policies on the environment and people. And they serve as a useful mechanism for involving a wide range of stakeholders affected by trade and investment flows. For example, these analyses consider how trade may impact the ability of countries to maintain high standards and regulate for environmental and social protection, including the enforcement of national and international standards.

Sustainability assessments are one of the best ways in which governments can better inform their negotiating positions and their commitments to further trade deals. For the past two years, WWF has been working with government aid agencies, foundations, and representatives from various industries, to test the effectiveness of these assessments in various sectors and in different regions. In pilot studies in Brazil and in the Philippines, the soya bean and reef fish industries are examined. WWF helps to promote the adoption and implementation of this tool, as well as building local capacity amongst stakeholders and governments interested in carrying out these assessments. In order for such tools to be effective, however, developed countries need to provide technical cooperation and support capacity-building in developing countries.

Growing international demand for environmentally sound products could, in principle, help safeguard valuable natural resources in developing countries. However, for now, these consumer-country demands are sometimes interpreted as intentionally excluding developing countries from the playing field. Again, developed countries driving this trade must assume more responsibility to ensure that middle- and low-income countries are given the technical assistance and support needed to meet these environmental standards.

National governments must take stock of the true costs of opening key sectors of their economies to international trade, including the environmental and social costs. A world of free trade without incorporation of the precautionary principle will never promote environmental and social benefits. A world of free trade without the transfer of sustainable financial and technical assistance to developing countries means that the majority of the global population will never fully benefit from free-market access. A world of free trade without transparency and public participation will never be sustainable, and the world's citizens will continue to call for more equitable rules that support the world's poor and the environment.

In sum, WWF believes that governments must improve the way in which natural resources are used to generate economic wealth, and that they must make efforts to ensure these resources are used more efficiently. Governments and policy-makers must also ensure that future regulations and negotiations deliver concrete developments and sustainability gains. In order to achieve these results, new initiatives and commitments are needed on all fronts.

OUR RESPONSE

Over the years, WWF has identified and promoted practical ways for people and nature to live in harmony, and it continues to search for long-term solutions to the challenges confronting humanity. By focusing on a limited number of environmental priority issues and on those parts of the Earth which hold the greatest biological diversity, WWF and its partners demonstrate their commitment to seeking more ethical and equitable environmental policies, which best protect the long-term interests of the planet and its people.

Since its foundation in 1961, WWF has continued to enhance its conservation network's ability to respond to the growing environmental demands, while sharpening its mission to focus on the principal threats to our global environment. WWF's activities have evolved from primarily protection of endangered species and spaces to a more strategic approach that covers all corners of the globe. Today, its mission is to stop the degradation of the planet's natural environment and to build a future in which humans live in harmony with nature by conserving the world's biological diversity, ensuring that the use of renewable natural resources is sustainable, and promoting the reduction of pollution and wasteful consumption.

As one of the world's largest non-profit conservation organisations, WWF operates in more than 90 countries, and comprises 28 National Organisations, 24 Programme Offices, (including two policy offices, in Brussels and Washington, DC), and four Associate Organisations (which have different names but a shared vision). Each of WWF's National Organisations has an independent legal status, and is accountable to its own Board of Directors and donors, but also regularly contributes expertise and funding to WWF International in Gland, Switzerland. The Board of Directors of WWF International includes many of the Board chairmen of the National Organisations. Over the past decade, the organisation has moved from a broad-based conservation agenda, according to which the National Organisations conducted a wide array of projects related to their own interests and expertise, to a much more focused and accountable network.

WWF International oversees the organisation's global conservation programme, identifying and monitoring trends, and developing WWF's coordinated positions on emerging issues for its network, as well as spearheading the organisation's global campaigns, communications, and fundraising. Throughout the network, WWF employs some 3,800 people and has nearly five million regular supporters. It invests about 310 million USD per annum in field projects, conservation advocacy, environmental education, and public awareness. In 2003, for instance, the organisation supported about 700 international conservation projects, in addition to thousands of local and national activities overseen by the National Organisations.

WWF regularly draws on its network of people on the ground, in local field offices, and in the nations' capitals to influence key environmental policies, often in close collaboration with other NGOs. It also calls on the media and its numerous supporters to participate in key environmental debates when needed. This knowledge and input at the national level has helped WWF to build credibility and influence at the international level, enabling its global campaigns to further contribute to critically important international negotiations on climate change and the use of toxic chemicals, to name two examples. While WWF works closely with like-minded, progressive governments to promote practical and necessary environmental regulations, it can also harness its international network to apply pressure to less friendly governments lagging behind. The fact that WWF bases its advice on the best science, drawing on its vast experience from working in many parts of the world, often helps illustrate to decision-makers the real severity of the environmental problems as well as some solutions. Owing in part to the many changes brought about by globalisation during the past decade, WWF International has undergone an organisational transformation, which enhances the way it does business.

Setting Global Targets and Timetables

By way of maximising its impact, WWF has identified a set of priorities, which include six globally important issues allied with some of the most biologically diverse and important places in the world. The organisation is also focused on several crosscutting issues, such as trade and investment, the rights of indigenous and traditional peoples, and the national implementation of broader treaties, such as the Convention on Biological Diversity. Moreover, WWF is deeply concerned about the root causes of biodiversity loss (poverty, migration, macroeconomic policies), and the poor enforcement of environmental legislation. Together, these priorities create a comprehensive umbrella and synergy for WWF's international conservation work.

The six global priority issues which WWF focuses on are deforestation, freshwater shortages, overfishing, species depletion, toxic pollution, and climate change. The biomes – forests, freshwater, and the marine environment – are critical for a healthy environment, which protects and provides for our future.[8] A finite number of flagship species have been included, because their protection benefits many other species. They also serve as important indicators of the level of threats to and the overall health of the ecosystems upon which they depend. Finally, toxic chemicals and climate change represent two of the most globally pervasive and insidious threats to biodiversity, and, as such, must be addressed. For each of the six global priority issues, WWF has established a programme with clear conservation objectives, and targets and actions, which contribute to WWF's mission. These six programmes are hosted by various parts of WWF's network, and work in a coordinated fashion with WWF International, its National Offices and Programme Offices, and its many partners throughout the world to deliver conservation and policy goals.

---

[8] Biomes are ecosystem types characterised by their distinctive vegetation and climatic conditions in a particular region, for example, temperate forests or arctic tundra.

Recognising that many local conservation problems have their roots in wider social and economic issues, which influence how people use and consume their resources, WWF also focuses on areas whose boundaries are defined by nature. These areas, or ecoregions, may be tropical forests or wetlands spanning one or more countries, or entire coral reef ecosystems, such as the Mesoamerican Reef, which extends 700 kilometres from the tip of the Yucatan peninsula in Mexico south to the Bay Islands off the coast of Honduras. WWF has identified some 200 such places – called the Global 200 Ecoregions – which contain the best part of the world's remaining biological diversity and must be protected if we are to leave a living planet for future generations. While WWF promotes the conservation of all of these areas, it has focused its support on 40 ecoregions, where it works with partners from all sectors to develop concrete action plans. Ambitious and comprehensive, these plans combine environmental, economic, and social actions to conserve and restore biodiversity throughout the entire region.

The vibrant synergy between WWF's six global priority programmes and the Global 200 Ecoregions is one of the organisation's strengths. WWF constantly tries to integrate its programme activities, bolster conservation efforts, and deliver stronger results. In addition, the various socio-economic and conservation approaches applied in the ecoregions provide a unique opportunity for WWF to learn more about the root causes of biodiversity loss. This, in turn, helps inform WWF's policy work on the six global priority issues, as well as helps it to link developments with crosscutting issues, such as that of trade and investment.

Expanding the Power of the Panda

For the past decade, WWF's overall membership has continued to be fairly consistent, but has varied considerably from country to country. For example, in the Netherlands, about 6 per cent of the total population belong to WWF National Organization, while WWF-US has maintained a relatively steady membership of more than 1 million individual supporters for more than 10 years. The size and breadth of experience of the organisation's staff has proven to be its most important asset, and this is particularly evident when WWF tries to influence environmental policies at the national and international levels.

Whether in front of the television cameras of the world's media or behind the scenes with senior government officials, WWF has established itself as a reliable source of conservation news and sound advice. It proactively promotes a conservation agenda in the media, and responds daily to journalists' requests from around the world for information on the latest environmental issues, as well as drawing on its global network of expertise to handle any unexpected environmental crisis. While the media has been supportive of the aims of WWF, its heyday of the late 1980s and early 1990s – when environment stories often made front-page news – is over. With a shrinking media pool, a diminished public attention span, and donor fatigue, it is much more difficult for WWF to publicise its positions. Nevertheless, having an international network has proven advantageous, as has its multi-faceted strategy to expand beyond traditional approaches. Behind the scenes, WWF enjoys unparalleled access to many government officials and decision-makers. It works hard to provide constructive input and to facilitate dialogue with all stakeholders responsible for deciding and implementing key environmental regulations. In the various regions, WWF regularly engages stakeholders, from resource users and management practitioners to community leaders and presidents.

Thanks in part to the technology advances brought about by globalisation, WWF is able to link its on-the-ground conservation efforts to its broader, global network, and to harness public pressure for critical environmental policy decisions. Its award-winning website has proved to be an immensely powerful tool for raising awareness and mobilising support for conservation action.[9] Some 200,000 people visit this website of WWF International every month. In addition, WWF's online campaign site, Panda Passport, attracts an average of about 25,000 visitors a month, who are willing to pressurise government leaders around the globe by signing petitions and emailing personal letters. WWF's National Offices use similar advocacy tools on their own national websites, in their own languages, influencing important policy debates in their home countries, and linking to WWF's international website when they need extra support. Such direct access to government leaders would have been unthinkable 10 or 20 years ago. News about governments' positions and decisions are often relayed instantly by the media and on the internet by organisations such as WWF. While politicians do not always appreciate receiving hundreds of emails, WWF has developed systems to effectively coordinate and deliver the public's responses, which ensure that their voices of concern be heard.

For example, in 2003, WWF's European Fisheries Campaign, which has fought to halt destructive EU subsidies that promote overfishing, even went so far as to broadcast internet messages from around the world to EU Fisheries Ministers prior to a critical decision on the reform of the Common Fisheries Policy. The impact of having real people – not just WWF – state their concerns directly to the politicians helped drive home the depth and breadth of public concern. This is a case in which WWF, as a global organisation, benefits from new technology, which is itself a product of

---

[9] WWF, 2004.

globalisation. The organisation enjoys using such technology to generate additional interest and lobbying power for conservation issues worldwide.

Speaking Out for Millions

The large-scale influence of certain NGOs has caused some right-wing, US conservative groups to question their legitimacy. In June 2003, one such group held an all-day conference entitled 'NGOs: The Growing Power of the Unelected Few', which explored the potential threats that international NGOs apparently present to the Bush Administration's foreign-policy goals and to free-market capitalism. On the other hand, there is much unease among the broad public. Many citizens do not feel represented by their governments, which have fallen into the trap of trade liberalisation without addressing the issues of sustainable development. Such sentiments refer not only to the anti-globalisation movement, but to NGOs such as ours as well, which are trying to bridge these extremes. NGOs are a legitimate voice for millions of concerned people, and WWF is one such group trying to make a real contribution to sustainable development.

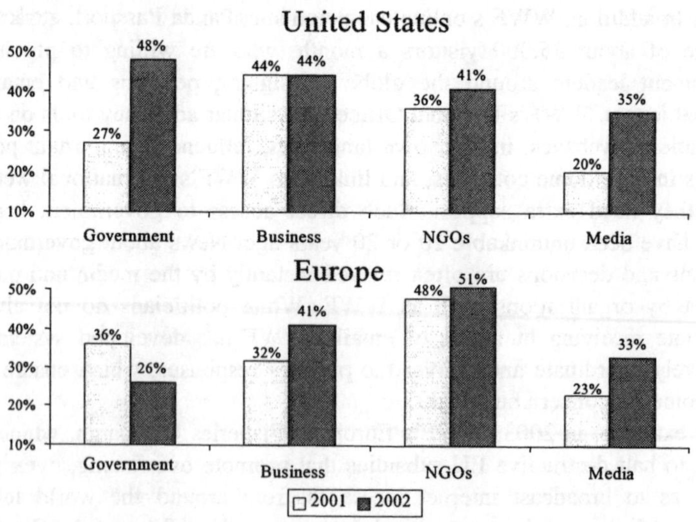

Figure 14.2 Trust in institutions

Following years of unfulfilled promises from government leaders to address contemporary environmental challenges, the European public often trusts NGOs more than their own governments, according to many opinion polls. Edelman[10] showed that opinion leaders' trust in the US government had increased, while trust in US businesses had remained the same; in Europe, however, trust in government institutions had decreased, while trust in businesses had increased between 2001 and 2002 (see Figure 14.2).

The end result of the fourth bi-annual survey was that the role, influence, and authority of NGOs had grown. In Europe, the trust in NGOs outscored that in all other sectors. Edelman also found that the credibility of companies declined the more they were perceived as global.

ORGANISING FOR SUSTAINABLE DEVELOPMENT

The Rio Earth Summit achieved a political and conceptual breakthrough, which created a new wave of optimism with respect to the world's ability to find a way to continue economic growth whilst safeguarding the environment. The Climate Change Convention and the Biological Diversity Convention were products of the Summit, and they provided a framework for creating new tools for conducting business. It appears, however, that some of these initiatives did not provide enough incentives for responsible government leaders and policy-makers to take the necessary steps required to implement these global treaties.

The Johannesburg Summit failed to address the harmful effects of trade liberalisation and subsidies, side-stepped the Climate Convention, and made lukewarm statements about the Biodiversity Convention, while compromising on the use of toxic chemicals to the extent that the outcome was weaker than the results of previous international agreements. More often than not, the world's largest consuming nations were the strongest opponents to setting any substantive and measurable commitments and indicators for monitoring progress.

Today, WWF feels somewhat disillusioned as many of the global negotiations pursued and facilitated over the last ten years resembled a 'race to the bottom', to the extent that environmentalists increasingly have to fight to prevent governments from going back on previous commitments, such as the Rio Principles and the Millennium Development Goals. Governments should reflect the legitimate voice of the people and should, therefore, recognise the great differences between nations, cultures, and economic circumstances. They should act in solidarity to address the global environmental problems rather than divide and rule.

---

[10] Edelman, 2003.

Promoting New Partnerships

Despite the governments' collective reluctance and failure to agree a sustainable way forward, the Johannesburg Summit had some positive outcomes. Over 300 new partnerships between governments, businesses, NGOs, and other civil-society organisations were forged. Among them, WWF, the Government of Brazil, the World Bank, and the Global Environment Facility announced a new venture to triple the area of Amazon rainforest under protection. Such initiatives have become a hallmark of WWF, which has found that working with a few willing partners is often more effective than working with many in promoting conservation. It also provides positive examples, which depict the benefits of effective environmental regulation and sustainable development. The Amazon initiative is extremely timely and will help address a crisis of unfathomable proportions there. Ironically, one of the apparent causes of the most recent forest destruction is land-clearing for soya bean farms which do not use GMOs. The farms have apparently been expanding to meet the growing demand in Europe. This is yet another example of the impacts of globalisation and the urgent need for new safeguards that assess and protect the threatened environment.

WWF believes that the potential for NGOs and businesses to work together is enormous. Not only can the private sector help generate much-needed funds for conservation, but it can also facilitate the adoption of new environment-friendly practices that are essential for pushing the environmental agenda. Moreover, some of the partnerships that have helped foster this change include government aid agencies and progressive foundations.

Using New Market Mechanisms

Changes in corporate practices are essential to measurable progress in reducing global warming, moving to renewable energy systems and clean technologies, phasing out toxic chemicals, and ensuring the sustainable use of commodities such as timber, fish, and agricultural products. Although companies are often part of the problem, they can undoubtedly contribute to the solution. According to the Institute of Policy Studies, the world's 100 largest economic entities consisted of 51 corporations and 49 countries.[11] Thus, corporate engagement is of critical importance if NGOs want to transform the current markets, influence national and international laws, and encourage best practice.

[11] Anderson and Cavanagh, 2000.

WWF has helped to establish, and continues to support, new market mechanisms, such as the Forest Stewardship Council (FSC) and the Marine Stewardship Council (MSC), which promote independent certification and eco-labelling schemes for forests and fish, respectively. Ten years since its inception, the FSC now has a membership of over 700 companies, including large producers and retailers such as B&Q and IKEA, subscribing to its principles and sustainable forestry-management practices. The MSC was started later under the auspices of a unique agreement between WWF and Unilever. It has become completely independent, with over 100 government, industry, and retail supporters, as well as about a dozen large- and small-scale fisheries that have earned the MSC eco-label.

Over the past decade, WWF has extended its outreach to the private sector, as well as the expectations that it brings to the table. In the past, WWF's problem was to decide whether to accept certain contributions and to decide if and how that entitled donors to use WWF's well-known panda logo. Now, with an increased awareness of the crucial role companies can play, both as influencers and as implementers, WWF has entered into more complex relationships with business, resulting in new sectoral standards and real changes in the way companies operate. This, in turn, furthers the conservation goals of WWF.

In addition to direct conservation benefits, some partnerships also deliver much-needed funding for conservation and help WWF to strengthen its brand and institutional capacity, and achieve its mission. To this end, and to ensure a more coordinated and strategic approach to engaging business and industry, WWF has established global guidelines which outline standards for engaging business and industry. They include WWF's right to maintain its independence and its right to criticise. While WWF continues to be vigilant in making these arrangements, no two agreements are ever exactly the same.

One of the most remarkable business arrangements that WWF has entered into was with Lafarge, the world's leading cement producer. After initial discussions with WWF France, WWF International agreed a five-year partnership in 2000 with three objectives: restoration of Lafarge's quarries, reinforcement of the company's environmental policies, and financial support for 'Forest Reborn', WWF's forest restoration programme. Since then, these agreements have been expanded to include agreements on energy use, a reduction in carbon dioxide emissions of 10 per cent below 1990 levels by 2010, and discussions about toxic chemical use.

Another partnership that WWF and other NGOs undertook together was with HSBC, one of the largest global banking institutions. In 2002, HSBC agreed to create a 50 million USD eco-partnership called 'Investing in Nature' to fund conservation projects around the world, and made donations to WWF and three other organisations (the Botanical Gardens, Conservation

International, and Earthwatch). WWF drives the conservation initiatives in Brazil, China, the United Kingdom, and the US, while HSBC helps to fund the overall management. In addition, WWF has agreed to work with the company on its own environmental policies.

While these partnerships are outside the realm of any official government regulation, the companies participating are obliged to observe national and local regulations and policies. As such, they have an enormous potential to influence that process. These partnerships represent a certain degree of risk, but they also have enormous potential to create real change in the way some of the world's largest multinational companies conduct their business, as well as to establish new standards in environmental practices.

FUTURE DEVELOPMENT AND PLANS

WWF's approach, which appeals to all stakeholders to find solutions to the world's environmental problems, is not new. However, the scope and severity of the problems plaguing our environment, fuelled in part by the globalisation of our economy, have never been more threatening.

More than 40 years ago, the founders of WWF saw an urgent need to reduce the effect of man on the natural world. At the time, the Earth housed around three billion people. Today, that figure has more than doubled and the human footprint can be seen everywhere. WWF's Living Planet Report developed projections of our ecological footprint based on various scenarios proposed by the UN, the IPCC, and the FAO, which all assume slowed population growth, steady economic development, and the use of more resource-efficient technologies between now and 2050. It suggests that the world's ecological footprint will continue to grow from a level estimated at 20 per cent above the Earth's biological capacity in 2000 to as much as 120 per cent above it by 2050.[12]

Once viewed as peripheral, such concerns for our environment and the future have now taken centre stage. WWF will continue advocating a sustainable future in which we can all live in harmony with nature. But governments must also pursue bold and comprehensive environmental policies at the national and international levels. Whereas before governments could afford to concern themselves with their own national problems and interests, the health of nations' economies, environment, and societies are now inextricably linked, in part owing to globalisation, and also because of the ubiquitous nature of many of the environmental problems plaguing the Earth today.

---

[12] WWF, 2002.

According to Panayotou, 'Economic globalisation changes the government-market interface; it constrains governments and enhances the role of the market in economic, social and environmental outcomes; on the other hand, it creates new imperatives for states to cooperate both in managing the global commons and in coordinating domestic environmental policies.'[13] Whether the governments and policy-makers of today will be able to meet this challenge is the question. The 'imperative' requires governments to work together on these global challenges, and has never been more pressing.

We envisage new constellations of enlightened governments, intergovernmental institutions, environmental and developmental NGOs, forward-looking companies, and creative thinkers, who will address these pressing issues. We foresee that such groups and alliances will engage in sustainable-development programmes and forge new policy alliances, which can mitigate the current flaws in the multilateral system.

REFERENCES

Anderson, S. and J. Cavanagh (eds) (2000), *Top 200: The Rise of Corporate Global Power*, Washington, DC: Institute for Policy Studies.

Edelman, R. (2003), *The Fourth Edelman Survey on Trust and Credibility*, Davos: World Economic Forum.

IUCN/UNEP/WWF (1991), *Caring for the Earth: A Strategy for Sustainable Living*, Gland: WWF International.

Mabey, N. and R. McNally (1999), *Foreign Direct Investment and the Environment: From Pollution Havens to Sustainable Development*, WWF Economics, Trade and Investment Reports, Godalming: WWF UK.

Panayotou, T. (2000), *Globalisation and the Environment*, working paper no. 53, Environment and Development Paper no. 1, Cambridge: Center for International Development at Harvard University.

WWF (2002), *Living Planet Report 2002*, Gland: WWF International.

WWF (2004), *http://www.panda.org*.

Zarsky, L. (1999), 'International investment rules and the environment: Stuck in the mud?', *Foreign Policy in Focus,* (4)22, Paris: Organisation for Economic Co-operation and Development.

[13] Panayotou, 2000: 1.

15. The Impact of European Non-Governmental Organisations on EU Environmental Regulation

John Hontelez

SUMMARY

The European Environmental Bureau (EEB) is a democratic federation of about 140 European environmental organisations. The European Union (EU) has fulfilled a prominent role in environmental legislation, though this may collide with the function of promoting a single market. The EEB, speaking on behalf of millions of citizens, defends environmental interests at the different EU institutions. The EEB's informal and formal EU contacts have had a significant impact in several areas, including the EU constitution, EU enlargement, the 6th environmental action programme, and policies on fiscal reform and chemicals. The EEB and the EU face major future challenges. The EU should fulfil a leading role in managing globalisation by promoting sustainable development, which should be anchored in the EU constitution and applied in environmental legislation. In this way, the EU would be able to better protect environmental and social interests and ensure good governance.

THE EUROPEAN UNION AND GLOBALISATION

In the view of the environmental movement, the EU is not just an answer to globalisation but also a promoter of this phenomenon. Governments, at least those of the larger industrialised nations, play an active role in promoting globalisation. The EU is, in this sense, a tool for European industrial states to play this role more effectively. For instance, Article 3 of the European Community[1] Treaty, lists the following Community activities: 'a common

[1] While the term 'European Community' is hardly used anymore, there is still a legal difference between the EU and the European Community. The Community is the successor of the

commercial policy; an internal market characterised by the ... free movement of goods, persons, services and capital; a common policy in the sphere of agriculture and fisheries; a common policy in the sphere of transport; a system ensuring that competition in the internal market is not distorted; ... the strengthening of the competitiveness of Community industry; encouragement for the establishment and development of Trans-European networks'.[2] This clearly shows a grouping of states not merely reacting to consequences, but actively accelerating regional and external trade, and taking concrete measures to do so, in the sense of removing practical constraints, in supporting advantageous external trade conditions, etc.

The EU, stimulated by lobbying from European companies and their organisations (in particular, the European Roundtable of Industrialists, the European Umbrella of National Federations of Business and Industry (UNICE), and European industry-sector organisations), has played an important enabling role in accelerating the process of globalisation. At the same time, as far as the EU region is concerned, it bolsters globalisation to limit subregional imbalances and environmental impacts, which is certainly positive. With regard to external trade (the 'common commercial policy'), the accompanying policies are much more timid, and the focus is clearly on promoting the interests of Europe's business. A very clear example is the role the EU is playing in strengthening the role of the World Trade Organization (WTO) and in pushing, for example, developing countries to open up public services (including drinking water) to – mainly European – multinational companies.

EUROPEAN ENVIRONMENTAL POLICIES

There are two logical reasons for internationalising environmental policies. First, many environmental problems cross borders. In Europe, both the pollution of rivers and coastal seas and acid rain are among the examples that have convinced the public and politicians most. Second, the interconnectedness of national economies can make it controversial and ineffective to take measures solely on a national basis. The business world never tires of complaining about the negative impact of unilateral national

European Economic Community and has its own Treaty still, whereas the EU is an umbrella including the Community but also the specific intergovernmental pillars dealing with nuclear energy (Euratom), coal and steel (a Treaty abolished in 2001), common foreign and security policy, and police and justice. I use the term 'Community' either when I relate to the time before the Maastricht Treaty (1992) or when the comments are specifically related to the European Community Treaty.

2 European Community Treaty, 1957.

environmental policies on their competitiveness. And when products rather than production processes are targeted for national environmental policies, complicated border measures are required to achieve the desired impact of such policies.

When the European Economic Community was founded in 1957, environmental problems were not an issue. The Community was set up to accelerate economic growth in the member states through cooperation and integration. Environmental issues first appeared on the agenda in the early 1970s, initially on the national government level; in 1972, the original six Community member states (with Denmark, Ireland, and the United Kingdom (UK) joining in 1973) agreed with the European Commission (EC) that environmental issues were a task for the Community as a whole. After the first United Nations (UN) Environment Summit in Stockholm in 1972, the EC presented proposals for what became the first Environmental Action Programme.[3] This Programme was both visionary and optimistic about the possibilities for effective responses to the new challenges. It elaborated on the mutual interdependence of economic development, prosperity, and environmental protection. Furthermore, it argued that the protection of the environment should belong to the essential tasks of the Community. At the time, it already stressed the need for a comprehensive assessment of the impacts of other policies, in order to avoid any damaging activities. However, three decades later, actual environmental policy integration is still in its infancy.

The reality of the 1970s was that Community policies focused on the protection of single environmental media, with water and air receiving the highest priority. In particular the water policies developed in those years were quite strict, probably because ensuring safe drinking water had strong public support. While environmental policies in the EC started with an environmental reasoning, the picture soon changed when the interests of achieving an internal market for goods and services became the main objective, where harmonisation of national environmental rules was one of the means. Such rules were obvious for products that were to be sold across the Community, but, to a certain extent, also for emission restrictions at production sites, as different standards could impact on competitiveness.

Until 1 July 1987, the legal basis for the EU, its Treaties did not include environmental policies, so legislation had to be agreed on the basis of internal market requirements or by consensus. The Single Act, an amendment to the Community Treaty, then became effective, and finally introduced an environmental chapter into the Treaty, which stressed the need for a common approach while leaving a limited freedom for individual governments with

3 European Commission, 1973.

respect to their national policies beyond the agreed-upon EU levels. This was a limited provision, as national policies would have to respect the overarching objective of the single market and do nothing that would result in trade barriers between member states.

Since 1987, the development of the role of the EU in environmental policies in its own region, elsewhere in Europe, and in the rest of the world has been impressive. For several member states, the EU is nearly the only source of national environmental policies. In those countries that traditionally have considerable internal drives for environmental policies, some 70-80 per cent of national legislation consists of the implementation of EU law. However, the EU is not only a producer of environmental legislation, but also a guardian of the principles of the single market against national environmental policies. The Commission frequently clashes with national governments that have approved national measures where, in their view, EU legislation is not effective enough. This often happens with respect to waste management, where the EC tends to defend the interests of industries that argue to be discriminated against by certain national measures. If intervention by the Commission itself does not help, industry can go to the European Court of Justice to ask for the removal or alteration of such national legislation.

How far this can go is illustrated by the recent case of the EC against the German government regarding a national law to enforce the sale of mineral water in returnable bottles. The Commission's argument is that this law would discriminate against foreign (French) producers of mineral water, who transport this water hundreds of kilometres to the German market. The German government's objective is not to promote consumption of mineral water from local sources but to reduce packaging waste. The possible side-effect of re-localising mineral water consumption is positive. However, it flies in the face of the ideology of the single market, regardless of the environmental impact of senseless mass transportation.

THE ENVIRONMENTAL MOVEMENT'S RESPONSE

The EU is crucial for environmental policies in its own region, and has been so for a long time. Environment and nature conservation organisations spotted this early on, which led to the foundation of the European Environmental Bureau (EEB) in 1974. The Bureau was set up to monitor developments in 'Brussels' (where the EC resides), inform its members about European developments, and organise coordinated responses to such developments in order to influence their outcomes. This basic mission of the

EEB has not changed over the years, but the way it has been implemented has changed dramatically.

Raison d'Être of the EEB

The founding members of the EEB were, on the one hand, established organisations working in nature conservation. Some, like An Taisce in Ireland, Elleniki Ettairia in Greece, and the RSPB and the CPRE in the UK, had a long history, while others were emerging environmental organisations, such as Amis de la Terre in France, Bundesverband Bürgerinitiative Umweltschutz (BBU) in Germany, and Stichting Natuur en Milieu in the Netherlands. Since its inception, the EEB's membership has increased to 142 organisations, in all but one EU countries, Algeria, Bulgaria, Norway, Romania, Serbia-Montenegro, and Turkey. Together, these organisations have a membership and supporter base of some 20 million citizens. Especially in the countries that recently joined the EU, membership has grown quite fast in the last five years.

The Way We Work

The EEB is a democratic federation of non-governmental organisations (NGOs). Its members determine the annual work programmes, and via the Board and working groups they contribute to and control the positioning and activities of the EEB. In this fashion, the EEB can connect the European agenda to the priorities and views of environmental organisations across Europe.

Currently, the Board of the EEB is composed of environmentalists from 23 countries. They are nominated by the EEB members from different countries and approved by the Annual General Meeting. They are expected to voice the ideas and wishes of the EEB members from the countries of origin on the Board, while promoting the EEB to the members and potential members in the home countries. The Board has the final responsibility over the political, organisational, and financial performance of the EEB. It takes direct responsibility for the bi-annual 'Memorandum for the Presidency', which presents the EEB's views on all issues appearing on the political agenda in the subsequent six months. The Board monitors the work of the Secretary General and his staff, and takes further political decisions if so required. Furthermore, Board members play a key role in the communication between the EEB and national Environmental Ministries. Contacts are needed to influence the positions of national governments in the Council of Ministers. However, the EEB also has a tradition of working with Presidencies, organising special events, setting up opportunities for meetings between the members of the Environmental Council and the European environmental movement, etc. In making those contacts, the national Board members are often essential.

Where the Board is most relevant for the horizontal issues, the network of EEB working groups forms the heart of the day-to-day communication and cooperation between EEB staff and the national member organisations. The working groups are open to all EEB member organisations. They serve to inform the membership about issues on the political agenda from very early on, prepare EEB positions that need to be approved in the end by the Secretary General or the Board, organise outreach activities to mobilise support for EEB positions, inside and outside the environmental movement, and coordinate activities towards the EU Institutions and their national actors: members of the European Parliament (EP) and the relevant Ministries. Currently, twelve working groups exist: on agriculture, air quality, biodiversity, chemicals, eco-labelling, ecological product policy, EU enlargement, environmental fiscal reform, noise, resource efficiency and waste management, urban environment, and water pollution. Together, they directly involve some 400 people across the EU.

In addition to the Board and the working groups, the EEB informs its membership at large through an ongoing stream of publications and internal memos. For a wider audience, the EEB publishes 'Metamorphosis', a newsletter with EEB comments on major issues on the agenda, such as the European Constitution, as well as the publication of the 'Ten Tests for the Presidencies' and a special insert on the EU enlargement.

The Green Eight

The EEB remained the only stronghold for national environmental organisations in Brussels until 1985. After 1985, international organisations such as Friends of the Earth International, Greenpeace, WWF, and Birdlife International began to open their own offices in Brussels. From a working group inside the EEB, a separate organisation on climate change emerged in the early 1990s. Furthermore, the European Federation for Transport and Environment was set up in the same period, with some of the largest European environmental NGOs, railways, and other organisations with a specific link or interest with the transport sector as members. The EEB is unique among these organisations in that it is the only organisation that exclusively monitors and influences EU policies, and thus is the only one which, in principle, looks at the whole picture. The other organisations focus on a limited number of issues, depending on their internal agendas.

Together with one more organisation, Friends of Nature International, the previously mentioned organisations and the EEB have set up an informal platform called the Green Eight. In this platform, the Secretary-Generals and Directors of the individual organisations organise practical coalitions on specific issues, exchange information and views on general developments, and take joint action on horizontal issues, such as the preparations for the European Constitution.

Structure and Strategy

Working at the EU level has never been easy for the environmental movement. We face a variety of problems. The complexity of EU decision-making and the overwhelming presence of lobbyists with agendas different to ours are two important but non-exhaustive factors.

Our member organisations work closely with the public, which is often not so aware of the relevance of activities in Brussels. What is happening locally and nationally is much more visible and of more direct concern. Even global issues such as climate change and the decline of the tropical forests are more visible. National politicians do not make much effort to educate the public. They prefer to stick to the impression that they still run the show in their own countries and that the EU is a negative outsider, a source of unpopular measures. Furthermore, the practical constraints of communicating in 21 languages are enormous. While English is being used to an increasing extent, many activists still feel excluded from European discussions, as the EEB cannot offer the range of interpretation facilities that the EP, for example, provides for its members. The distances and travel costs are another complication.

But we need the active participation of our members because the political decision-making process, although it starts in the EC, the central body best approached by professional lobbyists, needs national messages to be sent to members of the EP and certainly to the national governments which form the Council of Ministers. The secret of the EEB's success is that, to a large extent, we manage to involve our members to an increasing degree, despite these practical problems.

Environmental NGOs want to influence EU policies to ensure that they consistently work 'to improve the quality of the environment', as the Treaty of the European Community requires. This is a daunting task, as our area of focus stretches well beyond environmental policies in the traditional sense. In order to succeed, it is required that we look at such divergent areas as agriculture, external trade, development, transport, energy, structural funds, and fisheries.

The EEB has considerable experience with the 'traditional' environmental policies, the policies that emerge from the EC's Directorate General (DG) Environment and have as their sole or main objective the protection and improvement of the state of the environment. We have also been functional in furthering a number of the measures promoted by the DG Agriculture, in particular agri-environmental measures, though the DG Environment remains the main focus in our day-to-day work. Other DGs are targeted more consistently by WWF and Birdlife International (DG Agriculture and Fisheries, DG Regional Policy), the European Federation for Transport and Environment, and Climate Action Network Europe (DG Transport and Energy), as well as WWF and Friends of the Earth Europe (DG Trade and DG External Relations). Focusing on the DG Environment is a limitation. The existence of such a limit is primarily due to our limited resources, but can also be attributed to the DG Environment's level of involvement in horizontal issues and those where other DGs have the lead, its reception of our points of view, and a willingness to exchange (strategic) information. In any case, environmental policies will remain needed in the future.

In most cases, it is essential that we are aware of emerging proposals early on, and become involved in shaping them. Until the early 1990s, this was an entirely informal process. It depended on having the right contacts inside the EC and approaching them in the right manner. This was work that could be done only by people working permanently in Brussels, who understood the complex bureaucracy and had developed personal networks.

The EU is bureaucratic indeed. The Commission is a body with no equivalent at the national level. It is not comparable to a government, as it is not composed on the basis of the EP's political majorities. The EP does not have the same level of control over the EC as national parliaments have over their governments. However, the Commission has considerable power in certain areas, while it plays a marginal role in others. The decision-making rules are difficult to understand: even seasoned lobbyists need to consult the Treaties regularly to discern the exact division of power between the three main institutions in specific cases.

It is quite difficult to follow the work of the Council of Ministers, which represents the national governments. Officially, the Council has to await the position of the EP, but usually it takes that as a formality. It simply starts its internal negotiations and comes to political agreement independently of the debates among the Parliamentarians. The EC plays an important role here, as it can frustrate majorities in the Council by liaising with the minority, thus preventing changes in its proposals. Here, the Commission plays the role of executive and legislative body combined.

To have an impact on the outcomes of the Council, it is important to know details of the internal discussions. Most of these take place between civil servants and behind closed doors. In some countries, EEB members have close ties with the national ministries: for example, there are even regular meetings between the two parties; in other countries (including the Netherlands, Sweden, and the UK), such informal sources of information are poor or non-existent. In Brussels, the EEB's informal contacts can make up for this to a certain extent, though this mainly applies to issues being dealt with in the Environmental Council. It is much more difficult to get the information from the other Councils that would increase our impact.

In the early 1990s, the EC began to introduce forms of structured dialogue, but the environmental organisations were not their first target. Structured dialogue was primarily a response to firm pressure from business groups which opposed the EC's regulatory approach; these groups wished to see more voluntary agreements, enabling policies (funding of research, tax reductions, etc.) and promotion of positive examples. The business lobby was particularly strong and focused in the run-up to the Rio Earth Summit, and its impact was reflected in the 5th Environmental Action Programme. The message gained was that the EC understood that it needed to involve stakeholders much more in policy-making. Obviously, the Commission could not single out business as the only stakeholder (even though it did so in one of the first exercises in this new approach, the Auto-Oil Programme), so environmental NGOs became increasingly involved in ad hoc stakeholder processes, both before formulation of EU policies and in the framework of their implementation. The DG Environment initiated this policy and is today clearly the most experienced, though later the DGs Agriculture, Trade, Energy and Transport, Development, and Fisheries followed suit.

Taking part in such exercises is now part of our daily routine. The EEB is involved in many structured dialogues, usually with mixed feelings. In many cases, the main purpose of our presence is to supply balance to a predominantly business attendance. The environmental movement now has some 50 lobbyists/campaigners in Brussels at its disposal. Lobbyists from other sectors, with business and regional/local authorities being the largest, total some 13,000. In the discussions on the revision of the Packaging Directive, an important element of the EU's waste management policy, the one or two representatives from the EEB were the sole environmental lobbyists, surrounded by 60 to 70 representatives from business.

Formalised stakeholder dialogue can be useful, if the intention is to gather information and opinions in order to take better decisions while sticking with the objective of improving the quality of the environment. That means that the EC needs to exhibit political leadership. However, we have the impression that, in many cases, structured dialogue is used as an *alternative* to Commission initiative. Dialogue can also just become a ritual, a theatre to show transparency and a willingness to listen without having any impact on policies. The Advisory Committees to the DG Agriculture are dominated by farmers' associations, with environmentalists also serving as members. The discussions have very little impact on the EC's policies. The agricultural associations know that but do not mind. They have much better informal channels than these committees, and thus have no reason to challenge their existence.

Therefore, formalised dialogue is not good by definition, and can become a distraction. We continue to focus on informal dialogue, with technical experts and higher-level officials, including the Commissioners, at the right time. With this approach, we have had a good deal of success. We do make a difference, even though it is not always easy to determine exactly the magnitude of our influence.

The Impact of the EEB

When the EEB is successful, the EC, EP, and Council of Ministers adopt some of its demands and proposals. This happens as part of a complex process, spanning some one to four years, and decision-makers can have many reasons for not wanting to admit the EEB's role in this process. When we fail, it is often because our demands are based on what is needed for the environment, not on a realistic assessment of what can be achieved under the given circumstances. This is not a wrong approach per se. The bottom line is that we often have to balance the influence of industry, which almost invariably tries to weaken EC proposals. We regularly find ourselves in the position of defending a Commission proposal, even if it is not the ideal answer to our demands. I discuss a (non-exhaustive) number of issues on which the EEB has had a significant impact.

European Constitution

Inspired in particular by the Platform of European Social NGOs, the EEB began to get involved in the discussion on the Future of Europe. This discussion reached concrete form in the Convention, which presented concrete proposals for a Constitution to replace the existing EU Treaties in June 2003. The EEB mobilised other environmental organisations, working together in the Green Eight, and represented the Green Eight in hearings and direct discussions with the Convention and its members. At the national level, the EEB's efforts resulted in environmental NGO activities in several countries, including a number of EU Accession Countries. The challenge was to strengthen rather than weaken the anchors for environment and sustainable development in the Constitution and to introduce access to justice on the EU level.

Towards the end of 2002, it became clear that the challenge was serious: the Presidium of the Convention had presented an initial draft for the Constitution which omitted the language on sustainable development and environmental policy integration which the EU had adopted at the Amsterdam Summit in June 1997. The EEB launched a campaign which received the solid support of its members, other environmental organisations, and organisations working on other issues but convinced of the necessity of not reneging on sustainable development in the Treaty. We worked very hard, and we finally succeeded.

EU enlargement

With regard to the accession of 10 Middle and Eastern European countries, the EEB tried to influence the discussions on the conditions for accession through its EC and EP contacts. Our input was based on the comments we received from environmentalists in those regions, but this remained very difficult, given the complex and closed character of the negotiations. In the end, the results of the negotiations on the environmental chapters were quite satisfactory.

Another, perhaps more important long-term role for the EEB, is training and awareness-raising among environmentalists in the Accession Countries. A competent environmental movement in these countries on EU policies would be a contribution to effective environmental policy-making in the EU. To this end, the EEB organises yearly events to engage accession-country NGOs in issues like agriculture and regional policy. Accession-country NGOs have also become increasingly active in the EEB's work on biodiversity and sustainable development.

Environmental Action Programme

After protracted discussion, the 6th Environmental Action Programme was adopted in April 2003. The Council and the EP agreed on a compromise that was better than the original proposal by the EC. The EEB had considerable influence on the result; in 2001, it had made 92 concrete suggestions, of which 32 were adopted, 7 were reflected in some way, 18 were partially adopted, and 35 were refused. The EEB's success was due to very early involvement in the process and the concrete proposals it had made throughout the decision-making process. The EEB proposals were particularly well received by the EP rapporteur and some governments, whereas the EC refused to establish more clear targets and timetables, and to limit the work on thematic strategies in time. In the end, the EC had to give in on several issues.

Fiscal reform

In 2001, the EEB started a campaign on environmental fiscal reform. It provided a clear European framework for the work done thus far in isolated fashion in individual countries. It also encouraged EEB members in countries where the debate had not yet begun or had fizzled out. Events took place in Austria, Greece, Germany, Denmark, the Czech Republic, Poland, and Ireland.

The EEB organised, for the first time, coordinated pressure on the ECOFIN Council (Ministers of Finance and Economic Affairs) to agree on a directive on minimum ecotaxes on energy products. It launched the debate with social groups, trade unions, industry representatives, and scientists on desired reforms, and encouraged the Organisation for Economic Co-operation and Development (OECD) to reinstate this issue high on its agenda. Several Ministers came to speak on environmental fiscal reform at the EEB Annual Conference in 2003 and showed a good deal of interest in the issue: we simply have to keep on pushing for it.

Chemicals

Industry has heavily lobbied against the intentions of the EC, laid down in a 2001 White Paper, for a new chemicals policy. This has led to a considerable delay in the presentation of the draft regulation that will convert the options into legally binding measures. The EEB has worked closely with the European Bureau of Consumers organisations (BEUC) on this issue. Intensive discussions have been conducted with trade unions and animal protection groups in order to mobilise as much support as possible for our key demands. An EEB conference in Copenhagen in 2002 was successful in bringing together different stakeholders, while illustrating the EEB's ability to create pressure, conduct dialogue, and introduce innovative approaches.

The EEB cooperates with the European Coalition to End Animal Experiments, thereby increasing the awareness of animal protection NGOs for the need of a chemicals policy reform. The EEB insists on the minimisation of animal testing and is convinced that its use can be drastically reduced. Our success has reduced some of the lobby pressure on the EC against this policy reform, but did not really affect animal protection concerns in the UK.

As in other fields of its work, the EEB tries to build coalitions with individual companies that might speak out against the industrial federations in support of a new chemicals policy. The first companies of this kind to sign up to our aims were all British: Marks & Spencer, Kingfisher, and B&Q. These companies have direct contacts with consumers who want certainty about the safety of the products they buy, and discussions have taken place between these companies and the EEB.

Biofuels

EEB caused quite a stir when it announced its opposition to proposals for a major boost of biofuels production in the EU. Environmentalists concentrating on climate change were initially not happy, but those working for sustainable agriculture understood that the EC proposals would create new sources of soil and water pollution. In addition, the proposals did not make much sense from a cost-effectiveness perspective. We considered the plans to be a hidden new subsidy for farmers, rather than an environmental measure. The EEB had to defend its position many times, and to explain that it was not against the use of biomass for energy purposes, but against the specific focus on producing large amounts of liquid fuels from specialised agricultural crop cultivation. We could not convince the EP, but we had the Council on our side – though probably not for environmental reasons. In the end, no obligatory target for biofuels production was set, and VAT reduction on biofuels is optional and not obligatory.

The Future of the EEB

The coming years will not be easy. The enlargement of the EU means we will have to be able to convince the vast majority of no less than 25 national governments, issue by issue. Fortunately, we have invested a great deal in the expansion of the EEB's network into the 10 countries that have joined in 2004. Currently, we have 21 member organisations in these countries, and our membership is growing annually. These members take part fully in the work of the EEB. Owing to our capacity-building efforts, they are, by-and-large, even more involved and integrated than many of our 'older' members. In any case, the environmental movements in these countries expect quite a lot from the EU and, therefore, have already developed into experts on EU policy-making.

The highest risk is that the EU will abdicate its leadership role to its member countries and the rest of the world. Instead of effective legislation and other bold initiatives to steer the market, we might see increasingly softer policies (such as voluntary agreements), thematic strategies, and information exercises. There is also much discussion taking place on diversifying environmental policies in order to respond better to diverging environmental and economic circumstances within the larger EU. This is all very risky. As the markets of the EU countries are strongly integrated, it will be difficult for national environmental authorities to maintain high levels of achievement if this is not agreed on at the EU level. A race to the bottom might not be the immediate risk, but a standstill is certainly not unlikely.

In such circumstances, it is essential for the EEB that the environmental movement be strong and influential in all of the EU member states. With this in mind, the EEB has started a special training programme for national environmentalists on EU environmental policies.

MANAGING GLOBALISATION

Globalisation as it is currently defined is not going to ensure justice and prosperity to all of the world's population, and certainly not to future generations. If we continue to adhere to the ideology of economic growth, free trade, and deregulation, we will face serious problems concerning the depletion of the Earth's natural resources and growing tension between the rich and the poor, globally and within regions. The EU cannot change this all by itself, but it can certainly make a difference and set an example. In order to achieve this, however, it needs to reformulate its objectives and change its practices. Closely connected to this is the need to overcome its democratic deficit and transform itself into a true example of international participatory democracy.

Sustainable Development as the Central Objective

The notion of Sustainable Development was introduced in the Treaty of the European Community in 1997, at the Amsterdam Summit. The EU governments took on the concept of sustainable development without much thought concerning what it would mean for EU policies and measures. However, the combined initiatives of a number of governments (in particular, Finland and Sweden), the Environmental Council and Commissioner, and the EEB has led to further discussions on this point, resulting in the decision at the 1999 Helsinki Summit to develop an EU Sustainable Development Strategy within eighteen months.

The EC had some internal trouble in delivering a draft, which was published in May 2001, three weeks before the Gothenburg Summit. The strategy was not what it should have been; it is, in fact, an environmental amendment to the Lisbon Strategy which was adopted a year earlier aiming for the EU 'to become the most competitive and dynamic knowledge-based economy in the world, capable of sustainable economic growth with more and better jobs and greater social cohesion'.[4] The Summit accepted only a part of the EC proposals, given that there was little time to discuss the more complex elements – such as a deadline for the removal of environmentally problematic subsidies on the EU and national levels. The Summit did agree to merge the strategy's monitoring with the Lisbon process.

Since 2001, the implementation and updating of the Lisbon Strategy has been the topic of each Spring Summit of the European Council. In 2002 and 2003, the first attempts were made to integrate the environmental dimension as agreed in Gothenburg, but the results were poor. This is not surprising: the

---

[4] European Commission, 2000: conclusion 5.

Lisbon Strategy is motivated by the desire to catch up with the economy of the United States (US). This is not at all the right incentive for discussing a radically different approach to economic development, which concentrates on dematerialisation, quality of life, social justice, concern for future generations, and sufficiency. The EEB set up a coalition with the European Trade Union Confederation and the Platform of European Social NGOs to monitor and influence this process.

The European Convention: A Missed Opportunity

The Convention that prepared a draft for the European Constitution could have been a good place to discuss the need for a new leitmotiv for the EU. We tried to raise this issue, but we had to work hard to prevent even Article 2 of the Amsterdam Treaty from disappearing. There was very little interest among the Convention participants in this issue, all of the attention being focused on the relations between EU institutions and member states. We also pleaded for a reformulation of the objectives for the specific policies for which the EU has a mandate. The Convention did not do so, arguing that the mandate given by the European Council in 2001 did not include this exercise. The reason was probably a fear of opening a Pandora's box.

Pandora's box or not, the EU needs to reformulate its objectives; not just in general terms, but also with respect to those objectives that make a difference in policies on agriculture, trade, transport, research, cohesion, taxation, etc. The Green Eight published a set of specific proposals, directed at those involved in agreeing on the final content of the future EU Constitution.[5]

The EU could become a blessing for the region and the world, if it indeed developed a sustainable development strategy that encompassed all its policies on a permanent basis. However, defining what sustainable development means in terms of concrete targets and measures is an ongoing debate, which requires the involvement of civil society in the EU and those countries strongly impacted by the EU's external policies. Civil-society involvement is needed to come to decisions that receive popular support and cooperation, and to counterbalance the pressure of business and conservative political groups.

On democratisation and the involvement of civil society, the Convention did make progress, though less than we desired. Its proposal will increase accountability and transparency, and give the EP a more prominent role in several important policy areas. It introduced the concept of 'participatory democracy', to complement parliamentary democracy. Still, improvements

---

[5] Green Eight, 2003.

are necessary, in order to complete the EP's co-legislative role, to further clarify the minimum requirements for public participation, and to recognise access to justice as a logical and essential element of participatory democracy.

Pledges for Good Governance

The low turnout for the 1999 EP elections and the large-scale protests against the impact of national policies resulting from the economic conditions for joining the Eurozone, leading in 2001 to a 'No' vote in the Irish referendum on the Nice Treaty, made EU politicians aware that they had a real problem in their relationship with EU citizens.

As the EC's 2001 White Paper on Good Governance stated:

> Many people are losing confidence in a poorly understood and complex system to deliver the policies that they want. The Union is often seen as remote and at the same time too intrusive. ... Yet people also expect the Union to take the lead in seizing the opportunities of globalisation for economic and human development, and in responding to environmental challenges, unemployment, concerns over food safety, crime and regional conflicts. They expect the Union to act as visibly as national governments.[6]

The EC proposed a discussion on

> the question of how the EU uses the powers given by its citizens. It is about how things could and should be done. The goal is to open up policy-making to make it more inclusive and accountable. A better use of powers should connect the EU more closely to its citizens and lead to more effective policies.[7]

In its reaction, the EEB endorsed the White Paper's principles: openness, participation, accountability, effectiveness, and coherence.[8] But it had serious concerns with several other proposals, principally about the EC's idea that the EU would be more acceptable to EU citizens if it granted business a larger role instead of showing clearer political leadership. The discussion led to a set of measures, in particular, an internal EC code of conduct[9] and an integrated EC impact assessment process.

---

6 European Commission, 2001: 3.
7 European Commission, 2001: 8.
8 EEB, 2001.
9 European Commission, 2002.

Resisting Business Pressure

The EU institutions, including the EP, are too sensitive to business pressure. While the EU is potentially the most potent answer to globalisation, anticipating rather than responding to issues at hand, it acts as if it is dependent on the goodwill of European and US-based companies. The picture is not black and white, but the pleas for co-regulation (agreements between governments and industry sectors that do not have the character of law), cost-benefit exercises, the dogma that free trade and economic growth are preconditions for sustainable development, are illustrative of a dangerous lack of vision and political leadership. We see authorities and politicians claiming that they have fewer means to steer processes than they actually have at their disposal.

The EC regularly orders surveys on the attitude of the EU public with respect to the environment. It also investigates whom people trust when it comes to environmental issues. The figures differ per country, but the averages speak clearly: 48 per cent trust environmental NGOs, 23 per cent trust consumer organisations, 13 per cent trust the EU, 12 per cent trust national governments, and 1 per cent trust companies.[10] With this in mind, it would be a grave error to think that improving governance, reducing the distance between the EU citizens and the EU institutions, can be achieved by giving in to business pressure. Quite the opposite is true. The EU institutions should explore more fully visible political leadership, based on vision and respect for non-material values and signals produced by organised civil society with large public support, and which fights for justice, equal opportunities, respect for nature, and global responsibility. The approach I seek would certainly not be anti-business, and would find supporters inside the business sector, among those individuals and companies that also understand that current consumption and production patterns are unsustainable, and that innovation is not in the first place a matter of investing tax-payer money into research and development, but of the right mix of market instruments and regulations.

Legitimacy of the Environmental Movement

The EC's White Paper on Governance is making the degree to which NGOs are representative an issue. And this is not unique. Time and again, aggressive media articles appear about environmental organisations becoming too powerful. The American Enterprise Institute has set up an 'NGO Watch' to systematically collect information about NGOs in order to challenge their

---

[10] Eurobarometer, 2002.

legitimacy. The EC, the main sponsor of many European NGOs, has now established a set of questions on internal governance that NGOs need to answer yearly as part of their funding requests. Furthermore, it has set up a website where information can be found about NGOs with which the Commission maintains contacts.

I do not mind these questions and this transparency. We have nothing to hide and it is important that people understand that the EEB is not just an office of 14 individuals in Brussels, but an organisation with a democratic association structure. But the larger question, of course, is how we see ourselves in relation to authorities and elected bodies such as the EP. The EEB is here to improve the functioning of parliamentary democracy, not to undermine it. We think that politicians should take informed decisions. This means they need to know that the views from inside the administration, the US government, and business federations are not the only ones possible, and that a large share of the information that is presented as technical and neutral is in fact biased and based on political choices. We are there to add breadth to the picture, help politicians in asking the bigger questions, and remind them about the visions laid down in the EU Treaties and numerous policy statements. We insist on being heard, because we have the support of, and represent, millions of citizens.

However, people do support environmental organisations, through membership or financial support, more readily in some European countries than in others. Some politicians think that environmental NGOs that lobby should not receive any financial support from governments or the EC, using the US as an example. Luckily, these politicians do not dominate the scene in Brussels, so the EC does support NGOs to a certain extent. The EEB, for example, receives 50 per cent of its budget from the DG Environment.[11] Such support is justified, as it is used to improve the quality of parliamentary democracy by making it participatory democracy. Let us not forget that environmentalists function as a counterweight to an army of well-paid lobbyists working for industry. The latter are paid from the involuntary contributions of consumers. Consumers pay for a product or service, but not for lobbyists who defend positions with which these consumers may completely disagree.

---

[11] Of course, it remains important that environmental NGOs limit their dependence on government support. Direct public support is an essential proof of legitimacy.

CONCLUSIONS

The EU is facing a new era in both its internal and external relations. Internally, 10 new member states have joined the EU in order to be part of a strong economic block. They expect accelerated economic growth. Externally, the EU is facing an increasingly cynical US, who pursues a strengthening of its geopolitical position in international politics for the benefit of its national, highly unsustainable, economy. The logical result of these developments would appear to be an EU that is more aggressive towards trade liberalisation, moving with the US into forced liberalisation of services, intellectual property, investments, etc. However, such an outcome is not what environmental and social organisations seek. They seek a global society in which sustainable development, not free trade, determines political decisions, and where fair trade and protection of natural resources have a chance: a society, in short, in which the benefits of economic activities are better shared with local societies.

In the end, the EU is managed by the national governments. Thus, it is essential that the discussion on sustainable development continues at the national level, as well as at the geographical and institutional levels where the EU is active. We need to work towards a commonly shared vision on what sustainable development means for the EU, both for its own area and as a concept to guide its external actions.

This debate is not free from power relations. The national governments and the EC should show new leadership vis-à-vis the commercial sector. On the one hand, they should learn from progressive companies about what can be achieved with technological and management improvements, rather than listen to the conservative European industry federations. At the same time, they should always be aware that such innovative companies will always remain within the limits of their commercial mission. The key to change towards sustainable development lies, therefore, in strong public support. I am convinced that such support is easier to gain than many politicians think. Its essential preconditions are sincerity, fairness, reliability, transparency, and involvement of organised civil society in decision-making. In other words, good governance.

REFERENCES

EEB (2001), *A First Response from the European Environmental Bureau to the White Paper on Governance*, Brussels: European Environmental Bureau.

Eurobarometer (2002), *The Attitudes of Europeans towards the Environment*, The European Union Research Group (EORG) for the Directorate-General

Environment, Brussels: Directorate-General Press and Communication, Opinion Polls of the European Commission.

European Commission (1973), 'First Environmental Action Programme 1973-1976', OJ C 112, *http://europa.eu.int*, Brussels: European Commission.

European Commission (2000), 'Presidency conclusions, Lisbon European Council', 23 and 24 March, *http://ue.eu.int/ueDocs/cms_Data/docs/pressData/en/ec/00100-r1.en0.htm.*

European Commission (2001), *White Paper on Good Governance*, COM (2001) 428 final, Brussels: European Commission.

European Commission (2002), *General Principles and Minimum Standards for Consultation of Interested Parties by the Commission*, COM (2002) 704 final, Brussels: European Commission.

European Community Treaty (1957), *http://europa.eu.int/abc/obj/treaties/en/entoc05.htm.*

Green Eight (2003), 'Towards a green EU Constitution: Greening the European Convention Proposal', position paper, August, *http://www.eeb.org/publication/G8-IGC-EN-final-8-8-03.pdf.*

PART III

The Impact of Globalisation on Domestic Environmental Policies

16. National Room to Manoeuvre: The Dutch Position in EU Energy Policies

Maarten Arentsen and Theo de Bruijn

SUMMARY

We discuss interactions between supranational decision-making and national discretion in a context of globalisation and increasing complexity of political decision-making. The impact of energy policies in the European Union (EU) on domestic policies in the Netherlands, as well as the Dutch influence on common policies was studied for four dossiers: security of gas supply, emission trading, energy taxation, and sustainable electricity. In two cases (the gas dossier and the sustainable-electricity dossier), defensive national strategies were successful at the EU level. The offensive Dutch position in the emission-trading dossier had some impact on EU regulation, while the proactive strategy in the energy-taxation case was unsuccessful. We conclude that the relative power position of a country and the proximity of the viewpoints of different countries seem to be important determinants of national 'room to manoeuvre' in EU decision-making.

INTRODUCTION[1]

The process of European integration has undeniably affected national institutional settings of policy-making in Western Europe. Research clearly shows that Europe matters, but 'how and when is still controversial'.[2] A salient question is what margin of freedom remains in national policy-making. From a naive standpoint, one could argue that European integration imposes common approaches and harmonised policies on the member states of the

[1] The authors wish to thank Klaas Jan Koops of the Dutch Ministry of Economic Affairs for his stimulating remarks on an earlier version and for his contribution to the data collection.
[2] Börzel, 1999.

EU. However, empirical studies give a mixed picture at best.[3] Domestic policies are harmonising in some areas, while other policies remain nationally focused. In general, national policy contexts prove resistant to change despite the forces of Europeanisation.

We explored national positions in EU policy-making in the context of the expansion of economic and political processes. The continuous economic and financial integration within Europe ultimately affects the scale of political decision-making; the question is how. National decision-making and political autonomy will be affected if Europe is to become financially and economically integrated. But does an expansion of political decision-making necessarily follow expansion of economic processes? If the answer is confirmative, then how does the development of political decision-making affect national autonomy and how do countries evolve within the EU political process? We examined these questions analytically and empirically. We developed the argument from a Dutch perspective in EU political decision-making by examining four energy dossiers.

The tension between European and national levels makes the topic of energy relevant in this respect. Though energy was excluded from the Single European Act because it was considered a national topic, the European Commission (EC) pushed and still pushes the strengthening of a common EU energy policy for at least two reasons. According to the Commission, the EU region as a whole will become more dependent on imported energy resources (natural gas and oil) in the near future, inevitably increasing internal cooperation. Second, within the context of global climate change, the EU is considered a single continent speaking with one voice and forced to coordinate internal responsibilities at the European level. Given the tensions between national and European interests in energy matters, the relevant question becomes whether and to what extent the EU can allow for national discretion in energy policies. We addressed this question from an institutional and political perspective, not from a legal viewpoint. The aim of our analysis was to explore national 'room to manoeuvre' regarding energy policies in the European context.

The chapter is structured as follows. In section 2, we deal with the question of national room to manoeuvre from an analytical perspective. In section 3, we address the question empirically by focusing on the Dutch national position in four energy dossiers: security of gas supply, emission trading, energy taxation, and renewable electricity technologies. All four dossiers have been or still are on the EU political agenda. In the final section, we

---

3 Börzel, 1999; Cowles et al., 2001; De Bruijn, 2003; Haverland, 1999; Knill and Lehmkuhl, 1999; Olsen, 1997; Unger and Van Waarden, 1995; Weale et al., 1996.

evaluate the analytical perspective of the national position in EU energy policy and draw conclusions.

EU DECISION-MAKING AND NATIONAL STRATEGIES

The central rationale of EU cooperation is manifest in the principle of subsidiarity: the EU acts only if goal achievement is beyond the scale and scope of individual member states and if the EU can be expected to achieve the goals. Based on the principle of subsidiarity, energy has been excluded from EU decision-making for a long time; it was considered specifically a matter of national interest. At least two recent developments have changed this position. The first development is emerging globalisation. Europeanisation tends to be an autonomous process of economic, political, and cultural changes, which are not even typically European. The process takes place on a global scale and consists of several interrelated and simultaneous changes.[4] The institutional, economic, and political dynamics in the 'global village' are argued to be rapidly changing owing to technical, cultural, demographic, and economic changes. Europeanisation is recognised as a regional manifestation of this process[5] 'that refers to structural transformations in our economy through which economic activities are not only connected to their local base but are linked at a global level by means of various ramifications of a political, socio-cultural and economic nature'.[6] According to some scholars, this multidimensional process has restricted the political autonomy of the EU member states.[7] Others stress that 'it may exert pressure on the Member States to adapt to European rules and regulations and thus may affect the national institutional framework of policy making'.[8] Europeanisation affects the principle of subsidiarity because globalisation and Europeanisation imply political action on a larger scale.

The second development is the growing complexity of political decision-making. At all levels, decision-making has become a negotiating process between many actors; political action has thus become a process of co-production in network governance.[9] In such governance, policy choices emanate from highly organised social subsystems, such as production systems, rather than from a central state authority.[10] Efficient and effective governance must recognise the specific rationality of the subsystems. In order

4 Fuchs, 2002; Nijkamp and Verbruggen, 2003; Van Kersbergen and Van Waarden, 2001.
5 Addink et al., 2003.
6 Nijkamp and Verbruggen, 2003: 61.
7 Sorensen, 1999.
8 Van Kersbergen and Van Waarden, 2001: 29-30.
9 Van Kersbergen and Van Waarden, 2001.
10 Kohler-Koch, 1998.

to mirror the rationale and the creativity of the subsystems, policy processes need to be open and decentralised. In network models, states play a more limited role and concentrate on establishing supportive policy networks by bringing together the relevant actors.[11]

Member states face a myriad of interests that compete for priority, not only in national but also in EU decision-making. These multilevel and multi-actor dynamics tend to blur the idea of a single national position in EU decision-making. A diversity of national interests may be organised according to various social, economic, or political sectors. For instance, in EU environmental decision-making, the European car industry – a nationally based and privately organised sector – often has a strong voice in 'Brussels', where the EC resides. This occurs both through the inputs of national governments and as a result of being an organised international sector. EU policy-making is often a process of give and take between interests of national governments and economic sectors. Consequently, individual actors – including national governments – can find it difficult to hold a strong position in these complex bargaining games between highly disparate interests.

The impact these developments have on national positions in EU decision-making is considerable. They have changed the meaning of subsidiarity and have contributed to a diversification of national interest positions in the EU political process. As a result, individual countries are increasingly forced to position themselves and their policy preferences in a broader European political context consisting of various actors, both public and private. National governments must reassess their positions in the EU political process by answering questions pertaining to the meaning of this multi-actor context in terms of national gains and risks, the advantages and disadvantages of the European dimension to individual policy dossiers, and the (de)merits of coordinating policy. EU policy-making requires countries to constantly evaluate national and sectoral gains and losses. After agreeing on the strategic point of departure and considering potential gains, member states have to decide about timing an intervention to be successful in their strategy: either in the stage of policy-making or in the stage of policy implementation. Timing depends on the national interests at stake and the interrelatedness of these interests.

The combination of timing and interests may determine the kind of strategy to follow: offensive or defensive. In the first case, members may push their national policy model to the European level because of passionate preferences or because of specific interests. In the latter case, members may

[11] Kohler-Koch, 1998; Young, 2000.

have a passive attitude towards the particular initiatives, for example, if the national position is indifferent or strongly opposed to the EU policy initiatives. National governments may follow offensive strategies for several reasons. One such reason would be to ensure a close fit between the EU policy outcome and the domestic agenda. A driving force of this strategic position could be the endogenous institutional resistance to change. If national governments work with clearly defined national interests, their strategic focus usually narrows to getting the national policy model accepted as the EU model. In this way, nation states try to 'export' their own policy models to Brussels and thus to other EU member states.[12] Sometimes, the EC encourages this upgrading of effective national models to the EU level.[13] Another offensive strategy in EU policy-making is influencing the institutional setting with national policies. For instance, the key actors preparing the 5th EU Environmental Action Programme were Dutch civil servants and scientists strongly involved in and committed to the first Dutch National Environmental Policy Plan. They included many basic elements of their own policy, resulting in an EU plan closely resembling the Dutch programme. If national models are able to show their effectiveness or have the support of other EU institutions and agents, they have a good chance of rising to the international policy level. Horizontal information exchange between member states and active lobbying have been suggested as effective facilitators for this strategy, because states and the Commission are willing to copy successful national policy models.

Similarly, national governments can adopt certain strategies to implement EU legislation in national legislation.[14] When EU and national legislation coincide, member states tend to implement rapidly. For instance, in the case of the Integrated Pollution Prevention and Control (IPPC) Directive,[15] the Netherlands moved quickly. The difference between Dutch environmental regulation and the IPPC initiative was relatively minor. Except for some technical details, the fit between IPPC requirements and Dutch environmental licensing of industrial sites was convenient. Spain adopted a 'delaying strategy' in national implementation of the IPPC requirements because of significant differences with national legislation. For Spain, the IPPC Directive was 'revolutionary' and its implementation required dramatic modification of the Spanish regulatory system.[16] Denmark adopted an 'ignoring strategy' by

---

12 Lowe and Ward, 1998.

13 Pellegrom, 1997.

14 De Bruijn, 2003.

15 The IPPC Directive was initiated to harmonise environmental regulation of industry with the idea that the European environment could benefit from harmonised requirements for environmental licensing of substantive sources of industrial pollution (European Commission, 1996).

16 Rodriguez de Sancho, 2000.

restricting public participation in licensing procedures to the legal threshold of IPPC and by scarcely participating in the horizontal information exchange between member states on Best Available Techniques (BAT). Both public participation and the use of BAT were new to Denmark and, therefore, difficult to integrate into Danish environmental legislation.

In conclusion, the analysis thus far supports our initial assumption that even though EU integration has affected national institutional settings and national political decision-making, room to manoeuvre still exists. We modelled this space by distinguishing between offensive and defensive strategies that EU member states may choose given a specific configuration of actors, timing, and their perceived interest positions in individual policy dossiers. Based on this framework, we analysed the Dutch national position in four EU energy dossiers, as described in the next section.

NATIONAL ROOM IN EU ENERGY POLICY

Emerging Common Energy Policy

Energy – from coal and nuclear sources – was an issue at the very beginning of the European Community, but the energy sector was excluded from the Single European Act signed by the member states in 1985. The creation of an internal energy market for electricity and gas was the subject of a separate, long, and complicated bargaining process involving many actors with diverse interests. The process took more than a decade and only produced a minimal set of common rules on liberalisation and harmonisation of gas and electricity markets across Europe.

Despite inert political decision-making and compromised outcomes, both the Electricity and the Gas Directives changed the European energy market. Market liberalisation, dismantling of monopolistic electricity supply, privatisation of ownership structures, and consumer choice of electricity supplier became the major points of reference in electricity supply. In the face of economically driven competition and rivalry, the European electricity industry adapted a more commercial business strategy.[17] The 'utility' image of electricity was transformed in favour of a commodity. At different locations in Europe, power exchanges developed, where electricity was traded at prices reflecting market conditions. Electricity became a tradable commodity under competition-based supply-and-demand conditions. From August 2000, the legal gas order in the EU region changed along the lines prescribed by the Gas Directive. Countries enacted rules for the opening of

[17] Midttun, 2001.

the national gas market, third-party access to the pipeline infrastructure, and the unbundling of integrated companies and public service obligations (PSOs). In some countries, such as the Netherlands, the EU Directive induced public gas legislation for the first time in history. In other countries, such as the United Kingdom (UK), the EU Directive only codified already established competitive market conditions.

In the context of this new level playing field in European energy, we analysed the Dutch national room to manoeuvre in four energy dossiers, each with a strong national and European dimension: security of gas supply, emission trading, energy taxation, and sustainable electricity technologies. In the following subsections, the national position of the Netherlands in EU policy-making is empirically illustrated.

Dossier 1: Security of Gas Supply

The Netherlands is gifted with huge gas reserves, discovered in 1960. From the beginning, the Dutch adopted a public service orientation in production and supply, giving priority to both domestic gas consumption and long-term domestic supply security. To pursue this public service orientation, the Dutch developed a national gas market model based on three principles: centralised coordination of demand and supply, public-private partnership in upstream activities, and commercial exploitation of the national reserves for the benefit of the national economy. These core elements of national gas policies were adopted in a 1963 White Paper and guided national gas policies for the last four decades.[18] Since 1974 (the first oil crisis), security of the domestic gas supply has been added as a guiding principle of the Dutch gas policy. From 1974, the national long-term security of supply was given strict priority and sales strategies became conditioned by long-term planning (25 years ahead) of domestic supply and demand, to be assessed and forecast every two years by Gasunie.[19] This long-term planning was conservative, accounting only for proven reserves and excluding future finds. The national reserve position was further improved by the small-field policy (i.e., preferential depletion of small off-shore gas fields with the Groningen field in a swing capacity) adopted in 1974.[20]

The 1995 White Paper on Energy[21] did not make any significant changes

[18] Ministry of Economic Affairs, 1963.

[19] Gasunie is the central coordinator of the Dutch gas industry. For more than 40 years, it was the agency of Dutch gas policies. Until the market was liberalised, this public-private company coordinated supply and demand, exported Dutch gas, and owned and managed the high-pressure pipeline system in the Netherlands.

[20] The depletion of the Groningen gas field, the largest Dutch gas field, is easily manageable at low costs, because of the unique pressure-related features of the field.

[21] Ministry of Economic Affairs, 1995.

in the basic principles of the Dutch gas policy, but calibrated them in the context of the emerging liberalisation of EU gas markets. Until 1995, the long-term security of supply had been the ultimate bottom line for national gas strategies. Long-term security affected national depletion policies, reserve policies, sales policies, and exports and imports. The 1995 White Paper added to this nationally focused gas policy a strengthening of the commercial trade position of Gasunie and Dutch gas in Europe. It was expected that the liberalisation and opening of the Dutch gas market would affect the position of Gasunie in the domestic arena, and require compensation by increased gas exports. The company's obligation to forecast supply-and-demand developments 25 years ahead was relieved; the forecast now focused on the Dutch reserve position and strategic depletion instead of long-term security of supply. It was decided to keep the production of natural gas at a constant production level, which would satisfy long-term domestic demands and allow additional exports if Gasunie lost shares on the domestic market owing to liberalisation. The small-field policy was expected to continue under liberalisation.

During the early 1990s, the Dutch government took a passive position in EU negotiations regarding the liberalisation of the gas market. The Dutch position in the EU debate on electricity was different. The passive Dutch position in gas negotiations was partly motivated by protection of national gas reserves, but mostly by national security policy. Dutch gas sales in Europe had been recalibrated in the 1989 update of the national energy policy, but this had not affected the priority of longer-term security of domestic gas supply. The Dutch wanted to continue this security policy as well as its national industrial organisation as long as possible. Until 1995, neither the Dutch government nor the Dutch gas industry showed any interest in gas-market liberalisation. In 1995, this passivity vis-à-vis liberalisation and the longer-term gas security policy significantly changed.[22] The Dutch security policies were felt to be compatible with the idea of a competition-based internal gas market and could continue unchanged. By setting an annual production ceiling and continuing the small-field policy, it was possible to deplete the Dutch gas fields in the same time frame as in the old regulatory regime. With the suggested production quota of 80 billion cubic metres annually, the Dutch gas reserves would be exhausted in 2025. Until that year, domestic gas consumption would be secured by reserves. The change in the liberalisation debate did not change the national interest of the Netherlands in the security issue.

In the negotiations on the common Gas Directive, the Dutch reference

[22] According to Stern (1998), this change in the Dutch position greatly affected EU bargaining on the Gas Directive. It was a change of coalition in the Dutch government that caused the change of strategy.

points were: continuation of national strategic gas policy (including the small-field policy); continuation of the integrated structure of the national gas industry with Gasunie in a central coordinating position; and no third-party access to domestic gas-storage facilities, because these were considered part of the annual and seasonal load management system of the Netherlands. These national points of interest were safeguarded in the negotiations leading to the Gas Directive of 1998 and in the new Dutch Gas Law of 2000. The Dutch strategic gas policy and small-field policies were processed as a PSO of Gasunie, which was forced to unbundle its trade and transport activities in terms of accounts and access to storage facilities. Gasunie was excluded from the Third Party Access regime in the Gas Directive and continued as a transport service; consequently, it continued to control gas storage. The initial Gas Directive of 1998, as well as its national implementation in the Gas Law of 2000, coordinated with and reflected Dutch natural gas interests. The Dutch also managed to hold their position in negotiations on the second Gas Directive. In these negotiations, the Dutch interest was twofold: no tougher unbundling regime for Gasunie and no third-party access to gas-storage facilities. The first point, unbundling Gasunie, disappeared as a national point of reference in the EU negotiations owing to a change in the national political coalition in mid-2003. The newly established government no longer objected to a further unbundling of trade and transmission activities of Gasunie. The second point, third-party access to storage, was also excluded from the second Gas Directive. This Directive was also in line with the Dutch interests. The initial liberalisation course set by the first Gas Directive did not restrict national room to manoeuvre in negotiations on the second Gas Directive or the continuation of a nationally focused gas policy.

In 2003, Dutch gas interests were seriously challenged by the EC's initiative to develop an EU security-of-supply policy under the central coordination of the Commission itself. The initial proposal would have given the EC power to control depletion of Dutch gas fields as a safeguard for European gas supplies in times of crisis. The idea of EU-controlled energy resources met strong opposition by all member states. Members unanimously questioned the need for an EU-coordinated supply policy with far-reaching powers for the EC. The Dutch strongly opposed the initial proposal and were supported by many other member states, and the chance of acceptance was assessed as minimal.

In conclusion, the emerging European gas market has not restricted or conditioned the Dutch interests in the security of gas supplies thus far. In negotiations leading to the first and second Gas Directives, the Dutch position was clearly defined and remained intact. The Dutch were also able to weather the EC's initiative to develop a union-wide security-of-supply policy. In the stages of both policy development and implementation, the Dutch had ample room to manoeuvre in their security-of-supply policy.

Dossier 2: Emission Trading

Emission trading as an instrument to reduce greenhouse gas emissions has been debated for some time in the context of climate-change policy and the Kyoto agreement. However, scholars and practitioners disagree about emission-trading schemes.[23] The EU initiated a Directive on Emission Trading, of which the general objective is to create an EU-wide market for emission allowances so that the Kyoto target can be reached in the most cost-effective way. Emission trading will ensure that these reductions are made where they are least costly. At the same time, emission trading is environmentally effective by achieving a pre-determined emission reduction from the activities covered. The proposal ensures the proper functioning of the internal market and avoids major distortions of competition. The Directive is particularly important to ensure that legal commitments to reduce greenhouse gas emissions pursuant to the Kyoto Protocol, ratified by the EU and its member states in 2002, are economically feasible. The Dutch position in negotiations on the Directive and whether the Directive allowed for national room to manoeuvre in climate-change policy are discussed below.

From the start of the negotiations, the Dutch adopted an offensive strategy. The Netherlands had pushed for stricter climate-change policy at the European level for some time and envisioned the EC's initiative on emission trading as a timely opportunity. The Dutch wanted a stringent EU climate-change policy for domestic reasons, because they needed European backing for a stronger national policy. Without European support, the competitiveness of Dutch industry would be hamstrung by a strict national policy. The Dutch were strongly in favour of an EU-wide emission-trading scheme and, with Germany and Denmark, took a prominent position in the negotiations. In comparison to other policy dossiers, negotiations on the Emission-Trading Directive proceeded swiftly. As a result of Kyoto commitments, all member states were interested in a common instrument for emission reduction and, owing to its presumed (cost-)effectiveness, a trading scheme was regarded as the most promising alternative. The instrument had not been proven in practise, so a pilot programme would be required before its full operation in the EU region. This was another reason for the quick progress in negotiations; time was limited if member states were going to keep to the EU Kyoto commitments.

The Dutch tried to push their emission-trading model to the EU level, but

[23] Woerdman, 2002.

partly failed. They were successful in influencing the EU debate to review some questionable decisions. The Dutch were able to provide valuable information on the environmental and economic impacts of different modalities of emission-trading schemes, owing to their relatively long and extensive domestic research tradition on these topics. The EU debate benefited from Dutch expertise and information, and the Dutch were able to bring their own preferences to the EU floor. The Netherlands proposed the inclusion of carbon dioxide rights from two foreign implementation mechanisms of national commitments (the Clean Development Mechanism and Joint Implementation) in the emission-trading scheme. It also supported the introduction of a single trading scheme for the entire EU region, without national deviations. On the latter point, the Dutch strongly advocated zero national room to manoeuvre. The Dutch failed to convince the other parties on the significant point of an emission cap. At the time of the EU negotiations, an important Dutch advisory board suggested that the Dutch government should promote emission trading with relative instead of absolute emission caps, because the relative model was better able to account for the interests of sectors of Dutch industry exposed to international competition. In fact, the relative emission cap was the model favoured by Dutch industry; even after the rejection of the relative model by the other member states, Dutch industry continued to lobby for it. The Dutch government, on the other hand, changed position during negotiations to the absolute cap model, resulting in a discrepancy between different domestic interests on this point. The Dutch government joined the largest EU coalition in favour of the absolute model and Dutch industry tried to convince the EC of the advantages of the relative model by active lobbying. Most member states finally opted for the absolute cap model, contrary to initial Dutch preferences. In 2003, member states were in the process of implementing the requirements of the Directive in national legislation. In the stage of implementation, the Dutch position was challenged by the national distribution of carbon dioxide rights in other EU countries. There was substantial concern that internal market competition would be disturbed by member states using the distribution of carbon dioxide rights as a means of supporting suffering national industries.

The emission-trading dossier provides a second empirical example of how the Dutch manage interests and strategic positions in EU negotiations. The analysis shows considerable room to manoeuvre because of converging Dutch and EU interests. Strict EU climate-change policy was needed to support a strict national climate policy and, therefore, the Dutch took an advanced and offensive position in the negotiations. On many topics, the Dutch preferences were adopted at the EU level, with the exception of one significant point: the emission cap model. The Dutch opted for the relative model, but the EU decided in favour of the absolute model.

Dossier 3: Energy Taxation

The EU Directives on Energy Taxation date from 1993 and addressed harmonisation of tax rates on gas and electricity. The level of gas and electricity consumption in the EU region was exceedingly high and taxation was considered a cost-effective means of reducing consumption levels. For that reason, energy taxation had already been on many national political agendas for some time, and several countries had introduced national tax regimes. For reasons of competition and equal treatment of energy industries, the EU initiated a harmonising Directive in 1993, preceding legislation on this point in the Netherlands.

The Dutch position in negotiations on energy taxation was similar to its position in the emission-trading dossier. In this case, EU legislation was also considered necessary for the introduction of a feasible and effective tax regime in the Netherlands. The introduction of energy taxation had already been debated. However, without serious European backing, the introduction of an effective domestic regime would probably fail. The introduction of a separate Dutch tax regime would be counterproductive for the open Dutch economy and Dutch industry operating in European and global markets.

For that reason, the Dutch delegation joined the offensive strategic alliance pursuing a stringent EU-wide tax regime during the negotiations. To develop support for its position, the Dutch closely collaborated with this alliance of member states. They also offered exchanges of national positions in other policy dossiers in favour of an advanced EU tax regime and closely collaborated with the rotating chair of the Council of Ministers during the negotiation period.

Despite all these efforts, the progress in the negotiations on a harmonised European tax regime was minimal. Countries feared losing the competitive advantages of national industries and this resulted in a long period of inert decision-making. Together with other countries, the Dutch made a final attempt to move forward by suggesting the introduction of a stringent fiscal regime restricted to the group of relatively advanced EU countries. The assumption was that the suggestion of a partial introduction of a stringent regime would challenge the others to join. This final attempt was not successful either. Even the group of forerunner countries failed to agree about the details of a common tax regime. The negotiations eventually resulted in a complicated Directive providing a rich menu of national exemptions. These deviations and degrees of freedom led to a national tax regime far removed from the initial idea of a harmonised regime. Each country could take from the Directive what it pleased.

The Directive contained only two minor points of harmonisation: a common date for the introduction of a tax regime and a common minimum tariff. On all other points, the Directive allowed for national schemes. The Directive did not meet the initial Dutch interest of European backing for a stringent domestic tax regime. For that reason, the Dutch introduced a restricted tax regime only in 1996. This national scheme charged small industrial and household consumption at the minimal rate of 10 per cent of the initial Dutch tariff, as prescribed by the Directive. Medium and large industrial consumption was exempted from taxation for reasons of competition, also in case industry decided to join the Dutch Covenant of Benchmarking.[24]

In this conclusion, in this dossier, the Dutch were not successful in holding their initial interest position in the EU negotiations. On many points, the Directive was disappointing considering the initial Dutch ambitions for common energy taxation.

Dossier 4: Sustainable Electricity Technologies

In 2001, the EU accepted a Directive to promote electricity production from renewable energy resources in the internal electricity market.[25] The Directive aimed at harmonising the introduction and continued support of electricity production from renewable resources in the EU region. As in the case of energy taxation, the renewable-energy dossier proved to be extremely difficult. In the end, member states agreed only on a longer-term common target to increase the share of renewable energy in the internal electricity market to 22 per cent by 2010. Furthermore, the Directive formulated only indicative – not binding – national shares in this common EU ambition.[26] Finally, the Directive did not include any incentives for member states to continue designing national policies, despite long and serious negotiations on common policies. Like energy taxation, renewable-electricity negotiations testify to how sensitive a topic EU energy policy remains.

In the renewable-energy dossier, the Dutch interest position was straightforward. The Netherlands wished to safeguard a national share in the common EU goal for the long-term increase of renewable resources equal to Dutch political ambitions. The Dutch firmly held this position; the final Directive did not affect national targets for renewable electricity.[27] On market-share increase, the requirements of the Directive converged with Dutch goals. This was important because Dutch renewable-energy policies

---

24 The Covenant on Benchmarking implies exemption of energy-related legal obligations in case industry ranks in the worldwide top ten of most efficient industries.

25 European Commission, 2001.

26 National shares took account of differences between member states in physical and geographical conditions for the production of electricity from renewable resources.

27 The national target was 5 per cent sustainable energy, implying a 9 per cent share of sustainable electricity at that time.

were already ambitious given the physical and technical constraints in the Netherlands for renewable electricity production. The small size of the country did not allow for large-scale onshore wind parks. Its geography was incompatible with large-scale solar energy. Biomass, a serious option for the Netherlands, was technically still in research and development. And hydropower was not feasible given the flatness of the country. Dutch access to renewable resources and the production potential of renewable electricity are limited compared to those of most other European countries. For these reasons, a modest share in the common EU ambition equal to the Dutch national target was given top priority in negotiations.

The strenuous negotiations were extensive and focused on several controversial issues.[28] Countries could not agree on the definition of renewable resources. There were lengthy and tense discussions on whether to earmark electricity produced from waste incineration and large-scale hydropower as renewable electricity. The positions of member states in this debate reflected the differences in their access to these resources. The Dutch favoured a wide definition of waste and waste incineration, but objected to the inclusion of large-scale hydropower as a renewable resource.[29] The second debate focused on a timeframe for goal attainment, and the third on harmonising policy schemes for the support of the use of renewables. In the end, the EU countries agreed on a minimal set of corresponding rules for market-share increase and timing, with no agreement on a common support policy. The countries were obliged to certify renewable electricity and compelled to accept another country's certification in the case of imports.

After the acceptance of the Directive, member states continued to support the use of renewable resources in their own ways. Consequently, a large variety of support schemes exist across the EU.[30] In the Netherlands, a fiscal regime to support the production of renewable electricity was introduced. Initially, the tax was charged for all types of electricity consumption, including that of renewable electricity, although the charge on renewable-energy consumption was passed to generators. In 1998, renewable electricity was exempted from the tax regime and generators were directly supported by the tax funds charged on the consumption of conventional electricity. Because the tax exemption equalised tariffs for renewable and conventional electricity, the Netherlands saw a steep increase in the consumption of renewable electricity. This policy change did not support the domestic production of renewable electricity as the tax regime intended. Because of the lack of domestic generation capacity, Dutch distribution companies simply imported renewable electricity to satisfy the steep increase in demand. Since the Dutch

---

[28] Arentsen and Dinica, 2003.
[29] In the final Directive, large-scale hydropower was included as a renewable resource.
[30] Arentsen and Dinica, 2001; Dinica, 2003.

government accepted imported renewable electricity under the tax exemption, substantive parts of the Dutch tax funds for the support of domestic renewable electricity production went abroad. In 2003, a restriction of financial support to domestic renewable electricity production closed this gap in the Dutch support scheme.

In conclusion, the dossier on renewable electricity shows that the Dutch managed to keep their initial interest position in the negotiations. This was a position with one clear ambition: a Dutch share in the common EU political ambition equal to the one already agreed on domestically. For the domestic support of renewable electricity, the Dutch introduced a tax regime that supported the consumption instead of the domestic production of renewable electricity, resulting in a drastic increase of renewable-electricity imports and the subsidy of foreign production facilities.

CONCLUSION: HOW TO MANOEUVRE EFFECTIVELY?

In this chapter, we addressed several interrelated questions. First, how does emerging Europeanisation affect national room for political manoeuvre in EU decision-making? Second, how can EU member states manage their positions in this process? We modelled this space for member states by distinguishing between offensive and defensive strategy choices given specific constellations of actors, timing, and perceived interest positions in individual policy dossiers. With the help of this framework, we analysed the Dutch position in four energy-policy dossiers. Table 16.1 summarises the core findings of the empirical analyses described in the previous section.

It shows the Dutch have been successful in the two policy dossiers where they followed a more defensive (i.e., waiting) strategy. They were also successful in one case where they followed an offensive strategy (the emission-trading dossier), although this was tempered by the significant exclusion of the emission cap. The Dutch also followed an offensive strategy

Table 16.1 Summary of findings from four energy dossiers

Adopted national strategy		National interest position
Offensive	Defensive	
	Security of gas supply Renewable electricity	Completely held
Emission trading		Partly held
Energy taxation		Not held

in the energy-taxation dossier, but they were less successful here in keeping their initial interest position. How can we account for these differences? The analyses in this study allow for some tentative thoughts on this question.

An obvious difference between the two defensive strategies is their noticeably different backgrounds. In the dossier on gas security, the Dutch began in a strong position because their substantial reserves made them the major European supplier. Without Dutch support, it was difficult to move forward with the EU gas dossier. The strong EU-wide resistance to regulatory change in European gas was helpful for the Dutch position. Until 1995, nobody was willing to move, including the Dutch. The change of political coalition in the Netherlands in 1995 initiated a simultaneous change in the Dutch preferences and, eventually, changes in EU negotiations. In the case of renewable electricity, the Dutch also followed a defensive strategy, but from a different position. This position was not as strong, and they began the negotiations with a single clearly stated ambition. Other than this goal, they were indifferent to the issues. The Dutch defensive strategy focused only on assuring the preferred proportion in the common burden.

The Dutch position in EU negotiations was also distinctive in each of the two examples of offensive strategies (emission trading and energy taxation). In both cases, the strong EU context and backing was needed for national ambitions to introduce strong policies at home. In the case of emission trading, the Dutch were in the privileged position of 'most informed country' and were, therefore, able to significantly influence negotiations. The Dutch had nothing comparable to offer as regards energy taxation, only support for a stringent tax regime. Only a minority of countries were willing to accept this position, while a majority considered the Dutch ambitions to be only a single ingredient in a complex recipe.

Both observations show that the success of national strategy is affected by the power position a country holds vis-à-vis other countries. This brings us to a third point regarding the proximity of national interests and the dominant interest in the negotiations. If this distance is great, it becomes difficult for even a well-positioned country to influence the outcome to conform to national goals. If the distance is small, then a country can rely more easily on a defensive strategy, as the dossier on natural gas showed. In this dossier, the dominant preference had long been against liberalisation, which was concurrent with Dutch interest. A change of the Dutch position moved the negotiations to a legitimate and workable outcome. In the emission-trading dossier, the distance between the Dutch and the dominant preferences was negligible. Here, the Netherlands used its informed position to facilitate EU decision-making and benefit Dutch preferences on almost every point.

The above observations on the four policy dossiers analysed in this study compel us to suggest two modifications of the analytical framework described in section two. The first modification is that national strategy tailored for specific dossiers should also account for the relevant advantage countries may have compared to other countries. Small countries sometimes have the advantage of valuable resource surpluses, or can develop an advantage by accumulating knowledge and expertise. The second modification concerns accounting for the relative distance between national and dominant interests. The degree of success in utilising national room to manoeuvre is decided by the extent to which countries can integrate their national strategies into EU decision-making.

REFERENCES

Addink, H., B. Arts, and A. Mol (2003), 'Climate change policy in changing contexts: Globalization, political modernization and legal innovation', in E. Van Ierland, J. Gupta, and M. Kok (eds), *Issues in International Climate Policy: Theory and Policy*, Cheltenham: Edward Elgar.

Arentsen, M. and V. Dinica (2001), *Green Electricity in the Netherlands*, Oslo: Norwegian School of Management.

Arentsen, M. and V. Dinica (2003), 'Sustainable electricity supply in the European Union: Reconciling different scales of governance', in H. Bressers and W. Rosenbaum (eds), *Achieving Sustainable Development: The Challenge of Governance across Scales*, Westport: Preager.

Börzel, T. (1999), 'Towards convergence in Europe? Institutional adaptation to Europeanisation in Germany and Spain', *Journal of Common Market Studies*, **34**(4), 573-596.

Cowles, M., J. Caporaso, and T. Risse (eds) (2001), *Transforming Europe: Europeanization and Domestic Change*, Ithaca: Cornell University Press.

De Bruijn, T. (2003), 'The impact of policy style on policy choice across scales: The EU experience', in H. Bressers and W. Rosenbaum (eds), *Achieving Sustainable Development: The Challenge of Governance across Scales*, Westport: Preager.

Dinica, V. (2003), *Sustained Diffusion of Renewable Energy*, Ph.D. thesis, Enschede: Twente University Press.

European Commission (1996), Directive 96/61, OJ L 257, Brussels: European Commission.

European Commission (2001), Directive 2001/77/EC, OJ L 283, Brussels: European Commission.

Fuchs, D. (2002), 'Globalization and global governance: Discourses on political order at the turn of the century', in D. Fuchs and F. Kratochwil (eds), *Transformative Change and Global Order*, Münster: Lit Verlag.

Haverland, M. (1999), *National Autonomy, European Integration and the Politics of Packaging Waste*, Amsterdam: Thela Thesis.

Knill, C. and D. Lehmkuhl (1999), *How Europe Matters: Different Mechanisms of Europeanization,* European Integration online Papers (EIoP) 3(7), *http://eiop.or.at/eiop/texte/1999-007a.htm.*

Kohler-Koch, B. (1998), *The Evolution and Transformation of European Governance*, Political Science Series 58, Vienna: Institute for Advanced Studies.

Lowe, P. and S. Ward (1998), 'Britain in Europe: Themes and issues in national environmental policy', in P. Lowe and S. Ward (eds), *British Environmental Policy and Europe: Politics and Policy in Transition*, London: Routledge.

Midttun, A. (ed.) (2001), *European Energy Industry Business Strategies*, London: Elsevier.

Ministry of Economic Affairs (1963), *Gasnota (White Paper on Natural Gas)*, Dutch Parliament 1961-1962, 6767, 1, The Hague: Ministry of Economic Affairs.

Ministry of Economic Affairs (1995), *Derde Energienota (Third White Paper on Energy)*, The Hague: Ministry of Economic Affairs.

Nijkamp, P. and H. Verbruggen (2003), 'Global trends and climate change policies', in E. Van Ierland, J. Gupta, and M. Kok (eds), *Issues in International Climate Policy: Theory and Policy*, Cheltenham: Edward Elgar.

Olsen, J. (1997), 'European challenges to the nation state', in B. Steunenberg and F. Van Vught (eds), *Political Institutions and Public Policy: Perspectives on European Decision Making*, Dordrecht: Kluwer Academic Publishers.

Pellegrom, S. (1997), 'The constraints of daily work in Brussels: How relevant is the input from the national capitals?', in D. Liefferink and M. Andersen (eds), *The Innovation of EU Environmental Policy*, Copenhagen: Scandinavian University Press.

Rodriguez de Sancho, M. (2000), 'Spanish experience with the BAT info exchange', presented at the European Conference on The Sevilla Process, Stuttgart, April 6-7, *http://www.ecologic-events.de/sevilla1/en/documents/Sancho_en.pdf*.

Sorensen, G. (1999), 'Sovereignty: Change and continuity in a fundamental institution', *Political Studies*, **47**(3), 590-604.

Stern, J. (1998), *Competition and Liberalization in European Gas Markets: A Diversity of Models*, London: Royal Institute of International Affairs.

Unger, B. and F. Van Waarden (eds) (1995), *Convergence or Diversity? Internationalization and Economic Policy Response*, Gateshead: Athenaeum Press.

Van Kersbergen, K. and F. Van Waarden (2001), *Shifts in Governance: Problems of Legitimacy and Accountability*, The Hague: Netherlands Organisation for Scientific Research (NWO).

Weale, A., G. Pridham, A. Williams, and M. Porter (1996), 'Environmental administration in six European states: Secular convergence or national distinctiveness', *Public Administration*, **74**(Summer), 255-274.

Woerdman, E. (2002), *Implementing the Kyoto Mechanisms: Political Barriers and Path Dependence*, Groningen: University of Groningen.

Young, S. (ed.) (2000), *The Emergence of Ecological Modernisation: Integrating the Environment and the Economy?*, London: Routledge.

17. Strategies to Prevent Illegal Logging

Saskia Ozinga and Nicole Gerard

SUMMARY

A large proportion of tropical timber is logged illegally. We discuss different options available to tackle this major environmental problem, which has been enhanced by globalisation. Existing instruments are the OECD Convention on Bribery, legislation against money laundering, self-regulation of the financial sector, the Convention on International Trade in Endangered Species, and forest certification. These instruments face considerable implementation problems: environmental issues have low political priority and the cooperation of both exporting and importing countries is required. New, more targeted, legislation is needed, prohibiting goods that can be identified as illegal or goods that cannot be identified as legal. An eye should be kept on parallel systems of trade control such as the elimination of trade in conflict diamonds. Government should provide a clear framework of guidelines and regulations, and exert stringent controls. The role of non-governmental organisations (NGOs) is to inform and pressurise governments and companies. We conclude that addressing illegal logging requires independent monitoring, effective stakeholder participation, and dealing with corruption.

INTRODUCTION

Illegal logging is a pervasive and widespread problem, causing enormous damage to forests, forest peoples, and the economies of producer countries. No exact figures exist, but reliable estimates suggest that over 50 per cent of tropical-timber imports into the European Union (EU) are illegally sourced, as are between 10 and 20 per cent of timber imports from the boreal region. Costs to producer countries are high, ranging from 1.8 billion United States Dollars (USD) in the Philippines to 3 billion USD in Indonesia.

Concern about the extent of illegal logging around the world has grown significantly in recent years, with discussions taking place in many international forums, including the G8, the World Summit on Sustainable Development (WSSD), the World Trade Organization (WTO), and the EU. This heightened awareness has developed, in part, as a response to growing evidence of the destruction of forests and the accompanying serious loss of government revenues. In part, it is an offshoot of the growing emphasis on 'good governance' in international policy. And, in part, it reflects the increasing recognition of the role of consumer countries in fuelling demand for illegal products.

Despite this growing concern, there is a clear lack of immediate and well-coordinated action at the national and international levels to address the problem. The strengths and weaknesses of different options are listed below. The roles different stakeholders can and ought to play for any measures to be effective are described. Existing agreements that can be used to address this problem (such as the Convention on Bribery of the Organisation for Economic Co-operation and Development (OECD), the Convention on International Trade in Endangered Species (CITES), and money-laundering legislation), as well as forest certification schemes and standards for the financial sector, are discussed; however, additional legislation is needed.

Last, the chapter underscores that revision of forestry laws in most timber-producing countries is an essential condition to address the problem, as it has become clear that greater enforcement of forestry and conservation laws has the potential to affect rural livelihoods negatively. This is because existing legislation typically favours the timber industry over local people, and often prohibits forestry activities such as small-scale timber production, fuelwood collection, and hunting on which millions of poor rural households depend.

THE PROBLEM OF ILLEGAL LOGGING

The Scale of Illegal Logging

Illegal logging causes enormous damage to forests, forest peoples, and the economies of producer countries. Some estimates suggest that the illegal timber trade may comprise over a tenth of the total global timber trade, worth more than 150 billion USD a year.[1] Although exact figures are difficult to obtain, given the illegal nature of the activity, reliable estimates indicate that more than half of all logging activities in particularly vulnerable regions – the Amazon Basin, Central Africa, Southeast Asia, the Russian Federation, and some of the Baltic states – is illegal.[2] Activities constituting illegal logging

1 OECD, 2001.

2 For further details on illegal logging, see: Brack and Hayman, 2001; Brack, Hayman, and Gray, 2002.

include the harvest, transportation, purchase, and sale of timber in violation of national laws. The harvesting procedure itself may be illegal, including the use of corrupt means to gain access to forests, extraction without permission or from a protected area, the cutting of protected species, and the extraction of timber in excess of agreed limits. Illegal activities may also occur during transport, such as illegal processing and export, fraudulent declaration to customs, and the avoidance of taxes and other charges.

Although the clandestine nature of the illegal trade makes its scale and value difficult to estimate, extensive unlawful operations have been uncovered wherever authorities have tried to find them. As the World Bank (WB) observed, 'in many countries, illegal logging is similar in size to legal production. In others, it exceeds legal logging by a substantial margin … [P]oor governance, corruption and political alliances between parts of the private sector and ruling elites combined with minimal enforcement capacity at local and regional levels, all played a part'.[3] Illegal logging is the cause of widespread environmental damage and presents a grave threat to biodiversity. In addition, the scale of illegal logging represents an enormous loss of needed

Box 17.1 Some facts about illegal logging

- A joint British-Indonesian study of the timber industry in Indonesia in 1998 suggested that about 40 per cent of throughput was illegal, with a value in excess of 365 million USD.[4] More recent estimates, comparing legal harvesting with known domestic consumption plus exports, indicated that over 70 per cent of logging in the country is illegal in some way.[5]
- Over 80 per cent of logging in the Amazon may be in violation of government controls.[6]
- A World Resources Institute (WRI) comparison of import and export data for Burma in 1995 revealed substantial underdeclaration of timber revenues, accounting for foregone revenues of 86 million USD – equivalent to almost half of official timber export revenues.[7]
- Studies in Cambodia in 1997 by the WB suggested that illegal extraction, worth 0.5-1 billion USD, may be over 4 million cubic metres – at least ten times the size of the legal harvest.[8] If this level of extraction continues, the country's timber will all be logged within ten years of the industry's official beginning.

---

3 World Bank, 1999: xii.
4 ITFMP, 1999.
5 Scotland and Ludwig, 2002.
6 Scotland and Ludwig, 2002.
7 World Resources Institute, 1997.
8 Global Witness, 1999.

revenues to many countries. For example, a Senate Committee in the Philippines estimated that the country lost as much as 1.8 billion USD per year from illegal logging during the 1980s.[9] The Indonesian Government estimated in 2002 that costs related to illegal logging are 3 billion USD a year.[10] The substantial revenues from illegal logging sometimes fund, and thereby exacerbate, national and regional conflicts, as in Cambodia, Liberia, and the Democratic Republic of Congo.

The vast extent of the illegal timber trade distorts the entire global marketplace for a number of key timber products such as logs and sawn timber. It robs governments of revenues, and undermines both legal and sustainable management, which has to bear the additional costs of good husbandry and proper tax declaration. As the WB reported, 'widespread illegal extraction makes it pointless to invest in improved logging practices. This is a classic case of concurrent government and market failure'.[11]

Trade Liberalisation and Illegal Logging

The globalisation of trade, the elimination of barriers to trade, and the increase of incentives to export have all, to some degree, facilitated not only international timber trade generally, an intended effect, but also the 'laundering' of illegal timber by providing opportunities to disguise the true provenance of logs, and by facilitating transport. Trade in timber grew from

Box 17.2 Laundering illegal logs

One example concerns Indonesia's 2002 moratorium on exports of its logs, following the temporary ban it had imposed in 2001. Malaysia's role in smuggling illegally sourced timber from Indonesia has been clearly exposed.[12] To counter criticisms that it was laundering illegal logs, Malaysia ostensibly passed a ban on Indonesian logs.[13] However, in 2003, an investigation revealed that Indonesian timber listed under CITES was being smuggled into the port of Batu Pahat in Peninsular Malaysia. In addition to exposing Malaysia's disregard for CITES, the evidence proved that Malaysia's own import ban of Indonesian logs was being ignored: in the period of one hour, 32 Indonesian vessels loaded with illegal logs arrived at the port of Muar in Peninsular Malaysia.[14]

---

9 Callister, 1992.
10 ICG, 2001.
11 World Bank, 1999: 40.
12 EIA/Telapak, 2003.
13 EU Forest Watch, 2002.
14 Research carried out by the Indonesian NGO Telapak, jointly with the United Kingdom and Washington DC-based NGO EIA.

29 billion USD in 1961 to 152 billion USD in 1999.[15] It is widely argued that trade liberalisation in the form of lowering tariffs on timber and timber products has contributed to this increase.[16] Although specific research regarding the impact of trade liberalisation on illegal logging is lacking, it is clear that the increase in illegal logging has gone hand in hand with the increase in international trade in timber and timber products. Furthermore, the breaking down of border controls (such as in the EU) and the increased transport of logs, sawn timber, and paper and pulp from one producer country to another have all facilitated the laundering of illegally sourced timber.

The Urgency of Revision and Reform of Forestry Laws

An immediate problem facing any attempt to control the trade in illegal timber and wood products lies in defining what constitutes illegality. In many countries, forestry legislation is simply unclear and insufficient in terms of legal certainty. For example, a 1998 review of Cambodian forest legislation by the White & Case law firm found that the legislation was 'difficult to obtain, difficult to analyse, provides few objective standards for forest protection and provides no integrated guidelines or standards for forest management'.[17] An overview of Indonesian forest governance in 2003 revealed that 90 per cent of state forest lands have never been legally transferred from traditional landholders to the jurisdiction of the Forestry Department, meaning that most 'legal' forestry operations in Indonesia are in fact of dubious legality.[18] In Canada, also, the legal situation is complex. In large parts of British Columbia, indigenous peoples hold the rights and titles to their ancestral lands; in a recent decision, the Supreme Court of Canada upheld these rights. However, jurisdiction over resource management, including forest concessions, remains in the hands of the government, creating an unclear legal situation that has led to sometimes violent disputes over unceded lands.[19]

Lack of clarity in the legal framework can be linked to a second problem, which raises even more fundamental questions: the perceived legitimacy of the legislation. As noted in a recent study by CIFOR, 'Many existing forests and conservation laws have unacceptable negative impacts on poor people, ethnic minorities, and women, and in many places they are enforced in a fashion that is discriminatory and abusive'.[20] Ways must be found to address

---

[15] French, 2000.
[16] Rice et al., 1999.
[17] White & Case, 1999: 1.
[18] Colchester et al., 2003.
[19] The United Nations Human Rights Committee has twice condemned the Canadian government for undermining indigenous peoples' legal rights to land.
[20] Kaimowitz, 2003.

the problems associated with illegal forestry activities that at the very least do not aggravate the negative impacts of existing regulatory efforts on the rural poor. In many cases, that will mean a revision of forestry laws, taking into account local peoples' traditional and user rights. The examination of global options for addressing illegal logging should, therefore, be undertaken with the understanding that underlying issues of clarity and fairness of the national legal context, in addition to political commitment to implementing and enforcing such rules, are critical to the success of any action.

EXISTING MEASURES TO ADDRESS THE PROBLEM

Effective control of illegal logging will require action across many policy areas: the promotion of good governance, action to tackle corruption, land reform, industrial and fiscal policy reform, development assistance, and so on. We focus principally on the control of imports of illegally produced timber and the financial transactions surrounding the illegal timber trade. Even when thus narrowing the focus, governments and civil-society organisations face many hurdles when attempting to address the problem. These include proving illegality and cooperation with enforcement authorities in the countries of origin, which in many cases is poor or non-existent. Besides, the events of 11 September 2001 have had a resonant impact on global initiatives in almost every domain. Currently, the international focus is on security and anti-terror measures and now, more than ever, environmental issues do not appear to receive high priority. Under such circumstances, the greater wisdom may be to try to work within the current international tendencies, in an attempt to make these work to the environment's advantage. A number of international measures exist that could be used to address illegal logging and other environmental issues, despite the fact that they are not tailored to the environment.

Moving from general to more targeted initiatives, we briefly discuss the OECD Convention on Corruption, the EU Money-Laundering Directive, measures targeting the financial sector, forest certification schemes, and CITES as it relates to timber species. The list is by no means exhaustive; the examples are to be viewed as instruments that have the potential to be effective in addressing certain contributing causes and auxiliary effects of the illegal timber trade. For instance, other potential tools include stolen-goods legislation and Guidelines for Multinational Enterprises, such as the OECD guidelines.[21]

[21] For a more general overview of these tools, see Brack, Marijnissen, and Ozinga, 2002.

The OECD Convention on Bribery[22]

The OECD Convention on Bribery is a legally binding instrument whose requirements must be incorporated into or implemented in national legislation by its parties (OECD members and other signatory governments).[23] The Convention makes it a criminal offence to bribe a foreign public official. As illegal logging involves bribery in a number of instances, the OECD Convention clearly has a role to play in controlling illegal logging. The problem often lies in proving that bribery has taken place. The Convention's effectiveness would increase if all parties to the Convention implemented the recommendations made by the OECD, such as excluding companies that have been found guilty of bidding for public contracts.

The globalisation of international trade has contributed greatly to international initiatives to address corruption across governments and within industries,[24] and in the past decade-and-a-half such initiatives have gathered momentum. It is recognised that, to be effective, the fight against corruption must be undertaken on a multilateral basis. The self-interest of the entities involved is what generates this momentum. As stated in the Preamble to the OECD Convention, bribery 'distorts international competitive conditions'. The desire to compete on equal footing, a principal impetus behind the Convention, may ultimately prove to be a significant factor in its success. Anti-corruption measures will be more assiduously applied if the perceived interest of industry lies, for competitive reasons, in eliminating the corrupt practices.

The purpose of the Convention is to deter bribery in international business transactions and to criminalise the act of bribing a foreign official,[25] to give 'bite' in the national penal system to provisions punishing international corruption. It approaches its goals with flexibility: state parties are required to implement the Convention's objectives, yet can choose the means most suited to their national legal traditions.[26]

---

[22] For further reference, see: Aiolfi and Pieth, 2003; Cockcroft, 1999; Watson, 1998.

[23] All EU member states have ratified and introduced amendments to their legislation to implement the Convention.

[24] Examples are the 1996 OAS Inter-American Convention Against Corruption and the 1996 International Chamber of Commerce Revisions to its Rules of Conduct on Extortion and Bribery in International Business Transactions.

[25] Article 1(1): Each Party shall take such measures as may be necessary to establish that it is a criminal offence under its law for any person intentionally to offer, promise or give any undue pecuniary or other advantage, whether directly or through intermediaries, to a foreign public official, for that official or for a third party, in order that the official act or refrain from acting in relation to the performance of official duties, in order to obtain or retain business or other improper advantage in the conduct of international business.

[26] In countries that already have the legal framework to prosecute their nationals for crimes committed abroad (for example, the United States), this existing legislation must be amended to include bribery offences. In countries where the bribery of civil servants is listed as a

Box 17.3 Links between bribery and illegal logging

> The OECD Convention clearly can play a role in controlling illegal logging and other illegal activities. Simply put, without addressing pervasive corruption, efforts in almost any domain to combat illegal logging will be thwarted. In the forestry sector, bribery and corruption occur regularly in several areas: in the allocation of forest concessions; in the setting up or operation of pulp and paper mills that do not respect standards of health and environmental protection or that cannot source sufficient legal timber for their operations; in the procurement of official documentation legalising 'illegal' timber, particularly export licences (e.g., CITES permits); in the illegal construction of logging roads; etc. A recent Greenpeace report on the timber industry in Indonesia is chilling: not only is corruption routine,[27] but the military is a tool of the illegal loggers, perpetrating human rights abuses in the process. Finally, other official interests are complicit in the failure to prosecute crimes and human rights abuses committed by timber companies.

The Convention entered into force in early 1999. As of February 2003, 34 countries had ratified it, including Chile, Brazil, and Argentina. Obviously, many countries in which illegal logging occurs (such as Russia, Malaysia, Indonesia, Burma, Cameroon, Gabon, and Congo) are not signatories to the Convention. Nonetheless, the OECD secretariat has undertaken outreach efforts to raise awareness regarding the Convention and its objectives,[28] and it is hoped that, if the Convention proves successful in addressing bribery generally, more members will be forthcoming.

As with any illegal activity, the problem often lies in proving that bribery has taken place. However, because this Convention addresses the *supply* of bribes, it adds a valuable dimension to efforts to eradicate bribery. It does not rely solely on the will of the government receiving such income to address the problem. For example, under this Convention, a western company providing a bribe can become a target of legal action.

The instrument is imperfect. A weakness of the Convention is that foreign subsidiaries of companies are not explicitly covered. This is a serious omission in that payments are often made through subsidiaries.[29] That said, the Convention casts a wide net in providing criminal liability for complicity

---

criminal offence under money-laundering legislation, this legislation must be amended to include bribery of foreign civil servants.

27 According to Greenpeace (2003: 10), 'very little of the illegally felled timber that goes through Indonesia's timber mills does so without some form of official sanction – at some point in the production chain, some official, or some official document is misused in order to legitimize an illegal shipment of wood'.

28 Cockcroft, 1999.

29 Cockcroft, 1999.

in bribery, and this could – and NGOs would argue should – be instrumental in covering parent companies. One of the most interesting features, however, is the monitoring and follow-up foreseen by the Convention. This includes a process of peer review to be carried out within the context of the OECD Working Group on Bribery in International Business Transactions, involving questionnaires that governments must complete.[30] Both OECD and non-OECD parties gather to examine, in the first phase, the harmonisation of national legislation with the Convention's obligations. During a hearing that involves a to-and-fro between examining countries and the state parties under examination, the quality of the transposition is assessed. In a follow-up, the country's response to remedy shortcomings noted during the phase-one assessment is assessed. The phase-two evaluation implies an assessment of the parties' efforts at a practical level, including the resources dedicated to the effort, the number and training of personnel, the structures in place for dealing with cases, and the identification of obstacles to initiating prosecutions. It may also include an on-site visit. The procedure is open to members of civil society 'who can, and have, contributed written comments … publication of the reports is mandatory and they will also be available on the Internet'.[31]

It is too soon to evaluate the effectiveness of the Convention's implementation.[32] Where the Convention's existing provisions are not effective, they can be amended given the political will to do so. The procedure is in place to remedy and adapt with experience. The Convention may yet prove an effective instrument in combating international corruption, with significant opportunities to combat international trade in illegal timber.[33] It will be difficult, however, for the parties to obtain sufficient evidence to prosecute nationals for bribery offences. The parties are obliged to cooperate and to provide each other with legal assistance in criminal investigations and prosecutions.

To support these investigations, guidelines that help identify the types of bribery or corruption that occur in the forestry sector should be developed. Civil society exerts a watchdog role that is critical to uncovering corruption-related offences. The confidence of whistleblowers in coming forward will be a determining factor in the effectiveness of applying the Convention's

[30] These questions cover issues such as public-procurement sanctions, export credit agencies, blacklisting, codes of conduct, and accountancy provisions (such as clear and accurate reporting).

[31] Aiolfi and Pieth, 2003: 23.

[32] Finland is the first country to have undergone the evaluation of both phases, but it has not yet prosecuted a criminal case concerning bribery.

[33] These include: Article 3(3), providing for the seizure of the bribe and proceeds, or property of a corresponding value; Article 8, prohibiting off-the-books accounts and other accounting chicanery; and Article 9, on mutual legal assistance.

provisions and the possibility of denouncing corruption anonymously would be extremely helpful; a complaint procedure such as that available in the EU, or an ombudsman, would be useful in this framework. NGOs must continue to emphasise their role in providing and promoting information regarding the hidden destruction of the forestry industry. A few high-profile criminal prosecutions early in the Convention's implementation could provide a critical boost to its effectiveness, as well as a deterrent to international bribery; these would be a vital indication that the will to prosecute such crimes does exist.

In sum, one of the Convention's strengths lies in the self-interest that generated it: the will of the industries affected to eliminate the unfair competitive advantage gained and/or the cost of bribery from business transactions. The initiative should be supported by the NGO community, since in the absence of addressing the rampant corruption in the forestry sector, any other measures adopted to do so will fail. Increased efforts to encourage more timber-producing countries to become parties could be useful. Furthermore, a solid peer-monitoring framework exists, covering implementation on paper and on the ground, in which civil society can hope to participate actively. With genuine monitoring and participatory procedures in place, even where substantive implementation is lacking, it can be hoped that lessons taken from experience will be incorporated into measures and used to improve performance.

In order to meet these challenges, we recommend that governments take the following measures. First, governments should develop – at the national, international, and regional levels – forestry-sector guidelines for tax inspectors and public prosecutors to help them identify the possible forms that bribery and corruption can take in the forestry sector. Second, governments should send a questionnaire to all public prosecutors to ensure that they report all cases involving the application of bribery legislation to forestry-sector-related crimes. Problems preventing prosecutions should also be reported.

Money-Laundering Measures

All EU member states and many other countries have legislation on money laundering – the disposal of the proceeds of criminal activity. This legislation has the potential to be effective in dealing with the problem of illegal logging, so long as the legislation is sufficiently broadly defined. If illegal logging and the trade in illegally sourced timber are criminal offences under member states' law, then the proceeds of these activities can be subject to money-laundering legislation, provided they are deposited or disposed of within the EU. The fact that the activities themselves may take place overseas and be carried out by non-EU nationals is not relevant. To date, no EU country has attempted to use this legislation to tackle the laundering of the proceeds of illegal logging. Governments should alert banks, lawyers, accountants, etc., to the possibility that clients with interests in the forestry sector, particularly in countries where illegal logging is widespread, may be engaging in money laundering.

As with efforts to address international corruption, anti-money-laundering efforts are a current international focal point, following the 11 September 2001 events; again, the challenge is to try to make an existing instrument with broad potential work for the specific purpose of addressing illegal activities such as illegal logging. Money-laundering provisions could provide an interesting tool with which to target the illegal timber industry indirectly. Enhancing such provisions could eventually encourage, or even require, financial organisations to take a more proactive approach to exercising due diligence in conducting research into their clientele. Companies known to be involved in the illegal timber trade, or indeed in any illegal activity, should be shunned by the legitimate financial and credit community.

Money laundering refers to the processing of the financial proceeds of crime in order to disguise their illegal origin. At the global level, money laundering is a problem of vast proportions: one recent estimate put worldwide money-laundering activity at roughly 1 trillion USD per year.[34] National legislation allowing authorities to tackle money laundering and seize the proceeds of criminal activity has traditionally focused on the illegal trade in narcotics. Over the past decade, however, the need to expand the focus was increasingly recognised.

Every EU member state possesses legislation on money laundering. At the EU level, the first Directive on Money Laundering, which applied only to the proceeds of drug-related crimes, dates from 1991.[35] In 2001, a second Directive was adopted[36] that extended the scope of the 1991 legislation; it entered into force in June 2003. To date, no EU country has attempted to use this legislation to tackle the laundering of the proceeds of illegal logging. Theoretically, such legislation has the potential to be effective in dealing with the illegal timber trade, so long as the legislation is defined in a sufficiently broad manner, and if the political will can be found at various levels, to give substance to these provisions. With sufficient awareness of the problem and the willingness to interpret provisions in a manner applicable to the trade in illegally sourced timber, the proceeds of this trade could be subject to seizure. A challenge for the national NGOs, therefore, is to undertake awareness campaigns aimed at both authorities and the affected institutions regarding

[34] FATF, 2002; FinCEN, 2003.
[35] European Commission, 1991.
[36] European Commission, 2001.

the destructive consequences of the illegal trade in timber, in terms of lives disrupted, loss of state revenues, and environmental devastation.

If the activity in question is illegal under member state criminal law – as is the case with illegal logging in the United Kingdom (UK) and the Netherlands – then the proceeds of the activity could be subject to recovery, provided they were deposited or disposed of within the EU. The fact that the activity itself may take place overseas and be carried out by non-EU nationals does not preclude application of the Directive's provisions.

The use of such legislation is not without difficulty. Setting aside the difficulties of proving that a shipment of timber is illegal,[37] there is the potential hurdle of political and institutional will. Not only must member state authorities be willing to take action to interpret the provisions in ways relevant to the forestry sector, and to act when they receive reports of suspect activities or clients, but the institutions and persons subject to the Directive's provisions must also cooperate actively. Furthermore, in the case of offences committed in foreign countries, the success of action taken under money-laundering legislation requires cooperation with enforcement and judicial authorities in the countries of origin, which may not always be forthcoming.

For the Directive to be an effective tool, institutions and persons subject to the Directive must alert the appropriate member state authorities of suspect activities. Therefore, at a practical level, the Directive's effectiveness also hinges on the level of awareness of the institutions carrying out those transactions; indeed, the Directive is concerned with raising awareness.[38] It is likely that many of these institutions are not familiar with the extent of criminal activity in the forestry sector. Information and guidance are needed. The institutions that may handle the proceeds of the crime – banks, accountants, lawyers, etc. – should be alerted to the possibility that clients with interests in the forestry sector, particularly in countries where illegal logging is widespread, may be engaging in money laundering.

Indeed, a key to success in using any money-laundering provision to combat illegal logging lies in inspiring banks and other relevant institutions and persons to exercise due diligence in carrying out reviews, with a full awareness of the likelihood of certain clients' involvement in such activities.

---

[37] Brack, Marijnissen, and Ozinga, 2002.

[38] Article 1(10), replacing Article 11(1)b: 'Member States shall ensure that institutions and persons subject to the Directive take appropriate measures so that their employees are aware of the provisions contained in this Directive. These measures shall include participation of their relevant employees in special training programmes to help them recognise operations which may be related to money laundering as well as to instruct them as to how to proceed in such cases'. Article 11(2): 'Member States shall ensure that the institutions and persons subject to this Directive have access to up-to-date information on the practices of money launderers and on indications leading to the recognition of suspicious transactions.'

They must also alert authorities where they have reason to believe that a client may be engaged in criminal activities. Given that the majority of logging activities in many countries are generally accepted to be illegal, this should be a strong signal to a bank that clients operating forestry businesses in those countries may be committing forestry crimes. Such due diligence measures would be good business practice, as illegality, by definition, means increased risk. Nonetheless, banks and other institutions may resist the widening of the Directive's reporting burden on them.[39] Here is an opportunity for NGOs to embark on an awareness-raising campaign targeting both government and financial-industry officials with information on the devastating consequences of the illegal timber industry. Publicity of identities of wrongdoers and information exchange regarding suspicious companies is of critical importance. As with anti-corruption measures, international efforts to address corruption (above) and money laundering can be expected to reinforce each other.

The Financial Action Task Force (FATF)[40] publishes an annual Blacklist of Non-Cooperating Countries and Territories, which have critical deficiencies in their anti-money-laundering systems or a demonstrated unwillingness to cooperate in anti-money-laundering efforts. In June 2002, four timber-producing countries were included on the list: Indonesia, the Philippines, Russia, and Burma.[41]

In sum, the main advantage of money-laundering legislation is that the international pressure to eradicate money laundering has increased recently. The significant disadvantage is that the application of such provisions to the laundering of the proceeds of the illegal timber trade not only involves problems of proof, but is based upon the willingness of political and institutional actors to take meaningful action. The will to act already exists in those member states that have provisions applicable to illegal logging in their penal codes. A positive point is that lack of awareness of the illegal logging industry may be an obstacle that is relatively easy to remove; in the right ear, targeted information provided by NGOs could bring concrete results. As with anti-corruption measures, a few successful prosecutions could have genuine impact.

The policy implications for EU governments are as follows. Since member states are likely to expand the list of offences in the third Directive on Money Laundering, it is important that governments ensure that illegal logging falls

39 To illustrate this, the FBE (2000) had sent a strong negative signal early in the process of developing the new Directive.

40 A body created by the G-7 Summit in 1989 to examine methods of money laundering and to devise recommendations for addressing this threat to the international banking system and financial institutions.

41 FATF, 2002.

within the revised definition and that the burden of proof be shifted so that banks must report any activity they should consider suspicious based on the information they have available. EU governments should also inform and guide institutions that may handle the proceeds of crime – banks, accountants, lawyers, etc. – about the possibility that clients with interests in the forestry sector, particularly in countries where illegal logging is widespread, may be engaging in money laundering. Finally, EU member states should investigate the possibilities of taking action against illegal imports under national money-laundering legislation; UK legislation in particular appears to allow for this.

Financial Institutions[42]

Finance from private sources – banks, investment and pension funds – can be an important source of revenues for logging companies and other sectors of the forestry industry. Some, notably ABN-AMRO, have already announced that they will not fund any forestry companies involved in, colluding with, or purchasing timber from illegal logging operations. Other financial institutions should follow this example. EU and member state authorities should take action to encourage, and in due course require, financial institutions to draw up policies and action plans to ensure that they do not finance companies involved in illegal logging practices. This would also facilitate the implementation of money-laundering legislation.

The timber industry, the pulp-and-paper industry, the extractive industries (oil, mining), and agribusiness (soya, oil-palm, etc.) all contribute to forest loss; linkages with illegal (logging) practices have been clearly documented in all these sectors. The more capital-intensive the industry, the more important is the involvement of financial institutions. Capital-intensive sectors (such as the pulp-and-paper industry, extractive industries, and agribusiness) rely, to some extent, on financial institutions to enable them to operate. These financial institutions can be either private financial institutions (including banks and institutional investors or pension companies) or (semi-) public financial institutions (including multilateral development banks, foreign development agencies, and export credit agencies). A 2002 study by Profundo identified 21 financial institutions prominently involved in financing logging operations in the Congo Basin in Africa (including ABN-AMRO, HSBC, Credit Lyonnais, and Deutsche Bank).[43] A study by FERN strongly linked illegal forestry activities to the lack of due diligence in researching proposed activities on the part of export credit agencies as well as private financial institutions.[44] The study related the involvement of ten

42 Large parts of this paragraph are based on Van Gelder et al., 2003.
43 Van Gelder, 2002.
44 FERN, 2002a.

Box 17.4 Financing destruction

A clear case of financial investors tumbling over each other to finance an environmental, social, and economic disaster concerns a fully unsustainable paper mill in Indonesia, Asia Pulp and Paper. Over 300 international financial institutions were heavily involved in providing finance and guarantees to Asia Pulp and Paper, inflating the company with a 13.4 billion USD debt. These institutions failed to recognise that there was insufficient supply for the paper mill: the plantations that were to feed the mill were not ready and by no means able to provide the supplies, leading to large destruction of primary forests and human rights abuses. As a study by CIFOR pointed out, Indonesian pulp producers may have obtained as much as 40 per cent of the wood they consumed between 1994 and 1999 from illegal sources.[45]

export credit agencies to illegal practices in Indonesia, Bolivia, and Peru. Without the financial backing of these institutions, many destructive and illegal activities would not be possible. Despite this, the role of these institutions has been given limited attention in intergovernmental debates on illegal logging.

Because of the lack of attention given to the financial sector in relation to illegal activities, few existing regulations provide effective options to address the concerns. An international framework for risk assessment, developed by the Basel Committee on Banking Supervision, was adopted in the Basel Accord, but does not include any reference to environmental risks or risks related to illegal activities.[46] Nevertheless, over the past few years, the Committee has moved more aggressively to promote sound supervisory standards worldwide.[47] It has been suggested that the Basel Committee on Banking Supervision will formulate criteria to guide credit ratings by credit-rating agencies and by banks. These criteria can play a crucial role in stimulating rating agencies and banks to give sufficient attention to risks associated with illegal activities in forest-related and other sectors.[48] NGOs have been active in convincing export credit agencies to adopt binding standards, with some success: the OECD has developed a non-binding

[45] Barr, 2001.

[46] BIS, 2003.

[47] For instance, it published a report, of which key elements include thorough customer acceptance and identification practices, ongoing transaction monitoring, and a robust risk-management programme. In September 2002, the International Conference of Banking Supervisors, representing regulators from nearly 120 countries, recognised the report as the agreed standard (BIS, 2001).

[48] Van Gelder et al., 2003.

agreement and is revising this agreement with the intention of strengthening it.[49]

In cooperation with NGOs, Dutch commercial banks developed, in 2001, a set of minimum criteria for financial institutions servicing the Indonesian oil palm sector. They specified that banks should not be involved in clearing land by burning, deforestation, 'illegal' activities, and activities that generate social conflicts. Related discussions led four banks (ABN-AMRO, Rabobank, ING, and Fortis) to adopt risk-assessment policies that took these criteria into account. Some of the banks expanded the scope of the proposed criteria to other sectors, such as the pulp-and-paper sector, while others weakened the scope by focusing exclusively on illegality.[50] A more comprehensive set of minimum criteria was developed by ABN-AMRO for forests and plantations. Again, their genuine effectiveness on the ground remains to be seen, but these initiatives should, in principle, be supported and carefully monitored by NGOs and the banks. Adequate participation by NGOs could help to give substance to such initiatives, encouraging their adaptation where needed to ensure that they make a positive contribution to halting illegality and supporting sustainability.

Box 17.5 Making financiers pay

The first regulatory attempt to cast a wide net in making polluters pay – and financiers co-responsible for the environmental degradation caused by the (illegal or legal) activities of their clients – was the US's forward-looking Comprehensive Environmental Response, Compensation and Liability Act or 'Superfund' of 1980. The Act does not hold financiers explicitly liable, but, in the famous 'Fleet Factors' court case in 1990, a bank was nevertheless held responsible for the environmental pollution of its client. The outcome of this trial sent a shockwave through the US and the international banking community. Banks found liable were obliged to pay remediation costs and some banks even went bankrupt.[51] Some years later, modifications limited the liability of financial institutions.

The EU has a difficult history of attempts to adopt a stringent Directive on Environmental Liability; the scope of proposals has steadily shrunk and liability has shifted from strict to fault-based. It is unlikely that EU legislation will soon make financial institutions liable for financing damaging activities. Though political will seems to stop increasingly short of imposing liability on financial backers, the accountability of financial institutions in a less legal sense must continue to gather momentum.

---

[49] NGO demands for binding environmental and social standards for Export Credit Agencies (ECAs), signed by over 60 NGOs (FERN, 2002a).

[50] Van Gelder et al., 2003.

[51] Van Gelder et al., 2003.

In sum, when addressing illegal practices, such as illegal logging, the role played by financial institutions cannot be overlooked. Private commercial banks as well as (semi-)public export-credit agencies appear to be looking more seriously into risk assessments of their lending practices; however, more is needed. NGOs have been at the forefront in confronting financial institutions with the impact of their financial activities, and must continue to monitor the positive steps certain banks have taken. Without continued NGO interest and pressure, such initiatives risk becoming paper tributes to a cosmetic public-relations exercise.

In this respect, we make the following recommendations to governments and financial institutions. First, financial-sector regulators should issue specific industry guidelines for forestry-sector activities, specifying that companies wishing to raise equity on financial markets must disclose potential risks linked to forestry crime; this should encourage all financial institutions to adopt specific policies and guidelines for investments in the forestry sector. Second, governments should ensure that export credit agencies apply the best available (environmental and social) rules and procedures to all their operations. They should also increase the information disclosure practices of their export credit agencies regarding basic project information and environmental, social, and human rights impact assessments and economic analyses. And they should implement independent third-party monitoring of the projects in respect of the above-mentioned rules, once they are in force.[52]

CITES

The 1973 Convention on International Trade in Endangered Species (CITES) aims to protect endangered species from overexploitation by controlling international trade, employing a system of import and export permits. Species are placed on different lists: Appendix I includes all species that are threatened with extinction; trade in these species 'must be subject to particularly strict regulation' and is only authorised in exceptional circumstances. Appendix II includes species that are 'not necessarily threatened with extinction, but in which trade must be controlled in order to avoid utilisation incompatible with their survival'; this also includes other species that must be subject to regulation in order to control the targeted species effectively. Appendix III includes species that parties identify as being subject to regulation within their jurisdiction for the purposes of preventing or restricting exploitation, and where they need the cooperation of

[52] A coalition of over 60 European NGOs has presented a list of environmental and social guidelines for ECAs (FERN, 2002a).

other parties in controlling trade. Amendments to Appendices I and II are implemented by the Conference of the Parties, whilst parties themselves can place species on Appendix III.[53] At present, nineteen tree species are listed on CITES appendices. However, an evaluation of 255 tree species carried out in 1998 against the CITES listing criteria found that about 15 new species could be added to Appendix I and almost 100 to Appendix II, if there were the political will to do so.[54] Such additions to the appendices need to be agreed by the Conference of the Parties, and any proposal to add substantial numbers of new species, particularly those important in international trade, would almost certainly rouse strong opposition. Even unilateral additions to Appendix III can produce perverse effects.

CITES is currently the only worldwide legal agreement that can be used to control a part of the trade in illegally logged timber. It is also the only legal agreement to have been used by some member states to halt the import of illegally sourced timber. The main advantages of CITES are, therefore, that it is already in existence and is widely, if imperfectly, implemented. Yet, the effectiveness of CITES is hampered by weaknesses in the verification of trade documents. A key weakness is that the export and import permits effectively acquire a value, opening up possibilities of fraud, theft, and corruption. In theory, for an export permit to be issued, the competent authority of the exporting state must be satisfied that the specimen was not obtained in contravention of the state's laws for the protection of fauna and flora. On a practical level, a lack of personnel and resources to verify compliance with state rules is an impediment to adequate implementation of CITES rules. Furthermore, corruption also frequently plays a significant negative role in the issuance of permits. A further weakness lies in the cross-checking of the documents against each other. The World Conservation Monitoring Centre (WCMC), part of the UN Environment Programme, monitors the legal trade taking place under CITES, receiving copies of all import and export permits issued. Simple inspection of the permits is sometimes sufficient to reveal fraud. However, CITES lacks a comprehensive and independent system of monitoring and verifying the issuance and use of permits. Cross-checking of permits against actual timber species presents further difficulties. Central reporting and cross-verification of data would enhance the possibilities of verification that fraud has not occurred.

---

[53] See CITES: Article II, Fundamental Principles; Article III, Regulation of Trade in Specimens of Species included in Appendix I; Article IV, Regulation of Trade in Specimens of Species included in Appendix II; and Article V, Regulation of Trade in Specimens of Species included in Appendix III (CITES, 1973).

[54] World Conservation Monitoring Centre, 1998. The species evaluated were chosen to provide 'a reasonable representation of tree species from various regions, climates and grades of commercialisation and conservation' (p. 2). The availability of information on individual tree species varied considerably.

Box 17.6 Disappointing judicial review

The question of the validity of export permits has arisen with regard to exports of big-leafed mahogany from Brazil. The species is listed under Appendix III of CITES, and, in 2001, the Brazilian government ordered a complete ban on logging and export. Nevertheless, exports to Europe and North America continued in the first months of 2002. Shipments reaching the US, Canada, and a number of EU countries, were seized by the authorities pending further enquiries. In March, the European Commission issued advice to EU management authorities that they should not accept imports of Brazilian mahogany since reasonable doubt existed over their legality.

The willingness of the European Commission to take action was welcome: in this case, the UK government nevertheless declined to take action. The arguments in a subsequent court case brought by Greenpeace against the UK revolved around whether the export permits had been validly issued and under what circumstances the authorities in the importing state were justified in delaying the shipments and requiring further information on the validity of their export permits. Greenpeace lost its application for judicial review in the Court of Appeal: two of the three judges concluded that to allow importing countries to query the validity of export permits, even when some doubt existed, would introduce too great a level of uncertainty into international commerce. The third judge, however, dissented, accepting the argument that the survival of endangered species should be given higher priority.

As elsewhere, the political will to devote resources to addressing these shortcomings is a critical consideration. Since the events of 11 September 2001 in the US, for example, border resources have been diverted from monitoring environmental rules to security concerns. To repeat a truism, in the clash between global commerce and international environmental safeguards, the environment seldom comes out on top.

In sum, CITES cannot be expected to address the problem of illegal logging as a whole, but rather with regard to certain tree species. CITES' track record has been proven over three decades, although difficulties persist concerning fraud in permits, the listing of timber species, and the lack of willingness of authorities to take action, even when aware of problems. As in the Greenpeace case, NGOs must continue to insist on the effectiveness of action taken under this Convention and maintain a high profile where this is being undermined.

We recommend that governments should adopt a more coherent approach to checking the validity of CITES export permits at the point of import. Besides, they should attempt to include more timber species on the CITES appendices generally, as well as to encourage producer countries to list more timber species under Appendix III of CITES.

Forest Certification

Forest certification is a tool to help consumers choose ethical and environment-friendly products from responsibly managed forests. The crucial link between forest certification and the verification of legality lies in the tracking of the production and movement of the timber and wood products. Only forest certification schemes that have a rigorous chain-of-custody control from the forest to the point at which the product is labelled, and that have a certification standard specifying legal compliance, will contribute to controlling the trade in illegally sourced timber. In targeting the trade in timber and timber products, governments have at their disposal a range of instruments that can be used to increase the market share for products that are positively identified as having been legally produced. These include certification and government procurement policies. Here, we highlight the possibilities and problems of using forest certification schemes to combat illegal timber.[55]

Forest certification is a tool to help consumers choose ethical and environment-friendly products from responsibly managed forests. The process of certification involves the assessment of particular forests against publicly available criteria, and only if the forests meet these standards is timber certified. For forest certification to work, consumers must be able to identify timber, wood products, or paper that come from well-managed forests. These products, therefore, need to be labelled Once forests are

Box 17.7 Legal is not sustainable and sustainable is not legal

The certification of responsible forest management is different from legal verification. Legally sourced timber may not come from well-managed forests; indeed, it often does not. It is clear that 'legal' does not mean 'sustainable', as many requirements other than simple legality are linked with sustainability. The award of a Forest Stewardship Council certificate, for example, requires 10 principles and 56 specific criteria of good forest management to be met. Only the first principle relates specifically to legal compliance; the others relate to other essential requirements of sustainability. Furthermore, illegally harvested timber does not necessarily mean unsustainably produced timber. In countries where existing forestry regulations are inadequate or unjust, many undesirable practices in the forestry sector – such as the allocation of concessions on indigenous peoples' lands – may in fact be legal, under existing laws. In these cases, harvesting of the timber in a sustainable manner by the local people is seen as 'illegal'.

[55] For a more detailed overview of these tools (including government procurement), see: Brack, Marijnissen, and Ozinga, 2002; Garforth, 2004.

certified, the forest owners obtain the right to label products from those forests with the names of certifiers or logos. There is, however, a long and often complicated path from the forest to the point of sale: the product supply chain. To be able to guarantee the consumer that a particular product comes from a well-managed forest, this supply chain needs to be certified as well. The ownership and control aspect of the product supply chain is referred to as the 'chain of custody'.

To use a label indicating that a forest product comes from a well-managed forest as verification that the wood is legally sourced, three conditions must be fulfilled: (1) the forest certification standard must clearly require compliance with national laws; (2) the standard must be implemented effectively; and (3) an effective chain-of-custody control, from the forest to the point at which the product is labelled, must take place. In order to exclude non-certified content effectively, a credible chain of custody should include three main elements: identification, segregation, and documentation. Segregation requires clients to keep certified wood physically separate from uncertified wood at all phases of transportation, production, distribution, sale, and export. Accurate records must be maintained for the production of certified products. To date, only the Forest Stewardship Council (FSC) scheme, and possibly the Canadian Standards Association's (CSA) scheme, meet these requirements.[56]

However, as noted by the WB,[57] 'these 'quality assurance' [forest certification] systems have not been designed as tools to enforce the law and to be made compulsory. They are not based on regular and unannounced audits and on continuous sampling and they rely on paper-based chain-of-custody systems that are possible to forge. Given this, certification schemes do not provide the level of confidence that is likely to be required to demonstrate legal origin'.

In sum: only forest certification schemes that have a rigorous chain-of-custody control from the forest to the point at which the product is labelled, and that have a certification standard specifying legal compliance – after a review of relevant national legislation – can contribute to controlling the trade in illegally sourced timber. Even then, additional measures are needed. Most forest certification schemes do not meet these requirements. Therefore, governments should ensure that all certification schemes operational within the EU meet stringent compliance conditions. Besides, a clear distinction should be made between certification and labelling for sustainable forest

---

[56] For a more detailed analysis, see Brack, Marijnissen, and Ozinga, 2002; Ozinga, 2004.
[57] SGS Global Trade Solutions, 2003: 2.

management, and certification (possibly without labelling) for legal compliance. The latter is not and should not be an eco-label.

THE NEED FOR NEW LEGISLATION

Despite the wide range of policies and measures briefly described above that *could* affect the trade in illegally logged timber, none of them is specifically targeted at this problem. Undoubtedly, more could be done with existing legislation, or with relatively straightforward adaptations of provisions. There are several drawbacks to this approach. First, existing legislation requires close cooperation with enforcement authorities in the producing and exporting countries, which may not always be forthcoming, owing to lack of capacity, corruption, intimidation, etc. Second, existing legislation requires the cooperation of enforcement and other authorities in the importing countries, which, similarly, may not always exist under the current framework, given other priorities such as the fight against terrorism and drugs trade. Third, action that relies on judicial enforcement may take several years to show results, if a case is ever brought.

While new legislation may suffer from similar difficulties, it can target the issue more directly. Therefore, further options for legislation directly targeted at the problem of importation of illegal timber should be considered. This new legislation may take the form of sanctions against goods that can be identified as illegal. Alternatively, sanctions may be applied against goods that cannot be positively identified as legal, thereby closing markets to all imports that are not proven to be legal (including those of 'unknown legality'). The first option is an EU version of the US Lacey Act, that would outlaw the import, transshipment, purchase, sale, and receipt of timber obtained or sourced in violation of the laws of foreign states or of international treaties. The second option is to establish a licensing system for imports of timber and wood products, similar in principle to the approach adopted under CITES, with the requirement of transportation documents (export and/or import permits) to accompany consignments of controlled species. The latter was being envisaged by the European Commission (EC), intending to develop legislation to establish such a system.[58] Importantly, this does not represent a unilateral imposition of trade controls by the EU on producer states. Not only would this be likely to fall foul of WTO rules, but it would be almost impossible to operate effectively, as it would lack cooperation from the governments of producer countries in setting up and operating the necessary identification systems.

---

[58] European Commission, 2003.

The new EU legislation establishing this system must, therefore, create a *requirement* that imports of timber and wood products from participating producer countries be accompanied by a certificate of legality (or licence) at the point of entry into the EU. The system is possible only with the cooperation of the producer countries; accordingly, bilateral or regional partnership agreements are an essential part of this system. Bilateral agreements between Indonesia and the EU were being developed, and a bilateral agreement between Indonesia and Japan has been finalised.[59]

Although it was expected that the EU would choose this path to address the influx of illegally sourced timber, at least two potential problems arise with this approach. First, the bilateral or regional agreements must be developed with full participation of all civil-society groups, including local and indigenous peoples, in order to avoid the problems outlined above: reinforcing unjust and/or unclear existing laws will not address the problem and may endanger local peoples' livelihoods. In many producer countries, a revision or reform of forestry law is needed before any enforcement can take place. It is as yet unclear if full civil-society participation is foreseen in the development of these bilateral or regional agreements. Second, some – possibly many – producer countries may not want to enter into partnership agreements with the EU, for a variety of reasons. The legislation currently foreseen would not be able to address this issue, thereby perpetuating the existing practice of laundering illegally sourced timber. In its Action Plan, the EC showed it is aware of this problem by stating that 'the Commission will review options for, and the impact of, further measures, including, in the absence of multilateral progress, the feasibility of legislation to control the imports of illegally harvested timber into the EU, and report back to the Council on this work during 2004. Member states should also examine how the trafficking of illegally harvested timber is addressed under national laws'.[60] For guidance, the EC is expected to look into the US Lacey Act and the Kimberley Process; both are described below.

The import of stolen goods into the EU is not in itself a customs offence. In many ways, the simplest way to control imports of illegally sourced timber would be to change the law to make these imports themselves illegal as a back-stop in preventing imports of illegal timber from entering from countries with which no licensing system has been agreed. In this respect, the US Lacey Act, which makes it 'unlawful for any person … to import, export, transport, sell, receive, acquire, or purchase in interstate or foreign commerce … any fish or wildlife taken, possessed, transported, or sold in violation of any law

[59] Illegal-logging.info, 2003.
[60] European Commission, 2003: 15.

or regulation of any State or in violation of any foreign law'[61] could provide an important model.

Apart from species listed under CITES or identified as endangered in a US state, timber is not covered by the Lacey Act. There is, however, some current interest in the US in expanding the Lacey Act to try to prohibit imports of illegal timber. A two-tiered penalty scheme exists, creating both misdemeanour and felony offences, partially dependent on the level of knowledge of the laws violated by the accused. The penalties can involve imprisonment and/or fines, and forfeiture of equipment involved in offences. In all cases, the defendants need not be the ones who violated foreign law; the fact that fish or wildlife were obtained illegally is the important point. The Lacey Act also requires that shipments of fish and wildlife be accurately marked and labelled on the shipping containers. Failure to do so is a civil offence punishable by a fine. In all cases, federal agents are authorised to seize any wildlife that they have reasonable grounds to believe was taken, held, transported, or imported in violation of any provisions of the underlying laws. This may be done even if defendants can show that they were not aware that wildlife was illegally obtained. US prosecutors make frequent use of the Lacey Act. In 1999, for example, the US Fish and Wildlife Service was involved in almost 1,500 cases.[62]

The model provided by the Lacey Act is of obvious relevance to illegal logging or related illegal practices, and may inform legislation outlawing the import, transshipment, purchase, sale, and receipt of timber obtained or sourced in violation of the laws of foreign states or of international treaties. Although Justice and Home Affairs is not a matter of supranational competence, the basic prohibition can be established at the EU level. Proving illegality is not always straightforward, not least because of a lack of knowledge of, or a lack of clarity about, the foreign laws in question. US courts have interpreted the term 'any foreign law' broadly, including regulations as well as statutes, and not restricting the laws in question to those aimed directly at wildlife conservation. In Lacey Act proceedings, courts are given broad discretion because of the general lack of availability of foreign law materials and expert opinions. Sources used by courts have included affidavits, foreign expert testimonies, foreign case law, law review articles, information from foreign officials, and own research by courts.

As a cautionary note, this type of provision may be vulnerable to producer countries lowering the threshold of illegality to get round the law. Also, reliance on successful court cases to prove its deterrent effect could take years, though the act of giving additional powers to customs to make seizures

[61] USC, 2004: 2a.

[62] Animal Welfare Institute, 2000. It is difficult to acquire precise figures, as cases may often be coded as import violations or CITES violations.

may have a rapid deterrent effect. Despite these caveats, there is obvious value in giving willing authorities a clear and logical mandate to stop illegal products entering their jurisdiction. It provides a strong signal to participants in the market, and marks new limits of what is perceived to be acceptable behaviour.

In sum, new legislation aimed more directly at the problem of illegal logging should be envisaged. The option currently under consideration by the EC may prove effective, if only in the cases where bilateral agreements have been adopted with full participation of civil society. Something must be envisaged for dealing with the problem of importation of illegal timber from countries that reject bilateral agreements; such countries are likely to be the worst offenders. The US's Lacey Act provides an interesting model for new legislation to give authorities a mandate to stop illegal products from entering the market.

KEEPING AN EYE ON PARALLELS

Parallels with Other Trade Control Systems

In addition to CITES, several multilateral environmental agreements (MEAs) have been agreed in order to impose various controls on international trade in cases where the unregulated trade caused, or was likely to cause, significant environmental damage. Parallels exist with, for example, the Montreal Protocol on ozone-depleting substances, which uses a system of import and export licences, adopted primarily in order to reduce illegal trade; and the Basel Convention on transboundary movements of hazardous wastes, which uses a system of prior notification and consent.[63]

The purpose of these trade instruments is to establish a system in which exporting and/or importing countries (and in some cases countries of transit) must agree to the trade taking place before it can proceed. Unregulated trade is theoretically eliminated – or at least made more difficult. The shortcomings of CITES, involving susceptibility to fraud and inadequate cross-checking of permits, have been outlined above. The Basel Convention suffers from similar problems: most of the illegal trade in hazardous waste is believed to involve falsified documentation, and hazardous waste can often be difficult to distinguish from non-hazardous waste (indeed, the two are sometimes deliberately mixed together). Any system for controlling any part of the international trade in timber, therefore, needs to avoid these problems.

---

[63] About 20 of the 200 MEAs currently in existence contain trade measures (requirements, restrictions, or complete bans on trade).

The Kimberley Process

A relatively new process, the Kimberley Process, aims to identify and eliminate the trade in conflict diamonds.[64] Spurred by concern about the role diamonds played in fuelling civil wars in some African countries, the scheme had its origins in the decision of southern African countries, in early 2000, to take action to stop the flow of conflict diamonds to the market while at the same time protecting the legitimate diamond industry. The system came into effect on 1 January 2003, and focuses – like CITES – on the certification of exports; the intention is to establish minimum acceptable international standards for national diamond certification schemes. Participants undertake to establish internal systems to implement and enforce the certification scheme, including establishing suitable penalties for transgressions. The Process recommends, amongst other things, that the names of individuals and companies convicted of breaches of the certification scheme be made known to all other participants. Such transparency serves to flag suspicious actors and activities, and may reinforce efforts in the areas of anti-corruption and money laundering. The diamond industry has undertaken to introduce a system of self-regulation to support the Process, involving warranties underpinned by the verification of individual companies by independent auditors and supported by internal penalties set by the industry.

Obvious parallels exist between the aim of the Kimberley Process – to exclude conflict diamonds from the legitimate diamond trade – and moves to exclude illegally sourced timber from legal markets. There are also, of course, important differences: diamonds are traded in far lower volumes than is timber and can be sealed in tamper-proof containers; the number of countries involved in major imports and exports is lower; and the industry is largely united, worldwide, on the desirability of the system. Despite these differences, lessons can be learned from the Kimberley Process, and from the experiences of other MEAs with licensing systems. The Kimberley Process inspection scheme for certificates, particularly in the EU draft Regulation, is far stricter than those in CITES and other MEAs, and is expected to avoid many of the weaknesses of CITES import and export permits. The cooperation of the industry, essentially through giving favourable treatment to organisations that promise to fulfil particular criteria, is a useful precedent.

Developments should be watched. It was clear from the outset that

---

[64] It adopted the definition of 'conflict diamonds' as meaning 'rough diamonds used by rebel movements or their allies to finance conflict aimed at undermining legitimate governments, as described in relevant United Nations Security Council (UNSC) resolutions insofar as they remain in effect, or in other similar UNSC resolutions which may be adopted in the future, and as understood and recognised in United Nations General Assembly (UNGA) Resolution 55/56, or in other similar UNGA resolutions which may be adopted in future' (Kimberley Process Certification Scheme, 2002).

adequate independent monitoring should be an essential part of the scheme to ensure its effectiveness. NGOs working with the Kimberly Process have stressed that, without adequate monitoring, no way exists of ensuring that countries are actually halting the trade in conflict diamonds. 'Without effective and regular monitoring of the Kimberley Process it is difficult to evaluate that national regulations are worth the paper they are written on', stated Global Witness.[65] Thus far, a signal was sent when countries such as Brazil and Ghana, which failed to pass necessary legislation by the July 2003 deadline, were excluded from the Process: they will not be allowed to sell to large markets such as the US and Europe, or to diamond-processing countries such as Israel and India. Disturbingly, however, countries embroiled in internal conflict such as Congo have been authorised as participants, excluding only Liberia among African countries and, thereby, jeopardising the credibility of the scheme.[66]

In sum, existing environmental trade agreements offer useful insights into the potential pitfalls of efforts to address the illegal timber trade. The Kimberley Process in particular demonstrates that, when political will and industry backing combine, solutions can be devised with surprising speed. At the same time, political will seems to retreat before difficult decisions (for example, the decision to exclude Congo from the Process). Adequate independent monitoring and NGO participation are essential tools in keeping them honest.

CONCLUSIONS

As seen in section 1, illegal logging is a vast problem with considerable negative economic, social, and environmental consequences. Despite growing international concern, there is a clear lack of immediate and well-coordinated action at the national and international levels to address this problem. National governments do not speak with one voice: authorities in one ministry have priorities that differ from those of their colleagues in other ministries. Generally, environmental issues tend to rank low in the hierarchy of these priorities. Given political realities, it may be fruitful to attempt to include environmental issues within more general frameworks that currently receive attention and resources.

As a general consideration, many existing forest and conservation laws have unacceptable negative impacts on poor people, ethnic minorities, and women, and, in many places, they are enforced in a fashion that is

65 Global Witness, 2003.
66 Degli Innocenti, 2003.

discriminatory and abusive. There is a clear danger, then, that focusing blindly on law enforcement will backfire and have a negative impact on local peoples' livelihoods. Ways must be found to address the problems associated with illegal forestry activities that, at the very least, do not aggravate the negative impacts of existing regulatory efforts on the rural poor. In many cases, that will mean a revision of forestry laws, taking into account local peoples' traditional land and user rights. Efforts to ensure that loggers and other entities in the forestry industry comply with the laws and forestry regulations must go well beyond the chain-of-custody system and examine the general legal environment.

An observation that can be made with regard to all the initiatives considered in this chapter – from ensuring an equitable legislative framework to alerting officials to corruption, providing targeted insider information to authorities, evaluating efforts frankly, and encouraging reform in the light of experience – is that independent monitoring and effective participation of stakeholders and civil society are essential factors in ensuring that these efforts have practical results. Without watchdogs and true participation, international initiatives may simply become elaborate and costly public relations exercises.

Another general consideration is that, without addressing the existing pervasive corruption in the logging industry, efforts in almost any domain to combat illegal logging or any other illegal activity will be thwarted. Efforts to ensure that loggers and other entities in the forestry industry comply with the laws and forestry regulations must examine the general legal environment, and go beyond tracking systems or chains of custody. The forestry industry does not exist in a vacuum. If the underlying system is corrupt, it is likely that forest governance is corrupt as well.

Anti-corruption initiatives are presently an overriding international focus, which may bode well for the effects of anti-corruption measures on the illegal timber industry. Also, anti-corruption measures will be more assiduously applied if the perceived interest of industry lies, for competitive reasons, in eliminating the corrupt practices; this appears to be the case with the OECD Convention on Bribery. This Convention takes a fresh approach by addressing the supply side of such offences, and it seems to have laid the foundations of an interesting peer-review process that includes practical implementation and permits civil-society participation. Although it is too early to judge its performance, the framework to adapt to experience appears to exist. Governments and civil-society actors should be encouraged to give this framework substance.

The desire to compete on an equal footing has been the principal impetus behind not only the OECD Convention on Bribery, but also the OECD process to develop standards for export credit agencies, and possibly guidelines to be developed for commercial banks. When addressing illegal practices, the role played by financial institutions cannot be overlooked. Private commercial banks as well as (semi-)public export-credit agencies appear to be looking more seriously into risk assessments of their lending practices; however, more is needed. NGOs have been at the forefront in confronting these institutions with the impact of their financial activities and must continue to monitor the positive steps some banks have taken. Without continued NGO interest and pressure, such initiatives risk becoming paper tributes to a cosmetic public relations exercise.

The activities of the financial sector are clearly linked with international and national attempts to improve money-laundering legislation, as the key to success in using money-laundering provisions to combat illegal logging and other illegal activities lies in inspiring banks and other relevant institutions and persons to exercise due diligence in carrying out reviews, with a full awareness of the likelihood of certain clients' involvement in such activities. Given that the majority of logging activities in many countries are widely acknowledged to be illegal, this should be a strong signal to banks that any of their clients operating a forestry business in those countries may be committing forestry crimes. Such due diligence measures would in fact be good business practice, as illegality, by definition, means increased risk. Targeted information provided by NGOs could be instrumental in initiating a few high-profile prosecutions. A clear task for the NGO community is to provide this information.

CITES is the only existing international convention that can directly target the timber trade, albeit in a piecemeal fashion. Concerns surround political opposition to expanding the list of species protected. Experience with CITES also offers useful lessons for similar endeavours: reliance on paper certificates makes its provisions susceptible to fraud, theft, and corruption. The permits are not adequately cross-checked against each other or against the goods they accompany. Any system for controlling any part of the international trade in timber must be mindful of these lessons.

The recent Kimberley Process has taken on board some lessons from the problems that affect CITES and the Basel Convention. The Kimberley agreement established a certification scheme to control the international trade in rough diamonds, and thereby to prevent 'conflict diamonds' from entering the legitimate trade. Although too new to judge adequately, it nevertheless is worth watching as it may yield useful lessons in achieving similar objectives. However, the Kimberley Process may also present problems. Independent monitoring and transparency are often where political will runs out – as the Kimberley Process, as young as it is, may already illustrate.

The EC has proposed the development of new legislation to allow for a licensing scheme. This new legislation must create a requirement that imports of timber and wood products from participating producer countries be accompanied by a certificate of legality (or licence) at the point of entry into the EU. The system is possible only with the cooperation of producer countries; accordingly, bilateral or regional partnerships agreements are an essential part of it. Although a welcome first step in addressing the importation of illegally sourced imports into the EU, one issue that needs to be seriously considered is the requirement to develop these agreements with full participation of all civil-society groups, including local and indigenous peoples, in order to avoid the problems of reinforcing unjust and/or unclear existing laws. It is not yet clear if full civil-society participation is foreseen in the development of bilateral or regional agreements. It should also be noted that, in many countries, full participation of civil society is simply not possible.

None of these options is free of difficulty. The effectiveness of all the initiatives requires the vigilance and perseverance of NGOs and stakeholders in maintaining a frank dialogue with authorities. For their part, authorities must be willing to interpret and apply these provisions in a manner that is generally meaningful and, more specifically, that includes illegal logging within its scope. Authorities must be open to exchanging information with NGOs and stakeholders. All committed parties must expect practical difficulties and occasional breaches of political and institutional will to occur, and be ready to re-engage their efforts to meet such challenges. Ultimately, even if political will falters and concrete results are long in coming, efforts to bring the destruction caused by illegal logging to light can be considered partially successful if the actors involved – sometimes innocently – in the illegal timber trade are made aware of their role in financing and laundering the proceeds. If the consumers – not only of timber, but also of financial and banking services – take note of the link between their actions and destruction abroad, we will move incrementally forward.

REFERENCES

Aiolfi, G. and M. Pieth (2003), How to Make a Convention Work: The OECD Recommendation and Convention on Bribery as an Example of a New Horizon in International Law, OECD Working Group on Bribery in International Business Transactions (CIME), JT00122279, Paris: Organisation for Economic Co-operation and Development.

Animal Welfare Institute (2000), 'Lacey Act turns 100', Animal Welfare Institute Quarterly, **49**(4).

Barr, C. (2001), The Financial Collapse of Asia Pulp & Paper: Moral Hazard and Implications for Indonesia's Forests, CIFOR, Asian Development Forum-3, Bangkok, 12 June.

BIS (2001), 'Customer due diligence for banks', Basel Committee Publications no. 85, Bank for International Settlements, *http://www.bis.org/publ/bcbs85.htm*.

BIS (2003), 'The Basel Committee on banking supervision', Bank for International Settlements, *http://www.bis.org/bcbs/aboutbcbs.htm*.

Brack, D. and G. Hayman (2001), 'Intergovernmental actions on illegal logging', Royal Institute of International Affairs, *http://www.riia.org/sustainable development*.

Brack, D., G. Hayman, and K. Gray (2002), 'Controlling the international trade in illegally logged timber and wood products', Royal Institute of International Affairs, *http://www.riia.org/sustainabledevelopment*.

Brack, D., C. Marijnissen, and S. Ozinga (2002), 'Options to control the import of illegally sourced timber', *http://www.fern.org*, FERN/Royal Institute of International Affairs.

Callister, D. (1992), Illegal Tropical Timber Trade: Asia-Pacific, Cambridge: TRAFFIC International.

CITES (1973), Convention on International Trade in Endangered Species of Wild Fauna and Flora, Washington, DC, 3 March, *http://www.cites.org/eng/disc/text.shtml*.

Cockcroft, L. (1999), Implementation of the OECD Convention – The Conditions for Success, presented at seminar on corruption and bribery in foreign business transactions, Vancouver, Transparency International Canada, 4-5 February.

Colchester, M., M. Siriat, and B. Wijardjo (2003), Implementation of FSC Principles 2 and 3 in Indonesia; Obstacles and Possibilities, Jakarta: WALHI (Friends of the Earth Indonesia)/AMAN (Alliance of Indigenous Peoples of the Archipelago).

Degli Innocenti, N. (2003), '54 countries pass "Conflict Diamonds Test"', Financial Times, July 31.

EIA/Telapak (2003), 'Timber traffickers: How Malaysia and Singapore are reaping a profit from the illegal destruction of Indonesia's tropical forests', *http://www.eia-international.org/cgi/reports/report-files/media57-1.pdf*.

EU Forest Watch (2002), Issue 65, *http://www.fern.org*.

European Commission (1991), Directive 91/308/EEC, OJ L 166, Brussels: European Commission.

European Commission (2001), Directive 2001/97/EC, OJ L 344/76, Brussels: European Commission.

European Commission (2003), Communication from the Commission to the Council and Parliament; Forest Law Enforcement, Governance and Trade, Proposals for an EU Action Plan, COM 251 final, 21 May.

FATF (2002), 'Third FATF review to identify non-cooperative countries or territories: Increasing the worldwide effectiveness of anti-money laundering measures', *http://www.fatf-gafi.org/NCCT2002_en.htm*.

FBE (2000), Observations on the Proposal for a Directive Amending the 1991 Directive on Prevention of the Use of the Financial System for the Purpose of Money Laundering, Brussels: Fédération Bancaire de l'Union Européenne.

FERN (2002a), 'Key reforms needed for European export credit agencies', *http://www.fern.org*.

FinCEN (2003), '2003 National money laundering strategy', *http://www.fincen.gov/int_main.html*.

French, H. (2000), Vanishing Borders: Protecting the Planet in the Age of Globalization, Washington, DC: WorldWatch Institute.

Garforth, M. (2004), To Buy or not to Buy: Timber Procurement Policies in the EU, Gloucestershire: FERN.

Global Witness (1999), The Untouchables: Forest Crimes and the Concessionaires – Can Cambodia Afford to Keep Them?, London: Global Witness.

Global Witness (2003), Kimberley Process Finally Agrees Membership List but Lack of Monitoring Undermines Credibility, press release, 31 July, London: Global Witness.

Greenpeace (2003), Partners in Crime: A Greenpeace Investigation of the Links between the UK and Indonesia's Timber Barons, London: Greenpeace.

ICG (2001), Indonesia: Natural Resources and Law Enforcement, Asia Report no. 29, Brussels/Jakarta: International Crisis Group.

Illegal-logging.info (2003), 'Japan and Indonesia make a joint announcement of cooperation on illegal logging', *http://www.illegal-logging.info/news.php?newsId=23*.

ITFMP (1999), Illegal Logging in Indonesia, ITFMP report EC/99/03, Jakarta: Indonesia–UK Tropical Forestry Management Programme.

Kaimowitz, D. (2003), 'Forest law enforcement and rural livelihoods', International Forestry Review, 5(3), 199-210.

Kimberley Process Certification Scheme (2002), section 1, definitions, *http://www.kimberleyprocess.com/news/documents.asp?Id=14*.

OECD (2001), OECD Environmental Outlook, Paris: Organisation for Economic Co-operation and Development.

Ozinga, S. (2004), Footprints in the Forest; Current Practice and Future Challenges, Gloucestershire: FERN.

Rice, T., C. Marijnissen, and S. Ozinga (1999), 'Trade liberalisation and the impact on forests', *http://www.fern.org, FERN*.

Scotland, N. and S. Ludwig (2002), Deforestation, the Timber Trade and Illegal Logging, paper for EC workshop on Forest Law Enforcement, Governance and Trade, Brussels, 22–24 April.

SGS Global Trade Solutions (2003), Legal Origin of Timber as a Step Towards Sustainable Forest Management, final draft, World Bank/WWF Alliance.

USC (2004), 'Title 16, Chapter 53, Section 3372', *http://www4.law.cornell.edu/uscode/16/3372.html*.

Van Gelder, J. (2002), The Financing of African Logging Companies, Castricum: Profundo.

Van Gelder, J., E. Wakker, and W. Richert (2003), Sources of Investment for Forestry: Preventing Flows of Finance to Illegal Activities, discussion paper, London: Royal Institute of International Affairs.

Watson, G. (1998), 'The OECD Convention on Bribery', American Society of International Law, *http://www.asil.org/insights/insight14.htm*.

White & Case (1999), Report to Senior Officials of Royal Government of Cambodia and International Donors: Summary of Recommendations, New York: White & Case.

World Bank (1999), Forest Sector Review, New York: World Bank.

World Conservation Monitoring Centre (1998), Contribution to an Evaluation of Tree Species Using the New CITES Listing Criteria, Cambridge: World Conservation Monitoring Service.

World Resources Institute (1997), Logging Burma's Frontier Forests: Resources and the Regime, Washington, DC: World Resources Institute.

18. Globalisation and Crop-Protection Policy

Joost van Kasteren

SUMMARY

Farmers have long protected their crops. Pesticides have significantly enhanced food productivity, but have also engendered negative effects on the environment and public health. Integrated pest management (IPM) constitutes an alternative protection method that avoids these side effects. At the global level, the high costs to develop and commercialise new products have led to scale effects: six major suppliers, focusing on five major crops, which are sold on four large markets. Another global development is the Western setting of maximum residue levels (MRLs) of pesticides on crops, which is regarded by many (Third-World) exporters as a non-tariff trade barrier. European pesticide legislation is still fragmented, despite efforts to harmonise MRLs. The pesticide policy of the Dutch government is aimed at reducing the use of pesticides. However, its discretion is constrained by global pressures to abstain from regulation that interferes with the free market and by European harmonisation requirements. Against this backdrop, Dutch national pesticide policy should stimulate IPM, encourage self-regulation, and promote information transparency.

INTRODUCTION

Globalisation can be defined as the increase of connections between states and societies in the world. This means that decisions, actions, and events in one part of the world have consequences for individuals and societies elsewhere. Globalisation takes place in many areas, including agriculture (which is one of the main themes of the Doha round of trade negotiations). As crop protection is inextricably linked with modern agriculture (including organic farming), globalisation has an impact on the policies of national governments as to the authorisation and use of pesticides. A pesticide is a ready-to-use product that has only to be diluted in order to use it. It contains an active substance that kills the insect, weed, or fungus, but it also contains other substances for stabilisation and droplet formation. The recipe for a pesticide is called a formula. The central question of this chapter is how globalisation has affected the ability of national governments to conduct an effective crop-protection policy, taking into account local ecological circumstances.

To address this question, I first provide general background information on the use of pesticides and their impact on the environment. Then follows a description of three important trends that will severely limit the role of national governments in developing and executing a crop-protection policy. The first trend is the globalisation of pesticide development, production, and marketing, partly brought about by the high costs of authorisation. Six large agro-chemical companies hold three quarters of the world market for synthetic pesticides. High costs force them to focus their developing efforts on pesticides for five large crops (wheat, cotton, soya, maize, rice) and their marketing efforts on the major, easily accessible markets. This narrow focus might hamper the spread of more advanced methods of crop protection (such as integrated pest management) and hence the development of more ecologically sound agriculture. Limitations may become even more severe in the future when the now protected markets for agricultural products are opened up, which is the second global trend. As food chains become more global, a tendency to impose stricter pesticide regulations can be observed. Although set to protect the health of the consumer, such regulations might be considered a form of protectionism through non-tariff trade barriers, leading to objections by the World Trade Organization (WTO). Apart from the possible introduction of a health hazard, a lack of regulation may also lead to a competitive advantage for imported products, because farmers elsewhere are allowed to use – cheaper – pesticides that are forbidden in Europe. A third major trend, which directly influences the national crop-protection policy of the member states of the European Union (EU) is European harmonisation, which started in 1993. It will lead to similar authorisation procedures of pesticides in 2008. It has already had a severe influence on Dutch pesticide policy, showing the limitations of national policy in a supranational framework.

European harmonisation combined with the two global trends mentioned above seems to deny national governments the capacity to balance agricultural competitiveness against environmental quality and public health. This is harmful for both agricultural competitiveness and the environment. As developments in the Netherlands suggest, it is possible to find a solution.

New and more creative policies are needed to replace the traditional command-and-control approach. These are also discussed below.

PROTECTING CROPS

History

More than 10,000 years ago, people in different regions of the world started planting seeds in addition to the hunting of animals and the gathering of berries, seeds, and other edibles. In the Middle East, wheat was planted; in Asia, rice plants were used; and maize became the preferred staple food in Mexico. The social and ecological changes were enormous. Instead of containing thousands of different plants, one acre would now contain a single crop. Such monocultures are a paradise for insects, viruses, bacteria, fungi, nematodes, and other organisms that feed on plants. Through the ages, these crops also became more domesticated, losing their natural capacity for chemically defending themselves against their attackers.

Populations grew and became more and more dependent on these crops. As a result, society became more vulnerable to plagues and diseases attacking their crops. People tried to protect their crops by all available means. Around 1200 BC, Chinese farmers started using ashes and lime in the fight against parasites attacking their winter stocks. Certain plant extracts, like pyrethrum, were sprayed in pre-historic times by the Chinese to keep the insects at bay. Later, the Romans used arsenic for the same purpose, as Plinius described in 70 BC. From his writings, we also know that Roman farmers burned sulphur and bitumen to protect their vineyards from leaf roller.[1] Around the 16th century, new substances became available in Europe for plant protection, such as nicotine and the aforementioned pyrethrum, and a mixture of copper sulphate and lime, better known as Bordeaux Mix, which is still used today to protect grapes from mildew and grey rot. By the 18th and 19th centuries, new plant extracts such as rotenone and veratrine were used. Inorganic herbicides such as sodium chlorate and sodium arsenate were also available at this time.

The 1930s were the dawn of a new era of pesticide development. Several new classes were developed, including the carbamates (e.g., dithiocarbamate used as fungicide), the chlorinated hydrocarbons such as DDT (discovered in 1939 and used as insecticides), and the organophosphates such as Parathion and malathion (which were also used as insecticides). Modern herbicides such as the defoliants 2,4 Di-, and 2,4,5 Tri-chlorophenoxyacetic acid were developed later. These synthetic chemicals were to become the active

[1] Pelerents, 1989.

substances of pesticides that became successful before and especially after World War II. The development of pesticides and synthetic fertilisers (plant nutrients such as nitrates, phosphates, and potassium) stimulated an impressive transition in Western agriculture. Yields per hectare for most crops increased by a factor of 8-10. Labour productivity also increased, by a factor ranging between 100 and 200. As a result, the area now needed to feed one Westerner is around 500 square metres, compared to the 1 squared kilometre needed at the dawn of agriculture. The amount of labour required to work the land and feed a Westerner has decreased from around 300 hours in the Middle Ages to 75 minutes now. As a result, the percentage of the population of industrialised countries working in agriculture dropped from 50-60 per cent at the beginning of the 20th century to 2-3 per cent nowadays, and is still decreasing.

Box 18.1 Dutch laws on pesticides

Already in the 1920s, the Dutch Parliament had asked for a law to prevent fraud and deceit in the trade of pesticides, but only in 1948 did the first Pesticide Law come into force to protect farmers against bad trading practices. In the 1950s, people started to worry about the effects of pesticides on human health, i.e., both the health of farm workers who had to work with pesticides and the health of the public, who were exposed to pesticide residues on fruits and vegetables. This led to the Pesticide Law of 1962. The effectiveness of new pesticides was evaluated, and they were also tested with regard to safety and their possible effects on (public) health.

In 1975, the Pesticide Law of 1962 was adapted to include the 'harmful side-effects' of the pesticides and their metabolites. But it was not until 1989 that the criteria for these harmful side-effects were introduced in a government White Paper. And only in 1995 was the Pesticide Law of 1962 extended to include a paragraph containing criteria for harmful side-effects. These criteria were based on the 1991 European Directive 91/414/EEG for pesticides and Directive 94/43/EG, which was published in 1994, containing the Uniform Principles for the evaluation of pesticides.

In the 20th century, the Netherlands developed into an agricultural superpower. The country is the largest exporter of agricultural products after the United States (US) and France. The high yields are brought about by a relatively heavy use of pesticides. The total amount used in 1999 was about ten million kilogrammes of active substance, which amounts to five kilogrammes per hectare. Almost half (45 per cent) of the active substances that are used in the Netherlands are fungicides to fight fungi, including the infamous potato blight. Herbicides or weed killers are the second-largest category, at 28 per cent. Herbicides are used not only by farmers but also by local governments and private companies to eradicate weeds. Substances to disinfect soil (mostly nematicides, which kill nematodes, tiny worms) account for 14 per cent of total use, and insecticides – surprisingly – only account for 3 per cent. The category 'other' amounts to 10 per cent. About 230 substances are available for use. Before they can be applied, pesticides must be authorised by the CTB, the Board for the Authorisation of Pesticides, an independent administrative body, supervised by four departments responsible for agriculture (which is the coordinating department), public health, occupational health, and the environment. The department responsible for water management has an advisory role. The CTB has been independent since 1993, because the Dutch government wanted to separate the actual authorisation of pesticides from its crop-protection policy.

Ecological and Health Effects

The enormous increase in agricultural productivity came at a price in the form of ecological damage and, in some cases, also damage to people's health. When they reach the soil, pesticides can affect organisms other than their target species. Soil fauna (such as earthworms, springtails, and mites) that contribute to the formation of humus can decrease, thus influencing the soil structure and fertility. In turn, their predators (for example, centipedes) decrease in numbers. As many of these are generic predators, their demise can lead to an increase in plague organisms such as greenfly. Secondary plagues occur when broad-spectrum pesticides are used that also kill the natural predators directly.

Chemically stable, fat-soluble pesticides, such as DDT, can accumulate in the food chain to levels that are harmful to birds and mammals, including humans. Sometimes, the smaller mammals can be killed directly by an overdose (for instance, just after spraying). Usually, the effects are more long term (for instance, greater susceptibility to disease because the immune system is affected, or a decrease in fertility). The effects of DDT were described in Rachel Carson's book 'Silent Spring'.[2] She indicated how DDT enters the food chain and accumulates in the fatty tissue of animals and birds. The most haunting chapter in the book, 'A Fable for Tomorrow', depicts an unnamed town in the US where all life has been silenced by the insidious effects of DDT. Luckily, we will never know whether a 'Silent Spring' could have happened. What we do know is that the numbers of birds of prey have increased since DDT was banned in the Western world. Although it is not formally banned, stopping the use of DDT in Africa and Asia, on the other hand, has led to a dramatic increase in malaria deaths.

---

[2] Carson, 1962.

Rainfall can cause pesticides to wash out into ditches, brooks, rivers, and lakes. Careless spraying when the wind is too strong or when the nozzle of the sprayer is poorly adjusted can also cause this to happen. The contamination of surface water can equally occur when farmers clean their equipment after spraying or when disaster strikes. On 1 November 1986, at the Sandoz factory in Basel, a fire broke out and various chemicals used for pesticide production streamed into the river Rhine with the water from the fire extinguishers, killing most of the organisms in it. Many zooplankton species are sensitive to insecticides, while algae and other plants are, obviously, sensitive to herbicides. Many higher organisms such as arthropods are sensitive to insecticides in surface waters. It is not clear how fish are influenced by pesticides. Large doses, as in the Sandoz case, will kill most fish. Smaller doses can have sub-lethal effects, such as reduced fertility, sex change (imposex), and effects on the immune system.

Box 18.2 Drinking water supply in the Netherlands

Open water and groundwater are used as sources of drinking water. About 40 per cent of Dutch drinking water comes from open water and 60 per cent from groundwater. Both sources are contaminated by pesticides or their residues. According to the Dutch Association of Drinking Water Suppliers, groundwater contains detectable levels of 60 pesticides, while open water contains around 70 different types. Not all of the pesticides found in water are linked to agriculture. In some areas, more than half of the pesticides found are herbicides used to control weeds in urban areas.

The costs incurred by the drinking water suppliers connected with pesticides in their raw material amounted to 240 million euros (EUR) between 1990 and 2000. A large part of this money (60 per cent) was spent on installations and chemicals (activated charcoal) to remove pesticides from the water before distribution to consumers. Another 20 per cent was spent on monitoring water quality, while the rest of the money was spent on protection of sources (for instance, by stimulating farmers to use fewer pesticides and/or pesticides that quickly degrade into harmless substances), on displacement of wells, and on lobbying.[3]

Integrated Pest Management

Ecological problems in industrialised countries and economic problems in developing countries – synthetic pesticides were too expensive – led to the development of different strategies for crop protection, together called Integrated Pest Management (IPM).[4] They were developed thirty years ago

[3] Puijker et al., 2001.
[4] Pinstrup-Andersen, 1999.

and are practised on a large scale in Indonesia and other rice-growing regions in South-East Asia. In the Netherlands, IPM strategies have been adopted in greenhouse horticulture and in orchards. Although there are differences of opinion about IPM, most people agree that it must be science-based and economically viable for farmers. The emphasis is on anticipating pest problems and preventing them from reaching damaging economic levels. Both anticipation and prevention should be based on ecological knowledge and modelling, with the awareness that there is still much to do in that area.

IPM strategies include:

- Biological control such as protecting and releasing pests' natural enemies (including insects, spiders, nematodes, bacteria, and viruses). In Dutch horticulture, for instance, predatory mites are released in glasshouses to keep spider mite under control, and solitary wasps are used against white fly. In the field, nematodes can be used to keep snails under control.
- Cultural practice such as reducing field size and thereby the distance between the habitats of natural predators, crop rotation, crop residue management, pest monitoring and modelling, and new methods of weed control (including precision farming).
- Genetic strategies such as the use of naturally resistant varieties of crops bred for resistance either via classical routes or via genetic engineering. Another 'genetic' strategy is the introduction of sterile males to prevent reproduction; it has been used with some success against onion fly.
- Chemical control such as the use of biopesticides (for instance, bacterial toxins or synthetic mimics), insect growth regulators, and hormones and chemicals that modify pest behaviour and reproduction (such as sex pheromones that attract insects). A well-known and successful method is to attract plague insects using pheromones and then kill them with an insecticide.

Many growers and farmers are keen to use IPM strategies, because they are often cheaper than pesticides and using them challenges their craftsmanship. This does not mean, however, that synthetic pesticides have become obsolete. First, synthetic pheromones and insect hormones have to be made. Second, there is still a need for 'traditional' synthetic pesticides as a second line of defence, when the ecological strategy fails; for instance, when plagues or diseases are imported and natural enemies are not (yet) available. Third, non-chemical methods of weed control are labour-intensive. As labour is scarce in developing countries, because it is mainly women who work the land in addition to many other jobs, there is still a need for synthetic herbicides.

GLOBAL DEVELOPMENTS

With respect to crop protection, there are two major trends on a global scale. The first is the globalisation of the development, production, and marketing of synthetic pesticides. The second is the urge to liberalise trade in agricultural products. Apart from abolishing agricultural subsidies, this also means growing support for the abolishment of so-called non-tariff trade barriers. One of these barriers is the use of pesticide residue levels to stop imports.

Development, Production, and Marketing

The last ten years have seen an enormous reshuffling among producers. Six major companies are now responsible for three quarters of all agrochemical sales.[5] These are, in order of magnitude: Bayer Crop Science, with sales amounting to 5,500 million US dollar (USD) in 2002; Syngenta (5,300 million); BASF (3,000 million); Monsanto (2,800 million); Dow Agro Sciences (2,500 million); and DuPont (2,200 million). The first three have their headquarters in Europe, while the last three are US-based. Considering the size of the US-based companies, at least one merger is expected between a US-based company and one of its European or American competitors, leaving only five major producers.

New pesticides have been, and are still, developed to reduce the environmental effects of synthetic pesticides and to adapt to changing farming practices. Insecticides, for instance, are becoming more specific for use

Box 18.3 Global sales of pesticides

> The global sales of pesticides amounted to 28 billion EUR in 2002; in real terms, this was 1.5 per cent less than in 2001. The agrochemicals industry might consider this an improvement compared to 2001, which saw a drop of 4.4 per cent compared to the year before. Since 1998, the pesticides market has been reduced by 12 per cent. The only region that saw an increase (of 7 per cent) in sales of agrochemicals in 2002 was Western Europe.
>
> Dividing the use of pesticides by region, North America uses 30 per cent, Asia/Pacific 26 per cent, Western Europe 23 per cent, Latin America 12 per cent, and the rest of the World (including Africa) 9 per cent. Divided by type, 47 per cent of agrochemicals sold are herbicides, 29 per cent are insecticides, 18 per cent are fungicides, and 6 per cent are other types, including nematicides.

---

[5] Agrow, 2003.

against plague insects, while herbicides are made more degradable so they do not accumulate in soil, water, or living organisms. To accommodate the development of IPM, even more specific pesticides are needed, often in very small quantities.

According to a study commissioned by the European Crop Protection Association, both the costs and the time necessary to develop a new pesticide and bring it to the market have increased considerably over the last decades.[6] In 2000, it cost on average 184 million USD to discover, develop, and register each new crop-protection product. In real terms, this is 8.5 per cent higher than the costs in 1995. The rise is partly caused by increased research costs. In 2000, nearly 140,000 molecules were screened in order to discover and bring to market one new crop-protection product. In 1995, just over 50,000 molecules had to be tested to find one marketable pesticide. The lead time (i.e., the time it takes to bring a product to the market) extended from 8.3 years in 1995 to 9.1 years in 2000, hence decreasing the period during which the product is protected by patents.

The greatest rise in development costs is caused by field trials. Although there are legal methods of parallel authorisation (data used for authorisation in one country can be used in another country) and derived authorisation, the cost increase in registration, according to the ECPA, can be attributed to a rise in efficacy data required by regulatory bodies. On the other hand, the companies themselves are doing more and more field trials voluntarily, because they want to adapt one active substance to the requirements of a variety of crops and pests.

As development costs are increasing, there is a tendency among the large players to focus on crop-protection products for five major crops in the world (i.e., maize, soya, cotton, rice, and wheat). Another tendency is to focus on the major markets. These are North and South America, China, and India. Europe is also a major market (23 per cent of total sales), but owing to differences in authorisation procedures between countries, the market is fragmented and access to the market is expensive. The focus on the major crops could mean the development of fewer products for crops such as potato and onion that are big in the Netherlands and Europe, but quite small on a global scale. The fragmented European market may also make companies reluctant to present new, often more environmentally friendly, pesticides for authorisation. Together, these developments may lead to a situation in which not enough pesticides will be available in Europe.

6 McDougall, 2003.

Maximum Residue Levels

It is often suggested that the EU and its member states use MRLs of pesticides as a non-tariff trade barrier to block imports of fruits, vegetables, and cereals, especially from Third World countries. The 'zero residue' attitude of environmental and consumer organisations, and in their wake the retail organisations, seems to point in that direction. In 1995, on behalf of its member states, the EU officially approved the 'Final Act Embodying the Results of the Uruguay Round of Trade Negotiations', including the Agreement on the Application of Sanitary and Phytosanitary Measures (the SPS Agreement). The Principles of the Agreement are scientific proof, international harmonisation, risk assessment, and transparency. This means that MRLs cannot be used to block imports, unless they are a – scientifically proven – health hazard.[7]

There are some problems, however. International harmonisation of MRLs, for instance, should be based on the Codex Alimentarius list of pesticides and on MRLs. The acceptance of the Codex MRLs remains low, especially in the EU and the US. Also, each country can set its own level of protection (i.e., a lower MRL than listed in the Codex Alimentarius). Officially, this higher level of protection must be scientifically justified, but in the absence of any 'agreed' evidence, countries are allowed to restrict imports anyway. This comes close to the Precautionary Principle, adopted by the EU, which states that restrictions are justified even if the full scientific proof of damage cannot be given.

Box 18.4 Capacity building

In 2001, the Dutch Agricultural Economics Research Institute, together with the Plant Protection Service and the Institute of Food Safety, conducted a research project into the effects of EU pesticide regulation on fresh food exports from developing countries into the EU. The study was done on behalf of the Dutch Ministries of Agriculture and of Development Cooperation, and focused on agricultural products from Zambia, Ghana, and Ethiopia. The study showed that, although some problems arise because pesticides are used for which there is no MRL in the EU, most problems arise because existing MRLs are exceeded. The reasons are twofold. Often, farmers do not know how to use the pesticides in a proper way, ending up with too much residue on the product. This can be prevented by training farmers to meet the requirements of good agricultural practice. The second reason why MRLs are often exceeded is the lack of adequate laboratory facilities to check residue levels.[8]

---

7 Mahé et al., 1999.
8 Buurma et al., 2002.

Another reason Third World countries are suspicious of the use of MRLs as trade barriers is EU Directive 2000/42/EC. This states that, for any pesticide in combination with a commodity, there should be an MRL based on either residue trials as part of the authorisation procedure or residue trials by the exporting country. If no MRL has been set for a product, it defaults to the detection limit (LOD, Limit of Determination) of the material, which means that no residue is allowed at all. For large imports such as banana and pineapple, exporting countries and/or importing firms have filed a dossier with residue trials. For small imports, such as papaya from Ghana, the drawing up of such a dossier often costs too much money. Hence, the EU regulations tend to hurt small growers and small crops.

MRLs and their role in international trade are another illustration of the limitations national governments face and will face in the near future. On the one hand, people do not want any residue left on their apples, pears, papayas, and peppers. This is an ethical norm inspired by the precautionary principle 'better safe than sorry'. On the other hand, national governments are bound by treaties which allow for MRLs based on toxicity tests, meaning that some products can have some residue. A possible way out of this dilemma is to let consumers make their own choice by informing them about residue levels. The accompanying bureaucracy and even the information itself might be seen as a non-tariff trade barrier.

Box 18.5 DDT and malaria

A special case is the use of pesticides that have been banned in the industrialised world. A well-known example is DDT, which is still used to fight malaria by eradicating the parasite's carrier Anopheles. In South Africa, the government recently started using it again in Maputoland, the region near the border with Mozambique. In the mid-1990s, the use of DDT was stopped because of its toxicity and its environmental effects. This led to the incidence of malaria soaring in this poor region from 8,800 in 1995 to 62,000 in 2000. Hundreds of people died in Maputoland. After the government decided to begin using DDT again, the incidence of malaria dropped to the 1995 levels.

The South African government is now urging neighbouring Mozambique to start using DDT so the incidence of malaria can be further lowered. But the country is unwilling to do so. As large parts of it are regularly flooded, the government fears that DDT will spread to water wells and farmers' fields. Apart from the effects on health, this might also affect the export of fish, fruits, and vegetables to other countries.

EUROPEAN DEVELOPMENTS

Harmonisation

In 1993, EU Directive 91/414 was adopted. It regulated the authorisation and use of pesticides within the European Community. The aim of the Directive has been to limit the damage caused by pesticides to ecological systems, and public and occupational health. To realise this goal, the EU compiled a list of active substances for use in pesticide formulas: the Annex I list. According to the European Directive, the scientific and technical assessment of active substances is done by national authorisation bodies on behalf of the European Commission (EC) and is based on a European protocol. The regulating bodies of other countries can act as 'peers' to review the work of their colleagues.

When an active substance is included in Annex I, it can then be used in pesticide production. The authorisation of the actual pesticide takes place at a national level, albeit with some harmonisation between the EU member states. The producer applies for authorisation in, for instance, Germany. A national body in that country then evaluates the pesticide and its proposed use (pesticide/crop combination), and decides whether to authorise its use. If the producer wants an authorisation for the same pesticide but for use on a different crop, he must apply again. The evaluation is based on the Uniform Principles, which will supposedly lead to a uniform process in the member states. When a producer applies for authorisation of the same pesticide/crop combination in another country (for instance, the Netherlands), the pesticide/crop combination can be authorised straightaway (i.e., without evaluation, based on the mutual recognition mechanism). Often, however, a country needs to take specific circumstances into account and decide for itself if, and under what circumstances, the pesticide can be used. In the Netherlands, for instance, the high groundwater level and the use of this water as drinking water is such a specific circumstance. These country-specific factors, as they are called, must be reported to the EC, which then asks the European Food Safety Authority (EFSA) to assess these factors. If they are not acknowledged, the country is forced to authorise the use of the pesticide for the crop(s) specified.

The EU Directive was not implemented straightaway, but contained a transitional regulation, originally for a period of ten years, until 2003. During that period, substances that had been authorised before 1993 were to be reviewed according to the new standards. For several reasons, the re-evaluation took more time than expected and the period was extended by five years to 2008. It is expected that the re-evaluation of the most important substances will be completed by 2005 or 2006. New substances developed

Box 18.6 Fewer chemical substances in the EU

In July 2003, 320 of the more than 800 substances in the EU had been withdrawn because the producers did not want to 'defend' them by providing the necessary information regarding the agricultural, health, and environmental aspects needed for the re-evaluation of the products. By December 2003, another 110 substances had disappeared from the market, bringing the total to 430. Of these substances, a few will still be allowed in some member states for essential uses. This means that, although the products were not defended, they can still be used for a few years owing to the lack of readily available alternatives.[9]

Of the 480 existing substances that were submitted for re-evaluation and subsequent extension of authorisation, 48 had been reviewed in March 2003. Of these, 26 were added to Annex I. By March 2003, 30 new substances had also been evaluated, of which 29 were admitted to Annex I. This meant that Annex I contained 55 approved substances in March 2003. It is intended that every substance on the list be re-evaluated after ten years. The slow evaluation process may create some problems in this respect. The fungicide imasalil, for instance, was added to the list in 1997, so it should be re-evaluated in 2007. However, not all substances developed before 1993 will have been re-evaluated at that time, thus increasing the work load of the national authorisation bodies.

after 1993 are reviewed on the basis of the criteria derived from EU Directive 91/414 before being added to Annex I.

Maximum Residue Levels

European policy on MRLs was developed 25 years ago with the 1976 Council Directive 76/895/EEC. It established MRLs for 43 active substances in selected fruits and vegetables. This Directive was later superseded by directives for cereals and cereal products (1986), for products of animal origin (1986), and – again – for products of plant origin, including fruits and vegetables. At the moment, more than 17,000 MRLs have been set, 133 of which concern residues of pesticide/crop combinations.

Establishing MRLs is a shared responsibility of member states and the EU. Usually, an MRL is set by a national authorisation body for pesticides and is then submitted to the EC. The dossier is reviewed by a committee of scientific experts, the Committee for Plant Health, Plant Protection, and Residues, which is now part of the European Food Safety Authority. Following this, a European MRL can be set. Although the directive dates back to 1976, there are still no European MRLs for most pesticide/crop combinations. This means that member states can still set their own national

[9] European Commission, 2003c.

MRLs. It is possible for the import of grapes from Greece into the Netherlands to be blocked because the residue levels on the grapes exceed the Dutch MRL.

To prevent this from happening in future, in March 2003, the EC adopted a proposal to harmonise the MRLs properly at the European level.[10] The European Food Safety Authority will be responsible for risk assessment, while the EC will set the actual MRLs. The harmonisation will take four or five years; until that time, trade within the EU can be obstructed by differences of opinion regarding MRLs.

Box 18.7 Procedure of establishing maximum residue levels

MRLs are based on a comparison of consumer intake and toxicological data about the substance. When an active substance or a pesticide/crop combination is evaluated by the national authorisation board for pesticides, a residue level is established in or on an agricultural crop. This is generally done in supervised trials whereby the pesticide is used under the specified conditions of Good Agricultural Practices (GAP). Based on consumer intake models, the daily intake of residues can be estimated for the population in general or for sub-populations, such as children or elderly people.

Data from toxicological tests on the pesticide itself are used to derive an Acceptable Daily Intake (ADI). Usually, this involves finding the highest dose that would produce no adverse effects over a lifetime of ingesting these substances, and then applying a safety factor. The ADI is often expressed in milligrammes per kilogramme of body weight, which means that it is lower for children. Nevertheless, the EU has issued a Directive (99/39 EC) which places extra restrictions on the use of pesticides in the production of food for infants and young children. Most producers of infant food already have a 'no residue' policy.

The estimated consumer intake based on the use of the pesticide under specified conditions (GAP) is compared with the ADI based on toxicological tests, including a safety factor. If the consumer intake is lower than the residue level established under GAP, this is set as the MRL. If the calculated intake is higher, the conditions of use of the pesticide must be changed so that the residue level in or on the product is reduced. If this is not possible, the pesticide cannot be used on that crop, and the MRL is set at zero.

MRLs in the EU are based on risk assessment. In that respect fruits, vegetables and other food products differ from drinking water. For that commodity, there is an 'ethical' standard of 0.1 microgrammes per litre for any pesticide and 0.5 microgrammes for the total pesticide count. The environmental movement would like to see this type of ethical standard also applied to food products. Food should not contain any detectable pesticide

[10] European Commission, 2003b.

Box 18.8 Contamination levels

MRLs are monitored both at a national level and by the EC, more specifically, by the Food and Veterinary Office. Results are published every year, with a year's delay, so that the results of 2001 were published in April 2003.[11] Of 46,000 samples of fruits, vegetables, and cereals analysed, 59 per cent contained no detectable residues at all, while 37 per cent of the samples contained residues at or below the maximum level. On average, 4 per cent of the samples exceeded the MRL, ranging from 1.3 to 9.1 per cent. The national programmes also revealed that 18 per cent of the samples analysed contained more than one pesticide residue, an increase compared with the year before (15 per cent).

In a special programme that is coordinated at the EU level, the same pesticides and products have been monitored since 1996. These products are apples, tomatoes, lettuce, strawberries, and table grapes. In 51 per cent of the samples analysed, no residues were detected; 47 per cent contained residues at or below the MRL; and the MRL was exceeded in 2.2 per cent of the samples. Lettuce and strawberries are the commodities in which MRLs are most often exceeded (3.9 per cent for lettuce and 3.3 per cent for strawberries). A small proportion of lettuce (0.06 per cent) and apples (0.07 per cent) showed excess residues of endosulfan and triazophos, respectively, which gave cause for concern about acute exposure. Triazophon was withdrawn last July and endosulfan is currently under review, so it may or may not be included in the Annex I list.

residues. A few times per year, members of this movement campaign against the acceptance of pesticide residues in food. Recently, for instance, the 'fruits in school' programme was attacked because organic fruits were not used, thus exposing school children to 'poison'. Some retail organisations have picked up on this and have announced that, in the near future, they will sell only fruits and vegetables containing no pesticide residues at all.

NATIONAL DEVELOPMENTS IN THE NETHERLANDS

As mentioned above, Dutch agriculture is a relatively heavy user of pesticides. To reduce the dependence on synthetic chemicals, the Dutch government adopted a Long-Term Plan for Crop Protection in 1993. The goals of the Long-Term Plan were threefold: reduction of the dependence of agriculture on synthetic pesticides; reduction of the amount of pesticides used (in kilogrammes of active substance) by half; and reduction of the release of pesticides into the environment. An important instrument for realising these goals was the sharpening of the criteria for the authorisation and use of pesticides. For instance, the persistence of substances and their potential for

[11] European Commission, 2003a.

accumulation in the food chain became more important factors in the authorisation process. Also, the release of substances into ground- and surface water was to be limited. The criteria for authorisation and use were based on the European Directive 91/414, which was adopted at the same time.

Regarding existing pesticides, the government, the agricultural sector, and the pesticide industry agreed that several indispensable substances would remain available until the year 2000. This so-called 'canalisation procedure' was needed, because farmers' organisations claimed there were no alternative substances for methods of crop protection available at the time. The farmers' and industry's side of the bargain was that they would submit the substances for a partial re-evaluation in order to limit their uses to ways that were not harmful to the environment or to human health.

As the end date (2000) of the first Long-Term Plan for Crop Protection came near, it became apparent that the main target of less dependence on synthetic pesticides had not been realised. Although the use of synthetic pesticides had been halved over the previous ten years, farmers still needed them for crop protection. With that in mind, a new Long-Term Plan for Crop Protection was made. Instead of reducing dependence on synthetic pesticides and halving their use (which were important goals of the first Long-Term Plan), the second plan focused on the careful application of these substances, preferably as part of an IPM system. In connection with that goal, a system was proposed whereby farmers would have to be certified before being allowed to use pesticides.

The political debate regarding the second Long-Term Plan for Crop Protection was severely hampered by juridical and political disputes over the indispensable pesticides that were supposed to be made obsolete by 2000. The environmental movement and a large part of the public were of the opinion that farmers had had enough time to change their practices and use less harmful pesticides. The farmers, on the other hand, claimed that they were at a disadvantage compared to their competitors elsewhere in the EU if they were not allowed to use these indispensable pesticides.

To break the deadlock, the Dutch government proposed a law in early 2000 which would allow the continued use of a dozen agriculturally indispensable, but formally forbidden, pesticides. The proposal was published just before the growing season, but as the discussion in Parliament took some time, farmers already started using these pesticides illegally. The government turned a blind eye, but the environmental movement did not; it took the farmers to court and came out victorious in several cases.

The government proposal was eventually adopted and it became law in February 2001 under the provision that, for the indispensable substances, a full dossier for authorisation would be submitted before July 2001. If not, the provisional authorisation would be withdrawn. If a complete dossier was submitted, the substances could be used until July 2002. Before that date, the Authorisation Board for Pesticides would evaluate the dossiers and decide on the official authorisation of the pesticides. For several substances, no dossier was submitted, resulting in their formal ban in July 2001. This angered the farmers because, again, the prohibition came in the middle of the season, and it concerned substances and pesticides that were freely available in neighbouring European countries. A number of farmers drove to the adjacent Belgium to buy pesticides that were illegal in the Netherlands for use on their crops. Some even imported pesticides from countries as far away as Spain.

As the debate about indispensable pesticides raged on, little progress was made in respect of the new Long-Term Plan for Crop Protection. In 2002, the new Minister of Agriculture convinced farmers, industry, and environmental organisations to sign an agreement aimed at 'sustainable' crop protection. The parties agreed to realise existing environmental goals in an economically sound manner (i.e., with respect for the economic position of Dutch farmers). More parties became involved and, in the spring of 2003, the agreement (the covenant) was formally signed. An essential part of the agreement was the mutual trust between partners. To gain that trust, the government developed a procedure to solve the problem of the indispensable substances that had formally lost their authorisation. For a majority of pesticide applications, the problems were solved before the summer of 2003. This, however, did not mean the agricultural community was given a free rein. It was agreed in the covenant that the number of pesticides with a provisional authorisation should be halved every year until 2006. From 2004, all farmers are obliged to keep track of their use of pesticides.

LIMITATIONS TO NATIONAL POLICY

The Effects of Globalisation

The pressure on national crop-protection policy will increase further when, as a result of the Doha rounds, markets have to open up to agricultural products from around the world. Although many Third World countries follow the US or the larger European countries, there is no harmonised authorisation procedure for pesticides worldwide. This makes it difficult to keep out products which have been grown using pesticides that have not been authorised in the EU. If these products are cheaper because of that, it is a form of unfair competition, this time with European farmers in the role of the victim.

The same Doha rounds of trade negotiations will make it increasingly difficult to use public health as a non-tariff trade barrier for food products. On the other hand, the public in industrialised countries demands food products that are 'free of poison.' It will probably be some time before MRLs have been set on worldwide scale (Codex Alimentarius), and it will probably take more time before these MRLs have been adopted by the partners in the trade negotiations. Nevertheless, in the not-too-distant future, an ethical standard (no detectable level of residue) will no longer be accepted as a reason to block imports of food products. In a related case, concerning hormone-treated beef, the EU has already been punished by the WTO. An appeal to the precautionary principle did not help. Another possibility is that future exporters of food, such as China, will retaliate by blocking imports from Europe. Again, this could lead to a problem for the government with, on the one hand, an obligation to conform to international agreements, based on scientific evidence, and, on the other hand, a public that does not want 'poison' on vegetables and fruits.

The Effects of European Harmonisation

The first Long-Term Plan of the Dutch government aimed at reducing the environmental and health effects of crop protection through an authorisation procedure with criteria that were not used in most of the other EU member states. This was motivated by the fact that (a) the criteria would be harmonised in the near future, and (b) this was required by the specific circumstances in the Netherlands, in particular a high use of pesticides and an environment that is vulnerable to pesticide pollution owing to high groundwater levels, the interconnectedness of surface waters, and relatively small conservation areas. However sound these environmental reasons may be, rushing ahead of the other member states created a difficult situation for the government of the Netherlands. If Dutch farmers are not allowed to use the range of pesticides available to their competitors in other European countries, they are at a disadvantage. On the other hand, aiming for agricultural competitiveness by authorising pesticides that are authorised elsewhere is harmful to environmental quality.

A possible solution is to develop a type of authorisation that is neither national nor supranational, but that is based on the actual physical characteristics of regions. Therefore, instead of depending on lines on a map, authorisation would be based on groundwater level, soil type, average rain fall, biodiversity, and, naturally, type of crop and residue tolerance. This could result in some pesticide/crop combinations being allowed, for instance, on claysoil along the river Rhine in the Netherlands, the Po River in Italy, and the Oder in Germany/Poland, but not on the sandy soils in each of those countries. With this type of fine-tuning, it would be difficult to enforce the regulations. Farmers who wanted to break the rules to gain financial advantage would probably get away with it, because the pesticides they would not be allowed to use on their soil would be readily available elsewhere. Certification and some form of regulation in the production chain might be a way out of this problem. Nevertheless, the European Court of Justice might not allow this type of regional authorisation. Although not quite the same, it has rejected the Dutch manure policy, which deviated from the European policy on the grounds of local environmental differences.

Another problem is that authorisation procedures would become too fragmented, thus creating new barriers for industry in submitting new pesticide/product combinations for authorisation. As mentioned above, the development and production of new pesticides is mainly done by a few large companies operating on a global market. They tend to focus their development and marketing activities not only on five large crops, but also on three or four large producing regions: North and South America, India, and China.

To meet these developments, it is suggested that Europe will be divided into three 'climate zones' (North, Middle, and South), each with its own criteria for authorisation. Although this might stimulate industry to develop pesticides for the European market, it goes against the regional variation in physical characteristics and, hence, might lead to a deterioration of environmental quality. Whatever the outcome, authorisation in zones or on the basis of physical characteristics, the capacity of governments to weigh environmental quality and health against agricultural competitiveness will be reduced.

IN SEARCH OF FEASIBLE POLICY OPTIONS

The use of synthetic pesticides in agriculture has effects on ecology and human health. It can lead to water pollution, loss of biodiversity, and acute and/or chronic toxic effects on man. On the other hand, synthetic pesticides help to protect crops, both in the field and in the barn, and are important tools in enhancing agricultural productivity (i.e., the yield per hectare and the yield per unit of labour). It is unlikely that the world could sustain over six billion people without the use of pesticides and fertilisers. The agricultural sector and governments are faced with the task of maintaining, or preferably increasing, the level of crop protection, while at the same time reducing the effects on ecology and human health.

A possible solution is the development and use of IPM strategies based on biological control, cultural practice, genetic engineering, and smart synthetic pesticides for chemical control of plagues and diseases. Although it is already widely used in agricultural systems as diverse as rice growing in Indonesia and tomato growing in Dutch glasshouses, much more knowledge is needed to develop IPM for other crops. This ranges from fundamental knowledge of agro-ecological systems and food webs to practical knowledge of how to apply pheromones or use predatory spiders effectively in the field.

With regard to crop-protection policy, European and other governments rely heavily on a command-and-control approach consisting of extensive authorisation procedures, detailed instructions for use, and inspection and enforcement. This approach does not leave much room to stimulate the development of IPM strategies, as these strategies are – by nature – very specific: they have to be finely tuned to agro-ecological circumstances, type of crop, soil, and other factors. An extensive authorisation procedure can further block the development and marketing of specialised synthetic pesticides (for instance, pheromones that are used only in small quantities under special circumstances). On the other hand, international developments at both a global and a European level do not leave much room for a crop-protection policy based on the command-and-control approach. Extensive authorisation procedures, among other things, have led to a concentration of development, production, and marketing of synthetic pesticides. High investment costs force the six large global players to focus their efforts on large crops (cotton, rice, maize, soya, and wheat) and on major markets (North and South America, India, and China). In general, it seems likely that fewer active substances will be developed for crops that are big in Europe but small on a world scale, such as potato and sugar beet. Although Europe is supposed to be one large market, it is fragmented in marketing terms.

A second international development which will limit the command-and-control approach is the world trade system and its focus on eliminating trade barriers, including non-tariff ones such as the safeguarding of public health and concern for the environment. The SPS Agreement, which the EU member states have signed, stipulates that MRLs of pesticides will be harmonised through the Codex Alimentarius. This could mean that agricultural products will have to be accepted that contain a (higher) pesticide residue level than is now acceptable in Europe or in the Netherlands. It also means that products will have to be accepted containing pesticides that farmers in Europe are not allowed to use because of their possible effects on the environment.

A third development that affects national crop-protection policy is the ongoing European harmonisation. Although limited to 'active substances', the country-specific factors on the basis of which member states can refuse authorisation of pesticides will have to be scientifically assessed by the European Food Safety Authority.

Although the opportunities for national authorisation of pesticides are limited, this does not mean that there is no room for a national crop-protection policy, provided that governments are willing to step out of the authorisation-regulation-inspection-enforcement loop and develop other ways to achieve their goals. With regard to authorisation, so-called 'light' procedures could be developed for active substances that have been authorised for one crop (for instance, one of the big five) but can also be used on other crops, albeit with a small change in the recipe. A 'light' procedure can be – and sometimes is – used for substances that are identical or almost identical in structure and function with natural substances, such as synthetic sex pheromones.

Another possibility is to develop a prescription system as part of an authorisation procedure to regulate the use of certain hazardous pesticides. Like a medical doctor who prescribes medicines for a patient who is ill, a qualified phytopathologist could prescribe certain pesticides and supervise the way they are used. Other less harmful pesticides could be sold 'over the counter', similar to the self-help medicines sold in the chemist's shop.

Beyond the regulation-enforcement loop is the option of promoting self-regulation, for instance, by challenging farmers to adopt IPM systems that combine ecological knowledge and risk assessment of pests with the use of specified synthetic chemicals and biological control agents. The role of the government could be to facilitate the research and development of IPM systems, and the dissemination of the knowledge thus gained. In reward, farmers could be certified either by the government or – even better – by their clients, the food industry, and/or the retail organisations.[12]

An example of self-regulation is EUREPGAP, developed by the Euro Retailer Produce Working Group (EUREP). This group developed criteria for GAP (including the use of pesticides) for fruits and vegetables, and translated these into a certification procedure for growers. In December 2002, almost 4,000 growers in Europe had been certified, with more than half of these in the Netherlands. These certification systems do not have to be limited to Dutch or European farmers, but can also be extended to farmers in developing countries.

Still another option, which has not yet been fully developed, is to promote transparency by informing consumers about the use of pesticides during the production of both home-grown and imported agricultural products. As information and communication technology develops, and both growers and food-processing companies introduce quality systems, it will be possible to trace products in the shop back to their origins almost on an individual basis. Governments can promote the availability of detailed product information in shops and certify inspection procedures to guarantee that the information is correct. As not all of the information can be put on the packaging, retail shops could make it available via internet or information screens in the shop.

---

[12] Brouwer and Bijman, 2001.

Although international developments limit their options, governments can still develop and implement a crop-protection policy which takes into account local economic and ecological circumstances. In order to do this, they must rely less heavily on authorisation, regulation, and enforcement, and develop new policies to promote IPM, certification procedures, and transparency of production.

REFERENCES

Agrow (2003), 'Global agrochemical sales flat in 2002', *Worldcrop Protection News*, February, Richmond: PJB Publications.

Brouwer, F. and J. Bijman (2001), *Dynamics in Crop Protection, Agriculture and the Food Chain in Europe*, report 3.01.08, The Hague: Agricultural Economics Research Institute (LEI).

Buurma J., D. Boselie, J. de Jager, A. Smelt, and E. Muller (2002), *Impacts of EU Pesticide Legislation on Fresh Food Trade between the South and the EU*, The Hague: Agricultural Economics Research Institute (LEI).

Carson, R. (1962), *Silent Spring*, New York: Houghton Mifflin Company.

European Commission (2003a), 'Monitoring of pesticide residues in products of plant origin in the European Union, Norway, Iceland and Liechtenstein', Sanco/20/03, March, *http://europa.eu.int/comm/food/fs/inspections/fnaoi/reports/annual_eu/monrep_2001_en.pdf*.

European Commission (2003b), *Pesticides: Consumer Protection to be Boosted via Harmonisation of Maximum Residue Levels*, press release (IP/03/383), March.

European Commission (2003c), *Commission Close to Completion of Pesticide Review: 110 Additional Substances to be Withdrawn*, press release (IP/03/957), July.

Mahé, L., F. Ortalo-Magné, and J. Doussin (1999), 'Food safety and sustainability: The role of agents, the state and international organizations', in G. Meester et al. (eds), *Plants and Politics*, Wageningen: Wageningen Pers.

McDougall, P. (2003), 'The cost of new agrochemical product discovery, development and registration in 1995 and 2000', European Crop Protection Association, *http://www.ecpa.be/library/reports.html*.

Pelerents, C. (1989), 'Pesticiden en alternatieve bestrijdingsmethoden' ('Pesticides and alternative crop-protection methods'), in M. De Coster (ed.), *Milieuzorg in de landbouw (Environmental Management in Agriculture)*, Kapellen: DNB/Pelckmans.

Pinstrup-Anderson, P. (1999), 'The future world food situation and the role of plant diseases', *http://www/apsnet.org*, Glenn Anderson Lecture, Canadian Phytopathological Society.

Puijker, L., E. Beerendonk, and C. Van Beek (2001), *Door drinkwaterbedrijven gemaakte kosten als gevolg van het bestrijdingsmiddelengebruik: Inventarisatie over de periode 1991-2000 (Costs incurred by drinking-water companies resulting from the use of pesticides: Assessment of the period 1991-2000)*, Nieuwegein: KIWA.

19. Free Trade in Agricultural Products and the Environment

Henk Massink, Gerard van Dijk, Niek Hazendonk, and Jan van Vliet[1]

SUMMARY

We address the impact of a liberalised trade in agricultural products on the Dutch environment. The characteristics of the Dutch physical environment, biodiversity, and landscapes are described. Assumptions are specified and the main consequences of free global trade for Dutch farming are discussed: an increase in dairy farming, few changes in arable farming, and a decline in intensive livestock farming. Together with a significant intensification and scale enlargement of dairy production, this will lead to a moderately positive change in the physical environment, reduced biodiversity, and changed landscapes. While it is hard to assess causal relations between free trade, agricultural practices, and government policy, several – both positive and negative – effects of national policy on the Dutch environment can be identified. To reduce the negative effects, government policies should focus on agri-environmental management, the creation of landscape features, diversification of agriculture, and spatial planning. In conclusion, the government still has possibilities to conduct a domestic environmental policy under a free-trade regime.

INTRODUCTION

Since 1986, the trade in agricultural products has been included in negotiations on world trade (the General Agreement on Tariffs and Trade or

[1] This contribution is based on Van Dijk et al., 2003 and (the assumptions in) Massink and Meester, 2002. The present chapter does not represent any official position taken by the Dutch Ministry of Agriculture, Nature and Food Quality.

GATT and the World Trade Organization or WTO since 1995). This resulted in the 1994 Agreement on Agriculture. It includes a gradual decrease of the support for agriculture, while trade is liberalised further. Various agricultural sub-sectors with a high degree of protection keep a close eye on the trade-liberalisation process. In 2002, the then Dutch Ministry of Agriculture, Nature Management, and Fisheries (now Agriculture, Nature and Food Quality) published a policy document outlining what Dutch agriculture might come to look like following various forms of trade liberalisation.[2] The projected outcome of total trade liberalisation provides a particularly clear picture of the situation if all support for agriculture disappears. The basic template of the research underlying the 2002 study was as follows. Policy shocks, which were stylised into a scenario of total trade liberalisation, resulted in changes of costs and revenues. To cope with these changes in the economic environment, 18 strategic options for farms were mentioned.[3] The most profitable strategies were selected for each of those farm types. The use of these strategies would change the structure of Dutch agriculture, including changes in land use. One of the striking results of this exercise was the finding that, with total trade liberalisation, intensive dairy farming is likely to expand considerably. Assuming no changes in prevailing policy measures to mitigate the adverse effects of agriculture, changes in the structure of Dutch agriculture would be the only source of change in environmental impacts due to trade liberalisation.

The question was then prompted what the consequences of such a scenario might be for the environment, including biodiversity and the landscape. To enable a clear analysis, it is important to maintain a sharp distinction between the physical environment, biodiversity, and the landscape, since trade liberalisation can affect these three areas in different ways. In order to assess the effect of trade liberalisation, it is first of all important to define a baseline. Our baseline was the current situation, including the effects of prevailing policies that are assumed to be continued (for example, in the areas of health and animal welfare). The difference between this baseline and the expected developments in the agricultural sector as a result of trade liberalisation was

---

2 Massink and Meester, 2002.

3 The strategies were quitting, continuing unchanged, continuing unchanged with additional external family income, land exchange between livestock farmers and arable farmers, alliances between arable and livestock farmers, larger production units, intensification (crops/livestock farming), high-quality guarantees (certification), diversification, developing non-food crops, switching to organic farming, extensified farming combined with additional sources of income, switching to organic feed crops, small-scale production (especially for human health care purposes), diversifying the farm (conservation, recreation, health care, etc.), strengthening integrated production chains, and internationalisation of the processing industry. Distinguished types of farming were arable, dairy, beef, lamb, pig, and poultry farming.

used as an input for estimating the consequences of free trade for the environment.

The structure of this chapter is as follows. The current situation of the environment in relation to agriculture and its historical background are discussed in section 2. Changes in land use are described in section 3.[4] The expected changes to the environment are dealt with in section 4. The main points of possible changes on a European scale are also addressed. The consequences for the physical environment can be described for transport, use of energy, air quality, soil quality, water, waste, and materials, but we focused on nitrogen, phosphate, and plant-protection substances. Biodiversity is described using three main categories: breeding meadow birds, wintering water birds, and the flora and fauna of landscape features (including ditches). Landscape changes are discussed in terms of scale, landscape features, land use, 'urbanisation', and disused farm buildings. In section 5, the possibilities of controlling the developments are discussed, and some conclusions are presented in section 6.

THE ENVIRONMENT IN RELATION TO AGRICULTURE

In this section, we deal briefly with the current state of the physical environment, biodiversity, and the landscape in the Netherlands. The information on the physical environment is based on reports by the RIVM.[5]

The Physical Environment

The impact of agriculture on the physical environment mainly relates to emissions of nitrate, phosphate, ammonia, and plant-protection substances. The Netherlands has the highest degree of overfertilisation with nitrogen and phosphate of all countries in the European Union (EU). The minimum standard according to the Nitrate Directive of 50 mg of nitrate per litre in the upper groundwater is not being met. In 2000, the nitrate content of sandy soils was over twice this level. There has been some improvement, however, because the average content decreased from 150 mg/l in the years 1992-1995 to 125 mg/l in the years 1997-2000; the number of cases where the limit of 50 mg was exceeded dropped from 90 to 80 per cent of cases measured.

Phosphate concentration in Dutch agricultural land is high. Under current and proposed loss standards for phosphate (20 and 25 kg per ha), the

---

4 As regards the changes in land use, we used Massink and Meester (2002) as our basis. To provide a more in-depth analysis as to the effects of trade liberalisation, we drew on: Berkhout et al., 2002; De Bont et al., 2003a; De Bont et al., 2003b; Van Berkum et al., 2002.

5 RIVM, 2002, 2003.

accumulation and leaching of phosphate will continue. The long-term target of the Dutch government, such as formulated in the 'National Environmental Policy Plan 4' (NEPP4) is to reduce the phosphate surplus to 1 kg/ha by 2030.

Since 1980, the atmospheric impact of nitrogen (including from ammonia) has changed little, but it has been decreasing slightly in the last few years. The average deposition in the Dutch countryside was 2,900 mol/ha in 1990 and 2,300 mol/ha in 2001. The NEPP4 target for 2010 is 1,550 mol/ha. However, the deposition of acidifying substances has halved since 1980. This was mainly due to reduced sulphur emissions. Since the deposition of 45 per cent of potential acidifying substances and 35 per cent of nitrogen comes from surrounding countries, the effect partly depends on developments in those countries. It is estimated that the deposition of nitrogen and acidifying substances will on average be 0-20 per cent above the NEPP4 deposition targets in 2010.

The potentially negative effects of plant-protection substances have been reduced in the past four years by about 30 per cent in the soil and by 50 per cent in the surface water. Calculated over the past 15 years, the negative effects have been reduced by as much as about 70 per cent. Expressed in terms of emissions into soil, air, and surface water, the reduction in the past 15 years was some 75 per cent, 50 per cent, and 80 per cent, respectively. However, quality targets for surface water are still not being achieved. In the 'Policy Agreement Calendar for Plant Protection Substances', which was completed in 2003, a reduction of 95 per cent in effects on the environment is the target for agriculture as a whole, compared with the situation in 1998.

In 2000, about 3 per cent of the total desiccated area had recovered, whereas the target was 25 per cent. With the implementation of established policy, the total acreage of nature areas protected against manure pollution, acidification, and desiccation may increase to about 20 per cent by 2010.

Biodiversity

Wild flora and fauna in Europe depend on a large number of different habitats. Natural open habitats have almost disappeared in Europe, but many species continue to be found in man-made habitats such as farmed grasslands, which are in fact a substitute for natural habitats. Therefore, a large part of the flora and fauna of open habitats now depends on so-called semi-natural grasslands, which are managed by man but rich in species thanks to low or absent fertilisation. In other words, agriculture has not only eroded ecological values but also offered alternative habitats to a great number of species. However, these semi-natural habitats have come under serious pressure in the course of the twentieth century, owing to intensification of agriculture, especially fertilisation, but also drainage, early mowing, conversion of grassland to arable land, etc.

Agricultural areas of high ecological value in the EU (so-called High Nature Value areas) comprise the following categories: semi-natural habitats (mainly semi-natural grassland but also heath land and the like) with numerous wild species; areas that are no longer semi-natural but that are important to breeding birds (meadow birds, steppe birds) or migratory birds (geese, ducks, swans, waders, cranes, etc.); areas (inter alia bocage) rich in (semi-)natural elements (hedges, bushes, groves, ditches, etc.); and complex landscapes (including Iberian dehesas, montados, and old olive groves).[6] In the Netherlands, in spite of its small area and very intensive agriculture, we still find some internationally important categories, namely, meadow-bird areas and areas hosting migratory or wintering water birds. In addition, many plant and animal species are dependent on landscape features such as hedges and groves as well as ditches and their banks, in the latter case in particular in places where water quality is still high. Semi-natural grassland has practically disappeared from the farmland and remains only in nature reserves. The total of semi-natural grasslands (old and under restoration) today is approximately 30,000 ha, which is only 1.5 per cent of the agricultural landscape.[7]

In order to give some idea of the importance of the Dutch meadow-bird areas, we present in Table 19.1 some figures concerning the proportions of the Dutch populations in relation to those in the EU and Europe as a whole.[8] The number of black-tailed godwits has almost halved in the past ten years.[9] In other words, the current situation is already showing a decline. Netherlands and the decline of a number of other species. Although there are indications

Table 19.1 Importance of the Netherlands to a number of bird species

Species	The Netherlands relative to the EU (%)	The Netherlands relative to Europe (%)
Black-tailed godwit	86	48
Oyster catcher	50	34
Shoveler	40	10
Lapwing	29	3
Redshank	24	6
Curlew	6	4

6 This is a provisional specification. The European Environment Agency is working on HNV indicators.

7 RIVM et al., 2003.

8 Beintema et al., 1995.

9 SOVON, 2002.

that the Dutch agri-environmental management programme does not work as it should,[10] the government tries to stop the decline. In addition, there is now large-scale nest protection by volunteers throughout the country. The total area covered by agri-environmental management agreements was 87,900 ha in 2001,[11] although not all of this area was devoted to meadow-bird management. In addition, we should mention the management of meadow-bird reserves where sometimes spectacular densities occur. The gap is likely to become larger between nature reserves and areas under management agreements, on the one hand, and other agricultural areas, on the other hand.

The importance of the Netherlands to wintering and migratory water birds may be illustrated by the following figures: three quarters or more of the North-Western European flyway population of Bewick's swans, pink-footed geese, white-fronted geese and barnacle geese stop over in the Netherlands during the optimal time of the year.[12] For the greylag and brent geese, the figures are on average 47 and 37 per cent. It is clear that the size of the flyway populations largely depends on the Netherlands. In contrast to meadow birds, wintering geese are thriving. Apart from the species mentioned, there are a number of others for which the Netherlands is important outside the breeding season, including the wigeon, golden plover, and lapwing.

The arable land is less spectacular from a biodiversity point of view. The diversity in arable weeds is currently small and is still declining.[13] Nevertheless, a number of red-list species still occur,[14] and the management of some (parts of) arable land on 'plaggen soils' (artificially raised with heathland sods and sheep dung) focuses on arable weeds. The arable-farming area accommodates a number of bird species that are much less common elsewhere, such as the blue-headed wagtail, the rare Montagu's harrier, the corncrake, the skylark, and the meadow pipit.

By set-aside, ecological values such as arable weeds and food for birds of prey can be stimulated locally. This largely depends on EU agricultural policy. Special measures for protecting arable weeds are also possible in the framework of management agreements. As regards the importance of landscape features (ditches and their edges, hedges, bushes, pools), no quantitative data are currently available on their relative importance to plant and animal species. However, some calculations for the agricultural landscape as a whole are available: approximately 36 per cent of the regularly monitored vascular plant species and 34 per cent of the regularly monitored

[10] Kleijn et al., 2001.
[11] RIVM and DLO, 2002.
[12] Koffijberg et al., 1998.
[13] Hall et al., 1998.
[14] These are official lists of animal and plant species under threat.

fauna species in the Netherlands occur in – and partly depend on – agri-ecosystems.[15]

As regards the landscape features, the decline also continues. For example, the province of Zuid-Holland found that there had been a decline between 1976 and 1995 of nearly half the ecological value of ditch vegetation as well as ditch-bank vegetation in agricultural areas.[16] However, in the last few years, the decline in ditch-bank vegetation has stopped and the quality of ditch vegetation is clearly recovering.[17] The Netherlands has a relatively large concentration of 'wet elements'. According to a recent estimate, the overall length of ditches is in the order of 160,000 km, though another estimate doubled that figure.[18] Hedges, scattered groves, and rows of alders are important features for the fauna. Many of those features have disappeared in the past few decades. However, well-developed, small-scale landscapes still exist in some parts of the country, and contribute greatly to the countryside experience.

Although Dutch agriculture is very intensive, and though a large part of the biodiversity of Europe's agricultural areas depends on areas with extensive agriculture, the Dutch countryside has values that do not occur elsewhere to the same degree: meadow birds, wintering water birds, and an extremely large concentration of ditches.

Landscape

Traditionally, the Dutch landscape was very varied and reflected the geographical situation (soil, hydrology, reclamation history). Two extremes were the low sandy areas and open polders. The sandy-soil areas were characterised by a complex of nature areas (mainly heath and woods), as well as mixed farming. Fields were found in areas of arable land on plaggen soils and sandy ridges, and in transition areas from high to low ground. The landscape was mostly small scale, with numerous hedges, alder rows, and groves. At the other extreme were the open grasslands of the 'Green Heart' (low grassland polders surrounded by urban areas in the West of the country), the provinces of Noord-Holland and Friesland, and the areas bordering the former Zuiderzee. Many of these areas are rich in water, with ditches up to several metres wide.

The centuries-old parcellation and settlement patterns still reflect the country's reclamation history. In broad outline, this is still true for the whole country, but in sandy areas re-allotment projects have sometimes drastically

[15] OECD, forthcoming.
[16] Province of Zuid-Holland, 1996.
[17] Province of Zuid-Holland, 2000.
[18] Dijkstra et al., 2003; Zonderwijk, 1971.

changed ditch patterns. In polders with broad ditches, this has happened less. In the river areas, there was the contrast between backlands and border levees. The arable farming areas in the North and the South-West had a completely different character again, while the South of the Province of Limburg is in a class of its own. Many of the features mentioned above were levelled in the course of the twentieth century.[19] Small-scale areas became more open as a result of the disappearance of many landscape features. Rye fields were replaced by maize, and intensive livestock farming increased very much. In contrast, open areas have become less open owing to farm relocations (with plantations), urbanisation, intersecting new roads, etc. In spite of this, there are still clear differences, and these differences can be preserved and strengthened by means of targeted spatial planning and management policies (management agreements for landscape features). Geographic and cultural values are part of countryside values (agricultural cultivation, parcellation and settlement patterns, etc.) and of the way that people view the countryside (peacefulness, spaciousness, darkness, etc.).

The landscape changes of the past few decades were the result of intensification and increases in scale (enlargement of farming areas, but also of parcels and plots). The consequences of these processes, which mainly involved re-allotment, may be characterised as follows: less peacefulness, increased noise, and 'light pollution' in country areas; disappearance of historical parcel and plot borders; straightening of irregular shapes; levelling of land surfaces; disruption of soil profiles; and changes in infrastructure. The increase in scale has also led to changes in historical farm buildings in the countryside. Of the 192,000 farms in 1938, only 91,000 were left in 2000, and the decline continues unabated.[20] When buildings are taken over by city dwellers, the buildings themselves as well as the arrangement of the property change (city gardens, city plant layouts).

From a European perspective, its man-made character and the role of water are the most striking features of the Dutch landscape. And it is precisely the cultural landscapes that are affected most by changes in agriculture. Four groups of Dutch landscapes can be distinguished on the basis of international characteristics.[21] They are peaty grassland polders (peat reclamation areas), the old lake reclamation areas, the old marine clay polders, and (former) drift-sand areas. Only the first three are affected by agriculture and increases in scale. More than half of the North-Western European area of these types of landscapes is situated within the Netherlands. The old marine clay polders have been affected more strongly than have water reclamation areas and peat

---

[19] RPD, 2000.

[20] SHBO, 2001.

[21] In the analysis, North-Western Europe was used as the reference framework (Farjon et al., 2001).

reclamation areas. The old marine clay polders and the peat reclamation areas are landscapes that are vulnerable to the effects of increases in scale. Irregular block parcellations and long-strip parcellations have come under pressure.

The other characteristic landscapes of international significance are also situated mainly in the lowest part of the Netherlands: the Waddenzee, the coastal dunes, the salty to brackish estuaries, bogs, salt marshes, Zuiderzee polders, river areas, and the young marine clay polders. The agricultural areas are in the Zuiderzee polders, the river area, and the young marine clay polders. The negative effects of increases in scale will mainly occur in the river areas. The bog reclamation areas are the only agricultural landscapes in the higher part of the Netherlands that have been judged to be characteristic internationally. However, increases in scale have only limited effects on the structures of those landscapes.

IMPACT OF FREE TRADE ON AGRICULTURE

This section serves as a basis for the discussion of effects in the next section. In this context, it is important to note that as a result of current policies (for example, regarding fertilisers and minerals, plant protection, and animal welfare and animal health), as well as because of economic developments, there are already significant effects on the environment.[22]

Before stipulating the factors that served as inputs in our assessment of effects on the physical environment, biodiversity, and landscape, described in the next section, we mention some general aspects of trade liberalisation:[23]

- The efficiency of Dutch agriculture will grant it a strong market position under trade liberalisation.
- If trade barriers are removed on a global scale, Dutch agricultural products will gain access to more markets than at present.
- World market prices will no longer be influenced downwards as a result of supportive measures, which is likely to result in considerably higher world market prices.
- Dutch agricultural products that do not receive price or income support (as applies to two-thirds of overall production value, including horticulture, intensive livestock farming, seed and consumption potatoes, and onions) will experience more advantages than disadvantages from trade

---

[22] Furthermore, we assumed that the request for derogation by the Netherlands regarding the Nitrates Directive would, for the time being, result in a ceiling for the use of livestock manure on grassland of 250 kg/ha of nitrogen.

[23] Massink and Meester, 2002.

liberalisation (though these sectors are also subject to the effects of changes in land prices, animal feed, and manure contracts).

- A large proportion of the two million hectares of highly fertile and easy-tillable farmland will not find other economically attractive applications, even under total trade liberalisation.
- Removing price and income support will reduce the economic value of capital tied up in land and quotas, meaning that the capital costs of farming will decrease.

The Development of Dairy Farming

In a study on the developments in dairy farming and related developments in other sectors, one scenario was based on the abolition of milk quotas, a fall in milk prices of 30 per cent, and partial compensation (no compensation for production above the historical milk quota).[24] This scenario served as the basis for the analysis of the environmental consequences. The relevant quantitative effects are summarised in Table 19.2.

Table 19.2 Estimated effects of free trade on dairy farming

Product / area	Increase	Decrease
Dairy cows	306,300 (21%)	
Beef calves		63,000 (9%)
Beef cattle		146,000 ABU[25] (28%)
Milk	31.9%	
Grassland area (ha)	34,000	
Maize area (ha)	21,000	
Roughage crop area (Total, ha)	55,000	

The most significant impacts on the environment due to changes in the dairy sector (grasslands) are those on biodiversity and the landscape. Roughly 75 per cent of production will take place on a very large scale, with zero-grazed cattle. There is, however, still some uncertainty about the effects on the management of the land. The remaining quarter of the production will take place at land-tied farms with about 2 livestock units per ha and farmers having their own roughage crop production and outdoor grazing. This situation is to be found, among other places, in the buffer zones around the 'National Ecological Network', in protected landscapes, and in the 'Green

[24] Berkhout et al., 2002.
[25] Adult Bovine Unit.

Heart'.[26] This suggests that the location of this type of farm can be controlled to a certain extent. The increase in the maize area of 21,000 ha mentioned earlier will also affect biodiversity and the landscape, especially if maize cultivation replaces (existing) grassland.

The Development of Arable Farming

Trade liberalisation is estimated to lead to starch potatoes (48,000 ha in 2002) no longer being grown in the Netherlands.[27] As a result of strong competition from neighbouring countries, the size of the food-potato area will also decrease. Sugar beet and cereals will be maintained in the crop plan. The horticultural area will increase, although the possibilities are limited from a market perspective (at most 15,000 ha). A reduction of 11,000 ha is expected in the sugar beet area; a reduction in the food-potato area of 7,000 ha might also occur.[28] The size of both these areas is assumed to remain the same; however, a substantial reduction in the cereal area, amounting to as much as 39,000 ha, is expected. On balance, there would be an increase of 55,000 ha in the area of feed crops for the growing dairy-farming sector.[29] The effects on arable farming are represented in Table 19.3.

Table 19.3 Estimated effects of free trade on arable farming

Product area	Increase (ha)	Decrease (ha)
Starch potato		48,000
Food potato	0	0
Sugar beet		11,000
Cereals		11,000
Roughage crops	55,000	
Other (horticulture) crops	15,000	
Total	70,000	70,000

Intensive Livestock Farming

The prospects of intensive livestock farming are determined mainly by the industry's competitive economic position and policy developments in the areas of animal welfare and health. It is estimated that this will lead to a reduction of 25 per cent in intensive livestock farming of pigs and poultry.

[26] Massink and Meester, 2002.
[27] Massink and Meester, 2002.
[28] Bruins et al., 2003.
[29] Berkhout et al., 2002. The total agricultural area of The Netherlands is around 2 million hectares.

This would imply 1.5 million fewer fattening pigs and 23 million fewer poultry than in a basic scenario.[30]

Beef Production and Sheep Farming

Further trade liberalisation will bring few changes to beef production compared with the changes predicted in Agenda 2000.[31] Total trade liberalisation will put heavy pressure on the suckler-cow industry, unless support is replaced by payments in the framework of the agri-environmental programme. Abolition of the ewe premium will also place this (sub-)sector under great pressure. If sheep farming disappears, this may also apply to private flocks of traditional breeds in nature areas (heathland).

ENVIRONMENTAL IMPACT OF AGRICULTURAL CHANGES

Changes in the Physical Environment

Approach

For the assessment of environmental impacts, we applied a framework for sustainability assessment, in which the following distinctions are made in relation to the ecological component of sustainability: transport, energy, air, water, soil, waste, materials and environmental awareness.[32] Because the effects on the use and waste of materials, as well as changes in production methods due to an increase in environmental awareness, are negligible, they are not reported in more detail.

Impact on dairy farming

Transport, with all its associated environmental effects, will increase owing to the need to transport more milk products (up by 32 per cent), more feed (concentrate), more manure, and more calves. The transport of beef cattle, however, will decrease. In addition, dairy farming will become less land-tied, and the optimisation of milk production per cow will increase the need for transport. We did not quantify these effects.

Energy consumption will also increase as a result of further mechanisation (optimal feed recovery) and automation (milking robots), as well as increased livestock housing. However, in absolute terms, consumption in this sector will

30 Berkhout et al., 2002.
31 European Commission, 2004.
32 Meeusen and Ten Pierick, 2002.

be low, ultimately resulting in limited environmental effects (higher carbon dioxide emissions).

Changes in ammonia emissions into the air by the dairy sub-sector will be significant.[33] More dairy cows and beef calves will lead to an increase of ammonia emissions into the air.[34] This effect will not be counterbalanced by fewer emissions of ammonia due to the decreasing number of beef cattle. As a result, total emissions will increase by 3.6 million kg, which is 3.1 per cent of the level of the target for 2010, as laid down in the EU NEC Directive (115 million kg).

Emissions into the air of plant-protection substances will increase only slightly as a result of an increase of the maize area (21,000 ha) and the grassland area (34,000 ha). Because the use of these substances for maize and grass is small (0-1 kg active substance/ha/year), this effect will be outweighed by a decrease in the use of plant-protection substances due to a shift from arable farming to grass and maize production. As a result, the *total* use of plant-protection substances will decrease. We deal with this in more detail in the section on arable farming.

Emissions of phosphate and nitrogen into water and soil will increase significantly. The increase in the number of dairy cows and beef calves, and the decrease in the number of beef cattle,[35] will on balance result in the production of 6.9 million kg more phosphate[36] and 26.6 million kg more nitrogen.[37] It will not be possible to deposit these additional quantities of nitrogen and phosphate from livestock manure on the larger feed-crop area, with the result that expansion in the dairy-cattle sector will also lead to greater deposits and processing of manure outside this sector.

Emissions from plant-protection substances into groundwater and surface water will increase only very slightly. As regards emissions of minerals and plant-protection substances into the soil, the same applies as already observed for water.

---

[33] The Dutch dairy sub-sector emitted 7.4 million kg of ammonia in the year 2000, which is about 50 per cent of total agricultural emissions.

[34] Dairy cows (up by 306,300) and beef cattle (down by 146,000) are assumed to emit 21.6 kg ammonia; beef calves (up by 63,000) emit 2.5 kg. As a result of the further increase in the average milk production per cow, ammonia emissions per cow will also increase. But this effect will be marginal. The same conclusion applies to methane emissions, which are directly related to the number of cows and the amount of the manure output.

[35] According to Appendix A of the Fertilisers Act, dairy cows and beef cows produce 41 kg of phosphate and 161 kg of nitrogen per year, while the figures for calves are 5.2 and 12.0 kg, respectively (Staatsblad, 1986).

[36] This accounts for approximately 3.7 per cent of the total production of phosphate from livestock manure in 2000.

[37] This accounts for 4.9 per cent of the total production of nitrogen from livestock manure in 2000.

Shifts in arable farming area and crop plans

Transport volumes are likely to change only marginally. On the one hand, we expect an increase due to intensive crops requiring more treatment, and harvesting operations that may take place outside the farm. On the other hand, in order to compensate for a decline in profit margins, more will be invested in specific harvesting and sorting equipment and in storage accommodation. This will result in larger crop-growing areas at shorter distances from each other. In addition, approximately 55,000 ha of arable farmland will disappear.

As regards energy, few net changes are expected. The growing of energy crops (bio-diesel, bio-ethanol) is unlikely to become prevalent in the Netherlands. Vegetables and other horticultural crops will require more accommodation space in conditioned storage systems. In order to realise more added value, growers will prefer to store and keep their products at home. However, the ultimate effect will be small, partly because the total use of energy on arable farms, compared with energy use in field vegetable production, will be at about the same level: there will be a shift from the use of diesel in arable farming to the use of mainly electricity in vegetable growing.

Emissions from crop-protection substances into the air, water, and soil will decrease by 350,000 kg, representing 3 per cent of total crop-protection substances used in 2000. Smaller areas for starch potatoes, sugar beet, and grain imply less use of these substances.[38] These effects will be partly outweighed by the effects of the expected increase in the areas used for horticulture,[39] whereas feed crops, which will increase in area, require limited quantities of crop-protection substances (0-1 kg active substance/ha/year). The (effects of) using these substances is also likely to decrease in the period to 2010 because of a stricter admission policy, as well as the need for larger spray-free zones that apply to more intensive crop growing.

More intensive farming requires more minerals and will probably result in greater mineral losses. Winter wheat will decrease as a green crop. On the other hand, arable farming on sandy soil, more prone to leaching, will shift to clay soil, which will lead to less leaching. This latter shift will probably have a greater effect than shifts between the various crops. Direct spray irrigation, mostly using groundwater, will increase as a result of intensification. As regards the soil, intensification will lead to a search for the limits of crop alternation. However, high-quality soil structure is of great importance to

---

[38] Starch potatoes require the use of about 15 kg of active crop-protection substance per ha/year, whilst sugar beet requires 5.2 and grain 3.5 of active substance per ha/year.

[39] The quantities of crop-protection substances used for horticultural crops vary greatly. For field vegetable crops, the amount is about the same as for potatoes. Flower-bulb cultivation, by contrast, requires the use of considerably more substances.

arable farmers as well, and will, therefore, not substantially decline. Soil biodiversity might decrease slightly.

Decline in intensive livestock farming

The decrease in livestock numbers will lead to less production of nitrogen (44.4 million kg, representing an 8.2 per cent reduction in 2010 as compared with 2000), and phosphate (19.6 million kg, a 10.6 per cent reduction) from livestock manure; lower ammonia (3.7 million kg, a 3.2 per cent reduction) and methane emissions; less use of energy (and, therefore, slightly lower carbon dioxide emissions). Feed requirements will also be lower. Initially, this will probably lead to less feed being imported from countries outside the EU and, consequently, to a reduced burden on the environment in those countries and less transport. As a result of lower grain prices, the use of grain from domestic production in the Netherlands is expected to increase, leading to greater closure of the feed and fertiliser cycle at the national level. As a result of the decrease in nitrogen and phosphate production, this reduced quantity of surplus manure and minerals will no longer enter the markets and need not be deposited elsewhere. Table 19.4 represents the main environmental effects on the Netherlands of trade liberalisation.

On a global scale, fewer feed crops will have to be produced abroad to satisfy Dutch demand. On the other hand, since final demand for pork and chicken will not change because of lower Dutch production, it will have to be met by an increase in production abroad.

Table 19.4 Physical environmental effects of trade liberalisation[40]

Causes	Effects (in %) Ammonia	Phosphates	Nitrogen	Crop-protection substances
Dairy farming	3.1	3.7	4.9	
Arable farming				– 3
Intensive livestock farming	– 3.2	– 10.6	– 8.2	
	negligible	– 6.9	– 3.3	– 3

[40] The emissions of ammonia are compared with the target for the year 2010, laid down in the EU NEC Directive; phosphate, nitrogen and crop-protection substance emissions are compared with emissions in the year 2000; crop-protection substance emissions are not related to the best-performance obligations agreed in spring 2003 in the Crop Protection Policy Agreement; nitrogen emissions have not only decreased, but available land on which to deposit manure has also increased.

Changes in Biodiversity[41]

Grassland areas

Most of the nature values discussed above are concentrated in grassland areas. This is why changes in land management in those areas will have extra significance. Nature values have been declining for years under the present conditions. First, the species-rich semi-natural grasslands disappeared and, currently, field margins are also declining in quality. Among meadow birds, the species most under threat (ruff, snipe) were the first to decline (by 90 per cent and 75 per cent in 25 years, respectively). The formerly numerous skylark decreased by 90 per cent in 25 years. The black-tailed godwit, the grassland species for which the Netherlands bears the greatest international responsibility, declined by approximately 50 per cent in the last decade.[42] As indicated above, the decline is being mitigated by means of management agreements (agri-environmental programme) and, to a lesser extent, by the creation of reserves. For management agreements to be effective, initial (field) conditions must meet certain criteria. For example, management agreements directed at increasing breeding success make sense only if grassland is still attractive to breeding birds. Also, reducing fertiliser pressure on field margins for the benefit of vegetation will make more sense if some form of management (for instance, grazing) is maintained.

The expected changes in the use of grassland discussed in section 3 point to a considerable increase in the scale of dairy farming. Three quarters of the total production is expected to take place on large farms. However, it is uncertain to what extent such changes might also apply to land management, as land-tied farming is expected to decrease. Note that meadow-bird populations are decreasing, in spite of agricultural management agreements (the agri-environmental programme) and the management of nature reserves. Further rationalisation of land use will, among other things, lead to further levelling of grasslands, given the fact that mowing will be the only operation left (with cows remaining in stables all year). Young stock grazing after the breeding season might ensure some variation in sward conditions, but the degree to which this might occur is uncertain.

The biotope of meadow birds is expected to deteriorate further. Therefore, the basic conditions of the agri-environmental programme and nest protection will also deteriorate, and these measures will yield less as a result. However,

---

41 The changes described are based on a combination of the experiences of the Ministry's Reference Centre, and an e-mail discussion (concerning meadow birds) with a number of external specialists (SBB, SOVON, Wageningen UR, Bureau Terwan, and Bureau Altenburg & Wymenga), who are not in any way responsible for the contents of this study.

42 Bird data from: SOVON, 1987; SOVON, 2002; Teixeira, 1979.

exact figures for these areas are not available. The following negative developments may be mentioned in this connection:

- Large-scale grassland management: dominance of mowing management, ever-increasing uniformity in swards and surface levels.
- Increasing uniformity of swards as a result of a decline in (late) grazing (loss of 'clumpiness' and unevenness).
- Interventions to optimise large-scale management, such as further land drainage and possible filling-in of ditches in areas where this would be possible (loss of foraging areas for certain species).
- Continuing reduction in mowing periods for large areas, resulting in reduced foraging opportunities for meadow-bird chicks as well as reduced chances of survival.[43]

Finally, we should point out that the large-scale use of grassland is not by definition only negative to meadow birds. It would be beneficial if production increases were achieved only through the enlargement of farm areas and by means of sophisticated management agreements about mowing data ('mowing classes'). On the other hand, the simultaneous deterioration in the biotope (uniformity) would have the opposite effect. The experiments with the black-tailed godwit, which started in 2003, may provide greater insight here. Finally, a development towards a combination of large-scale land-tied farming and extensification, which would be extremely favourable from an ecological point of view, is not likely.[44]

A quarter of milk production is expected to be produced by farms that hardly change or even extensify.[45] If this quarter could largely be located in the best bird areas, the net effect on meadow birds might be better than otherwise. However, for the time being, the 'controllability' of various decisive factors seems uncertain. The consequences of the expected increase in grassland area at the expense of arable land will probably be of little benefit to meadow birds, given the fact that this increase will not occur in wet areas. For wintering water birds, the future with liberalised farming seems less problematic. Most goose species are not choosy. The expected larger area of monotonous grasslands might still be attractive as a foraging area. Bewick's swans and pink-footed geese deserve special attention, however, because of their strong preference for certain regions and the primary importance of the Netherlands to these species.

Vegetation along ditches, dependent on a combination of wet and nutrient-poor conditions, on the one hand, and grazing, on the other hand, might

[43] Terwan et al., 2002.
[44] Bruins et al., 2003.
[45] Massink and Meester, 2002.

decline sharply if grazing disappears entirely. In addition, the filling-in of ditches might take place on a large scale in sandy and gradient areas, given that these areas need much less earth fill than do peaty polder areas. This would result in the disappearance of the entire habitat, and all the species associated with it. In the event of increases in scale, other landscape features (including hedges, alder rows, groves, and water pools) would come under pressure in those areas where they are still widespread. The expected increase in the maize-growing area will also adversely affect vegetation along ditches. These will no longer be grazed and the vegetation will become poor in species, or even be ploughed. Rotation of maize growing over farms might result in even larger areas being affected.

Arable farming

As mentioned in section 3, arable-farming areas will decrease slightly in size in favour of feed crops. In addition, the penetration of horticultural crops and associated provisions is predicted, in rotation with arable-farming crops. This concerns approximately 15,000 ha, but, as a result of rotation, the consequences will affect an area six times that size.

The disappearance of more than 50,000 ha of arable farming land could have consequences for bird species that are relatively dependent on arable farming, such as the corncrake (Oldambt), Montagu's harrier, the grey wagtail, the skylark, and the meadow pipit.

Without direct income support and under full trade liberalisation, there will probably be no future for traditional forms of arable farming with its characteristic crop plans.[46] However, without detailed information on the preferences of the relevant plant and animal species for certain crops, little can be said about the consequences. The same holds true for the consequences of horticultural crop penetration.

Sandy areas

Sandy areas comprise a complex of grasslands, arable land, intensive livestock farming, and many wooded areas and other nature areas, as well as concentrations of small landscape features (hedges, groves, etc.). Ditches in the transitional areas (gradients) from sand to peat often have species-rich ditchbanks. The possible consequences of trade liberalisation for separate landscape components were partly dealt with earlier. Meadow birds and ditches in the grasslands, as well as landscape features, could suffer from increases in scale. Increases in the size of maize-growing areas will also prove harmful. A reduction in intensive livestock farming will lead to reduced

[46] Massink and Meester, 2002.

pressure on nature and forest areas, but this will be due less to trade liberalisation than to environmental policy.[47]

After the near disappearance of most sheep flocks in the Netherlands around 1900, there are currently more than 30 shepherded flocks, which have an important function in the maintenance of nature areas.[48] With the abolition of the ewe premium, these flocks will be in difficulty. The disappearance of these flocks would represent a loss from the point of view of agricultural biodiversity (breeds), site management (nature reserves), and tourist attractions.

The enlarged EU

It is important to briefly discuss the situation in the EU prior to and after the extension from 15 to 25 member states. In the EU-15, there are, despite a serious decline, still large areas of extensive agriculture. Those areas of livestock farming are particularly important to biodiversity. There are two main threats to these areas, both as a result of trade liberalisation (including the abolition of support for agriculture). On the one hand, there may be intensification in certain areas, and, on the other hand, there is a threat of land abandonment, especially in these usually marginal areas. Both these situations would have important consequences for the landscape and biodiversity. Particularly in the new and candidate member states, land abandonment is already widespread (up to 30 per cent of the agricultural area). This will produce problems, particularly for the formerly extensively farmed areas. Ideally, abandoned areas important to biodiversity would have to be re-farmed, or managed in other ways, as soon as possible to prevent or reverse a decline in nature values.

Changes in the Landscape

The development of large-scale livestock farms, as expected in the event of total trade liberalisation, might make it more difficult to preserve typical landscape features, unless the government's regional policy manages to put a halt to such developments. The fitting of large-scale farms into a small-scale landscape will present a challenge. In contrast, the building of large new farms in open areas will reduce openness and the loss of function of traditional regional farm buildings will be accelerated.

In section 2, we highlighted the fact that, from an international perspective, the peaty grassland polders in the Netherlands, the bog reclamation areas, the old lake reclamation areas, and the old marine clay polders are of great

---

[47] Massink and Meester, 2002.

[48] Elbersen et al., 2003.

importance. Changes in these areas, therefore, count double. In addition, increases in scale and mechanisation will lead to fewer and fewer animals and people in the landscape. Cows will almost disappear from the Dutch landscape. Road and waterway infrastructures will also have to be adapted. In addition, intensification of and changes in land use (for example, increase in horticulture and possibly in arboriculture in arable-farming areas) will have consequences for the landscape. There is also the consideration that urban functions (paddocks and the like) are the most profitable use for landowners, which may also have consequences for a large part of the Netherlands.

We already mentioned the accelerated redundancy of farm buildings. These will be given new functions, which may lead to changes in the character of the countryside. Properties with living or leisure functions lead to a totally different image. In fact, such developments involve a spread of urban conditions, and everything associated with such a spread (traffic, 'light and noise pollution', as well as horticulturist-type management of plantations and gardens). This tendency will not only affect the landscape, but may also restrict the development of agriculture itself as well as that of the countryside.

POSSIBILITIES TO CONTROL DEVELOPMENTS

Government Control and Agriculture in the Past

The description of the recent past in section 2 clearly shows that, even in times in which globalisation was not yet identified or discussed, certain processes took place that are currently judged to be harmful. These processes occurred particularly in the period of European unification, during which time European agriculture was highly protected. It has become clear that protection and support of agriculture can have undesirable consequences. For that reason, the EU agricultural policy has been subject to much criticism from, among others, environmental organisations. The idea behind this criticism was that subsidised farming intensified production, which had negative consequences for the environment.[49]

The discussion in section 2 showed the following indications of causal relationships between policy and increased pressures on the environment:

- The possibility of importing relatively cheap raw materials for livestock feed stimulates the growth of intensive livestock farming, with all the associated effects on the environment.

---

[49] Other matters could be mentioned here, including animal welfare, working conditions, and food safety.

- The growing of starch potatoes, which takes place in the Netherlands as a result of EU subsidies, involves considerable use of plant-protection substances.
- The maize premium stimulates maize cultivation, with consequences for grassland habitats (birds, field margins) and the landscape.

In short, there is a relationship between policy as a form of government control and a decrease of environmental qualities. At the same time, the discussion in section 2 showed that government policy ensures a reduction in several negative effects:

- Emissions of plant-protection substances into the ground, air, and surface water have been substantially reduced.
- Ammonia depositions have halved.
- Desiccated areas are subject to restoration, though still far behind schedule.
- Setting-aside land leads to the recovery of wild-bird populations in arable-farming areas.

The discussion in section 2 further showed that there is a remarkable relationship between biodiversity and (government-promoted and -permitted) human intervention. While Western Europe now lacks natural grasslands, wild flora and fauna presently exist on semi-natural grasslands and are dependent on extensive agricultural management. Another interesting fact is that the number of black-tailed godwits increased substantially at the beginning of the last century thanks to increased fertilisation, which ensured an abundance of food. However, this trend has now been reversed, while the above-mentioned semi-natural grasslands now remain only in Dutch nature reserves.

The relationship between government policy – national as well as European – and intensive farming is difficult to trace. It is possible to argue that there is a relationship between farming subsidies and intensification. However, it can also be argued that farming subsidies put a brake on intensification. In any event, there is a demonstrable relationship between intensification brought about by automation and increases in scale, on the one hand, and changes in the landscape, on the other hand. It is common knowledge that ditches, hedges, alder rows, groves, water pools, field margins, and levelled surfaces have disappeared and that waterways and paths have been straightened. These developments were partly due to land-consolidation (re-allotment) projects, financially supported by the government, but also occurred spontaneously. However, policy control has proved to be able to change developments in this area as well, both in the sense of 'greening' land-consolidation (development) projects and in the sense of payments for the maintenance of landscape features and biodiversity (management agreements). An outcome of the discussion in section 2 was also that recovery of bank and ditch vegetation is possible.

Furthermore, the discussion in section 2 showed that man-made habitats and landscapes can be deemed worthy of protection. This is particularly true of the Netherlands, with large parts of the Dutch landscape being man-made, but also of all other Western and Central European countries, where biodiversity-sensitive management is promoted by means of agri-environmental programmes.

On the basis of the data in section 2 concerning government controls, we conclude the following:

- It is highly probable that there are forces other than government policy that determine the development of agriculture. Technological and economic developments are the most likely such forces.
- In respect of certain aspects, there is a demonstrable relationship between policy and its external effects.
- Possibilities exist to slow down, and control and mitigate developments by means of government measures.
- It is useful to put into perspective the call for government protection by placing it in a cultural context.

Government Control and the Future of Agriculture

The above results show that care should be taken not to call too quickly for government intervention in order to maintain the status quo, even against the background of trade liberalisation. It is important to realise that increases in scale can have causes other than trade liberalisation, and that the process of trade liberalisation takes place in a situation in which the internal market is protected. The decline in intensive livestock farming, described in section 4, also shows that there are other factors involved: rather than the abolition of import restrictions, it is measured in the areas of animal welfare and the environment that cause reductions. On the other hand, two observations should be made. In the first place, there is no real status quo, but a continuous decline in biodiversity and landscape values in the current situation. It is the government's policy to take active measures against the decline.[50] The second point is that, given this policy, a speeding-up of developments in case of full trade liberalisation may entail the need to speed up the implementation of compensating measures as well.

[50] Ministry of Agriculture, Nature and Food Quality, 2000a.

The foregoing discussion makes it clear that the government has the power to control developments to some extent; not in the sense that it can stop increases in scale, but in the sense that it can determine the conditions under which such processes take place. Below, we discuss possible negative effects of developments in agriculture in situations of total trade liberalisation, with some observations about possible government control. One important sector in this connection is dairy farming, which will grow and intensify, with consequences for meadow birds, landscape features (including ditches), and the landscape itself.

There is nothing wrong with intensification as such. In order to face increasing competition successfully, we need efficient modern farms. It is the task of the government to facilitate this process of growth. In this connection, consider (re)training in business methods, (re)allocation of land, infrastructure, farm takeovers, and support through research, research stations, and new technologies. We suggested earlier that the effect of intensification could be limited if the estimated quarter of the dairy production not being intensified were to take place in the areas with the highest nature and landscape values. In practice, this would mean that these farms should be concentrated in areas that are less suitable for agriculture, and in areas with agri-environmental management agreements, or in both at the same time.

To achieve that objective, we suggest the deployment of a combination of several policy instruments:

- The agri-environmental programme and other forms of 'countryside stewardship schemes'.
- The 'Quality impulse' for 400,000 ha of countryside (the creation of 40,000 ha of landscape features) and National Landscapes.[51]
- Application of the proposals from the Belvedere Memorandum on cultural history and spatial planning,[52] and diversified agriculture (to be stimulated by the Rural Development Plan 2007-2013).[53]
- Spatial town-and-country planning and architecture policies.

Whether all of this would be sufficient to place the quarter referred to 'in the right place' remains uncertain, however. It is also debatable whether that quarter would be sufficient. In any event, the divide in the countryside between highly rationalised and other farming areas will become more pronounced.

---

[51] Ministry of Agriculture, Nature and Food Quality, 2000a.
[52] Ministry of Education, Culture, and Sciences et al., 1999.
[53] Ministry of Agriculture, Nature and Food Quality, 2000b.

Farm designs can, to a certain extent, be controlled through planning and welfare regulations. The redundance of old farms for other purposes cannot be stopped, while, at best, a number of conditions can be set for new buildings. The government will have to be creative in finding new uses for vacant farm buildings without doing serious damage to the image of the countryside. Success in that area would also benefit the local economy.

The expected changes in dairy farming will be the most significant, also with regard to their negative environmental effects. The first effect of trade liberalisation will be intensification in dairy farming. However, in the longer term, livestock numbers per ha could decrease if milk production per cow continues to increase with a constant overall level of milk production. Some of the negative effects (including transport and deposition of manure at a distance) could be counterbalanced through the stimulation of land-tied farming and through further increases in house parcels. Ammonia emissions could be reduced by adjusting feed rations and using low-emission livestock buildings. The environmental effects of dairy farming on areas with the highest nature and landscape values could be counterbalanced by stimulating dairy-farming extensification in or near those areas. Countryside stewardship schemes (including management agreements) could then ensure both adequate management and supplementary income possibilities.

In the European context, the future of extensive livestock farming areas is of crucial importance for the biodiversity of agricultural areas. Two factors are significant in this regard: future possibilities for agriculture in the extensive-farming areas and the nature of management. The first factor is connected with net farming income, the second with the use of a high-quality system of management agreements (agri-environmental programmes). The more intensively used areas will, to some extent, have the same problems as the Netherlands.

In the new member states in Central and Eastern Europe, where land abandonment is already a widespread phenomenon, professional site-managing organisations could theoretically offer an alternative, but this is hardly affordable for those countries. To some extent, solutions may be found, both in renewed use as extensive livestock farms and, on a limited scale, through site-managing organisations or other actors. In addition, agri-environmental programmes are important tools to promote biodiversity-sensitive management where agricultural management still has prospects.

CONCLUSIONS

It is not possible to point to a simple causal connection – either in a positive or a negative sense – between trade liberalisation (as one aspect of globalisation) and changes in the environment. In some respects, further trade liberalisation will be beneficial to the environment in the Netherlands. This applies to the contraction of intensive livestock farming. By contrast, in so far as trade liberalisation provides an incentive for (faster) intensification and increases in scale, the consequences are expected to be less positive, particularly for biodiversity and the landscape. In addition, the penetration of horticultural crops into existing arable or grassland areas could also result in greater pressure on the Dutch environment.

The national government could influence developments, for example, by ensuring that the large farms that will be created in dairy farming at least have to meet a number of minimum sustainability requirements. It would also be possible to offer farms financial incentives in this regard. Moreover, the government could try to control the quarter of dairy farming that will continue to produce more extensively in such a way that it remains in ecologically valuable areas.

As regards the development of EU policy on rural development, this is an important factor for addressing environmental problems in all EU countries, not least the new member states. It is also to be expected that the developments in other countries (intensification of production, on the one hand, and marginalisation, on the other hand) will continue and that the resulting impacts on biodiversity and the landscape will offer no compensation for losses incurred in the Netherlands. This highlights the importance of such 'non-trade' issues, not only in the EU context but also in the framework of the world trade negotiations.

REFERENCES

Beintema, A., O. Moedt, and D. Ellinger (1995), *Ecologische Atlas van de Nederlandse weidevogels (Ecological Atlas of Dutch Meadow Birds),* Haarlem: Schuyt & Co.

Berkhout, P., J. Helming, F. van Tongeren, A. de Kleijn, and C. van Bruchem (2002), *Zuivelbeleid zonder Melkquotering? (Dairy Policy without Fixing of Milk Quotas?)*, report 6.02.03, The Hague: Landbouw Economisch Instituut.

Bruins, W., T. Edens, J. van Esch, L. Loseman, J. Reinders, and G. Schroën (2003), *Boeren op Pad naar Vrijhandel (Farmers on their Way to Free Trade)*, report 2003/203, Wageningen/Ede: Expertisecentrum LNV.

De Bont, C., C. van Bruchem, W. van Everdingen, J. Helming, and J. Jager (2003a), *Mid Term Review: Gevolgen van de Voorstellen van de Europese Commissie voor de Nederlandse Landbouw (Mid-Term Review: Consequences of the European Commission's Proposals for Dutch Agriculture),* report 1.03.01, The Hague: Landbouw Economisch Instituut.

De Bont, C., W. van Everdingen, J. Helming, and J. Jager (2003b), *Hervorming Gemeenschappelijk Landbouwbeleid 2003: Gevolgen van de Voorstellen van de Europese Commissie voor de Nederlandse Landbouw (Reforming Common*

Agricultural Policy 2003: Consequences of the European Commission's Proposals for Dutch Agriculture), report 6.03.05, The Hague: Landbouw Economisch Instituut.

Dijkstra, H., F. Bianchi, A. Griffioen, and F. van Langevelde (2003), *Typering Landschapseenheden in Nederland naar Groen-Blauwe Dooradering (A Green-Blue Veined Typology of Dutch Landscapes),* Wageningen: Alterra.

Elbersen, B., A. Kuiters, W. Meulenkamp, and P. Slim (2003*), Schaapskuddes in het Natuurbeheer: Economische Rentabiliteit en Ecologische Meerwaarde (Sheep Flocks in Nature Management: Economic Return and Ecological Added Value)*, project number 12245, Wageningen: Alterra.

European Commission (2004), 'Agenda 2000', *http://europa.eu.int/comm/agenda 2000.*

Farjon, J., G. Dirks, A. Koomen, J. Vervloet, and G. Lammers (2001), *Nederlandschap Internationaal: Bouwstenen voor een Selectie van Gebieden bij Landschapsbehoud (Building Blocks for Area Selection to Conserve Landscapes)*, report 358, Wageningen: Alterra.

Hall, M., J. van 't Hoff, R. de Koning, J. Meijering, and K. van Scharenburg (1998), *De Toekomst van Natuur en Landschap in de Provincie Groningen (The Future of Nature and Landscape in the Province of Groningen)*, Groningen: Province of Groningen.

Kleijn, D., F. Berendse, R. Smit, and N. Gilissen (2001), 'Agri-environmental schemes do not effectively protect biodiversity in Dutch agricultural landscapes', *Nature*, **413**, 723-725.

Koffijberg, K., J. Beekman, L. van den Bergh, C. Berrevoets, B. Ebbinge, T. Haitjema, J. Philippona, J. Prop, B. Spaans, and M. Zijlstra (1998), 'Ganzen en zwanen in Nederland 1990-95' (Geese and swans in the Netherlands 1990-1995), *Limosa,* **71**(1), 7-32.

Massink, H. and G. Meester (2002), *Boeren bij Vrijhandel (Farming and Free Trade),* The Hague: Ministry of Agriculture, Nature and Food Quality.

Meeusen, M. and E. ten Pierick (2002), *Meten van Duurzaamheid: Naar een Instrument voor Agroketens (Assessment of Sustainability: Towards an Instrument for Agricultural Chains)*, report 5.02.11, The Hague: Landbouw Economisch Instituut.

Ministry of Agriculture, Nature and Food Quality (2000a), *Nature for People, People for Nature: Policy Document for Nature, Forests and Landscape in the 21st Century*, The Hague: Ministry of Agriculture, Nature and Food Quality.

Ministry of Agriculture, Nature and Food Quality (2000b), *Rural Development Programme: The Netherlands 2000-2006*, The Hague: Ministry of Agriculture, Nature and Food Quality.

Ministry of Education, Culture, and Sciences, Ministry of Housing, Spatial Planning and the Environment, Ministry of Agriculture, Nature and Food Quality, and Ministry of Public Works and Water Management (1999), *The Belvedere Memorandum: A Policy Document Examining the Relationship between Cultural History and Spatial Planning*, The Hague: VNG Uitgeverij.

OECD (forthcoming), *Environmental Indicators for Agriculture – Volume 4*, Paris: Organisation for Economic Co-operation and Development.

Province of Zuid-Holland (1996), *Staat van de natuur 1995 (State of Nature 1995)*, The Hague: Province of Zuid-Holland.

Province of Zuid-Holland (2000), *Staat van de natuur 2000 (State of Nature 2000)*, The Hague: Province of Zuid-Holland.

RIVM and DLO (2002), *Natuurbalans 2002 (Nature Balance 2002)*, Alphen aan den Rijn: Kluwer.

RIVM (2002), *Milieubalans 2002 (Environmental Balance 2002)*, Alphen aan den Rijn: Kluwer.

RIVM (2003), *Milieubalans 2003 (Environmental Balance 2003)*, Alphen aan den Rijn: Kluwer.

RIVM, CBS and DLO (2003), *NatuurCompendium 2003 (NatureCompendium 2003)*, Utrecht: KNNV Uitgeverij.

RPD (2000), *Balans van de Ruimtelijke Kwaliteit 2000 (Balance of Spatial Quality 2000)*, The Hague: Ministry of Spatial Planning, Housing and the Environment.

SHBO (2001), *Historische Boerderijen in Nederland: Een Onderbouwde Raming van het Resterende Bestand aan Historische Boerderij-Complexen Gebouwd voor 1940 (Historical Farms in The Netherlands: A Founded Estimation of the Remaining Historical Farms Built before 1940)*, Arnhem: SHBO/Stichting 2003 Jaar van de Boerderij.

SOVON (1987), *Atlas van de Nederlandse Vogels*, Arnhem: SOVON.

SOVON (2002), *Atlas van de Nederlandse Broedvogels (Atlas of Dutch Breeding Birds)*, Leiden: Naturalis/KNNV Uitgeverij/EIS-Nederland.

Staatsblad (1986), *Meststoffenwet (Fertilisers Act)*, 598, The Hague: Sdu Uitgevers.

Teixeira, R. (1979), *Atlas van de Nederlandse Broedvogels*, 's-Graveland: Natuurmonumenten.

Terwan, P., J. Guldemond, and J. Buijs (2002), *Toekomst voor de Grutto? (Prospects for the Black-tailed Godwit?)*, Zeist: Vogelbescherming.

Van Berkum, S., A. Westerman, and C. Wolswinkel (2002), *De Internationale Locatie van de Tuinbouw bij Handelsliberalisatie (Impact of Trade Liberalisation on International Horticulture Sites)*, report 6.02.11, The Hague: Landbouw Economisch Instituut.

Van Dijk, G., J. van Vliet, and N. Hazendonk (2003), *Vrijhandel, Milieu, Natuur en Landschap (Free Trade, Environment, Nature, and Landscape)*, report 2003/226, Wageningen/Ede: Expertisecentrum LNV.

Zonderwijk, P. (1971), *Verantwoord Gebruik van Onkruidbestrijdingsmiddelen in Berm, Sloot en Beek (Responsible Use of Pesticides in Banks, Ditches, and Brooks)*, Wageningen: Plantenziektenkundige Dienst.

PART IV

The Role of Government in International and Supranational Forums

20. Different Countries, Different Strategies: 'Green' Member States Influencing EU Climate Policy

Sietske Veenman and Duncan Liefferink[1]

SUMMARY

Environmentally proactive member states of the European Union (EU) influence European legislation in different ways. Green 'pushers' seek direct influence at the EU level, whereas 'forerunners' try to get national elements incorporated in an indirect fashion. Together with a second dimension, the nature of national policies (purposeful or incremental), four national positions towards EU legislation can be identified: pusher-by-example, constructive pusher, defensive forerunner, and opter-out. We applied an amended version of this framework to the influence of three 'green' member states (the Netherlands, the United Kingdom, and Denmark) on the European Commission and the European Council when preparing and deciding on new legislation as to carbon dioxide taxation and emissions trading (related to climate policy). Different aspects of influence were tested: timing, nature of contacts, alliance building, and uploading of national elements. We concluded that one type of position should be revised (late starter is preferable to opter-out); besides, countries may simultaneously follow different strategies rather than adopt one single approach.

[1] The authors would like to thank Frank Wijen, Mariëlle van der Zouwen, and Alkuin Kölliker for numerous suggestions and comments. Thanks are also due to Jan Pieters for his inputs and observations and to Mar Robillard for her assistance in the English language. Naturally, any shortcomings are those of the authors.

INTRODUCTION

Studies of the strategies that EU member states employ to influence the European environmental policy process can either cover all member states, or focus on certain types of member states. The focus of this study was on member states that are, as far as environmental policy is involved, relatively ahead of the others. Almost by implication, such proactive member states are also the ones that most actively shape the EU environmental policy process. Generally speaking, 'laggard' countries try to delay or stop the policy process.[2] The objective of this study was to provide a better understanding of the processes and mechanisms behind the making of EU environmental policy. As is set out in the next section, the basis for this exercise was the 'pusher-forerunner' theory, developed by Liefferink and Andersen in the late 1990s.[3] In this chapter, the theory is extended so as to be more specific with regard to the strategies employed in Brussels, where the European Commission (EC) resides, by what may be called the green member states.

To reach the objective of the study, the empirical focus was narrowed down to two case studies within the area of climate policy: carbon dioxide taxation and emissions trading (ET). Carbon dioxide taxation was chosen because it allows a clear distinction to be made between member states in favour of and member states opposing the idea of a carbon dioxide tax, and, in addition, because it provides a clear view of the distinct roles within the supportive camp. Since this case extended over more than ten years, two episodes were chosen. To include a more recent issue in this study, the ET case was selected. Also here, a clear group of propagators can be discerned, whose strategies are scrutinised in section 3.

For both cases, furthermore, we focused on the same group of countries in order to achieve a valid comparison. The cases give a fairly good picture of the different roles that several countries have played.[4] Regarding carbon dioxide taxation, Germany, the Netherlands, Finland, and Denmark are often mentioned as driving forces. Concerning ET, Denmark, Germany, the Netherlands, and the United Kingdom (UK) played important and supporting roles. Member states active in both cases were Denmark, the Netherlands, and Germany. The UK played a particularly active role in the ET case. Germany was left out of focus owing to the theoretical perspective: although generally supportive, the country did not play a truly stimulative role during either of the cases. In carbon dioxide taxation, Germany distanced itself from the supporters, because it was unwilling to accept that some countries were 'first-movers' in the introduction of a carbon tax. It insisted on having a binding

---

2 Börzel, 2002.
3 Liefferink and Andersen, 1998a.
4 Zito, 2002.

deadline by which all member states should have applied a carbon-energy tax. In the ET case, Germany was hesitant because of internal conflicts. The UK, on the other hand, was included in the study, even though it hardly behaved 'green' in the carbon dioxide case. Its inclusion is justified by its key role in the ET case. In sum, the member states concentrated upon in this study are the Netherlands, Denmark, and the UK. In these countries, interviews were held with 20 officials involved in the specific cases.[5]

The cases are presented and their theoretical implications discussed in section 3. The conclusions are presented in section 4.

THE STRATEGIES OF PUSHERS AND FORERUNNERS

Theorising Relations between Domestic and European Politics

The pusher-forerunner theory closely links the domestic motives of states to their behaviour at the international level. As such, the theory is part of a long tradition. Although far from a complete review of the relevant literature, the present section provides a discussion of the position of the pusher-forerunner theory within this tradition, enabling its strengths and weaknesses to be discerned more clearly.

A now classic contribution to the field was made by Putnam.[6] The main contribution of Putnam's work was that it provided a systematic perspective on the relationship between domestic institutions and politics, on the one hand, and the international negotiation table, on the other. Using a game theoretical approach, Putnam defined the situation of an overlap between the two levels as a 'win-set'. Although not designed with a specific view to the EU, it can be applied to this context.[7] Putnam's conceptualisation of the two levels, however, remains at a general level and does not provide a useful starting point for a more detailed analysis of the different strategies employed by states acting at the EU level.

---

5 In the Netherlands, respondents from the Ministries of the Environment, Economic Affairs, Finance, and Foreign Affairs were interviewed, as was an environment attaché at the Dutch Permanent Representation to the EU in Brussels. In Denmark, interviews were held with officials at the Energy Authority and the Environmental Protection Agency, the Ministry of Taxation, and the Permanent Representation. The UK was represented by officials from the Departments of Food and Rural Affairs, and Trade and Industry, as well as Her Majesty's Customs and Excise, and the Permanent Representation. Finally, three interviews were held with civil servants at the EC. All interviews were carried out between the beginning of May and mid-July 2003.

6 Putnam, 1988.

7 Andersen and Liefferink, 1997.

From the 1990s, various typologies of the negotiating styles and strategies of EU member states were published. Héritier,[8] for instance, distinguishes 'leaders' and 'laggards'. In her typology, 'leaders' are the member states that are keen on following or even shaping EU legislation. 'Laggards', in contrast, tend to resist EU pressure. A subsequent typology by Börzel can be regarded as a refinement of the leader-laggard dichotomy.[9] Based on the policy preferences and the action capacity of member states, Börzel distinguishes three basic types of member states' strategies. First, 'pace-setters' have a relatively strong economy and a well-developed environmental policy and, therefore, a strong incentive to strive for harmonisation and the 'uploading' to the EU of their own national policies. The second category consists of 'foot-draggers', who aim at stopping or at least delaying the attempts of other member states to upload their domestic policies. Because of a relatively poor economic situation, they have neither the policies nor the action capacity to upload themselves. The last category is that of the 'fence-sitters', who take a neutral position by neither promoting policies, nor preventing others from doing so. Although Héritier's and Börzel's schemes are no doubt helpful for understanding the mechanisms of policy-making in Brussels at large, they are not specific enough for the present purposes. The green member states which we focused on in this study can all be included in the 'leaders' or the 'pace-setters' categories, without a basis of differentiation between the strategies within these categories.

This problem was addressed in another article by Héritier.[10] Here, she formulates a typology of so-called 'first-movers' strategies (i.e., the strategies of member states who are ahead of the others and try to upload their own policies to Brussels). Four steps for uploading are distinguished: (1) a clear home run, (2) a 'saddled' home run, (3) a 'moderated' home run, and (4) no 'home runs'. The first step implies a regulation exactly as the initiating member state proposed. An increasing number of concessions is made in steps (2) and (3), while step (4) implies no uploading at all. The problem with this approach is that it suggests that all first movers want to upload their own policies. This is evidently not the case. An example of this is the Danish bottle case, where Denmark had a national policy 'ahead' of the EU, but did not want to upload it to the EU level. Denmark would not have fitted into Héritier's framework.

The pusher-forerunner typology by Liefferink and Andersen offers more room for differentiating between the possible roles of green member states.[11] Like Héritier's 1996 framework, it focuses on first movers, but a distinction is

---

[8] Héritier, 1994.
[9] Börzel, 2002.
[10] Héritier, 1996.
[11] Liefferink and Andersen, 1998a.

made between 'pushers', who want to influence EU environmental policy directly (for instance, by uploading domestic policies) and 'forerunners', whose first priority is to maintain their national policies and whose influence on EU policy may be more indirect. The Danish bottle case falls under the latter type. The pusher-forerunner theory is presented in more detail below and some amendments are proposed.

The Pusher-Forerunner Model

Before discussing the pusher-forerunner model and its applicability to the present study, a preliminary remark has to be made. The aim of this chapter is to outline and explain actual strategies adopted by member states in order to influence the EU policy-making process in Brussels. In the original article by Liefferink and Andersen, the word 'strategy' is also used, but here it refers to distinct 'roles' or 'positions' of green member states in relation to EU environmental policy in a more general sense. The two are linked: being in the position of either a pusher or a forerunner with regard to a given policy issue leads to a certain form of strategic behaviour in Brussels. However, this is not necessarily a one-to-one relationship. It is conceivable that different strategies are used to propagate a particular position in Brussels, dependent on the circumstances. The central objective of this study was precisely to find out more about this relationship. For that reason, we reserve the term 'position' for defining the way a green member state's domestic policy on a given issue relates to the relevant EU policy or policy process (i.e., pusher, forerunner, etc., in the sense of Liefferink and Andersen); in other words, for defining the basic 'role' of a member state in a given policy play. The member state's actions based on this position, or the precise way the 'role' is performed on the stage in Brussels, is referred to as 'strategy'.

The pusher-forerunner theory investigates the positions of the 'leaders' in EU environmental policy-making (i.e., those member states that want to go further in environmental policy than most of their partners). The pusher-forerunner theory is based on two distinctions. First, a distinction is made between countries that deliberately seek to influence EU environmental policy directly and countries whose impact on EU environmental policy works in an indirect way, mainly via internal market policies. The former are referred to as pushers, the latter as forerunners. Pushers are member states that emphasise the importance of international and EU cooperation. They work from the conviction that their interests are in the end better served by common policies, even if they are at a somewhat lower level. It is considered more important to have the certainty of what the neighbouring countries do than to have what might be optimal from a domestic point of view. High standards are maintained at the domestic level, but basically within the limits of EU and international law. It is, therefore, clear that harmonisation within the EU is important to pushers. Forerunners, in contrast, aim at maximising freedom to develop and implement their own national policies. In their view, international policies should not restrict the authority of a country to design its own policies and maintain stricter standards. International and EU policies are welcome as far as they help to achieve domestic policy goals. Forerunners have a preference for minimum harmonisation, leaving room for alternative options and exemptions.[12]

The second distinction in the theory is between countries which have developed their advanced policies purposefully (i.e., with a view to EU policy) and those which have done so more gradually and incrementally (that is, without consideration of EU policy). The former refers to a planned initiative to go one step further than the rest. In the latter case, the position of being ahead of the others emerges over time, fuelled essentially by domestic considerations.

When the two distinctions (direct/indirect and purposeful/incremental) are combined, four possible positions of green member states in EU environmental policy emerge, as set out in Table 20.1. Field (a) describes member states eager to influence EU environmental policy directly by 'giving a good example'. The goal of uploading this 'good example' to the EU is inherent in this position. In field (b), progressive domestic policies are developed more gradually and basically as part of a domestic process. Only when it has appeared that the national position is ahead of the position of most other member states, the possibility of 'elevating', or uploading, the policy to the EU level is fully worked out. Uploading may be considered desirable to raise the environmental effectiveness of the policy (e.g., in view of transboundary flows of pollution) and to alleviate the impact of – possibly costly – national measures on competitiveness. Typical features of the

Table 20.1 Positions of green member states in EU environmental policy[13]

	Policy	
Influence	Purposeful	Incremental
Direct	(a) Pusher-by-example	(b) Constructive pusher
Indirect	(c) Defensive forerunner	(d) Opter-out

[12] Liefferink and Andersen, 1998b.
[13] Adapted from Liefferink and Andersen, 1998a.

constructive- pusher strategy are an orientation towards building alliances (for instance, with the EC or other member states) and, if necessary, the willingness to accept or even broker compromises. Field (c), like field (a), represents a deliberate forerunner position, but without the explicit goal of acting as an example. Nevertheless, a considerable impact on the EU may occur via internal market policy, as national measures may disrupt conditions of trade and/or competition. This may even lead to legal procedures before the European Court of Justice (ECJ) and thus initiate debates about the issue at the EU level. The classic example of the defensive forerunner role is the Danish ban on cans which led to a seminal Court case in the 1980s, but also helped to prepare the ground for the Packaging Waste Directive of 1994. Field (d), finally, describes the situation in which a forerunner position is developed gradually, as in field (b), but without the wish to create a conflict in Brussels. If such a conflict nevertheless occurs, particularly because national measures turn out to impinge upon the functioning of the internal market, it may lead member states to opt out and go their own way. The EU Treaty provides limited scope for following this road in the form of Articles 95, paragraphs 4-9 (sometimes referred to as 'environmental guarantee') and 176.

It is possible that positions shift in time. Germany is an example in the case of the introduction of catalytic converters in cars in the 1980s. Germany successfully linked its role of 'pusher-by-example' with the threat of becoming a 'defensive forerunner' by making advanced preparations for the unilateral introduction of catalytic converters.[14] A flaw of Liefferink and Andersen's original theory is its claim that positions may be combined. This is not very plausible, however. Member states cannot develop their policies purposefully (as pushers-by-example and defensive forerunners do) and yet 'find themselves' to be out of step with the others (as constructive pushers and opters-out do) at the same time. Similarly, the focus is either predominantly domestic (as with defensive forerunners and opters-out) or there is a clear wish to influence the EU as a whole (as with both types of pushers). It is difficult to see how the roles can logically be combined. In sum, it is possible for member states to change positions over time, but it is not possible to play two roles simultaneously. We now turn to the question of how the different roles or positions may lead to actual strategies in Brussels.

[14] Liefferink and Andersen, 1998a.

Strategies of Pushers and Forerunners: General Expectations

It is often emphasised that the different strategies of member states must be linked to different phases of policy-making.[15] However, the literature suggests many different ways to distinguish phases in policy-making.[16] In the context of the EU, the most obvious way of constructing phases is to follow the – more or less – chronological order of the EU policy process.[17] In the first phase, the EC (or 'Commission') formulates new proposals. In the second phase, the member states gathered in the Council negotiate about the proposal. It should be kept in mind that phasing in relation to the EU policy process is always a simplification. In reality, the phases are not necessarily consecutive and policy issues may move back and forth. For analytical reasons, in this research, we made a distinction between the 'Commission phase' and the 'Council phase'. We examined these phases in detail and formulated a set of expectations regarding the strategies of different types of green member states in those phases. We also tried to extend the original theory so as to accommodate these expectations in a consistent way.

The Commission phase

The EC and the member states are mutually dependent on each other during the Commission phase. The member states need the Commission to listen to their wishes and the Commission needs the member states for support and legitimation.[18] In spite of the EC's formal right of initiative, member states have many possibilities of influencing the Commission's proposals (for instance, by developing informal contacts).[19] Pushers and forerunners are likely to act differently in the Commission phase. Pushers aim at harmonisation, which implies that they have an interest in creating wide support for a new common policy. In order to achieve this goal, they try to influence the formulation of EC proposals in detail. The interest of forerunners in the Commission phase, in contrast, is essentially limited to safeguarding the room for national discretion in the proposals. The involvement of pushers in the Commission phase, therefore, is likely to be more intense and encompassing than that of forerunners. The same difference is clear during the meetings of national experts in the Commission phase, which form a breeding ground for new steps.[20] All member states concerned with environmental policy may be assumed to send well-prepared and

---

[15] For example: Héritier, 1996; Pellegrom, 1997.
[16] For example: Dunn 1994; Kingdon, 1984.
[17] Schout et al., 2001.
[18] Héritier, 1996.
[19] Pellegrom, 1997.
[20] Pellegrom, 1997.

competent experts to such meetings to get an impression of the political support and seize opportunities to influence the EC's decisions. Experts from both types of pusher countries have a special interest in gaining early influence. They tend to downplay their national interests (have a 'neutral' attitude) and look for common solutions.[21]

As regards their relations with the EC and their behaviour in expert meetings, a slight difference may be expected between constructive pushers and pushers-by-example. Whereas the former seek to achieve their goals by building alliances with the EC and other member states, the latter try to influence the policy process mainly by setting and showing 'good examples', usually in the form of allegedly successful national policies. Pushers-by-example thus focus somewhat less on actively working together with the EC than do constructive pushers. Defensive forerunners, in contrast, are focused on sovereignty and subsidiarity. They prefer that each member state be able to do whatever it wishes at the national level within the European limitations and that these limitations be kept to a minimum. This national leeway does not necessarily have to be set in proposals. The history of many European directives shows that it can also be reached during Council negotiations, either with or without the help of the European Parliament (EP). This strategy is also reflected during expert meetings. Defensive forerunners send well-prepared experts to the meetings, but they tend to focus on the protection of national interests and national policies. Opters-out, finally, show no specific interest in becoming heavily involved in informal contacts with the EC or contributing actively to expert meetings. Because initiating new, common environmental policies is not the focal point for defensive forerunners or opters-out, the Commission phase is not considered to be as important for these as it is for constructive pushers or pushers-by-example.

The Council phase

During the Council phase, priorities can be different. It should be emphasised that in the Council all member states try to insert elements of their national interest, pushers no less than forerunners. The fact that draft European legislation is discussed in the Council at all indicates that harmonisation is aimed at. The basic form in which this harmonisation is to take place is laid down in the proposals and defended by the EC. In this situation, neither pushers nor forerunners fail to take good care of their key national interests, but there may be differences in how they do so.

The pusher-forerunner theory fails to formulate specific expectations as to the different strategies of pushers and forerunners in the Council phase. The literature suggests that the following aspects are particularly relevant: the

21 Schout et al., 2001.

negotiating style in the Council, the strategy of alliance building, and the approach to the uploading of pre-existing national policies to the EU level.[22] Some member states use a pragmatic and consensus-seeking approach, others have a harder negotiating style. For pragmatic negotiators, achieving consensus is considered more important than to be fully in the right. It is acknowledged that expertise can be a major resource in working parties such as the Environment Group,[23] but in the end it is seen as part of the larger political game.[24] The pragmatic style is appropriate to constructive pushers as well as opters-out. The latter do not push towards new directives and have nothing in particular to defend, which makes a pragmatic strategy most likely. The ultimate aim of constructive pushers is the adoption of the proposals at stake. Therefore, constructive pushers are willing to make compromises. The style of pushers-by-example is constructive too, but more aggressive than that of constructive pushers, as they want the directives but are also keen on uploading domestic elements into the directives. This leads to a harder negotiating style, which tends to stress the technical and scientific bases of the positions taken.[25] For defensive forerunners, finally, being able to retain existing national policies is considered more important than achieving harmonised directives. This role is not focused on compromising, because their goal is to protect national environmental policies, to safeguard sovereignty, and to have minimum harmonisation within the EU.

Like-minded countries tend to find each other in the course of negotiating in Brussels. Alliance building in the Council working groups is to a large extent an implicit process. If countries decide actively to coordinate their strategies, this mainly happens at the daily work level, for instance, among environmental attachés at the Permanent Representations in Brussels.[26] In addition, regarding issues of major importance, bilateral contacts between capitals may occur, particularly in order to win doubters for one side or the other. These alliances are formed mostly on ad hoc basis, because structural alliances and the formation of 'cliques' tend to be in contradiction with the open and case-by-case character of the EU policy process.[27] Constructive pushers may reasonably be expected to be most active in building alliances with the EC and/or with other member states. Their eventual aim is to construct directives, which is most likely to happen when as many member

---

[22] Héritier 1996; Jordan and Liefferink, 2004; Klok 2002.

[23] The Environment Group consists of environmental attachés of the member states' Permanent Representations in Brussels. It serves as the main preparatory body for the negotiation of environmental issues in the Committee of Permanent Representatives (COREPER) and, eventually, the Environment Council.

[24] Liefferink and Andersen, 1998a.

[25] Jordan, 2002.

[26] Pellegrom, 1997.

[27] Liefferink and Andersen, 1998a.

states as possible take part. Being pushers-by-example, it is not likely that many other member states will push similar national 'examples' in Brussels, but alliances with other member states are still helpful in pushing effectively. Defensive forerunners focus primarily on retaining their national standards, rather than on harmonisation and compromises. Therefore, they are not likely to be much of alliance-builders. Opters-out, finally, are out of step with the EU mainstream, which logically pre-empts the possibility of building alliances with other member states.

Finally, on the subject of the uploading element, the following can be expected. Pushers-by-example strive to fully upload their national policies because they work, almost by definition, from pre-existing national policies which they hope will be copied by the EU. Constructive pushers may also be interested in uploading certain national elements, but are less eager to do so, as their chief objective is to bring about harmonised policies. Therefore, they are willing to make significant concessions if necessary. Defensive forerunners and opters-out, in contrast, are not specifically interested in uploading policies at all. They both focus on sovereignty, rather than harmonisation.

Summing Up: The Pusher-Forerunner Model Extended

The above were our expectations as to the strategies of the four types of pushers and forerunners in respect of five different aspects. Box 20.1 summarises these expectations with reference to the four different roles or positions derived from the original pusher-forerunner model. The above expectations can also be summarised in a more systematic way. For this purpose, we must go back to the basic distinction between the Commission phase and the Council phase, and draw together the five aspects of Box 20.1 under the following two headings: (1) the level of activity during the Commission phase, and (2) the negotiating style during the Council phase.[28] Linking this once again to our expectations and to the categories of the original pusher-forerunner theory, two things draw attention. First, both types of pushers appear to invest more in the Commission phase of the environmental decision-making process than do forerunners. Second, incrementally developed national policies seem to induce a more pragmatic

---

[28] The aspects of cooperation with the EC and input into expert meetings (Expectations 1 and 2 in Box 20.1, together making up the Commission phase) are arguably connected. This is also the case for the aspects of negotiating style and strategy of alliance-building (Expectations 3 and 4 in the box, together characterising to a large extent the Council phase). The aspect of uploading (Expectation 5), finally, is essentially implied in the basic categories of pushers and forerunners. Combining these aspects for reasons of analytical clarity, therefore, is not likely to result in an unacceptable loss of information.

Box 20.1 Expected strategies of pushers and forerunners in Brussels

Expectation 1 (cooperation with the Commission):
Pushers concentrate more on the Commission phase than do forerunners and opters-out, because pushers aim to maximise harmonisation and cooperation in the EU. Constructive pushers primarily strive to formulate proposals together with the EC. Pushers-by-example concentrate less on working together with the Commission, but focus on being fast movers.

Expectation 2 (expert meetings):
All 'green' member states supply well-prepared and competent input to the various EU expert committees during the Commission phase. However, forerunner member states' experts advocate their national interests, whereas pushers look for common solutions.

Expectation 3 (negotiating style in the Council):
A pragmatic style is a common characteristic of constructive pushers and opters-out. Constructive pushers in particular focus on consensus. The style of pushers-by-example is pragmatic and consensus-oriented as well, but more aggressive in order to get certain national elements through. Defensive forerunners have a harder negotiating style.

Expectation 4 (alliance building):
Alliance building in the Council generally has a pragmatic and ad hoc character. Nevertheless, pusher member states – and especially constructive pushers – are more open to forming alliances than are defensive forerunners or opters-out.

Expectation 5 (uploading):
Pushers-by-example strive for full or almost full uploading, while constructive pushers are willing to concede to a more moderate form of uploading. Forerunners and opters-out have no particular interest in uploading.

Table 20.2 Strategies of green member states in EU environmental policy

	Council phase style	
Commission phase activity	Consensual and constructive	Adversarial and tough
Much	(a) Constructive pusher	(b) Pusher-by-example
Little	(c) Opter-out	(d) Defensive forerunner

approach than do deliberate, purposeful forerunner policies. These two insights can be set out in another 2*2 matrix, which may be regarded as an amendment to the original pusher-forerunner theory, focused on the actual strategies in Brussels (rather than the starting positions) of green member states (Table 20.2). The empirical material presented in the next section was used to see if this amendment makes sense.

PUSHERS AND FORERUNNERS IN EU CLIMATE POLICY

The objective of this section is to give an overview of the strategies of the three member states during the carbon dioxide and ET cases. First, the roles that the Netherlands, the UK, and Denmark played during the policy processes of the two cases are outlined. Then, the information is linked to the set of expectations developed above.

Carbon Dioxide Taxation in Three EU States

We describe the roles of the three member states during two episodes of the carbon dioxide taxation case. The first episode focuses on the Commission phase and deals with the formulation of the first proposal for a carbon dioxide tax in 1992. The second episode concentrates on the Council negotiations in 1993, when the Danish Presidency tried unsuccessfully to force an agreement on the issue.[29]

The Netherlands

In 1992, during the preparation of the EC proposal for a carbon dioxide tax, an ad hoc committee in the Netherlands presented a report analysing several options for carbon dioxide taxation in an international perspective. The report was mainly intended for domestic purposes, however, and was not actively used in Brussels. Additional reasons for not using this report included the existence of a certain amount of confusion as to which ministry was responsible for the issue and the lack of a well-functioning Dutch network in the EU. A proper network was built up only when the Netherlands tried to establish a 'club of countries' in favour of the carbon dioxide tax in 1995 (see below).

During the Council negotiations in 1993, the Netherlands was in favour of a carbon dioxide tax and willing to make compromises. These compromises went quite far from the Dutch point of view: the Netherlands was prepared to

[29] During the collection of the data for this case, which took place ten years ago, it turned out to be difficult to find respondents who could clearly remember the events, particularly those of the episode covering the Commission phase. In order to compensate for this problem, some additional information was collected and used in this case study about the re-formulation of the carbon dioxide tax proposal in 1995 and 1997.

give up the Dutch system of fuel taxation. Uploading the national system to the EU was never considered a serious option. The Dutch realised that this system was unique as it strongly relied on gas, which can, unlike other fuels, be measured using a meter at every outlet. The clear willingness to compromise makes the Netherlands a genuine constructive pusher: it wanted a tax to be introduced and actively worked towards this goal in Brussels.

During the 1993 Council negotiations, but also in later stages, several alliances were built. These were ad hoc alliances in the sense that they were issue-specific. In practice, however, the same countries that were basically in favour of a tax found each other again and again. In 1995, the Netherlands made some efforts to establish a club of countries in favour of a carbon dioxide tax, consisting of, among other countries, the Netherlands, Germany, and the Nordic member states. It should be noted, however, that the EC was never invited to the meetings. After two meetings, the initiative died.

The UK

The UK did not participate much in the early phases in 1992, because it did not expect the carbon dioxide taxation proposal to be accepted. As there was no national policy in the field, uploading was not at stake. When the issue turned out to persist, the UK made it clear that it did not want a harmonised tax. It did so, for instance, during the Danish Presidency in 1993. All member states were aware of the British position. Even when the Danes tried to force an agreement on the issue, sometimes referred to as Danish 'arm-twisting',[30] the UK refused to compromise. In this context, it should be remembered that carbon taxation is defined as a fiscal issue and thus comes under the unanimity rule in the Council, so the UK did not require the support of other countries which opposed the tax. Although some countries held views more or less similar to those of the British, there is no evidence of serious bilateral efforts to take a common line. On the other hand, Danish attempts to fully isolate Britain failed.

In conclusion, and as pointed out in the introduction, the UK cannot be regarded as a green member state in this case. As it did not want to develop a carbon dioxide tax at all, its style was adversarial, while it acted neither as a pusher nor as a forerunner.

Denmark

Denmark had a good relationship with the EC during this case because it had introduced a national carbon tax in 1992/1993[31] and continuously supported the EU proposal. A Danish respondent described the relationship with the

---

[30] Klok, 2002.
[31] Klok, 2002.

Commission as follows: 'The Danish prepare things at a convenient point in time in a good manner so they can have influence. The Commission will ask for advice when problems arise.' And, 'How the Commission thinks it is good is the way to go for the Danish. Denmark did its homework early.' On the one hand, Denmark was a first mover in this issue, which makes it a potential pusher-by-example; on the other hand, Denmark worked closely together with the EC, implying that it also acted as a constructive pusher.

Particularly after the unsuccessful 'arm-twisting' during the Presidency of 1993, the Danes were willing to make compromises. As one respondent phrased it, 'Rather a minimum tax than no tax at all.' They tried to convince their colleagues with arguments and to build alliances. At the same time, however, Denmark strongly persisted in uploading its own tax system. Some tax-friendly countries were willing to understand, but the Danish national system was too complicated and was unacceptable to most other countries.

Emissions Trading in Three EU States

The policy process in relation to the introduction of an emissions-trading system for carbon dioxide (ET) took place in a relatively short period between 1998 and 2001. Therefore, the process could be looked upon as a whole without choosing certain episodes. This enables us to give an overall view of the positions and strategies of the three member states in our sample.

The Netherlands

The Commission started working on the ET draft in 1998. In this early phase, Dutch interdepartmental coordination was deficient. First, the two Ministries involved (Environment and Economic Affairs) were still arguing about who was going to be responsible for the ET issue. As a respondent commented, 'We should have done something during the early phases, but we did not know that something was going on; there was no communication.' One Dutch respondent mentioned an alliance, while another assured us that such an alliance had never existed, because 'If ever it had existed, I would have known about it.' Second, at the same time as the EC was preparing the EU proposal, another advisory committee was established in the Netherlands, charged with developing a national ET system. Although the two ET designs were different, the Dutch decided to continue preparing the national system. This strategy made the Netherlands a (belated) defensive forerunner. As a result, the Dutch participation in Brussels was minimal.

The Dutch style during the Council negotiations was described in different ways by different respondents. Some argued that the most important target for the Netherlands was to construct the European ET directive. They claimed that the Netherlands, without fully denouncing national interests, had been flexible during the negotiations and willing to make compromises: 'The Netherlands set the trend during the negotiations by saying: "we want this proposal and we want it quickly!"' This view of the process coincides with the perception on the part of the same respondents that the Netherlands was prepared to give up the national ET system, as it had been mainly intended for national use anyway. Seen this way, the Netherlands acted as a constructive pusher during the ET negotiations. In sharp contrast with this, other respondents asserted that the Netherlands had been inflexible during the negotiations. These respondents furthermore felt that the national system had been written with an explicit view to uploading it to the EU level, which would have made the Netherlands a pusher-by-example, although starting late.

The UK

The UK developed its own ET scheme from 1997 and started trading in 2002. The national scheme included voluntary elements and covered not only carbon dioxide but all six gases mentioned in the Kyoto Protocol. The British took part in the early phases as soon as they knew the EC was writing a proposal. The UK is known for its well-resourced and efficient internal coordination system. In addition, in view of its national trading scheme, the UK had highly relevant knowledge and experience available. During these early phases, the UK had bilateral contacts with the EC to preconcert the EU system and the UK national scheme. These dialogues made clear that the UK was in favour of an EU-wide ET system, but that its view of how it was to be conceived was different from that of the Commission. The British tried to explain their national system to the other member states, which had no experience in this matter, and to the EC during both the expert meetings and the Council negotiations. The ultimate goal of this was not to upload the UK system to the EU entirely, but rather to be able to maintain it at the domestic level. To achieve this, only certain elements had to be uploaded. The British did not manage to convince the Commission, however.

In the beginning of the Council phase, the British continued their efforts to protect their national ET system, if necessary by uploading parts of it to the EU level. This was done by way of a tough negotiating approach. As one respondent mentioned, 'We make clear to the other member states what we want and what our fallbacks are.' Overall, the strategy of the UK so far can be described as that of a defensive forerunner. This defensive attitude changed, however, when the British saw that no other member states were on their side and that they were not going to be able to fully preserve the national scheme. The British changed their strategy and, given that they remained basically in favour of an EU-wide ET scheme, retreated to a more pragmatic negotiating style. But they still focused on maximising their national freedom. One respondent claimed, 'As long as the UK gets what it wants, it does not matter what other member states get.' By this change of strategies, the UK moved slightly towards a constructive-pusher style. After the UK had changed its strategy, it had a number of meetings with the Danish Presidency to discuss the UK's position and fallbacks. This information about what Britain could and could not accept was apparently taken into account by the Presidency. In addition, several ad hoc alliances were formed by the UK during the process.

Denmark

Denmark was the first country within the EU to launch a national trading scheme. The national ET system was part of a political agreement of March 1999 and passed Parliament in June 1999. Owing to a delay in the EU approval, the system started only in January 2001. Although the system was small, since merely 15 companies participated, Denmark established its position as first mover and potential pusher-by-example.

In the early phases of the ET process, Denmark participated on what respondents referred to as 'a normal basis'. This normal basis implied some lobby activities, suggestions for the green paper, sending well-prepared experts, and having informal discussions with the EC. A respondent mentioned that there were 'lots of lunches, coffees, and beers'. It should be remembered that ET was new to the EC also, so the Danish knowledge and experience was relevant to it. More important, Denmark inspired the Commission with its (small) national system. Like the UK, the Danes had several bilateral meetings between mid-1999 and mid-2000. It turned out that Denmark and the EC had the same ideas about certain aspects of ET, including the focus on carbon dioxide only, instead of all six Kyoto gases. Denmark thus took its chance to exert considerable influence in the Commission phase. As one respondent remarked, 'During the time the Commission writes a proposal, all member states are equal. In the Council, the votes count.'

When the ET proposal was officially submitted to the Council in late 2001, Spain held the Presidency. The Spanish mainly investigated the member states' positions and did not try to come to an agreement. In the first half of 2002, the Danes took over the Presidency. This forced them to assume a more neutral role and thus they ran the risk of losing the benefits of their successful preparatory work in the Commission phase. In our interviews, we found indications that the Danes, in order to avoid this risk, entered into a basically tacit, but relatively long-standing strategic alliance with the UK. As Council President, they were interested in keeping the British aboard, but at the same time they probably tried to make sure that the UK acted generally in line with the Danish position. Denmark, in sum, put more emphasis on building an EU system for ET than on uploading its pre-existing national scheme and thus acted as a constructive pusher in this case.

Evaluating the Strategies

This section reviews the processes described above in the light of the expectations developed earlier (see Box 20.1). For the convenience of the reader, the expectations are reproduced before discussing them.

Expectation 1: Cooperation with the Commission

> Pushers concentrate more on the Commission phase than do forerunners and opters-out, because pushers aim to maximise harmonisation and cooperation in the EU. Constructive pushers primarily strive to formulate proposals together with the EC; pushers-by-example concentrate less on working together with the Commission, but focus on being a fast movers.

For all member states, it is more effective to influence the Commission while writing proposals than during the Council negotiations, when the basis for the directives is already on the table. The ET case shows that not only pushers-by-example and constructive pushers, but also defensive forerunners try to influence the EC. It should be added, however, that forerunners are less likely to succeed, as the Commission concentrates on harmonisation, while defensive forerunners do not. For example, the UK as a defensive forerunner in the ET case tried to influence the EC, but was superseded by the Commission. If this happens, the defensive forerunner has to carry on to the Council, as the UK did.

We found that, as expected, pushers focus intensively on the Commission phase. Pushers are in favour of harmonisation, as is the EC, and they can offer their knowledge and political support. The Commission phase, therefore, appears to be a more appropriate time to push for harmonisation than the Council phase. During Council negotiations, all member states, even pushers, try to get certain national elements accepted. Because of this focus on national interests, it is harder to draw attention to harmonisation issues. In both the ET and the carbon dioxide taxation cases, the pushing member states showed their support and assisted particularly during the Commission phase.

Differences between pushers-by-example and constructive pushers in the Commission phase are subtle. Both make an effort to establish a good relationship with the Commission. But pushers-by-example have an example to show the EC, while constructive pushers try to cooperate with the Commission to formulate EU proposals. Member states can combine these strategies, as Denmark did in the carbon dioxide taxation case (see also below). In order to play the role of constructive pusher or pusher-by-example in the Commission phase, member states need to have a clear conception of their own strategic goals. They also need resources (including people, knowledge, and experience) as well as a good internal coordination system. The Netherlands was a relatively poor constructive pusher in both cases, participating little in the early phases because of a weakly coordinated internal system.

It should be noted, finally, that the EC also plays an active part in this phase. Essentially, the EC can choose which member states to listen to. If the British were superseded in the ET case, this can also be interpreted the other way round, in terms of the EC hearing only what it wants to hear. It can shop around among all knowledgeable, participating member states and listen to those with similar ideas.

Expectation 2: Expert meetings

> All green member states supply well-prepared and competent input to the various EU expert committees during the Commission phase. However, forerunner member states' experts advocate their national interests, whereas pushers look for common solutions.

As expected, we found that both pushers and forerunners generally sent well-prepared experts to expert meetings in the ET case. Unfortunately, little information about expert meetings concerning carbon dioxide taxation could be gathered during the interviews. In the ET case also, few differences between constructive pushers, pushers-by-example, and defensive forerunners with regard to strategies employed in expert committees could be observed. We sensed, however, that the forerunner (the UK) focused more on advocating its national interest than did the two pushers (Denmark and the Netherlands).

It was mentioned by the respondents that the opportunities offered by expert meetings to promote national interests should not be overestimated. Experts act on a personal basis, even though they generally have a good idea of what their governments want. Most of them do not, however, get any instruction from their governments. In some member states, moreover, experts delegated to Brussels do not write official reports afterwards. All in all, experts have a strong tendency to think in terms of common solutions rather than national interests. This tendency is further enhanced by the working routine of most expert meetings, in which the EC lets everybody have his say without much discussion or negotiation.

Expectation 3: Negotiating style in the Council

> A pragmatic style is a common characteristic of constructive pushers and opters-out. Constructive pushers in particular focus on consensus. The style of pushers-

> by-example is pragmatic and consensus-oriented as well, but more aggressive in order to get certain national elements through. Defensive forerunners have a harder negotiating style.

With regard to this issue, most respondents pointed out that every member state has to make compromises, because 'that is how the EU works'. However, as assumed, there are differences between member states. The defensive forerunner in the ET case (the UK) started the negotiations as a tough negotiator. This can be related to the basic aim of the defensive forerunner, which is sovereignty, not harmonisation. The ET case showed that as long as defensive forerunners are able to maintain their national policies, they do not care about what the other member states achieve. This is in clear contrast to pushers, who aim at harmonisation and pay attention to what other member states do and get.

A further difference between the two types of pushers was observed. Particularly in the carbon dioxide taxation case, the country combining elements of a constructive pusher and a pusher-by-example (Denmark) used a tougher negotiating style in the Council than did the genuine constructive pusher (the Netherlands). Pushers-by-example implement their domestic policies with an eye to the EU process and are ambitious to upload at least certain elements. Constructive pushers are more focused on making compromises. Their primary goal is to agree on directives and, therefore, they are more willing to adjust their incrementally grown (or growing) national policies.

As is discussed in more detail below, member states can change strategies over time. During the ET case, the defensive forerunner (the UK) changed strategies when the negotiations did not proceed as expected. The defensive forerunner changed its strategy to become more flexible. Even then, however, compromises were only acceptable under certain conditions. The defensive forerunner safeguarded certain aspects of the national scheme by making clear to the Presidency what its fallbacks were.

Expectation 4: Alliance building

> Alliance building in the Council generally has a pragmatic and ad hoc character. Nevertheless, pushing member states – and especially constructive pushers – are more open to forming alliances than are defensive forerunners or opters-out.

Contrary to what was expected in the second part of the expectation, both cases showed that all member states form alliances, forerunners to the same extent as pushers. Under qualified majority voting, as applied in the ET case, member states simply have to form alliances in order to construct a majority. But also during the carbon dioxide taxation case, which came under the unanimity rule, member states formed alliances; no member state wanted to become fully isolated.

During negotiations, like-minded countries tend to find each other. This depends mainly on the interests of the countries involved, but also on the persons in charge. Like-minded persons also tend to find each other. In addition, maintaining large informal networks increases the chances of actually finding coalition partners. Thus, effective networks include not only the Permanent Representations based in Brussels, but also civil servants in other capitals and in the EC.

The first part of the expectation also seems to have been, at least partly, disconfirmed. Apart from several ad hoc alliances made in both cases, the ET case showed one alliance of a more strategic, long-standing character. We come back to this in more detail below.

Expectation 5: Uploading

> Pushers-by-example strive for full or almost full uploading, while constructive pushers are also willing to concede to a more moderate form of uploading. Forerunners and an opters-out have no particular interest in uploading.

In the cases analysed here, all member states having a national policy wanted to see certain national interests taken into account in the directive and, at the end of the day, all member states preferred to upload as much as possible. Nevertheless, there were differences between the strategies employed by the different role models and there were also differences with regard to their fallback positions. Defensive forerunners, for instance, can act as tough negotiators because they are not eager to have directives at all. Their first priority is to maintain pre-existing national systems. As used by the UK in the ET case, uploading is a possible means to achieve this goal, but not the ultimate objective. Pushers-by-example are hard negotiators, too, but for different reasons. Contrary to defensive forerunners, they have a strong interest in a common policy and insist on having it on their own terms. It turned out that constructive pushers are the most pragmatic negotiators. They do not hold on to their incrementally grown policies as much as the others do. As long as satisfactory European policies are developed, they can live with concessions.

CONCLUSIONS

Having discussed the expectations individually and in quite some detail, we now draw together the main conclusions of this study. For this purpose, Table 20.3 first repeats the amended scheme of pusher and forerunner strategies in

EU environmental policy as developed at the end of section 2. As the reader may remember, the two parameters of the scheme were the level of activity during the Commission phase and the negotiating style during the Council phase.

A careful look at Table 20.3 reveals that the original category of the 'opter-out' has been replaced by the ideal-type of the 'late starter' (field (c)). Liefferink and Andersen had defined the opter-out as a country with an incrementally developed national policy which more or less unexpectedly turns out to be ahead of the rest of the EU.[32] It is doubtful whether this situation occurs often; it did not in the present cases. Whether it occurs or not, the category of the opter-out does not seem to make sense in the present scheme. There are countries that perform weakly in the Commission phase and turn to a more active, probably constructive strategy in the Council phase.[33] As we have seen with the Dutch in our cases, lack of resources and/or deficient internal coordination are plausible reasons for ending up in

Table 20.3 Extensions of strategies of green EU member states

	Council phase style	
Commission phase activity	Consensual and constructive	Adversarial and tough
Much	(a) Constructive pusher	(b) Pusher-by-example
Little	(c) Late starter	(d) Defensive forerunner

field (c). If this happens, the term 'late starter' appears to be more to the point in the present context than 'opter-out'.

As far as the Commission phase is concerned, it was confirmed by our cases that both types of pushers influence the EC more actively than do defensive forerunners. This was related, among other things, to the observation that harmonisation is, so to say, the keyword of the Commission phase. This explains why pushers rather than forerunners are likely to find a listening ear with the EC in this phase. In the Council phase, in contrast, harmonisation arguments are less popular and national interests are paramount, giving forerunners more opportunities to flag their message. This is not to say that forerunners do not engage in the Commission phase at all. As pointed out before, it is only less likely that forerunners will be heard by

32 Liefferink and Andersen, 1998a.

33 As is discussed in more detail below, the most likely strategy for a late starter in the Council phase is a pragmatic one.

the EC. Small member states in particular have a strong interest in making themselves heard in the early phases of the process. There is at least some truth in the slightly optimistic remark made by one of our respondents, quoted above, that all member states are equal in the eyes of the Commission, whereas it is the weight of the vote that counts in the Council. Most green member states recognise this, but are not always in a position to take full advantage of it. It is here that the new category of the late starter proves its value. As we have seen, a relative lack of domestic knowledge and experience, combined with some flaws in the working of the internal coordination system, prevented the Netherlands from forcefully engaging in dealings with the EC and acting as a full-fledged pusher of the ET issue.

Our expectation regarding expert meetings was not confirmed. As discussed above, member states are apparently not strongly inclined to use expert meetings for their own purposes. One reason for this might be that member states simply do not consider these meetings to be sufficiently important (which would not be surprising considering the enormous number of expert meetings held in Brussels every year!) and, therefore, do not develop firm strategies to exert influence during these meetings. The effect of this is enhanced by the character of the meetings themselves: meetings are considered to be 'hearings' rather than forums for discussion and negotiation.

The examination of the green member states' negotiating styles in the Council produced a number of interesting findings. As expected, the defensive forerunner appeared to be the toughest negotiator. The insistence on national autonomy and the low interest in common policies make it possible for defensive forerunners to play it hard. Similarly, and more than assumed beforehand, the pusher-by-example was found to employ a tough negotiating style. This can be explained by the fact that pushers-by-example, like defensive forerunners, purposefully implement national environmental policies that they do not wish to give up and in fact seek to upload to the EU level. Both apparently can make a credible effort to stand alone if necessary, rather than make too many concessions. However, pushers-by-example at the same time strive for harmonisation and may change, under certain circumstances, to a more consensual negotiating style in order to reach an agreement. This implies that pushers-by-example are more likely than defensive forerunners to shift their strategy during the negotiations (towards a more constructive and consensual style). The place of the pusher-by-example in Table 20.3, therefore, is less fixed than that of the defensive forerunner.

As described in the original article, the constructive pusher turned out to be a pragmatic negotiator, preferring a harmonised policy at the EU level and, therefore, willing to make compromises. It is interesting to note that this style of negotiating also implies keeping a keen eye on what other member states do and get. Whereas defensive forerunners (or other member states, for that matter) are interested in only making sure that everybody has a more or less fair share of the costs and benefits associated with the new policy, constructive pushers also want to ensure that the harmonised policies assume the preferred form.

Similarly, the late starter in our cases used a pragmatic approach in the Council. For late starters, this is arguably the most likely situation. It is in practice possible for late starters to have a tough negotiating style, for instance, when member states are very late in finding out that proposed EU measures may severely harm their existing domestic policies. Tough, adversarial strategies in the Council are to be expected rather from defensive forerunners or pushers-by-example, who are eager to maintain their pre-existing national policies. Genuine defensive forerunners and pushers-by-example, however, are aware of being ahead of the others and, therefore, are not likely to start late. The combination of a late start and a tough, defensive style in the Council, in other words, cannot logically be expected to occur.

No clear difference was found between the four roles as regards the eagerness to build alliances. In fact, all member states have an interest in doing so. However, two insights can be added to the original theory. First, as assumed by Liefferink and Andersen, having more or less similar positions in a particular situation was found to be the most important reason why countries form alliances. In addition to that, the personal basis, based on informal networks in Brussels, appeared to be a key factor as well. Like-minded persons also tend to find each other. Thus, the social skills of diplomats and civil servants working in Brussels are an important resource for member states, next to, for instance, expertise or a well-functioning domestic coordination system. Second, although the 1998 article firmly excluded the appearance of alliances during the negotiating process other than those with an ad hoc character, the ET case has shown one long-standing, structural alliance. The case suggests that the Danish Presidency built up a relationship with the UK extending over the entire period of the Presidency, perhaps in order to be sure that – through the British – the Danish voice be heard during the negotiations. The most interesting feature of this alliance is that the Presidency was involved.

It is difficult, if not impossible, to make generalisations about the circumstances in which structural alliances might appear. Further research is needed, but nevertheless an attempt is made here. First, long-standing alliances seem more likely to emerge if the Presidency is involved. The Presidency is in a better position to direct alliances and to assure that certain aspects be taken into account in a balanced way. Besides, the Presidency has a natural interest in constructing (qualified) majorities. Second, a long-standing alliance seems to be more likely if one or more large member states are involved. Large states are more dominant; it is easier for them to say no as their vote in the Council carries more weight.

Analysing the findings at a more general level, one must note that, contrary to the four distinct positions described in the 1998 article, strategies can both be combined simultaneously and change over time. This dynamic character of the member states' strategies stresses once again that countries are not always tied to playing the same role. The Netherlands is not always a late starter, nor is the UK always a defensive forerunner or Denmark a pusher-by-example. Which position and strategy member states take may differ from issue to issue, and even in respect of one issue, and from department to department. In addition, there may be external reasons for changing strategies, such as changes of government, economic crises, or major environmental events.

The possibility of combining two strategies at the same time implies that one position of the original theory can lead to two parallel strategies at the same time. For example, as we have seen with Denmark in carbon dioxide taxation, a member state basically performing a pusher-by-example role can combine this without internal contradiction with a constructive-pusher strategy by working closely with the Commission and assisting in the formulation of a proposal. If member states combine two strategies, it is important that the coordination within those states works well; different departments of the government are not likely to effectively maintain parallel strategies without a clear, over-arching view.

To change strategies over time is obviously a strategic decision. If the original strategies do not work out as wished, states may decide to change them. In the ET case, the UK moved from a defensive-forerunner role towards a constructive-pusher role. Other changes of strategy are also conceivable. Pushers-by-example, for instance, can shift from a tough negotiating style towards a consensual negotiating style if, during negotiations, they feel the need to put a stronger emphasis on the harmonisation aspect of their original position. Pushers-by-example, which from the outset combine the desire to shape EU policy with the wish to retain their pre-existing domestic policies (in practice leading to a strong ambition to upload the domestic models), may in general be more liable to changing their strategy during the negotiations than are constructive pushers or defensive forerunners, who both have more clear-cut priorities. Earlier in this section, we referred to the possibility of a late starter taking different directions in the Council phase. However, to what extent some strategies are more apt to change than others, and in which preferred sequence, so far remains unclear. Firmer claims in this regard can only be made on the basis of additional research. The important insight gained in this study is that strategies may be combined, both simultaneously and over time.

REFERENCES

Andersen, M. and D. Liefferink (1997), 'Introduction: The impact of the pioneers on EU environmental policy', in M. Andersen and D. Liefferink (eds), *European Environmental Policy: The Pioneers*, Manchester: Manchester University Press.

Börzel, T. (2002), 'Pace-setting, foot-dragging, and fence-sitting: Member state responses to Europeanization', *Journal of Common Market Studies*, **40**(2), 193-214.

Dunn, W. (1994), *Public Policy Analysis: An Introduction*, University of Pittsburgh, Prentice Hall: Englewood Cliffs.

Héritier, A. (1994), ''Leaders' and 'laggards' in European Clean Air Policy', in B. Unger and F. van Waarden (eds), *Convergence or Diversity? Internationalization and Economic Policy Response*, Aldershot: Avebury, 278-305.

Héritier, A. (1996), 'The accommodation of diversity in European policy-making and its outcomes: Regulatory policy as a patchwork', *Journal of European Public Policy*, **3**(2), 149-167.

Jordan, A. (2002), *The Europeanisation of British Environmental Policy: A Departmental Perspective*, Basingstoke: Palgrave Macmillan.

Jordan, A. and D. Liefferink (eds) (2004), *The Europeanisation of National Environmental Policy*, London: Routledge.

Kingdon, J. (1984), *Agendas, Alternatives and Public Policies*, Boston: Little, Brown and Company.

Klok, J. (2002), *Negotiating EU CO_2/Energy Taxation; Political Economic Driving Forces and Barriers*, Copenhagen: AKF Forlaget.

Liefferink, D. and M. Andersen (1998a), 'Strategies of the green member states in EU environmental policy-making', *Journal of European Public Policy*, **5**(2), 254-270.

Liefferink, D. and M. Andersen (1998b), 'Greening the EU: National positions in the run-up to the Amsterdam Treaty', *Environmental Politics*, **7**(3), 66-93.

Pellegrom, S. (1997), 'The constraints of daily work in Brussels: How relevant is the input from national capitals?', in D. Liefferink and M. Andersen (eds), *The Innovation of EU Environmental Policy*, Oslo: Scandinavian University Press.

Putnam, R., (1988), 'Diplomacy and domestic policies: The logic of two-level games', *International Organization*, **42**(3), 427-460.

Schout, A. et al. (2001), *Environmental Integration and Coordination of European policies: The National and European Administrative Challenges in Perspective*, final report, 2000.04.045, Maastricht: European Institute of Public Administration.

Zito, A. (2002), 'Integrating the environment into the European Union: The history of the controversial carbon tax', in A. Jordan (ed.), *Environmental Policy in the European Union*, London: Earthscan.

21. The Dispersion of Authority in the European Union and its Impact on Environmental Legislation

Ludwig Krämer[1]

SUMMARY

The achievements and limitations of environmental legislation by the European Community (EC), as well as the possible effectiveness of a central environmental authority, are the central topics of this chapter. After a description of the history of EC environmental legislation, the institutional framework and achievements of EC legislation pass in review. This is followed by an analysis of factors that influence European environmental legislation. Next, I discuss the relationships between EC and national legislations, and between environmental and other policies. The question of whether EC standards are sufficiently stringent is answered negatively and major shortcomings are identified. A powerful central environmental authority would not have been more effective than the present regime; four directives are used as examples. I conclude that the stringency of EC environmental legislation is mainly determined by political will – dispersion of authority is not so important – and that, notwithstanding major achievements at the European level, environmental interests have been insufficiently represented and considered, leaving concerns for the future.

INTRODUCTION

In the present study, it was examined whether the absence of a European Union (EU) government and, therefore, of a uniform, coherent environmental

[1] The author expresses his personal opinion only.

policy has influenced the stringency of EC environmental legislation.[2] I start by providing a short description of the situation prior to 1987, when a section on the environment was inserted into the EC Treaty. The institutional aspects of the EC system and their impact on the stringency of legislation are then discussed. Next, the possibilities and limits of EC member states in adopting more stringent legislation and environmental aspects in other policy areas are outlined. The discussion then turns to whether EC standards are strict enough and whether a more centralised authority would be more effective. Finally, I provide some concluding remarks.

EC ENVIRONMENTAL LEGISLATION BEFORE 1987

The environment was not mentioned in the EC Treaty of 1958. European environmental policy was conceived in the early 1970s. In the absence of explicit provisions in the EC Treaty, the European Commission[3] proposed to establish, in this sector, action programmes[4] which should fix objectives, principles, and priorities for action; it was followed in this approach by the Heads of State and Governments.[5] The EC principles for environmental action contained from the beginning, and long before a chapter on environmental policy was inserted into the EC Treaty, two basic conditions: first, European environmental policy should not put into question environmental achievements which had been reached within member states. Second, European environmental measures should take into consideration the diversity of the environment in the member states.[6] Neither aspect caused

---

[2] The EU is based on the European Treaty. The Treaty on the EC is an integrated part of the Treaty on the EU. However, as environmental legislation may, under the present legal system, be based on the EC Treaty, but not on the Treaty on the EU, the notion of 'EC' is used throughout the following text. For further reading on EC environmental law, see: Jans, 2000; Krämer, 2003.

[3] The European Commission is referred to as 'Commission'; the abbreviation EC is reserved for European Community.

[4] See the European Commission's (1972) proposal for an EC environmental action programme.

[5] It was for the first time at the Paris Summit of Heads of State and Governments in October 1972 that the EC institutions were invited to elaborate an EC environmental action programme, (European Commission, 1973a).

[6] See European Commission (1972: 6): 'However, the EC must take care to leave national, regional, and local bodies as much freedom of discretion as possible. Harmonisation must be aimed at only to the extent necessary for a *minimum* degree of protection for the whole EC and for the free circulation of goods as well as undistorted competition. Solutions must be sought which take into account the multiplicity of situations. The diversity in geographical and natural situations, and between the tasks of the different regions, may sometimes require the application of *different standards'* (translation by the author, emphasis added).

The First Environmental Action Programme states that an EC environmental policy should aim at coordinated and harmonised progress 'taking into account the regional differences existing in the Community' and 'without, however, hampering potential or actual progress at

problems, as none of the nine member states of that time was keen to work towards an environmental policy which would establish a uniform EC policy such as in the food and industrial products sectors. It was clear to everybody that between the south of Italy and the north of Scotland, or between Ireland and the Alps, differences existed as regards water scarcity, soil erosion, forest management, and the control of pollutants.

EC environmental policy thus deviated from the beginning from the principle of Articles 94 and 95 EC, which provided for the harmonisation of national legislations for the establishment and functioning of the internal market. EC legislation was developed which took into consideration the necessity not to hamper national environmental achievements, on the one hand, and to take into consideration the different economic and ecological situations within the EC, on the other hand.

As regards the first aspect, EC environment policy developed minimum directives.[7] This constituted, in the 1970s and 1980s, a considerable innovation at the EC level. Under the system of Articles 37, 70, 94, and 308 EC, minimum directives did not exist in the areas of agricultural and transport policy or the harmonisation of legislation, which were among the few sectors under EC legislation. In contrast, environmental legislators tried from the beginning to leave member states the possibility of maintaining or introducing more stringent national legislation, even after the adoption of EC legislation in the specific sector. A typical example is found in one of the first environmental directives, Directive 75/440 on the quality of surface water,[8] which provided in Article 6, 'Member States may at any time fix more stringent values for surface water than those laid down in this Directive'.

Similar clauses were progressively introduced in most directives in the areas of water[9] and air pollution, nature protection, and horizontal action such as environmental impact assessments. Such minimum clauses were not inserted into product-related directives, while the practice in waste-related legislation differed: some directives contained such a minimum clause, while others did not. This reflects the uncertainty as regards the classification of

---

the national level' (European Commission, 1973b: 7). See also European Commission (1983: 1), which mentions the importance of considering 'the need to take account of the different economic and ecological conditions and the differing structures of the Community'.

7 Such directives were also developed in the sector of consumer policy. It should be noted that, between 1973 and 1989, the Commission's environmental and consumer protection administration was placed in the Environment and Consumer Protection Service, which became a Directorate General in 1982.

8 Directive 75/440 concerning the quality required for surface water was intended for the abstraction of drinking water in member states (European Commission, 1975a).

9 For example, Directive 76/160 on bathing water, Article 7(2) (European Commission, 1976a); Directive 76/464 on the discharge of dangerous substances into waters, Article 10 (European Commission, 1976b); Directive 80/68 on groundwater, Article 19 (European Commission, 1980a); and Directive 80/778 on drinking water, Article 16 (European Commission, 1980b).

waste legislation: where the accent was placed more on the product-related aspects of waste, the legislation did not contain a minimum clause; where environmental aspects had priority, the minimum clause was retained.

With regard to the second aspect, the diversity of the environment in member states, EC environmental legislation was not outspoken. This was because the natural or geographical environment was rarely the subject of EC legislation. The two exceptions for which EC provisions were established also provided for regional diversity: Directive 79/409 on wild living birds[10] took account, within the different provisions, of the diversity within member states,[11] while Directive 78/659[12] went so far as to allow member states to designate themselves those fishing waters to which the Directive should apply; only in these waters were clean-up measures to be taken. The effect of this Directive was limited, as many member states either did not designate fishing waters that came under the Directive or designated waters which were clean anyway. For the rest, nature-protection legislation concerned matters such as the hunting of seals and whales, or trade in endangered species. Other environmental legislation dealt with pollution caused by humans or with human activities. In this legislation, the diversity of natural or geographical situations in member states rarely played a role.

The EC used a number of other methods in order to reach EC-wide consensus on environmental legislation despite the wide variety of environmental situations and the existence of an active environmental policy in only a minority of member states. One such method was alternative standard-setting. For example, as no agreement could be found between the United Kingdom (UK) and the other EC member states regarding whether the EC should fix, for the discharge of toxic, persistent, and bioaccumulative substances, emission standards or concentration (quality) standards,[13] the EC opted for emission standards, but allowed a member state to introduce quality standards where it 'can prove ... that the quality objectives ... are being met and continuously maintained'.[14] In Directive 80/779, which fixed air quality standards for sulphur dioxide and suspended particulates, a similar alternative

---

[10] Directive 79/409 (European Commission, 1979a).

[11] Directive 79/409, Art.3: '[Member States are to] preserve, maintain or re-establish a sufficient diversity and area of habitats'; Art.4: 'Member States are to designate habitats for the most endangered species "the most suitable territories"'; Art.9: 'Member States may derogate from several provisions of the Directive under certain conditions; Member States have the possibility to provide for rules on the hunting of birds which allow them to largely maintain traditional hunting practices (France, Belgium, Italy, Malta)' (European Commission, 1979a).

[12] Directive78/659 concerns the quality of fresh waters needing protection or improvement in order to support fish life (European Commission, 1978b).

[13] Prior to 1993, environmental directives had to be adopted unanimously.

[14] Directive 76/464, Article 6(3) (European Commission, 1976b).

approach was adopted.[15] As Germany stated that it was not able to meet the Directive's standards by using the measuring method which was favoured by the other member states, the Directive allowed a different measuring method to be used, though it admitted that the two methods were not completely equivalent.

The attempt to fix alternative standards did not exclude products. Directive 75/716 fixed the sulphur content of gasoils and introduced two types of gasoil: gasoil A was to be used everywhere; gasoil B, having a higher sulphur content, was to be used in specific, unpolluted areas.[16] By fixing areas more or less arbitrarily, member states could avoid taking measures to reduce the sulphur content of gasoil. Directive 78/611 fixed the lead content of petrol at a maximum of 0.40 gramme per litre, but allowed member states to fix another, lower standard, provided that this standard was not lower than 0.15 gramme; and Ireland was allowed to use a standard of 0.64g until 1986.[17] This differentiation was introduced because, in 1978, Germany already had a lead standard of less than 0.40g and was not prepared to abandon it; and Ireland successfully argued that it was lagging behind economically. Another example concerns Directive 73/404, which required detergents to be 90 per cent biodegradable.[18] However, the testing methods which were fixed determined that the biodegradability should not be less than 80 per cent, which de facto introduced different standards.

Other methods of taking into consideration the diversity of environmental situations in member states were the following: member states were allowed to determine themselves where they applied EC legislation; member states were allowed to determine areas where they would not, for a period of time, apply the requirements of EC legislation; and individual member states were allowed specific derogations from the requirements of an environmental directive.

The most relevant aspect of differentiation was the omission of the Commission to ensure a proper application of the provisions agreed. In the early years, the Commission was more or less of the opinion that member states were in charge of implementing and applying the rules that had been agreed at the EC level. It took action under Article 226 EC against a member state when that member state had (obviously) not correctly transposed the provisions of a directive into national law. But for the rest, it did not look into the details of national legislation and preferred to ignore the details of application of EC environmental law within member states.

---

[15] Directive 80/779 (European Commission, 1979b).
[16] Directive 75/716 (European Commission, 1975c).
[17] Directive 78/611 (European Commission, 1978a).
[18] Directive 73/404 (European Commission, 1973c).

Most of the above-mentioned methods were unsuccessful. The process of regulating discharges of dangerous substances into water, which progressed slowly after 1976, came to a complete halt in 1991. The alternative standards for pollution or for measuring methods were progressively abandoned. The same applies to the differentiation of product-related standards and of derogations. The only approach that succeeded, and was, in 1985/1987, incorporated in the new chapter on environmental policy in the EC Treaty, was the right of member states to maintain or introduce more stringent environmental protection standards once the EC has legislated (Article 176 EC). This clause is uncontested nowadays, and its deletion would inevitably lead to considerable conflict between member states and the EC as regards responsibility for environmental policy matters.

ENVIRONMENTAL LEGISLATION WITHIN THE FRAMEWORK OF THE EC TREATY

Institutional Aspects

Since the introduction of Articles 174 to 176[19] to the EC Treaty in 1987 and the introduction of majority voting in most environmental matters by the Maastricht Treaty in 1993, the procedures for standard-setting under Articles 95 and 175 EC are largely the same: a proposal for legislation is made by the Commission, on which the European Parliament and the European Council[20] decide jointly using the procedure of Article 251 EC; the Economic and Social Committee and the Committee of the Regions give an opinion on the Commission's proposal. The main differences lie in the residual powers which member states have after the adoption of the EC legislation: Articles 95(4) to 95(8) allow the maintenance or even the introduction of more stringent national legislation only under strict conditions, which are restrictively monitored by the Commission and which allow such a deviation to succeed only in exceptional circumstances. These conditions are absent from Article 176 EC.

---

[19] Article 174 of the EC Treaty fixes the objectives, principles, and conditions of EC environmental policy. Article 175 determines that decisions in environmental matters are normally taken by the European Parliament and the European Council jointly; the Council decides with a qualified majority. In certain, expressly enumerated cases, the Council decides unanimously. The implementation of measures that were decided at the EC level is in the hands of member states. Article 176 provides that, where an EC measure was adopted on the basis of Article 175, member states may maintain or adopt more stringent environmental measures, provided that these measures are compatible with the other provisions of the EC Treaty.

[20] The European Parliament and the European Council are referred to as 'Parliament' and 'Council', respectively.

The environmental legislation which is elaborated under Article 175 EC is usually not controversial in the Council. The first reason for this is simply that the environment knows no frontier and the great majority of solutions that are found by the Community are, from an environmental point of view, common sense. The Council's environmental working groups, in which the national environmental attachés of the Permanent Representations – supported by officials from the environmental departments of member states – gather, and the environmental ministers normally find consensus on basic questions. Controversies mostly concern details. The environmental administrations of the national governments and of the Commission – which frequently meet at conferences, international conventions' meetings (where the EU tries to speak with one voice), on the occasion of the six yearly Council meetings, and at other gatherings – are regularly provided with scientific data, information, and analyses of environmental issues. This practice has been strengthened by the numerous papers, reports, programmes, and communications issued by the Commission. This has led to analyses of the problems by various national administrations that are often similar. The policy question concerns the kind of common EC standards that should be developed, and the discussions inside the Commission's working groups and inside the Council often centre on this question. Environmental legislation normally does not have significant competition implications or involve trade-offs with other areas.

As Article 174 EC requires the EC to aim at a high standard of environmental protection, the Commission tends to suggest environmental legislation which is relatively strict. While the Parliament, driven by its influential Environmental Committee, regularly sides with the Commission, the question in the Council is whether the Commission's proposal is not too strict – which mainly corresponds to the question of whether the obligations it imposes on economic operators are too strict. The Council rarely amends a Commission proposal to make it more stringent, though the Parliament occasionally succeeds in introducing stricter amendments via the co-decision procedure of Article 251 EC.[21] In 1991, I stated that it 'is exceptional that a Commission proposal is significantly amended during the legislative process. It is the Commission which selects the topics for regulation and decides on shape and mostly on the content of Directives, fixes priorities and changes orientations'.[22] Despite the introduction of majority voting and co-decision procedures, this conclusion is still correct.

---

[21] The only significant success of the Parliament in the last years is to have inserted into Article 16 of Directive 2000/60, which establishes a framework for Community action in the field of water policy, provisions according to which the discharge of the most dangerous substances in water should progressively be stopped within the next fifteen years (European Commission, 2000).

[22] Krämer, 1992: 153.

EC environmental policy has been successful over the last thirty years. The most obvious signs of this success are: [23]

- The establishment of environmental administrations and elaborate infrastructures (ministries, agencies, inspectorates, research centres, environmental non-governmental organisations (NGOs) at national, regional, and local levels
- The adoption of some 250 pieces of EC environmental law (directives, regulations) which cover almost all sectors of environmental concern
- The introduction of numerous pieces of environmental law which were innovative, even with regard to the so-called 'green' EC member states and the majority of which would, without the EC, have been introduced only in a minority of EC member states [24]
- The capacity to perceive the environment in Western Europe as a joint venture, to progressively regard environmental threats as common, and to elaborate joint solutions – this capacity is reflected by the adoption of six EC-wide Environmental Action programmes, which span the period from 1973 to 2010, and fix objectives, principles, and priorities for EC environmental action; the fact that relatively little environmental legislation was adopted by the EC member states at the national level demonstrates this awareness that environmental protection policy is a common undertaking
- The serious and continuous effort to monitor the application of environmental law in all parts of the EC
- The continuous effort to participate in global environmental discussions as an EU that speaks with one voice and tries, as the only region in the world, to reconcile economic growth and environmental protection

Influences on the Stringency of Legislation

These 'signs of success' are more concerned with external factors than with the stringency of EC environmental standards. The fact that between 60 and almost 100 per cent of EC member states' environmental legislation is based on EC provisions[25] does not reflect the stringency of that EC legislation. This

---

[23] See Krämer (2002) for a more detailed assessment of EC environmental law.

[24] Examples are binding standards for drinking water, coastal (bathing) water, environmental impact assessments for projects, plans, and programmes, permit requirements for industrial and waste management installations, natural-habitats protection, access to information, industrial-accident prevention, and pollution-emission registers.

[25] This figure is based on the assessment of national administrations themselves. The 100 per cent mainly refers to the Cohesion-Fund countries (Ireland, Portugal, Spain, and Greece).

stringency is influenced by a number of factors, including:

- The objective of Articles 175 and 176 EC. As mentioned above, Article 175 EC does not require a harmonisation of national legislation; this is an important difference from the standards that are set under Article 95 EC. Unification or harmonisation of national laws is not an objective of EC environmental policy. The objective is rather to ensure that there are common minimum environmental standards throughout the EC which are respected everywhere. This objective allows EC environmental legislators to indicate to member states which have more sophisticated national provisions that they may maintain their provisions even where the EC as a whole is not (yet) able to introduce the same degree of protection.
- The consensus of environmental administrations. The EC and the national environmental administrations were in the past and are still largely in agreement. The only significant exception at present is in the waste sector, where the power struggle between national and EC administrations is not yet over. In this sector, the EC has not yet demonstrated, through its strategies and in particular through its waste-management legislation, that it is more effective than national administrations and policies.[26]
- The different political value of environmental policy in EC member states. This value indirectly influences EC standard-setting. Member states with a marked interest in environmental matters try to occupy high administrative or political posts in the Commission's administration, and in this way steer the Commission's policy towards particular approaches. This is one of the most effective ways of influencing EC environmental policy, though it needs strategic planning and great breadth.

Member states which intend to influence the orientation of EC environmental policy need to have a long-term strategy. For example, the shift from emission limits to quality standards for water pollution and for industrial installations[27] was largely influenced by the UK, which vetoed opposing decisions under the unanimity rule until 1991/93. When majority voting was introduced, the UK managed – with its concept of integrated pollution prevention and control – to persuade economic operators from other member states and, later, the majority of administrations that emission limits were too expensive and favoured strong operators. In another case, Sweden and its Nordic allies managed to have measures against acidification adopted by the

[26] For reasons of space, this controversy is not discussed here.

[27] Directive 96/61 concerns integrated pollution prevention and control (European Commission, 1996a).

EC[28] and will shortly make the EC follow its approach to chemicals and products. Germany successfully pushed for a loosening of standards for the contained use of genetically modified micro-organisms.[29]

Preventing the EC from adopting stringent measures is more difficult, in particular where the policies of member states do not have alternative concepts to offer. Examples of this kind can easily be found for most member states. Thus, the EC measures on the pollution of waters by nitrates from agricultural sources did not meet with much enthusiasm in the Netherlands and Denmark, yet were adopted. Germany was not in favour of measures such as environmental impact assessments and access to environmental information, but did not veto their adoption for political reasons. Generally, it may be argued that every EC government is in favour of the protection of the environment, provided that this protection does not cost money, jobs, or the competitiveness of economic operators. The legislation on habitat protection[30] was probably not popular in any member state, but owed its adoption to the consensus of the central nature-protection administrations and to the fact that no realistic alternative measure was visible in any member state or at the international level.

More Stringent National Environmental Legislation

Article 176 EC remains important in the sense that member states with more stringent environmental measures are normally free to maintain this legislation after the adoption of EC legislation. This does not apply to product-related legislation, as this is based not on Article 175 but on Article 95 EC and this provision requires – for reasons of free trade and undistorted competition ('level playing field') – uniform legislation in so far as possible. However, the practice of the last fifteen years shows clearly that Article 176 EC is rarely used to develop or maintain domestic environmental policy. One of the main reasons for this is the competitive disadvantage which the domestic economic operators would have with regard to competitors from other member states. Scandinavian member states, for example, signalled to the Commission that this was the reason why they did not intend to increase the carbon dioxide tax in their countries. Other examples concern the extension of environmental impact assessment to projects which are not covered by Directive 85/337, the extension of the permit requirements under

[28] Directive 2001/81 prescribes national emission ceilings for certain atmospheric pollutants (European Commission, 2001c).

[29] Directive 1990/219 concerns the contained use of genetically modified micro-organisms (European Commission, 1990); amended by Directive 98/81 (European Commission, 1998).

[30] Directive 92/43 deals with the conservation of natural habitats and of wild fauna and flora (European Commission, 1992).

Directive 96/61 to small and medium-sized installations which are not covered by that Directive, and stricter standards for drinking water, and air and water pollution. Under the present majority of conservative parties, the Austrian Parliament even went so far as to adopt a resolution not to introduce or maintain national legislation that went further than EC legislation.

Other reasons include the costs of investments: the assumption prevails in most member states that the degree of environmental protection agreed at the EC level is more or less an appropriate level; thus, there is no need to go beyond this level. Finally, the great majority of member states do not have autonomous national environmental policies, but follow the consensual evolution of EC environmental policy.

Environmental Policy and Other Policies

The influence of other policy aspects on EC environmental standard-setting is considerable and is greatest where a specific environmental issue touches on another sector of policy. The integration requirement of Article 6 EC,[31] which has existed in the EC Treaty since 1987, has not significantly changed this. The first reason for this is that standards which concern the environment and transport, the environment and energy, or environmental aspects of internal-market policy are not fixed by environmental policy-makers, but rather by transport, energy, or trade policy-makers and/or administrations. The priority in these policy sectors is often not the environment or even an appropriate balance between the environment and transport, etc., but rather specific transport, energy, or internal-market targets. A good illustration of this is the Sixth EC Environmental Action Programme: the Commission's proposal for the decision to adopt the programme contained few concrete targets and instructions for the transport and the energy sectors.[32] The Decision of the Parliament and (Environmental) Council indicated in Article 5 a number of concrete measures that were to be undertaken in these two sectors. However, it will be interesting to see to what extent the Commission will make the corresponding proposals, and the Energy and Transport Councils adopt them.

The second reason follows directly from this separation of responsibilities. Indeed, nobody has determined at the EC level what an environmentally sound transport, energy, or internal-market policy is or should be. The integration requirement of Article 6 EC has been, until now, an empty shell which has not much substantive content. There have been serious attempts in the agricultural and fishery policies to better take environmental requirements

[31] Article 6 EC: 'Environmental protection requirements must be integrated into the definition and implementation of the Community policies and activities referred to in Article 3, in particular with a view to promoting sustainable development'.

[32] European Commission, 2001a.

into consideration. These efforts were mainly influenced by the reduction of fish stocks and the fact that the EC agricultural policy of the past could no longer be financed; this might cause the long-term sustainability of the present approach to be questioned. In contrast, environmental considerations play almost no role in the legislation on nuclear energy, the performance of cars, and biotechnology.[33] The 'Cardiff process', which was intended to fulfil the requirement of Article 6 EC, led to the elaboration of a number of strategic papers for the integration of environmental requirements into the different sectors of EC policy.[34] However, these strategies were followed by few concrete implementation measures.

The limiting effect of other policy interests on the stringency of environmental legislation is particularly clear in the area of global activities. Here, the EC promotes stringent environmental rules in exceptional circumstances only. This is to a large extent because the international discussion is strongly influenced by trade priorities and by the fact that the United States is interested in 'negative global integration' (i.e., free trade), but opposed to 'positive integration' (i.e., stringent global environmental standards). In the last twenty-five years, the Montreal Protocol on ozone-depleting substances of 1988 was the only truly successful international environmental convention, and the few EC successes (the export ban on waste, the making of the Kyoto Protocol, and the banning of twelve chemical substances in the POP Convention) cannot lead to another conclusion, as the global standard-setting is, overall, desperately slow, general, and poorly enforced. This raises the question of whether the EU should not concentrate more on regional standards for the enlarged European continent.

As regards the initial question, whether the dispersion of authority in the EU has an impact on the stringency of environmental legislation, it is submitted that this question is wrongly posed. It may be possible to have one environmental authority, in particular in the form of an Environmental Ministry, in small countries such as the Netherlands or Denmark. However, in larger countries, such a central environmental authority is not an appropriate political form of organisation, as the examples of the UK, Germany, Italy, and Spain demonstrate. Even in France, a highly centralised country, many environmental standard-setting responsibilities are regionalised, such as basin management in the water sector and activities in the industrial-installation area. A centralised environmental authority in a region with 450 million people is neither possible nor desirable. Europe is far from being a consensual society: economic, social, cultural, religious, and historical

[33] In biotechnology, the concerns are almost exclusively those of human health.
[34] Such strategy papers were elaborated for transport, energy, agriculture, the internal market, development policy, fisheries, economic questions, and external affairs.

differences are still considerable, and influence attitudes towards the environment. Thus, a country which has a population density of more than 250 persons per square kilometre and where more than 15 per cent of the territory is below sea level has a different attitude towards the environment than a country with only 70 persons per square kilometre. The UK's and Ireland's attitude to discharges into waters has been influenced by their island status and the fact that their rivers are fast flowing. Spain, with its large land surface, may have a different approach to waste disposal on land than, for instance, Denmark. And attitudes towards the need to protect the environment continue to differ significantly in East and West Germany, fifteen years after the fall of the Berlin Wall.

Recent examples in Austria, Denmark, and the Netherlands also show that changes in national governments may impinge considerably on the attitude towards the stringency of environmental legislation, at the national and the European level. Fifteen Environmental Ministers or Prime Ministers of the stature of Svend Auken or Klaus Töpfer would make a different EC environmental policy and legislation than would fifteen ministers of the type of Lady Thatcher or Björn Lomborg.

The Stringency of EC Standards

Arising from the question of whether the environmental legislation established by the EC is strict enough is the question of whether this legislation is good. As the EU is not a state, and as environmental law, intended to protect an interest without a lobby that profits from such protection,[35] differs in this respect from other areas of law, the normal theories used to evaluate whether the extent to which legislation is good cannot be applied. However, the environment can be regarded and assessed. Assessment by measuring emissions, concentrations, pollutants, fauna and flora species, and noise levels, but also illnesses, nuisances, etc., constitute means to obtain results regarding the effectiveness of legislative environmental measures. This leads to the conclusion that EC environmental legislation is strict enough if it manages to protect, preserve, and improve the quality of the environment.

In its attempt to assess efficiency, the EU – in a misunderstanding of the Treaty provision that implementation of environmental measures decided at the EC level was the task of member states (Article 175(4) EC) – has been rather omissive. The EC does not report satisfactorily on the transposition and application of legislation. Nor has it proceeded, as in the early years of EC

[35] This comment cannot further be elaborated here. However, it seems obvious that spiders, flies, most other parts of the biological diversity, the groundwater, the landscape, babies, and future generations do not have a lobby in our society.

environmental policy, to lay down methods of analysis or measuring methods to report every two years and to integrate dispersed data. While the publications of the European Environment Agency appear to progressively fill some lacunae, at present, Decision 1600/2002 still states that 'monitoring, data collection and reporting requirements should be addressed efficiently in future environmental legislation',[36] a request that does not come too early.

Despite these imperfections of reliable data and facts,[37] it becomes obvious that, in many areas (climate change, stratospheric ozone depletion, air pollution, chemicals in the environment, waste and water management, soil degradation, prevention of technological and natural accidents and their consequences, biological diversity, and health aspects of environmental impairment), a considerable number of future measures are necessary in order to prevent further environmental degradation, and this despite the numerous environmental measures which were adopted in the past. The conclusion, then, must be that past legislation has not stopped environmental degradation and that, in general, the EC measures were not strict enough. This observation is confirmed by, for example, the statement in the Fifth Environmental Action Programme that there was 'a slow but relentless deterioration of the general state of the environment of the Community notwithstanding the measures taken over the past two decades'.[38] Another instance is Decision 1600/2002, which asked for a considerable amount of new legislation, because 'a number of serious environmental problems persist and new ones are emerging which require further action'.[39]

At the same time, this Decision, which listed 156 measures to be undertaken in the coming years, declared that 'legislation remains central to meeting environmental challenges', hopefully finishing the long-lasting discussion on whether voluntary commitments by economic operators should not or could not substitute legislation. Indeed, at present, economic operators are not prevented from voluntarily reducing their emissions or discharges into the environment, from producing using cleaner technologies and fewer resources, from applying less dangerous substances for toxic and dangerous chemicals, etc. If this were done, no legislation would be necessary, as, from an environmental viewpoint, it is irrelevant whether the environment remains unimpaired because of voluntary action or legal prohibition. However, day-to-day reality shows that there is no hope of humans behaving in an environmentally responsible way without rules that oblige them to do so.

---

36 Decision 1600/2002, laying down the Sixth Community Environmental Action Programme, Article 10(f) (European Commission, 2002).

37 Consider, for example, a recent report on Europe's environment (European Environment Agency, 2003).

38 European Commission, 1993.

39 Decision 1600/2002, 4th considerant (European Commission, 2002).

Economic greed which can give direct advantages to actors is a stronger spur than consideration for the environment, which does not yield direct profits.

The main lacunae of EC environmental legislation are:

- The lack of a systematic approach to reducing emissions into the air and the water. The attempt of 1976 to systematically reduce dangerous discharges into water was abandoned,[40] as was the attempt of 1984 to systematically reduce air emissions from industrial installations.[41] Attempts to have habitats (including those of endangered species) designated and protected, began in 1979,[42] but advanced slowly. Noise-abatement measures, announced at the end of the 1970s,[43] are only just starting. The replacement of dangerous substances in products and production is beginning and waste generation is increasing, while recycling and recovery measures show limited effects.
- The concentration of EC measures for discharges, permits, impact assessments, monitoring requirements, accident-prevention measures, waste- and water-management measures, etc., on large installations, projects, and operators, while 80 per cent of EC companies are considered small or medium-sized companies. This approach, based on subsidiarity considerations, would be reasonable if there were national or regional policies to take care of those installations, projects, and emissions that are not covered by EC legislation. However, this is the case in few member states.
- The inability to integrate environmental requirements into transport, energy, trade, and other EC policies. This policy requirement of 1987 continues to be wishful thinking rather than part of EC policy.
- The absence of an implementation-and-enforcement policy within the EC, where local, regional, national, and EC bodies cooperate to ensure application of the different provisions that exist in order to protect the environment. However, inside the EC, such implementation-and-enforcement policy is lacking; and the measures which the Commission undertakes in the frame of Article 226 EC and its informal complaint procedure cannot seriously make up for this deficiency.

---

40 Directive 76/464 (European Commission, 1976b). The water framework directive 2000/60 (European Commission, 2000) now tries to start a new attempt to systematically approach water impairment.

41 Directive 84/360 concerns the combating of air pollution from industrial plants (European Commission, 1984). Directives 96/61 (European Commission, 1996a) and 96/62 on ambient air quality assessment and management (European Commission, 1996b) made some attempt to systematise the efforts on fighting air pollution.

42 See, in particular, Directive 79/409 on the protection of wild living birds (European Commission, 1979a).

43 See the Second Environmental Action Programme (European Commission, 1977).

THE EFFECTIVENESS OF A CENTRAL AUTHORITY

As mentioned above, the question of whether a central environmental authority is more effective than the present system is rather hypothetical. A central EC environmental authority would be largely dependent on the general 'quasi-governmental' authority that existed at the EC level and on the political degree of priority which such a European 'central government' policy would attribute to environmental matters. At the EC level, some short discussions on this issue took place in the early weeks of the setting up of the European Environment Agency in 1989. At that time, it was considered whether such an agency should be established instead of the Commission's environmental administration and not, as was finally decided, as a complementary and strengthening body to the Commission. This idea was launched, however, not to improve the legislation, management, and enforcement of EC environmental protection policy, but rather to weaken it.

In order to examine more closely the potential role of a less dispersed environmental political authority at EC level, I took a closer look at different legal instruments, and considered whether the outcomes of the legal instrument's solution would have been different under a less dispersed authority. I selected at random four environmental directives.

Environmental Impact Assessments (Directive 85/337)

Directive 85/337 requires an environmental impact assessment before planning permission for an infrastructure project is given.[44] Projects which are listed in Annex I to the Directive need to undergo an environmental impact assessment in all cases; projects which are listed in Annex II must undergo such an assessment where, in view of the characteristics, the size, or the location, a significant impact on the environment may be expected. Major problems in the practical application of the Directive occur where: (a) member states abuse their freedom to exert their discretion and do not provide for an environmental impact assessment for an Annex II project, though it is likely to have a significant impact on the environment; (b) member states decide, for political reasons, to realise a project before an assessment of the direct and indirect negative impacts of the project on the environment is made; and (c) a project is realised without an environmental impact assessment, though Directive 85/337 requires such an assessment (should the planning permission procedure then be repeated or is the omission just a procedural error without any material consequence?).

---

[44] Directive 85/337 concerns the assessment of the effects of certain public and private projects on the environment (European Commission, 1985).

As regards (a), it seems self-evident, from an environmental viewpoint, that any project – even the construction of a private house – should be preceded by some sort of assessment of the environmental impact of the project; the intensity of the assessment should depend on the likely impact. Seen from this perspective, it is not logical to make an environmental impact assessment for some projects but not for others. This is, however, not the approach which was chosen by the Directive. Any central environmental decision-making body for an EC of 450 million people would be obliged to give local, provincial, regional, or national planning authorities some discretion to assess whether a given project should or should not undergo an environmental impact assessment. In theory, it seems that each project should be submitted to some form of environmental impact assessment. However, no member state or region has opted for such an approach; it is, then, unlikely that a more centralised EC authority would have come to a different solution than the one which was laid down in Directive 85/337.

The sanction for omitting the making of an environmental impact assessment (point (c) above) could have been fixed by Directive 85/337, but was neither fixed by it nor introduced following the two revisions of the Directive in 1997 and 2003.[45] A less dispersed EC authority might have fixed such a sanction, but nothing indicates that this would have occurred. Indeed, in practically all EC member states, the choice of sanction is left to the courts and not laid down in legislation.

Finally, the making of political decisions on projects before an environmental impact assessment has been made (see (b) above) is common practice in all member states, whether it concerns the laying of lines for high-speed trains, the enlargement of sea-ports or airports, or the construction of motorways. No member state has seriously laid down provisions to stop this practice. Again, therefore, it is unlikely that a less dispersed EC environmental authority would have been able to fix effective and efficient EC rules for this way of bypassing the objectives of Directive 85/337.[46]

Integrated Pollution Prevention and Control (Directive 96/61)

Directive 96/61, on integrated pollution prevention and control, provides for a permit requirement for all industrial installations which are based on the best available techniques.[47] The Directive contains some general indications

[45] Directive 97/11 (European Commission, 1997) and Directive 2003/35 (European Commission, 2003).

[46] One attempt by the EC legislator to reduce this kind of practice is constituted by Directive 2001/42, on environmental impact assessments for plans and programmes, effective from July 2004 onwards (European Commission, 2001b).

[47] European Commission, 1996a.

of what constitutes an available technique. The details are being elaborated, with strong input by vested-interest representatives, in non-binding EC reference documents for each type of industrial installation. The Directive contains an annex which determines, with the help of thresholds, which installations – in total, some 20,000 – are covered by these provisions.

First, it is not clear why only some industrial installations, and not all installations, should follow the requirements of the best available techniques. Under EC law, the installations which are not covered by Directive 96/61 do not need a permit at all. This leads, for example, to the absurd result that Italian incinerators for non-hazardous waste which recover energy do not need a permit, because Italy exempted them from such a requirement and was allowed to continue with this practice under Directive 96/61.[48] Second, it is not easy to understand why the standards for the best available techniques are laid down in non-binding documents with the active help of the sectors concerned; this is bound to lead to results that have as a primary objective not the protection of the environment but the economic interests of the concerned. Third, the Directive provides that, when a permit is granted, local environmental conditions as well as the specific economic situation of applicants are taken into consideration in fixing the conditions for the emissions. This approach leads to the result that the environmental standards for air, water, and soil emissions for the same type of installation (for example, a cement kiln or a refinery) vary from one installation to the other, and from one local authority to the other. Such an approach must inevitably lead to the bargaining of low environmental standards.

Overall, the approach chosen by Directive 96/61 deliberately moved away from the fixing of uniform emission standards for industrial installations of the same type, the approach which underlay Directives 76/464 and 84/360.[49] It is clear that the new approach taken by Directive 96/61 is based on a political choice. The same political option would have been placed before an EC central environmental authority. It is, therefore, not certain that Directive 96/61 would have applied to all industrial installations or would have opted for another approach that would have ensured better environmental protection, or that it would have prevented the 'race to the bottom' of local authorities which compete to provide the location of new industrial installations.

Water Quality (Directive 2000/60)

The Water Framework Directive 2000/60, adopted in 2000, establishes the

---

[48] Directive 96/61 (European Commission, 1996a); Directive 75/442 (European Commission, 1975b), amended by Directive 91/156, Article 11 (European Commission, 1991).

[49] European Commission, 1976b and European Commission, 1984, respectively.

objective of ensuring the 'good' ecological quality of waters.[50] The water quality is fixed by river basins, according to criteria which are laid down in the Directive, but which allow, nevertheless, a considerable margin of discretion and even derogation. The good quality of water and decisions concerning whether water has a high, good, or moderate quality status are determined according to river basins and may thus vary from one basin to the other. Once more, the competition between river basins is more likely to lead to a race to the bottom than to a race to the top.

An EC central environmental authority could have opted for a more uniform approach throughout the EC. However, this is a political choice, and nothing indicates that it would have done so. Indeed, the attempt of Directive 76/464 to establish common EC emission standards for toxic, persistent, and bioaccumulative substances failed, largely owing to opposition from the UK, which was joined by other member states. There is no indication that today another, more integrated water policy would find a majority, even under a non-dispersed EC environmental authority.

Habitat Protection (Directives 79/409 and 92/43)

Directives 79/409 and 92/43 jointly aim to protect threatened habitats within the EC.[51] While Directive 79/409 requested member states to designate habitats for endangered bird species, Directive 92/43 provided a system according to which member states established lists of appropriate habitats and the Commission then selected from these lists the 'EC-habitats'; member states had to take the necessary conservation measures.

Both systems managed with difficulty to ensure the protection of threatened habitats. Owing to controversy at the local and regional levels, and the opposition of vested-interest groups (farmers, fishermen, hunters, urban developers, constructors, and planners of roads and other traffic-infrastructure projects), the designation of habitats advanced slowly. It was estimated, for instance, that in 2003, 22 years after the date fixed by Directive 79/409, only about two thirds of the habitats which should have been designated under that Directive were actually designated; this figure does not include any assessment of the intensity (quality) of the conservation measures adopted by member states.

It is not clear how a less dispersed EC environmental authority could have reached better results. The permanent conflict between the natural environment and human activities leads to the natural environment losing out

---

[50] European Commission, 2000.

[51] European Commission, 1979a and European Commission, 1992, respectively.

under a central government system as in France, the Netherlands, and – to take a non-EC country – Russia. Regional authorities do not fare any better, as the examples of Germany, Belgium, Italy, and Spain demonstrate. In conflicts between economic and ecological interests, the environment is the loser in 99 out of 100 cases, under whatever political authority.

The conclusion of this look at different EC environmental directives is that the stringency of EC environmental legislation does not depend on the dispersion or concentration of authority in the EU. It depends, rather, on the political will to ensure an appropriate and fair balance between environmental concerns, which are a general interest, and other concerns, which are more vested interests. If all the words of the EC Treaty were implemented by action and not only by lip-service, the environment would be in better shape.

The following conditions should be met to realise such a situation:

- The objective of a high level of environmental protection.
- The necessity not only to maintain the status quo, but to improve the quality of the environment.
- Making the polluter and not the taxpayer pay for environmental impairment.
- Ensuring sustainable development – which means that at regular intervals (for instance, every five years), improvements to the environment should be clear.
- The integration of environmental requirements into the elaboration and implementation of other policies.
- Ensuring that environmental provisions are actually applied and do not only fill up the statute books.

Politically, the fact that there is but one environment, has not had a significant influence on most EC governments or on the EC institutions. The fact that the different parts of the EU have a chance of forming a union only if the rule of law – and not the rule of the more powerful – is applied, has not yet been accepted.[52]

Having said this, it seems clear that a number of the above-mentioned deficiencies could be reduced or remedied by an EC environmental authority which had more responsibilities. This applies in particular to the concentration of legislation on large installations, projects, etc., and to implementation and enforcement. The first aspect has already been mentioned. As regards implementation and enforcement, the European

[52] A good example is the approach adopted in 2003 by Germany and France to comply with the requirements of the Stability Pact; it is ironic that Germany, which had insisted on the fixing of precise criteria, was one of the first member states to consider departure from these rules. This recalls Hugo Grotius: 'All evil starts when man departs from law'.

construction and its repartition of responsibilities is based on the assumption that both sides, the EC authorities and the member states' authorities, assume their respective responsibilities. However, it is submitted that the majority of EC member states do not have an active environmental policy which would, more or less, pursue the objectives of Article 174(1) within the national boundaries. Even in member states which for a long time were active in environmental issues (Denmark, the Netherlands, Austria, and Germany), economic constraints push environmental issues back to the second or third rank of priorities. Even today, only a minority of EC member states have a policy which could be called an implementation and enforcement policy for the environment. It is simply not realised that the environment is an interest without a group and that, therefore, the traditional implementation-and-enforcement provisions do not work satisfactorily.

Furthermore, neither environmental science nor political science managed to prove that in the long and medium term a sound – sustainable! – economic policy is only possible where it is backed by sound environmental policies, within the EU and at the national and global levels. In a consumer society which is strongly influenced by discussions in the media, it is not enough to make good environmental policy; it is also necessary to demonstrate the economic and social benefits of such a policy and draw attention to the negative consequences of policies that neglect environmental concerns.

There should be no doubt that, despite its many imperfections, omissions, and deficiencies, EC environmental policy and legislation has been successful. It has achieved much more than most governments of the EC member states would have achieved had they acted at the national level only. Legislation on drinking water and coastal water (bathing water), on waste water and nitrates in water, as well as on the discharge of substances into water and the protection of groundwater, would not exist in the majority of member states today, had the initiative not come from the EC. The same argument can be used for provisions for access to information on the environment, environmental impact assessment, integrated pollution prevention and control, the prevention of accidents in industrial installations, air pollution from large combustion plants and waste incinerators, hazardous waste, the ban of waste exports, end-of-life vehicles, bird and habitat protection, chemicals and car catalysts, climate-change provisions, and many other pieces of legislation. The EC, often supported by some member states, was the driving force, initiator, and monitoring force; also, it was capable of offering – and requesting! – a level playing field in environmental matters for all EC member states.

CONCLUDING REMARKS

The search for a less dispersed environmental authority at the European level resembles the search for the Holy Grail: it is unattainable. More traditional measures are necessary: establishing European public opinion and then discussing openly how the balance between economic growth, social concerns, and environmental needs can be struck. This requires more factual knowledge of environmental impairment and its causes. There is in particular a need for more capacity to formulate the environmental needs and more determination to look for allies in the public and parliamentary debates. Where environmental issues are discussed in public, there is a chance of finding a better balance between environmental and other interests than when the discussions take place behind closed doors. Policy is the long and patient drilling of hard wood, and no other definition can be offered for environmental policy. European and national environmental policies lost much attention in the public political discussions over the last years, owing to economic distress, immigration and minority problems, and other concerns that kept media and politicians occupied. There is little sense in lamenting this, as these concerns exist and are legitimate.

What environmental policy has not yet managed to convey, or has forgotten in day-to-day wrangling, is its own importance to the economic and social well-being of citizens. This includes the message that the notion of 'future generations' first of all concerns our children and grandchildren, that investment in the environment is investment in a sustained economy, and that lifestyle changes are possible without impairment of the quality of life. The values of an open society, quality of life, and the natural environment are becoming appreciated as they become rare. Albert Camus once stated that, under the Nazi regime in Europe, the Pope might have raised his voice in favour of the Jews, but that his voice was so silent that nobody heard it. Environmental policy and environmental politicians must be careful to speak up in favour of the environment, or run the risk that their environmental concerns will not be heard, noted, or taken into consideration. This risk is omnipresent in the undertaking of making the EU and will remain present in the years to come.

REFERENCES

European Commission (1972), *Proposal for an EC Environmental Action Programme*, OJ C 52, Brussels: European Commission.

European Commission (1973a), *Sixth General Report*, Brussels: European Commission.

European Commission (1973b), *First Environmental Action Programme*, OJ C 112, Brussels: European Commission.

European Commission (1973c), *Directive 73/404*, OJ L 347, Brussels: European Commission.

European Commission (1975a), *Directive 75/440,* OJ L 194, Brussels: European Commission.

European Commission (1975b), *Directive 75/442,* OJ L 194, Brussels: European Commission.

European Commission (1975c), *Directive 75/716*, OJ L 307, Brussels: European Commission.

European Commission (1976a), *Directive 76/160*, OJ L 31, Brussels: European Commission.

European Commission (1976b), *Directive 76/464*, OJ L 129, Brussels: European Commission.

European Commission (1977), *Second Environmental Action Programme*, OJ C 139, Brussels: European Commission.

European Commission (1978a), *Directive 78/611*, OJ L 197, Brussels: European Commission.

European Commission (1978b), *Directive 78/659*, OJ L 222, Brussels: European Commission.

European Commission (1979a), *Directive 79/409*, OJ L 103, Brussels: European Commission.

European Commission (1979b), *Directive 80/779*, OJ L 229, Brussels: European Commission.

European Commission (1980a), *Directive 80/68*, OJ L 20, Brussels: European Commission.

European Commission, 1980b, *Directive 80/778*, OJ L 229, Brussels: European Commission.

European Commission (1983), *Resolution on the Third Environmental Action Programme,* OJ C 46, Brussels: European Commission.

European Commission (1984), *Directive 84/360*, OJ L 188, Brussels: European Commission.

European Commission (1985), *Directive 85/337*, OJ L 175, Brussels: European Commission.

European Commission (1990), *Directive 1990/219*, OJ L 117, Brussels: European Commission.

European Commission (1991), *Directive 91/156*, OJ L 78, Brussels: European Commission.

European Commission (1992), *Directive 92/43*, OJ L 206, Brussels: European Commission.

European Commission (1993), *Fifth EC Environmental Action Programme*, OJ C 138, Brussels: European Commission.

European Commission (1996a), *Directive 96/61*, OJ L 257, Brussels: European Commission.

European Commission (1996b), *Directive 96/62*, OJ L 296, Brussels: European Commission.

European Commission (1997), *Directive 97/11*, OJ L 73, Brussels: European Commission.

European Commission (1998), *Directive 98/81*, OJ L 330, Brussels: European Commission.

European Commission (2000), *Directive 2000/60*, OJ L 327, Brussels: European Commission.

European Commission (2001a), OJ C 154E, Brussels: European Commission.

European Commission (2001b), *Directive 2001/42*, OJ L 197, Brussels: European Commission.

European Commission (2001c), *Directive 2001/81*, OJ L 309/22, Brussels: European Commission.

European Commission (2002), *Sixth Community Environmental Action Programme*, Decision 1600/2002, OJ L 242, Brussels: European Commission.

European Commission (2003), *Directive 2003/35*, OJ L 156, Brussels: European Commission.

European Environment Agency (2003), *Europe's Environment: The Third Assessment*, Copenhagen.

Jans, J. (2000), *European Environmental Law*, 2nd ed., Groningen: Europa Law Publishing.

Krämer, L. (1992), *Focus on European Environmental Law*, London: Sweet & Maxwell.

Krämer, L. (2002), *Thirty Years of EC Environmental Law: Perspectives and Prospectives: Yearbook of European Environmental Law 2*, Oxford: Oxford University Press.

Krämer, L. (2003), *EC Environmental Law*, 5th ed., London: Sweet & Maxwell.

22. Mutual Recognition in the Testing of Chemicals through the OECD

Rob Visser[1]

SUMMARY

The OECD's mutual acceptance of data (MAD) system plays an important role in the testing and registration of chemical products. OECD member countries and other states adhering to the system mutually recognise the outcomes of safety tests of chemicals conducted by a participating country. Standardised guidelines and principles are followed during such tests. Representatives from science, government, industry, civil society, and intergovernmental organisations are involved in the development of such guidelines and principles. The mutual recognition of test data has important advantages: enhanced protection of the environment and public health, increased cost efficiency, minimised non-tariff trade barriers, and the sharing of an increasing burden. Potential drawbacks are the lengthy process of obtaining consensus and the politicisation of the system. Globalisation has magnified the merits and drawbacks of the system.

INTRODUCTION

The Organisation for Economic Co-operation and Development (OECD) is an intergovernmental organisation established in 1960. It groups 30 industrialised countries: all 'older' European Union (EU) member states (Austria, Belgium, Denmark, Finland, France, Germany, Greece, Ireland, Italy, Luxembourg, the Netherlands, Portugal, Spain, Sweden, United Kingdom), some 'new' EU countries (Czech Republic, Hungary, Poland, Slovak Republic), some non-EU European countries (Iceland, Norway,

[1] Any opinions expressed in this chapter do not necessarily represent those of the OECD or its member countries. For general information on OECD's work in this field, see OECD, 2004.

Switzerland, Turkey), the NAFTA countries (Canada, Mexico, United States), and some Asia-Pacific states (Australia, Japan, New Zealand, South Korea); the European Commission participates in its work.

The mission of the OECD, as laid down in its Convention, is:

- To achieve sustainable economic growth and employment and rising standards of living in member countries while maintaining financial stability, and thus to contribute to the development of the world economy.
- To assist in achieving sound economic expansion in member countries and other countries in the process of economic development.
- To contribute to growth in world trade on a multilateral, non-discriminatory basis.

The OECD assists member countries in addressing the economic, social, and environmental challenges of interdependence and globalisation by providing comparative data, analysis, and forecasts to underpin multilateral cooperation. It does not fulfil functions that some other intergovernmental organisations carry out. What the OECD is, and what it is not, is summarised in Table 22.1.

Table 22.1 Functions of the OECD

OECD is:	OECD is not:
• a forum for policy dialogue and development	• a provider of technical assistance
• a centre for policy research	• a supranational rule-making body
• a facilitator for achieving harmonisation, cooperation, and cost sharing	• a bank

The Organisation's governing body, the Council, has the power to adopt legal instruments, usually referred to as 'OECD Council Acts'. The OECD's Convention states that 'in order to achieve its aims, the Organisation may: (a) take decisions which, except as otherwise provided, shall be binding on all members; and (b) make recommendations to members'.

Decisions are legally binding on all member countries. While they are not international treaties, they do entail the same kind of legal obligations as those subscribed to under international treaties. Members are obliged to implement Decisions and they must take the measures necessary for their implementation. Recommendations are not legally binding, but practice accords them great moral force as representing the consensus political will of member countries and there is an expectation that member states will do their utmost to fully implement a Recommendation.

In undertaking its work, the OECD does not only relate with its member countries. More than 70 developing countries and economies in transition are engaged in working relationships with the OECD and an active Centre for Co-operation with Non-Member Economies exists. The extent of the OECD's working relations across the globe is shown in Figure 22.1.

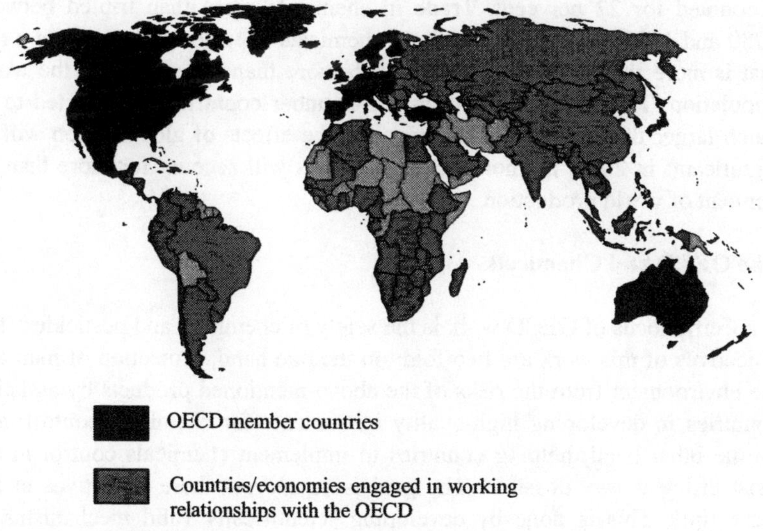

OECD member countries

Countries/economies engaged in working relationships with the OECD

Figure 22.1 The OECD's geographical scope

The OECD also has official relations with other intergovernmental organisations, such as the International Labour Organisation (ILO), the Food and Agriculture Organisation (FAO), the International Monetary Fund (IMF), the World Bank (WB), the World Trade Organization (WTO), the World Health Organisation (WHO), the International Atomic Energy Agency (IAEA), and a large number of other United Nations (UN) organisations and programmes.

The Chemicals Industry

The chemicals industry is essential for modern life and central to the global economy. The products of the chemicals industry, worth 1,500 billion US dollars (USD) annually, account for 9 per cent of the global trade in manufactured goods. The OECD Environmental Outlook for the Chemicals Industry provides some characteristics and projections about this rapidly globalising industry;[2] some of the key information is provided in Annex 22.1.

The chemicals industry is growing rapidly: in 1998, it was more than eight times as large in volume of output than it was in 1970. The OECD countries accounted for 83 per cent of world production in 1970; however, the contribution of non-member countries is growing: in 1998, their share accounted for 22 per cent. Trade in chemicals more than tripled between 1980 and 1998. World production of chemicals is projected to grow at a rate that is more than world GDP, and much more than the growth of the world population. The relative growth in non-member countries is projected to be much larger than in OECD countries and the effects of globalisation will be significant: in 2020, the non-OECD countries will account for more than 30 per cent of world production.

The OECD and Chemicals

A specific focus of OECD work is the safety of chemicals and pesticides. The objectives of this work are two-fold: on the one hand, protection of man and the environment from the risks of the above-mentioned products by assisting countries in developing high-quality instruments for chemicals control: and on the other hand, helping countries to implement chemicals control in the most efficient way possible. The goal is to address these objectives at the same time. This is done by developing scientifically valid mechanisms to minimise non-tariff barriers to trade, assisting governments and industry in avoiding duplicative activities, and facilitating work-sharing efforts across the OECD. Experts from industry, trade unions, academia, environmental non-governmental organisations (NGOs), and animal welfare NGOs participate in this work. The safety of the products of modern biotechnology is addressed in the same vein. The objectives of the OECD's work in this field are shown schematically in Figure 22.2.

---

[2] OECD, 2001.

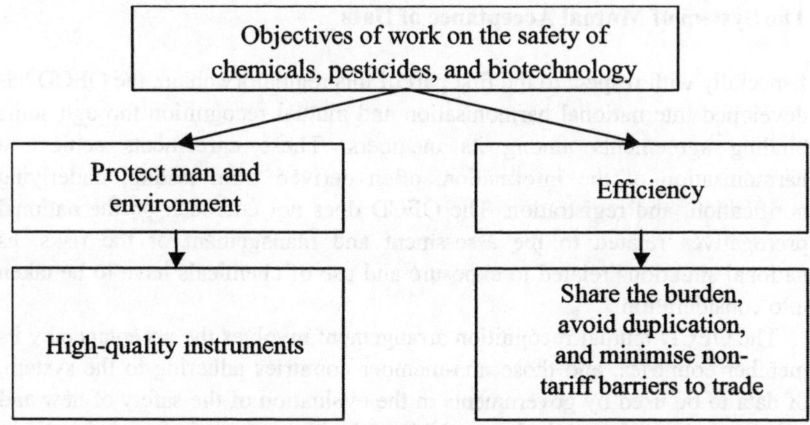

Figure 22.2 The OECD's objectives on chemical safety

MUTUAL ACCEPTANCE OF DATA

With respect to chemicals, pesticides, and the products of modern biotechnology, the regulatory chain appears as in Figure 22.3.

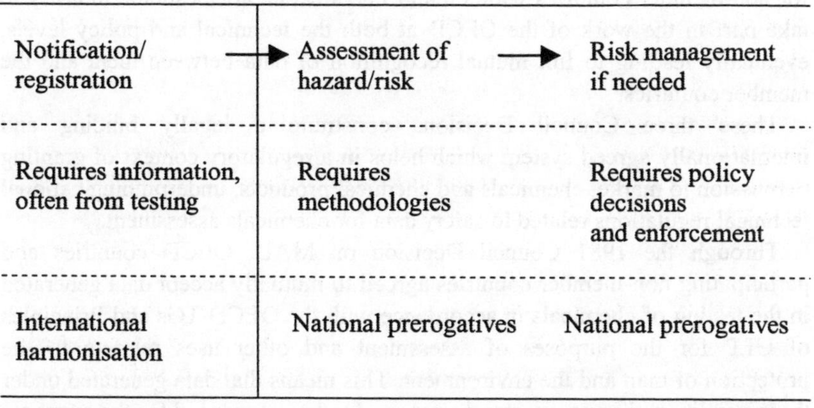

Figure 22.3 Regulatory chain of chemicals

The System of Mutual Acceptance of Data

Especially with respect to the first part of this regulatory chain, the OECD has developed international harmonisation and mutual recognition through some binding agreements among its members. These agreements achieve a harmonisation of the information, often derived from testing, underlying notification, and registration. The OECD does not encroach on the national prerogatives related to the assessment and management of the risks, as national situations related to exposure and use of chemicals have to be taken into consideration.

The OECD mutual recognition arrangement involves the acceptance by its member countries, and those non-member countries adhering to the system, of data to be used by governments in the evaluation of the safety of new and existing industrial chemicals, pesticides, biocides, etc. It is founded in three legally binding Council Decisions: the 1981 Decision on the Mutual Acceptance of Data (MAD) in the Assessment of Chemicals[3] with its Test Guidelines (TGs) and Principles of Good Laboratory Practice (GLP); the 1989 Decision-Recommendation on Compliance with Good Laboratory Practice[4] with its three Annexes which give substance to the Recommendation paragraphs; and the 1997 Decision on the Adherence of Non-Member Countries to the Council Acts related to the Mutual Acceptance of Data in the Assessment of Chemicals.[5] This last Decision sets out a step-wise procedure for non-member countries with a major chemical and/or pesticide industry to take part in the work of the OECD at both the technical and policy levels, eventually leading to full mutual recognition of data between them and the member countries.

These three Council Decisions constitute a legally binding and internationally agreed system which helps in a regulatory context of granting permission to market chemicals and chemical products, underpinning national technical regulations related to safety data for chemicals assessment.

Through the 1981 Council Decision on MAD, OECD countries and participating non-member countries agreed to mutually accept data generated in the testing of chemicals in accordance with the OECD TGs and Principles of GLP for the purposes of assessment and other uses relating to the protection of man and the environment. This means that data generated under these conditions in one country do not need to be reproduced for the purposes of assessing the chemical in another country. The MAD system has allowed OECD countries to avoid non-tariff trade barriers, which can be created by different national regulations while improving protection of human health and

[3] OECD, 1981.
[4] OECD, 1989.
[5] OECD, 1997.

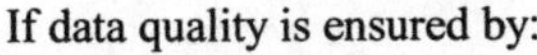

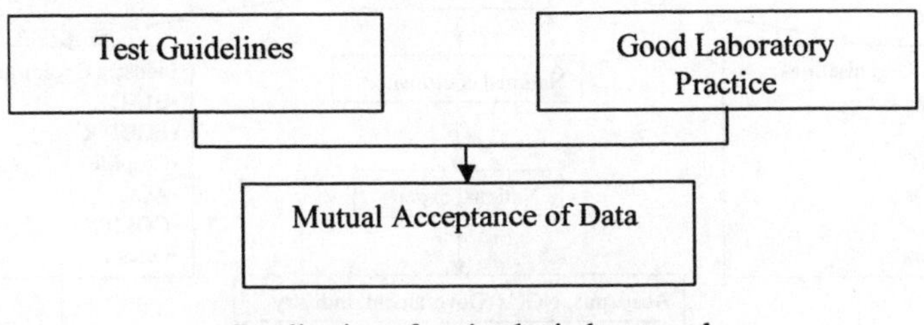

Figure 22.4 The MAD system

the environment. The MAD system also helps to provide a common basis for cooperation among national authorities. The system saves the chemicals industry the expense of duplicative testing of products marketed in many different OECD countries. Schematically, this can be represented as in Figure 22.4.

The OECD considers that the three Council Acts constitute a multilateral agreement among governments on the MAD, which is open to both member and non-member countries. In endorsing the Council Acts, all 30 OECD member countries are partners in the agreement. Twenty-five OECD countries have implemented the 1981 and 1989 Council Acts through national legislative and administrative instruments. Two other OECD countries are in the process of establishing monitoring programmes; and the remaining three have not yet implemented the Council Decisions, but are obliged to accept data from the other member states and adhering countries.

Test Guidelines

The OECD TGs are the first constituent part of MAD. The results of tests with chemicals are the basis for decision-making on the risks of such chemicals for human health and for the environment. OECD TGs exist in the areas of physical-chemical properties, human toxicity, ecotoxicity, and degradation and accumulation.

Approximately 100 TGs prescribe in detail the conduct of safety tests needed for regulatory purposes. They are continuously developed and updated in accordance with the scientific state of the art. A network of 7,000 experts in various areas ensures the peer review and validation of proposed new TGs or updates. The way this is organised is schematically represented in Figure 22.5.

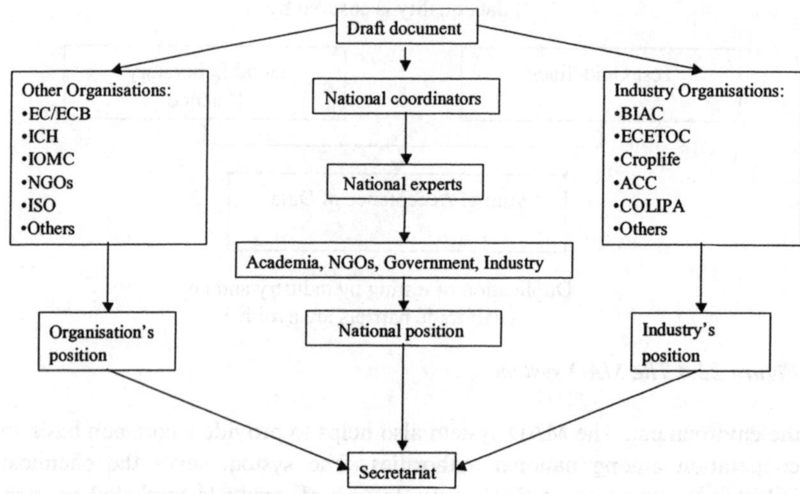

Figure 22.5 The OECD's test guidelines

Good Laboratory Practice

Governments which receive notification or registration dossiers for permission to market industrial chemicals, pharmaceuticals, or pesticides can easily verify whether the testing for which the results are reported in the dossier was carried out using OECD TGs. In addition, however, a system to assure the quality of the test undertaken is needed. After all, decisions which can have an impact on the health of man and on the environment have to be made based on these data, so their good quality has to be absolutely certain. The OECD quality control system is based on the Principles of GLP, which are annexed to the 1981 Council Decision. They address details of the responsibilities of a test facility and the requirements for its organisation and personnel, and the quality assurance programme of a facility (including the laboratory standard operating procedures, apparatus, materials, and reagents).

Through the 1989 Council Decision, it is required that data submissions be accompanied by a declaration by the test facility that a study was carried out in accordance with GLP Principles. Governments need a means of verifying such declarations.

In order to build confidence in the compliance with GLP, all participants in the MAD system have established GLP monitoring authorities. The heads of these monitoring authorities (GLP inspectors) meet on a regular basis in an OECD forum, the Working Group on GLP, in which the non-member countries (provisionally) adhering to the system also participate. Several countries have more than one (product-based) GLP compliance monitoring authority and are represented accordingly in the Working Group.

The Working Group was set up to implement the instructions laid down in the 1989 Council Decision. These instructions called for the OECD to put mechanisms into place to ensure continuing exchange of information on technical and administrative matters related to the application of the GLP Principles and on the implementation of compliance monitoring procedures. The Working Group meets regularly to reach consensus on issues leading to greater harmonisation in practice and to ensure international liaison among the monitoring authorities and, through them, among the national regulatory authorities which receive and evaluate safety data. It calls meetings with the chemical industry to reach agreement on the application and interpretation of specific parts of the GLP Principles, holds training courses for inspectors, sets policies for compliance monitoring issues, etc. It oversees the publication of Consensus Documents on interpretation of the Principles, Guidance Documents on aspects of compliance monitoring, and Advisory Documents on policy aspects of GLP. The Working Group is serviced by the OECD Secretariat.

Monitoring of Compliance with GLP

Implementation of the Council Decisions – especially of the national legal and administrative procedures transposing the OECD guidance for GLP compliance monitoring – is thus monitored by member countries through the OECD Working Group, and peer pressure takes the place of formal enforcement measures. In spite of this, there have been isolated cases of refusal by a member country to accept safety data for product evaluation on the grounds that it had insufficient confidence in the GLP compliance monitoring procedures of the country where the data had been generated. When such cases were notified to the OECD, the secretariat facilitated bilateral discussions between the countries concerned and the threat of bringing the issue to the OECD Council resulted in individualised ad hoc solutions being found. Nevertheless, the perception persisted in some member countries that a mechanism for more complete mutual confidence building as regards the implementation of the 1989 Council Act was necessary if countries were to be completely comfortable with mutual acceptance. It was clear that confidence in the quality of data could not be merely legislated or mandated, but must come from objective evidence about the application of the criteria set down in the Annexes to the 1989 Council Act.

From 1998 to 2002, therefore, the Working Group on GLP established an informal system to undertake a first round of evaluations of national compliance monitoring programmes through so-called ‘mutual joint visits’.

Over a three-year period, all 34 GLP compliance monitoring authorities in place at that time were visited by teams comprising representatives of three other authorities, who evaluated the programmes through review of programme documentation, discussions with staff, and observation of inspections and study audits on site. The reports from the visits were discussed in the Working Group, which made recommendations for necessary improvements and drew conclusions on the extent of implementation by each country of the Recommendations in the 1989 Council Act. This project remained informal, with peer pressure being sufficient to bring about the needed modifications in specific programmes. When the project was evaluated by the OECD Chemicals Committee in 2002, it was agreed that a continuing process based on the experience gained needed to be put in place and that the same procedures should be used in respect of the adherence of non-OECD members to the Council Acts. In this way, the Working Group on GLP is able to keep all of the national GLP compliance monitoring programmes under review and to exert pressure for change as appropriate in order to ensure mutual confidence.

MAD and Non-Members

By establishing the basis for harmonised national policies for data acceptance (the 1981 and 1989 Council Acts) and developing and maintaining internationally agreed instruments for their implementation (TGs, GLP Principles, and Compliance Monitoring Procedures), the OECD has created among its member countries a standards-based system which could be considered to meet the main technical barriers to trade (TBT) objectives of transparency and avoidance of the unnecessary barriers to trade created by technical regulations. By opening up this system to non-OECD countries through the 1997 Council Decision, the system has become, in principle, accessible to all WTO members. The MAD system fulfils the major WTO requirements for 'international standards' of transparency, avoidance of trade barriers, and openness to participation by all WTO members. A discussion of the relation of MAD with the WTO and ISO standards is given in Annex 22.2.

The process for adherence of non-members to the MAD system involves: acceptance of data from OECD countries (using GLP, TGs) by non-members; assistance provided to non-members by the OECD in developing the GLP compliance system; participation of non-members in OECD work; and acceptance by OECD countries of data from non-member countries (using GLP and TGs).

The first non-member countries to provisionally adhere to the 1981 and 1989 Council Acts and take part in the work of OECD in this area with regard to the development and updating of TGs and with regard to GLP and its compliance monitoring were South Africa (which became a full participant in the system in 2003), Slovenia, Israel, and India. Negotiations are advanced with Brazil, and starting with China and the Russian Federation.

BENEFITS OF HARMONISATION

Below, the benefits of harmonisation for the various stakeholders (governments, industry, trade unions, civil-interest NGOs) are analysed with respect to the quality of environmental and health protection, efficiency, trade, and possibilities for work sharing.

Quality of the Environment and Health

Harmonisation of TGs and GLP helps to improve the quality of the test results. During the development of TGs, the best expertise from government, academia, industry, and civil-interest NGOs is brought together to work at achieving consensus on a method which is scientifically acceptable to all. In particular for smaller countries, which might not have easy access to the varieties of expertise required for the different tests, the cooperative development of TGs might be helpful to ensure that the tests they require are based on wide and up-to-date scientific input. When all regulatory testing is done using the same test methods, consistent experience accumulates, which helps in the interpretation of the test results. Regulators in governments and experts in industry and civil-interest NGOs can compare results which have been achieved over the years. Harmonisation in GLP inspections helps to build confidence in the tests carried out in the countries participating in the OECD MAD system.

Efficiency

The OECD has made a quantitative estimate of the savings resulting from the use of harmonised TGs on GLP.[6] For the study, information on the costs of creating and maintaining TGs and GLP programmes was obtained from OECD records and from government industry experts who had participated in various development activities. The costs of the programmes were divided into secretariat costs (the costs for OECD Secretariat support) and country costs (the costs for governments, industry, and NGOs to participate in and contribute to the work of the programmes). The savings resulting from the use

---

[6] OECD, 1998.

of TGs and GLP were calculated based on information obtained from contract laboratories regarding the costs of tests and on information obtained from governments concerning the amount of notified or registered chemicals. Industry was asked to check these data to ensure that the analysis did not overestimate the savings.

Conservative assumptions were used regarding the number of substances tested, the number of tests required, the costs of the test package, the number of markets to which the substances were introduced, and the percentage of tests that would have to be repeated if the OECD MAD system did not exist. For example, based on discussions with government and industry experts, it is reasonable to assume that each (often multinational) company which conducted safety testing and notified or registered a new industrial chemical or pesticide in one market, also did this in other OECD markets. For the purposes of this report, the 30 OECD countries were generally not considered individually, but rather as part of three major regional markets: North America, Europe, and the Asia/Pacific region. It was assumed that each new product notified or registered in one region was also notified or registered in the two other regions. Considering the OECD to have only three regional markets is an underestimate, however.

After new substance clearance in one market, a company would have to repeat a considerable part of the testing to obtain clearance for other markets, in order to comply with different requirements, were it not for the OECD TGs, the Principles of GLP, and the MAD system. However, it cannot be assumed that, if OECD TGs and GLP did not exist, country A would not accept any of the data generated by country B. It is assumed that test methods would have been developed by countries independently and that they would not have been exactly the same. It seems likely, however, that country A would have accepted some of the data developed using country B's different methods. Based on discussions with experts in government and industry, it was estimated that 30 per cent of the foreign test data would not be accepted because of differing methods and that testing would have to be repeated. This was a conservative estimate, which involves as much as 70 per cent of the data being accepted in a second country, even if such data did not fully conform to the requirements of that country.

The health and environmental benefits of improved chemicals management are obviously great, but they were not considered for this quantitative estimation. Other non-quantifiable benefits (like improved quality of test procedures or reduced use of animals) were also not included in the calculations; neither were benefits resulting from reduced non-tariff barriers or reduced delays in the marketing of new products.

The results of the study on savings resulting from avoiding duplication of testing by application of the MAD system are summarised in Table 22.2.

Table 22.2 Financial benefits of the MAD system

Yearly costs and savings resulting from MAD (estimated in USD)			
Costs to government and industry of participation in the programme		Estimated savings for government and industry in the testing, evaluation, and approval of different chemical methods	
Secretariat	3.0 million	New industrial chemicals	5.8 million
		New pesticides	20.0 million
		New pharmaceuticals	36.0 million
Countries	6.5 million	High-production-volume industrial chemicals	1.7 million
Total:	9.5 million	Total:	63.5 million
Yearly net savings due to the EHS programme: 54 million			

Trade

The MAD system creates many advantages for trade. In an era of globalisation, with tariffs progressively decreasing, differences in regulations and (test) standards have become an increasingly important remaining trade barrier. Furthermore, differences in regulation discourage research, innovation, and growth, and increase the time required to introduce a new product to the market. Harmonisation helps significantly in addressing the above issues. The OECD has a procedure which can be used to notify non-acceptance of a test carried out according to the MAD criteria; the OECD will then investigate if the non-acceptance is correct according to the MAD decision or if there is a non-tariff trade barrier.

Burden Sharing

The harmonisation of test procedures and of the quality control of tests has important specific advantages for governments, because it provides a global basis for work-sharing efforts. If the same high-quality tests, using the same methods, are received by all regulators as a basis for their hazard/risk assessments, authorities can take better advantage of work done by others, which helps to reduce the resources needed for chemicals control. In practice, through the OECD, work-sharing schemes have been organised for the testing and assessment of new chemicals, high-production-volume chemicals, and pesticides – so the possibility of work sharing based on test results obtained through MAD is an increasingly used reality.

WHAT MAKES MAD WORK

Below is a short discussion of each of the elements that are involved in making the MAD system work, including consideration of why the various stakeholders are interested in playing their roles. This involves science, governments, industry, civil society, and intergovernmental organisations.

Science

The scientific basis of the MAD system is of crucial importance. It is our experience that, in any consensus-building discussion, in the end, sound science prevails. Politically inspired positions on certain aspects can be held for some time, but when proponents of such positions are asked to present the scientific basis for them, the discussion comes back to the scientific validity of the arguments. This can in many cases be established by relying on the opinions of a large majority of peer experts who review scientific information. We have had the same experience in other work in the OECD related to chemical safety (e.g., harmonisation of classification of hazardous substances), biosafety (such as genetically modified organisms), and food safety. While perception and policies related to the risks of chemicals and other products may vary greatly among countries, discussions on tests and data underlying assessments and risk management have always been possible in the OECD, because for these data the scientific quality of the arguments made is, in the end, the basis for coming to agreement.

Governments

The notification/registration of chemicals is a regulatory process, which is in the hands of governments. The role of OECD governments is to indicate the priorities at the start of the harmonisation process, to participate as active and equal discussion partners in the scientific debate, and finally, when a good measure of consensus on the scientific aspects has been achieved, to take the scientific results and commit – in the case of TGs and GLP through an OECD Council Decision – to use them in all member countries in the regulatory process. Having an internationally harmonised and efficient system for implementing regulatory requirements helps governments to respond to arguments concerning the costs of regulations, the distortion of competitiveness among industries in different countries by different regulations, and the unnecessary use of laboratory animals.

The monetary savings resulting from harmonised testing have been mentioned above. Some non-quantitative benefits for governments (and NGOs) of harmonising include: the creation of networks of government, industry, and NGO experts in member countries; having a forum in which new policies can be discussed, developed, and, where appropriate, harmonised OECD-wide (8 Council Decisions and 12 Council Recommendations exist in the overall chemical safety field); the development of technical instruments that improve the quality of chemical evaluations and regulations; access to information and advice from countries with different policy experiences; and greatly increased availability of safety data on high-production-volume chemicals.

Industry

An obvious benefit for industry of international harmonisation is the reduction of duplicative testing and the costs thereof. This helps all companies in OECD countries in the same way. Therefore, industry in OECD countries is united in supporting the harmonisation work.

Non-quantifiable benefits include the reduction of non-tariff trade barriers, the reduction of delays in marketing new products, the creation of a level playing field regarding regulations in member countries, the creation of networks of government, industry, and NGO experts in OECD countries, and the opportunity to obtain information about member countries' policies and regulations.

Because industry has so many direct benefits from harmonisation, it is also important for industry to play a credible role in the scientific discussions. In the OECD work, this has always been the case. Also, the arguments made by industry concerning the development of test methods are always underpinned by reviewed science.

Civil-Society NGOs

Civil-society NGOs actively participate in the process and, in addition to providing scientific input, their role is also to help ensure that the process remains transparent. The outcome of the use of reliable test methods is the assessment of the hazards of chemicals in the best possible high-quality way, which is the main interest of NGOs representing civil society. Civil-society NGOs include stakeholders with different interests: trade unions, environmental NGOs, consumer NGOs, and animal welfare organisations. The interests of these various groups are not always the same. For example, environmental and consumer NGOs might want to increase the number of animals required per test, while the animal welfare NGOs might want to reduce this. However, the peer-reviewed scientific evidence is the only convincing basis for consensus.

Other Intergovernmental Organisations

No other intergovernmental organisation has a system which compares with MAD or is working on TGs or GLP – the OECD's role is unique. Of course, the OECD works closely with other organisations which can provide input into the process and coordinate follow-up activities as well. For example, while the WHO works, in cooperation with the OECD, on the development of methods to assess the health risks of chemicals, it does not undertake work related to test-method development. A formal coordination mechanism, the Inter-Organisation programme for the sound Management of Chemicals (IOMC), exists to this end (see Figure 22.6). The WB and the UN

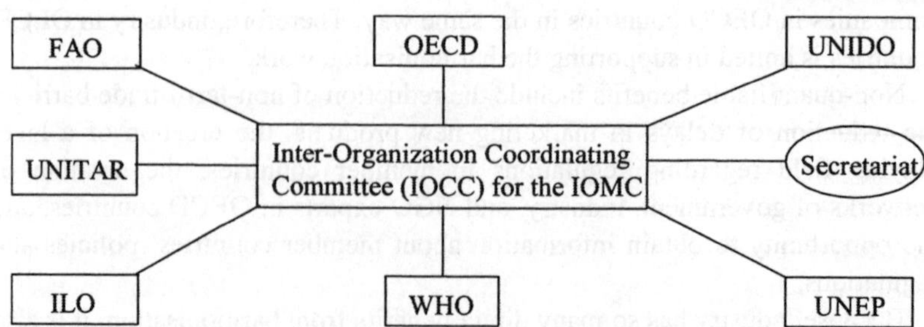

Figure 22.6 The IOMC coordination mechanism of chemicals management

Development Programme are expected to become members of the IOMC soon. Close working relations also exist with the WTO.

Because the MAD system is included in a legally binding OECD Council Decision, member countries have to implement it by including it in their regulatory practices. For European countries, this is done by Community legislation; all OECD TGs and the GLP principles are transcribed directly without making changes in EU Directives.

DEVELOPMENT OF MAD IN A GLOBALISING WORLD

Below, the developments in the implementation of the MAD system over the last 23 years as a result of globalisation are briefly discussed, as are opportunities and threats for the future.

Developments in the MAD Process

Changes in the MAD process as a result of globalisation include not only the increasing importance of the trade aspects of the system. With tariffs going down, it is becoming more relevant to address non-tariff trade barriers, because they are becoming the most important impediments to trade. However, globalisation has also brought the possibility of making information more widely available, which means that more people with a wider range of backgrounds are able to provide input. Currently, all relevant stakeholders are involved in the network of 7,000 experts that contribute scientific input to the development of TGs.

While globalisation has resulted in a wider participation in the work of the MAD system and greater transparency, a second consequence of globalisation is that the scientific scrutiny of proposals is much more rigorous. Often, validation programmes for new tests are required and the updating of TGs in order to adapt them to the latest scientific standards is more frequently requested. The result of this is that, while procedures have become more complicated (which can be compensated to some extent by the use of electronic means of communication and discussion), the quality of TGs has certainly improved over the years. Globalisation here means a race to the top, rather than a race to the bottom.

Another result of globalisation is the participation of non-member countries in the MAD system. Not only is chemical production increasing outside the OECD countries; the development of chemical safety regulation is also increasing in these countries. The logical result of this is that, in these countries, similar types of chemicals control systems to those in the OECD countries have been developed and that similar types of instruments (such as test methods and test quality assurance) are at the basis of this legislation. OECD countries and industry in OECD countries would like to see systems for chemicals control in non-member countries converge with the OECD system, rather than to have diverging systems come in place, which would lead to duplicative work and trade barriers.

Many non-members that are important in view of their chemicals production and/or chemicals control regulations would, in principle, be happy to use the instruments already developed and used by OECD countries, provided they could be part of the development, updating, and decision-making structures for these instruments. Through the earlier mentioned 1997 Council Decision, the part of the OECD work related to MAD has been opened up to non-members; however, only those UN member states whose requirements for the testing and assessment of chemicals are similar to those of OECD countries will have an interest in joining the OECD system. Nevertheless, the opening up of the OECD MAD system has been an important step to increasing global regulatory efficiency and avoidance of trade barriers.

Opportunities

The future for the implementation and maintenance of the MAD system looks quite promising. New endpoints for safety testing are becoming of international interest. A recent example is testing for chemicals which even in low doses can cause disruption to the hormonal systems of man and animal, thereby causing problems in the reproductive processes (endocrine disruption). The OECD MAD system has demonstrated that it is capable of rapidly addressing the requirements for new TGs resulting from these new interests. In addition, it has created innovative testing methodologies, because more tests using cultivated cells or tissues or computer models (the so-called in vitro or in silico tests) are being developed in order to address concerns related to animal welfare (in particular, the use of animals in tests) in chemicals testing. In order for such tests to be acceptable in a regulatory context, the development of agreed TGs is necessary.

Furthermore, in most countries, the resources available for the testing and assessment of chemicals and pesticides are very limited, while the number of chemicals and pesticides is rapidly increasing. Governments are, therefore, required to improve efficiency. Because many countries follow the same process for making assessments of chemical safety, a good basis for work sharing between countries is available. Work sharing is a major way of improving efficiency. However, in order to be able to make optimal use of work-sharing possibilities, it is crucial that tests be performed using the same method (i.e., following an agreed TG).

Finally, the number of players using the MAD system is increasing. Not only are more countries participating, but the increasing number of international voluntary initiatives for the testing of chemicals by industry makes the MAD system a crucial key pin, because tests performed in such initiatives must be internationally acceptable. Industry tests can only be accepted by governments if authorities can be sure of the scientific correctness of the tests, which means that OECD TGs and GLP have to be used.

Threats

The most important threat to the MAD system is also its greatest strength: the consensus basis. The openness and transparency of the system have much improved the quality of tests. However, it can also be a liability in the process of developing TGs. For example, one of the stakeholders may hold up consensus by insisting on extensive and costly validation procedures, which can very much delay the availability of a TG for regulatory use. Of course, the objective of the development of all TGs is to have a guideline of the best quality possible, but sometimes best is the enemy of good. If a large majority of experts agrees with a draft, it should be accepted that good is good enough and no further validation should be needed.

A second threat is the politicisation of the process. As long as scientific elements are discussed, the chances of finding consensus are quite good. Even though policies with respect to making decisions on allowing products to enter the market might differ across countries (for example, in the case of genetically modified crops), agreements can still be reached on the underlying science. We have been fortunate in the OECD that we have been able to work constructively on the basis of addressing science, even when faced with differing political views on market admission. Our work in the field of modern biotechnology is a good example of an area in which the OECD is able to achieve harmonisation in regulatory oversight, even though safety/risk policies differ. However, if countries diverge from the scientific discussions and let emotional or trade-related arguments play a major role in their willingness to reach consensus on TGs or assessment methodologies, the work on harmonisation in the OECD may become very complicated.

CONCLUSIONS

It is important to remember that, in the regulatory process for chemicals, when going from testing to assessment to management decisions, national prerogatives become more and more important. With respect to harmonisation, the level of ambition of international work will, therefore, decrease going down this regulatory chain. The MAD system relates to the first part of the regulatory chain (referred to in Figure 22.7).

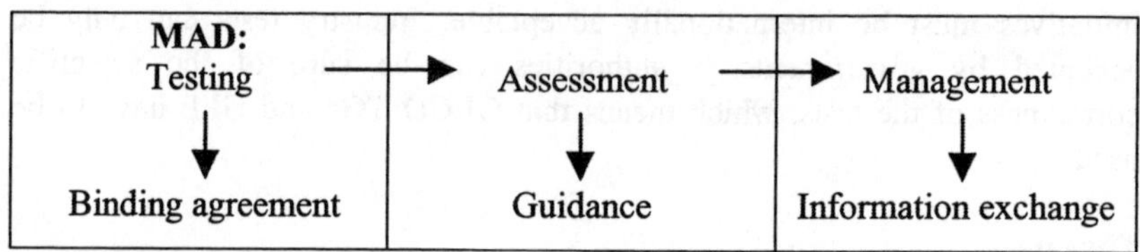

Figure 22.7 The MAD system and the regulatory chain

The key requirement for harmonisation to work well is the creation of a win/win/win situation for all stakeholders were harmonisation to be achieved. In the case of the OECD MAD system, all stakeholders achieve benefits in different areas:

- Governments: good-quality data and assessments, efficiency in assessments, a basis for work sharing.
- Industry: savings in testing, cost efficiency in notifications, trade.
- Civil-society NGOs: good-quality data, transparency, animal welfare.

The main characteristics of a successful modern process in the harmonisation of chemical safety policies are a solid basis of sound science, a system of quality control which inspires confidence, a transparent, open process, and an inclusive, participatory process.

REFERENCES

OECD (1981), *Decision of the Council Concerning the Mutual Acceptance of Data in the Assessment of Chemicals and Addenda 1-15 (Test Guidelines)*, C(81)30, Paris: Organisation for Economic Co-operation and Development.

OECD (1989), *Decision-Recommendation of the Council on Compliance with Principles of Good Laboratory Practice*, C(89)87, Paris: Organisation for Economic Co-operation and Development.

OECD (1997), *Decision of the Council Concerning the Adherence of non-Member Countries to the Council Acts Related to the Mutual Acceptance of Data in the Assessment of Chemicals*, C(81)30(Final), C(89)87(Final), and C(97)114, Paris: Organisation for Economic Co-operation and Development.

OECD (1998), *Savings to Governments and Industry resulting from the Environmental Health and Safety Programme*, Paris: Organisation for Economic Co-operation and Development.

OECD (2001), *Environmental Outlook for the Chemicals Industry*, Paris: Organisation for Economic Co-operation and Development.
OECD (2004), 'Chemical safety', *http://www.oecd.org/ehs*.

ANNEX 22.1: THE CHEMICALS INDUSTRY AND PROJECTIONS FOR FUTURE DEVELOPMENTS[7]

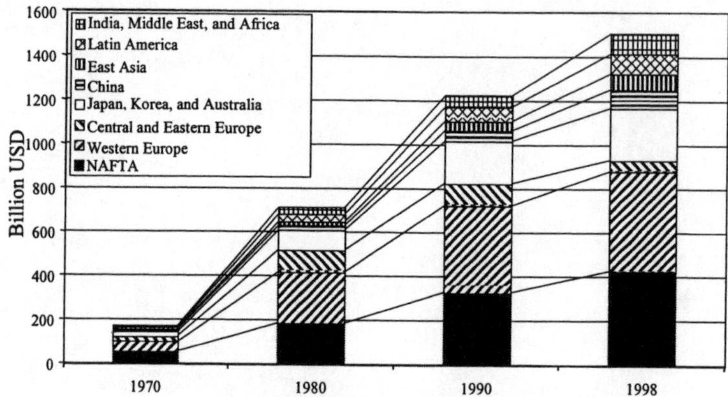

Figure 22.8 Volume of world chemicals industry output

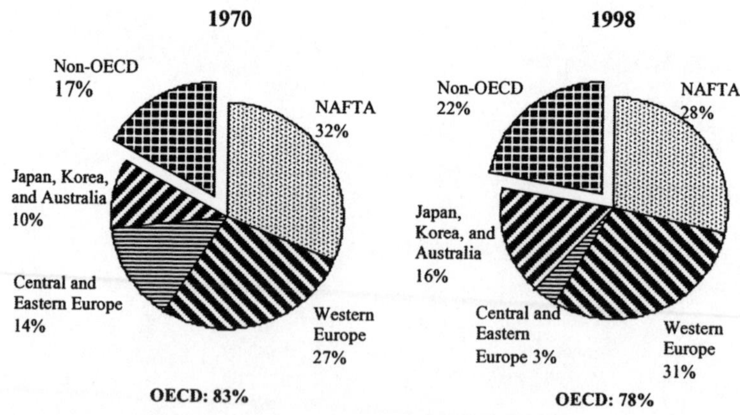

Figure 22.9 Share of world chemicals industry output

---

7 OECD, 2001.

Annex 22.1, continued

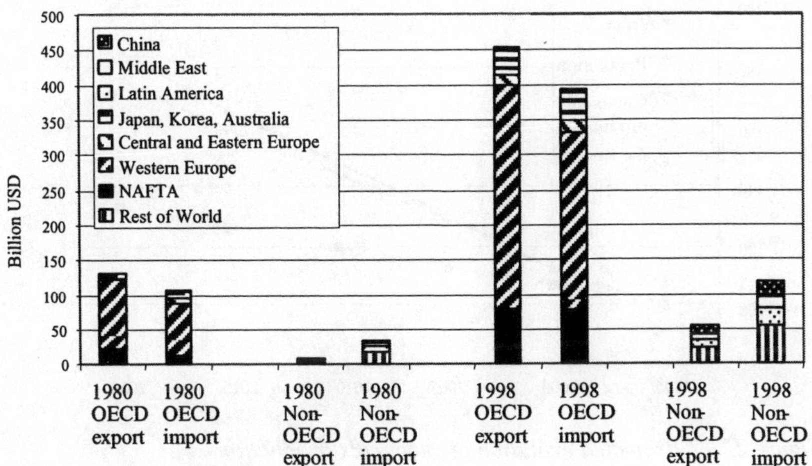

Figure 22.10 Volume of trade in chemicals

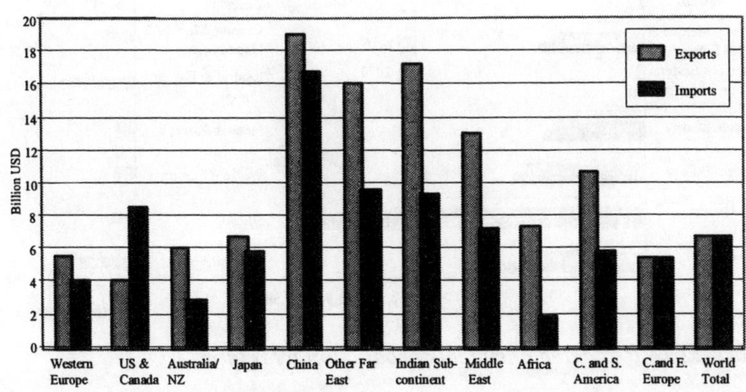

Figure 22.11 Real-term annual growth in trade in chemicals (1979-1996)

Annex 22.1, continued

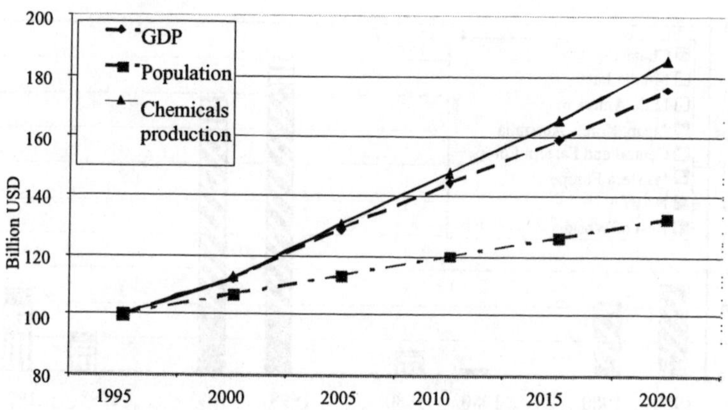

Figure 22.12 Projected evolution of chemicals production

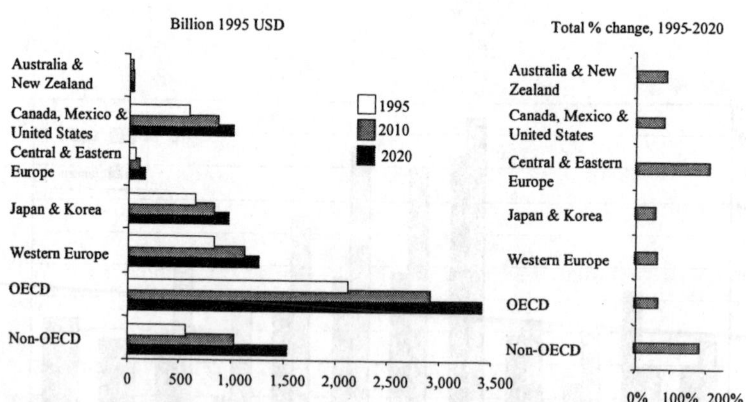

Figure 22.13 Projected chemicals production by region

ANNEX 22.2: MAD, THE WTO, AND ISO STANDARDS

The WTO

The WTO rules have been established to apply to measures affecting trade in goods, and the Agreements refer to product standards and assessment procedures for conformity with technical standards for products. The OECD TGS for chemicals and the OECD Principles of GLP are standards for testing the safety of products (chemicals, pesticides, etc.) rather than standards for specification of the products themselves. In the same way, the procedures for monitoring compliance with GLP are procedures for conformity assessment, albeit not assessment of the conformity of a product with a standard, but rather of the conformity of the testing of the product with a standard (the OECD Principles of GLP). However, since chemicals generally cannot be marketed without being tested for their safety to health and the environment, the standards used to do this and the procedures used to monitor compliance with these standards – the OECD standards, which are the basis for national technical regulations related to the data acceptance process – might be considered in some way 'product standards', since they ultimately affect trade in goods (chemicals, pesticides, etc.).

Although both the Technical Barriers to Trade (TBT) and the sanitary and phytosanitary measures or standards (SPS) Agreements include 'testing' procedures in their definitions of covered measures, it is not clear whether the definition of 'testing' procedures in the WTO Agreements is intended to include testing systems which are distinct from national or international product standards and assessment of conformity with these product standards. It could also be argued that the OECD standards fulfil the major WTO requirements (transparency, avoidance of trade barriers, and openness of participation) and that, therefore, were they to be challenged in a WTO dispute settlement procedure, they could be considered 'international standards'.[8]

In any case, the TBT Agreement allows the use of other standards as the basis for national technical regulations when the international standards which exist would be an 'ineffective or inappropriate means for fulfilment of the legitimate objectives pursued'. The fact that the Agreement names such legitimate objectives as including the 'protection of human health... or the environment' could justify the use of the OECD Principles of GLP and compliance monitoring procedures as the basis of technical regulations related to chemical safety data acceptance, which should not be 'more trade-

[8] It should be emphasised that the WTO is not itself a standards-setting body, but rather an arbitrator of disputes related to the implementation of standards.

restrictive than necessary to fulfil a legitimate objective, taking account of the risks non-fulfilment would create'. In conclusion, the WTO rules do not seem to apply specifically to the acceptance of safety data on chemicals for the purposes of assessment related to health and the environment, except insofar as this can be considered part of a product specification. In order to provide a definitive answer on this point, one would need either an agreed interpretation by WTO members or the question would have to be addressed by a WTO Panel. Since, in principle, all WTO members can take part in the development of OECD standards related to MAD, it is conceivable that, if challenged, the latter would be considered to be 'international standards'.

ISO Standards

In the context of the safety testing of chemicals for health and environmental assessment purposes, the issue of the acceptability of other international standards is often raised, specifically of the ISO standards. The fact that the TBT Agreement includes in its Annex a 'Code of Good Practice for the Preparation, Adoption and Application of Standards' which is based heavily on ISO principles for standards-setting shows the WTO's prejudice in favour of ISO standards where they exist. In the area of standards related to the acceptance of chemical safety data, ISO standards do exist for certain safety studies – primarily for toxicity to the aquatic environment – which are considered comparable with OECD Test Guidelines in this area.

There are no ISO standards – or any other standards – which are comparable with the OECD standards for quality and rigour of safety studies (OECD Principles of GLP). Thus, data generated solely under the ISO/IEC Guide 25 are not likely to be accepted by regulatory authorities in OECD countries for the purposes of the assessment of chemicals related to the protection of health and the environment. This was emphasised by the OECD Chemicals Committee in 1994, when it endorsed a position calling for the monitoring of compliance with the OECD Principles of GLP of data received in safety studies. Therefore, the OECD standards and OECD procedures for GLP and compliance monitoring of facilities presuming to use these standards appear to be the only standards acceptable to regulatory authorities in OECD countries.

For all the reasons and points indicated in the above paragraphs, it seems likely that, were a dispute to be raised in the WTO framework about the appropriateness of using the ISO/IEC Guide 25 instead of the OECD Principles of GLP, use of the OECD Principles in the technical regulations of a country would not be considered contrary to the WTO agreements, even if the Principles were not considered 'international standards'.

23. Architecture of the Kyoto Protocol and Prospects for Public Climate Policy

Frank Wijen and Kees Zoeteman[1]

SUMMARY

Options for governments' future climate policy are discussed as a function of the architecture of the present regime; the latter is anchored in the Kyoto Protocol, which is aimed at reducing the human impact on climate change. We describe the basic tenets of this agreement, and explain how it was realised despite the widely divergent interests. The strengths and weaknesses of the Kyoto regime, and related future opportunities and threats, are presented. The degrees of collective decision-making and international participation were the basis for exploring four scenarios (local market, local collectivity, global market, and global collectivity) and concomitant policy instruments and actors. The possibilities of enhancing participation by linking issues and creating bandwagons are discussed. We conclude that the main flaw of the Kyoto regime is its lack of appropriate incentives. To realise a more effective regime, future climate policy should be geared towards making participation more attractive and rendering compliance self-enforcing.

INTRODUCTION

The Kyoto Protocol, concluded in December 1997 to address anthropogenic (i.e., human-induced) climate change, is undoubtedly the most prominent

[1] We gratefully acknowledge the insightful comments by Daan van Soest, Paul van Seters, Jan Pieters, Xavier Martin, David Levy, Erwin Bulte, Sjoerd Beugelsdijk, Marcel Berk, and Theo Beckers on an earlier version, as well as research assistance by Wiebe Vos and Suzanne Verheij and financial support by the Dutch Ministry of Spatial Planning, Housing and the Environment (VROM/DGM/KVI) to conduct this study.

global environmental agreement. This status stems from the perception that climate change represents a major environmental problem without easy solutions, the high stakes involved for many parties, the controversial nature of the agreement, and the media coverage it has received.[2] While many uncertainties persist, increasing scientific evidence points to a large human influence on global atmospheric conditions.[3] This human impact is caused mainly by the combustion of fossil fuels, which gives rise to greenhouse gases (GHGs) that retain solar radiation, causing both progressive warming of the atmosphere[4] and the increased occurrence of extreme weather events. These changes may have far-reaching consequences: the progressive flooding or desertification of large areas, the melting of glaciers and ice caps, major shifts in patterns and levels of economic activities, and the disappearance of a significant number of living species. The possible consequences of climate change (including the disappearance of small island states) are thus enormous, as is the impact of addressing the causes. Many constituencies have important interests in the production, marketing, and use of derivatives of fossil fuels (oil, gas, and coal). Mitigating the human impact on climate would directly affect the financial resources of countries and companies in this business. Besides, most states and economic actors would be concerned indirectly, since a large share of economic activities are supported by fossil-fuel-generated energy. As a result, major changes of production and consumption practices would be required globally.

A major impediment to action has been the global-public-good nature of the climate issue: GHG emissions from anywhere on earth contribute to the global problem, while parties taking action enjoy only a small share of the fruits.[5] This misfit between loci of costs and benefits has given rise to free-rider behaviour (i.e., a calculative abstention from collective action out of personal interest), reinforced by time lags between efforts and results: future generations bear the consequences of present (in)action. Another complicating factor has been the North-South divide: low-income countries have not committed themselves to action, arguing that rich nations are responsible for most of the problems and should thus take the lead. Consequently, the Protocol, which ensues from negotiations between parties with divergent interests, has also been controversial in nature. Finally, regular public attention has been drawn for over a decade by the media coverage of

2 Grubb et al., 1999; Oberthür and Ott, 1999.

3 IPCC, 1995, 2001.

4 This does not imply, however, that the temperature rises around the globe. Owing to changes in gulf streams, certain regions may actually be confronted with lower temperatures (Bartsch et al., 2000; Oberthür and Ott, 1999).

5 Kaul et al., 1999.

the climate negotiations. As a result, the Kyoto Protocol has become better known among the public at large than any other environmental agreement.

Thousands of persons have been involved – directly and indirectly – in Kyoto's lengthy antecedents, laborious conclusion, and capricious aftermath: negotiators from virtually all countries, climate scientists, business lobbyists, environmental activists, and media representatives. Myriads of publications on the Kyoto Protocol have seen the light. The object of this study was not to reiterate the factual evolution of global climate policy or to provide an in-depth description of Kyoto's provisions and mechanisms; these have been excellently explained elsewhere.[6] The aim of this study was twofold. First, to assess the strengths and weaknesses of the climate policy regime that has emerged over the past few decades against the objective of sharply reducing the human impact on climate. Second, to explore the opportunities and threats of future climate policy given the prevailing regime – the idea being that if the present is a reflection of past developments, a future climate regime is shaped to an important extent by the present architecture.[7] The Kyoto Protocol has remained for years in a critical stage, with all options (entry-into-force, collapse, and renegotiation) still open.

While drawing on different theoretical strands, we made predominant use of insights from regime theory[8] and institutional theory.[9] Regimes are 'sets of implicit or explicit principles, norms, rules, and decision-making procedures around which actors' expectations converge in a given area of international relations'.[10] They are constituted by factors such as the interests and influences of different parties, common norms, established customs, and generated and shared knowledge. Effective regimes contribute to putting interconnected (environmental) problems higher on the political agenda, addressing (global) problems in a more comprehensive way, and formulating national policy responses. The regime approach has different (economic, legal, political, normative) dimensions, though political aspects tend to dominate. Therefore, regime theory provides a suitable lens to conceptualise the (global) interplay of forces, ambitions, and insights in relation to the climate-change issue. Institutional theory shows significant overlap with regime theory, though it focuses more on explaining the behaviour of (groups

6 Bartsch et al., 2000; Depledge, 2000; Grubb et al., 1999; Oberthür and Ott, 1999.

7 This is not to say that a future regime is fully determined by the present conditions, as it is also influenced by other factors, in particular the future political willingness to act.

8 Haas et al., 1993; Keohane and Nye, 2001; Krasner, 1983; Young, 1994, 1999.

9 DiMaggio and Powell, 1983; Meyer and Rowan, 1977; Phillips et al., 2000; Tolbert and Zucker, 1996.

10 Krasner, 1983: 2. While regime theory applies primarily to international relations, it can easily be extended to the national context – for example, to explain group processes that cannot be legally enforced.

of) individuals and organisations. Institutions can be defined as 'enforced rules, formal and informal, about what actions are required, prohibited, or permitted'.[11] They constitute the basis for institutionalisation, which is 'a core process in the creation and perpetuation of enduring social groups'.[12]

Empirical insights were obtained through interviews and an expert panel. Between May and December 2003, in-depth interviews were conducted with 30 representatives of (national and supranational) government, business, non-governmental organisations (NGOs), and academia in Europe and North America who had all been directly or indirectly involved in the Kyoto process. Respondents were asked about their roles, the relative importance of (domestic and international) actors, the realisation of the Kyoto Protocol, its (de)merits, and its prospects. Most interviews were tape-recorded and transcribed; detailed notes were taken of the rest. The transcripts, notes, and recent newspaper excerpts were coded in Atlas/ti, a qualitative software package.[13] In December 2003, a one-day workshop was attended by 20 experts from government, business, and academia. Brainstorm-and-analysis sessions were held on the present and possible future climate regimes. Salient insights from this workshop were represented in a detailed report.[14] The coded excerpts and the workshop report were subsequently analysed.

The structure of this chapter is as follows. In the second section, dealing with the past and the present, we indicate the basic tenets of the Kyoto Protocol. Factors explaining the 'success' of Kyoto are then highlighted,[15] followed by a description of developments since the agreement was concluded. These lead to an assessment of the strengths and weaknesses of the Kyoto regime. They constitute inputs for the third section, which focuses on future climate policy. We explore four possible climate policy scenarios, and for each scenario discuss policy instruments and the actors involved. Next, the possibilities of creating participation leverage are indicated. In the final section, we draw conclusions on the opportunities and threats of future public climate policy.

---

[11] Prakash, 2000: 17.

[12] Tolbert and Zucker, 1996: 180.

[13] Codes capture and bundle highlighted chunks of text around specific themes (such as 'strengths of Kyoto' or 'relative importance of stakeholders'), facilitating the subsequent analysis of these themes. See Weitzman and Miles (1995) for a description of the analytical possibilities of Atlas/ti.

[14] Wijen and Zoeteman, 2004.

[15] The 'success' refers to getting all parties to sign the Kyoto Protocol; it does not refer to its ratification or entry-into-force.

KYOTO'S ARCHITECTURE[16]

Kyoto's organisational embedding and substantive arrangements, such as agreed prior to and during the Protocol's conclusion and further elaborated during subsequent meetings, are described in this section. The contents were analysed against the backdrop of the different interests, and the regime's (de)merits were examined.

Institutions

An international group of climate scientists, which has participated from 1988 onwards in the Intergovernmental Panel on Climate Change (IPCC), has applied major resources to investigating the human impact on climate, as well as the socio-economic consequences of climate change. Notwithstanding the complexity of the causal chains of evidence and the uncertainty of future developments, the IPCC has increasingly come to the conclusion that the human impact on atmospheric GHG concentrations is significant and that the consequences of climate change are far-reaching.[17]

As a result of the growing scientific evidence of the human impact, the United Nations Framework Convention on Climate Change (UNFCCC) was signed at Rio de Janeiro in 1992. This Convention was aimed at stabilising GHG concentrations at levels that are compatible with present and future socio-economic development, thus preventing dangerous human interference with the climate system. Guiding principles were also formulated: common but differentiated responsibilities and capabilities, implying that while the same objective is pursued, individual actors assume actions according to their (financial and institutional) carrying capacity. A related principle referred to consideration of the vulnerable position of developing countries. Consequently, developed countries ('Annex I Parties') committed themselves to adopting adequate national measures. The UNFCCC was endowed with a Secretariat, established in Bonn, to coordinate and file the positions of different nations and monitor their actions. The Secretariat explored possible common grounds between states in a neutral way, thus facilitating Kyoto's preparatory negotiations, such as the annually revolving Conferences of the Parties (COPs).

The IPCC's findings greatly facilitated the conclusion of the Kyoto Protocol. After 1997, the IPCC continued to provide scientific evidence. The

[16] Unless indicated otherwise, factual information in this section draws on: Barrett, 1998; Bartsch et al., 2000; Grubb et al., 1999; Gupta, 2001; McGivern, 1998; Oberthür and Ott, 1999; Yamin, 1998. The design of the Kyoto Protocol was inspired by the Montreal Protocol, concluded in 1987, on the ban of ozone-depleting chlorofluorocarbons (Barrett, 2003).

[17] IPCC, 1995, 2001.

UNFCCC constitutes the political and legal infrastructure of the Protocol, with an active Secretariat in charge of its implementation.

Targets and Mechanisms

The overall target of the Kyoto Protocol is to reduce human-induced GHG concentrations by an average of 5.2 per cent over the period 2008-2012 ('first commitment period'), as compared with the base year, 1990.[18] The net greenhouse impact is targeted: the emissions of a basket of six GHGs (of which carbon dioxide is, in absolute terms, by far the most important)[19] minus the absorption of these gases by 'sinks' (mainly additional forests). In line with the principle of differentiated responsibilities, developed nation states committed themselves to individualised targets, ranging from a 28 per cent reduction to a 27 per cent increase.[20] No emission ceilings were formulated for developing countries. The height of national targets was determined on the basis of past emission records (with high past levels, both in absolute and relative terms, entailing more stringent targets) and negotiation power (with large and indispensable parties obtaining modest marks).[21]

While it has formulated precise targets and timetables, the Kyoto regime does not prescribe any policies and measures (PAMs). This implies that nation states have the discretion to choose domestic implementation modes. Yet, the Protocol provides international compliance options. Parties which exceed their emission caps may buy off the difference from countries which have unexploited emission room ('hot air'). By creating a market for GHG emission credits ('carbon trade'), the Protocol has caused GHGs to be monetarised for the first time. Joint Implementation (JI) is another international compliance mechanism. Countries meet their national commitments by paying for (the additional costs of) emission-reduction measures implemented in other Annex I countries. A similar tool is the Clean Development Mechanism (CDM), where (subjects of) states invest in project-based reduction measures in developing countries. The Kyoto Protocol has no

---

[18] It is generally recognised that this target is insufficient to achieve the Climate Convention's objective, which would necessitate more than halving anthropogenic GHG emissions (Metz and Berk, 2001; Oberthür and Ott, 1999). At the same time, unaltered practices ('business-as-usual') would entail double-digit increases (Löschel and Zhang, 2002; McGivern, 1998).

[19] Carbon dioxide accounts for 82 per cent of GHG emissions in industrialised countries. The other regulated gases are methane (12 per cent of emissions), nitrous oxide (4 per cent), and three halocarbons (2 per cent).

[20] Several Northern European countries agreed upon relatively large emission-reduction targets, while several Southern European states were entitled to the largest relative increases, following an internal EU 'bubble' agreement.

[21] An example of a large and indispensable party is Russia, which succeeded in negotiating stabilisation, although its economic collapse and production inefficiency would have justified a major reduction of its GHG emissions.

financial provisions for implementation in the sense of redistributing funds, apart from some assistance to developing countries (through the Global Environment Fund, GEF).

The Protocol is thus flexible as to the gases targeted (trade-offs among GHGs are allowed), the nature of measures (technical or accounting solutions), and the locus of implementation (domestically or abroad). It is also flexible as to timing: excess emissions in a particular year may be compensated afterwards, as only the cumulative emission record by 2008-2012 counts, while unused assigned quantities may be saved for future periods ('banking'). Non-complying parties must assume a 30 per cent reduction surcharge in the next commitment period. There are no direct financial sanctions or other enforcement mechanisms, as is the case with most international agreements.

Realisation

While realising a global agreement, involving numerous parties, on a public good with uncertain, long-term consequences is a major enterprise, the task becomes even more arduous when conflicting interests exist. Some parties (like producers of fossil fuels and cold-climate states) had articulated interests in thwarting the realisation of such a climate agreement; indeed, they actively resisted through the Global Climate Coalition.[22] It is a wonder that agreement was finally reached in Kyoto among all nation states involved in the negotiations.

A major reason for Kyoto's 'success' is that the Protocol consists of elements that are crucial to the different parties and/or that do not contain stipulations against which they have prohibitive objections. No party is completely satisfied with the present, heavily compromised agreement, but it perfectly reflects the different interests defended by blocks of negotiating countries. The European Union (EU) obtained firm targets and timetables, but had to give in with respect to its desired prescription of PAMs. The United States (US) succeeded in imposing market-like instruments and sinks as well as in avoiding PAMs, but had to compromise on the involvement of developing countries. Japan, which had mostly negotiated on the side of the US and other major non-EU industrialised countries (united in the 'JUSSCANNZ coalition' and the similar 'Umbrella group'),[23] benefited from the honour of having a prestigious agreement concluded on its territory, but

[22] Levy and Egan, 2003.

[23] The JUSSCANNZ coalition consisted of Japan, the US, Switzerland, Canada, Australia, Norway, and New Zealand, countries which for different reasons opposed a stringent climate regime. The Umbrella group was made up of the JUSSCANNZ countries except for Switzerland, together with Russia and Ukraine.

had to assume an emission-reduction commitment despite the relatively high energy efficiency of its economy. Other advanced industrialised countries succeeded in avoiding high emission-reduction targets, but had to accept emission ceilings. Eastern European countries and Russia, whose GHG emissions had dramatically dropped during the 1990s, also had to show political commitment, but had the prospect of political benefit (from improved relations with the EU and US) and financial pay-off (from selling hot air). The developing countries, which had joined forces in the G77,[24] managed to avoid any binding commitments, but had to demonstrate moral engagement and accept that industrialised countries obtained modest targets and could avoid domestic action through international mechanisms.

Common factors also drove the Protocol's realisation. The threat of human disasters and economic disruptions, induced by progressive warming and extreme weather events, were also a major explanation of Kyoto's realisation. An increased incidence of death from heat (virtually all warm summers at higher latitudes of the last two centuries have occurred since 1980), the exponential rise in damage from weather-related catastrophes over the last forty years, and even the outright disappearance of (parts of) low-lying countries highlight the necessity to act.[25] The IPCC's mounting evidence of human influence on climate, as well as indications of consequences of climate changes, had significantly raised this awareness.

The pressure to reach an agreement was also upheld by the media. Environmental NGOs, coordinating actions through the Climate Action Network (CAN), succeeded in securing ample media attention. No party wanted to become the public scapegoat of a mediatised negotiation failure; the propensity to compromise was thus enhanced. The willingness to assume dissimilar but 'fair' shares of the reduction burden was increased by scientific calculations of national and sectoral contributions to GHG concentrations. For example, the EU's burden-sharing agreement was underpinned by calculations of relative contributions, thus turning a political polemic into a scientific debate.[26] Finally, Kyoto's chair, Raúl Estrada-Oyuela, skilfully exploited the potential common grounds and hammered out a success where failure was imminent.

---

[24] The G77 is a group of over a hundred developing countries, negotiating as a solid block despite internal differences. For example, the OPEC nations wanted to avoid any significant actions because of their fossil-fuel interests, while the AOSIS (Association of Small Island States) countries advocated important actions because their survival was at stake owing to rising sea levels.

[25] Holdren, 2003.

[26] Phylipsen et al., 1998.

State of Affairs

By the year 2000, the aggregate level of net GHG emissions in Annex I countries was fairly well in line with the agreed target.[27] Major contributions to this relatively favourable performance came from: Eastern Europe and Russia, whose GHG emissions had dropped dramatically after the collapse of their economies in the early 1990s; the UK, which, for financial reasons, had reconverted its energy supply from coal into gas, entailing substantially lower GHG emissions; and Germany, which had modernised the energy-inefficient production installations in the Eastern part of the country following reunification. Other indicators left less room for optimism. By 2001, the EU, which was responsible for 24 per cent of GHG emissions in Annex I countries, had only slightly decreased its net emissions, particularly as a result of enhanced mobility, while aggregate stabilisation at the 1990 level was projected using existing domestic PAMs.[28]

Other Annex I parties were also likely to underperform. By 2000, GHG emissions had risen by 14 per cent in the US (accounting for 36 per cent of Annex I emissions) and 11 per cent in Japan (responsible for 9 per cent). The emissions of developing countries, which were formally without emission caps, had increased considerably; especially the rapid economic development of large countries such as China and India had entailed significant emission increases.[29] Therefore, while the overall performance by the turn of the millennium seemed to be in line with the Protocol's target, the prospects of sharply rising global GHG emissions did not give rise to optimism.

The Kyoto Protocol enters into force when 55 per cent of the signatories representing 55 per cent of the total carbon dioxide emissions by Annex I countries have ratified it. In late 2003, the first condition had been met (120 parties had ratified the Protocol), but the second hurdle remained (the ratifying parties represented merely 44 per cent of total emissions) because the two largest emitters (the US and Russia) had not ratified the agreement.[30] Given the agreement's outright political rejection in the US in 2001, the Protocol could then only enter into force if Russia (responsible for 17 per cent of Annex I emissions) ratified it. As the country had postponed this decision because of an internal conflict of interests – between those seeking to attract foreign energy-efficient investments and those benefiting from the sales of fossil fuels – and the desire to capitalise on its pivotal position – by

[27] UNFCCC, 2003a.

[28] European Environment Agency, 2003.

[29] China is the world's second largest producer of GHGs in absolute terms (Leal Arcas, 2001).

[30] UNFCCC, 2003b. Australia, which accounted for 2 per cent of Annex I emissions, had also failed to ratify the agreement.

inducing the EU and the US to bid against one another, Kyoto's entry-into-force had turned into a game of Russian roulette.

While the US and Russia frustrated the ratification of the Kyoto Protocol, the EU moved forward by creating the institutional framework to implement the Kyoto mechanisms (in particular, emission trading from 2005 onwards). It also reiterated its commitment to the Protocol, regardless of its future status.[31] The European Bank for Reconstruction and Development set up an investment fund for energy-efficient projects in Eastern Europe and Russia. At the end of 2003, the political dispute between the EU, on the one hand, and the US and Russia, on the other hand, continued, as a result of which the Kyoto Protocol faced a highly uncertain future.

(De)merits

When the institutional embedding, targets, and mechanisms of the Kyoto Protocol, along with developments since its conclusion, are combined, the following strengths and weaknesses of the Kyoto regime can be identified against the aim of mitigating the human impact on climate change.[32]

Strengths

- Despite the huge divergences of interests, the Kyoto Protocol has been signed by virtually all nation states, making it a truly global environmental agreement. This creates a broad basis and an important signalling function for engaging in climate-related actions around the globe. A broad climate coalition is necessary to overcome the problems inherent in a global public good such as the climate issue, and paves the way for entering and gaining momentum in learning trajectories on emission-poor activities.
- The Kyoto regime has a solid institutional infrastructure. It is based on the UNFCCC, to which all Kyoto signatories (including the US) still formally adhere. The IPCC's extensive and sustained research has considerably expanded the knowledge base of climate change. The UNFCCC Secretariat effectively coordinates the different positions and monitors the performance of countries. Annual COP meetings attended by all parties allow for the settlement of unresolved and upcoming issues.
- Innovative, flexible instruments to reduce the costs of implementation have been created. International carbon trade, JI, and the CDM are, at least in principle, ways of making the agreement cost-effective by creating international markets where transactions are settled at the least costs. An additional benefit of these financial instruments is the monetarisation of

[31] Wallström, 2003.
[32] Barrett, 2003; Grubb et al., 1999; Leal Arcas, 2001; Metz and Berk, 2001; Oberthür and Ott, 1999.

GHG emissions. This sensitises economic agents to the costs of emissions, which is a precondition for taking financially inspired actions. In addition, JI and CDM provide opportunities to reinforce international cooperation among participating countries.

- The Kyoto Protocol has specified concrete, binding targets and timetables for industrialised countries. This implies that individual commitments cannot be waived because of elusive wording.[33] While the overall reduction target of 5.2 per cent may seem modest, it represents, in the context of steadily growing national products, double-digit reductions in effective terms; it also implies some decoupling of GHG emissions from economic output.

Weaknesses

- Only a minority of nation states have committed themselves to emission ceilings. Some of these states (especially the US and Australia) have withdrawn or (as with Russia) have postponed ratification. Apart from delaying or obstructing the entry-into-force decision, this leaves a small basis for action.[34] Besides, the rapidly rising emissions of developing countries have remained outside the regulative scope of the Protocol.
- Even when all parties comply with their targets, the overall level of GHG concentrations is hardly affected. Halting the human impact on climate requires much more ambitious decreases. Besides, the present targets of some countries (for instance, Russia) have been set so low that they require no effort to be met, and thus do not lead to the envisaged behavioural changes.
- Kyoto's enforcement regime is weak. While the penalty of additional future reductions in case of underperformance is foreseen, it can easily be avoided – especially because future targets (for the post-2012 period) have not yet been agreed upon. The present regime is further handicapped by the absence of an effective global enforcement organism (such as a powerful world environment organisation) or mechanism (such as financial or trade sanctions), though national parliaments or NGOs may pressurise governments to meet their targets.
- Carbon trade, JI, CDM, and sinks are waivers to domestic action, because parties may comply through accounting measures instead of technical

---

[33] Vague or ambiguous phrasing is not uncommon in international environmental agreements, facilitating their realisation but subsequently entailing interpretation problems that hamper their implementation.

[34] While the US has rejected the Kyoto Protocol at the federal level, many local governments (states and municipalities) have taken significant climate-related actions. Australia has indicated that it aims at complying de facto.

actions.[35] They distract attention from the sources of GHG emissions and discourage behavioural changes and the advancement of climate-neutral technology.

- The time horizon of the first budget period (2008-2012) dissuades governments from developing long-term solutions. Parties are induced to implement short-term, incremental solutions (such as optimising existing techniques), rather than forcing costly breakthrough innovations to realise the required leapfrog improvements.
- The present regime stresses the bearing and sharing of the emission-reduction burden, rather than indicating novel (economic) opportunities that may arise from emission-poor products and processes.
- Incomplete international participation and differentiated targets entail an uneven economic playing field. This especially affects energy-intensive businesses which produce in committed countries and sell on highly competitive global markets. They are likely to insist on exemption from measures. Eventually, they may even relocate to countries with permissive climate regimes.
- The present regime is overly complex. It leads to a high administrative load, which certain (developing or emerging) nations cannot or do not wish to bear. Besides, complexity facilitates fraud. Bribery of officials is a serious risk, especially when stakes are high and opportunities for corruption are ample.
- Economic activities that cannot be easily attributed to national territories (in particular, international transport) are not covered by the Protocol, though many of these 'footloose' activities have a high climate impact.

PROSPECTS FOR CLIMATE POLICY

Effective strategies consist of turning strengths into opportunities and reducing the threats that weaknesses entail. Public climate policy should aim at creating and seizing the opportunities that the Kyoto regime provides, while minimising and managing its threats. This is a complex challenge because of the numerous intervening variables (such as economic, political, technical, and demographic developments), the future directions and magnitudes of many of which are uncertain. It is useful to consider different future options, which can be explored using scenarios.[36] Scenarios are internally consistent, challenging descriptions of possible futures. Their aim is

[35] Although the Kyoto Protocol states that international mechanisms are complementary to significant domestic action (the supplementarity principle), it fails to specify the minimum share of domestically taken measures.

[36] De Mooij and Tang, 2003; Fahey and Randall, 1998; Van der Heijden, 1996.

not to predict the future, but to sketch and understand feasible alternatives in order to be prepared for the contingent situations in which they materialise.

As described in the next sub-section, we developed four such scenarios. Government interventions that are compatible with the different scenarios, as well as the main actors involved, are reviewed. Finally, we discuss the possibilities of raising support for climate policy.

Climate Scenarios

Establishing scenarios involves the identification of future uncertainties (given the prevailing policy question), which – after correlated factors are merged – yields two critical dimensions. We adopted the key dimensions identified by the Netherlands Bureau for Economic Policy Analysis,[37] which show similarities with those of the IPCC and the United Nations Environment Programme (UNEP).[38] The first dimension is the extent of international cooperation, ranging from complete national sovereignty to full-fledged international cooperation. It indicates the degree to which nation states craft their climate policy in concert with other countries. The second dimension concerns the allocation of climate-related decisions. In the extreme cases, climate policy is entirely organised either by the private sector or through public responsibilities (i.e., collective action). Confronting the two dimensions yields four scenarios, which are depicted in Figure 23.1.

Local market

In this scenario, little international cooperation concurs with private-sector organisation. The lack of international support may occur if the Kyoto Protocol does not enter into force, after which parties that ratified the Protocol earlier may no longer feel committed to their initial targets. The failure or continued postponement of Kyoto's entry-into-force may result in such a complete loss of momentum. If the Kyoto regime materialises but nation states make no effort to meet their targets, international commitment will also fall short. Knowing that the Kyoto targets cannot be effectively enforced, nation states may indulge in free-rider behaviour and abstain from taking measures, especially when these are perceived as costly or untimely (for example, when they are to be taken during an economic recession).

When markets are the locus of decision-making, climate-relevant behaviour tends to be confined to economically attractive actions: 'no regret' measures that pay off (for example, higher fuel efficiency leading to lower energy costs) and new business opportunities (such as the development and

[37] De Mooij and Tang, 2003.
[38] IPCC, 2001; UNEP, 2002.

Figure 23.1 Climate policy scenarios

marketing of fuel-cell cars for areas struck by air pollution). Behavioural changes that offer no (financial) advantages to individual consumers or producers are not undertaken. Given the lack of international cooperation, costly measures involving competitive disadvantages for energy-intensive businesses exposed to international competition are particularly eschewed.

Local collectivity

This scenario involves the combination of public responsibilities and low international participation. The prevalence of national sovereignty may result from pessimism after the collapse of the Kyoto regime or the lack of incentives to comply in the absence of an effective enforcement mechanism. By contrast, some (EU) countries show a firm political and/or moral commitment to the climate cause. The asymmetry between the minority of countries that take concerted, possibly costly actions and the majority of nations that abstain from collective action leads to political tension. The proactive countries may consider the presence of 'free-riding' nations to be unfair and may experience adverse economic effects of climate-related measures. Industries facing international competition are hampered by the uneven playing field. Bilateral agreements among like-minded countries are concluded and regional regimes (i.e., political networks of nations with similar orientations as to climate policy) are likely to arise in the absence of one global regime.[39] Such regimes turn the climate issue from a global public good into a regional club good: proactive countries collectively shield their markets from adverse competition from laggards who do not face measures that raise their production costs.[40]

Countries may opt for collective actions owing to political awareness (for example, because they are directly threatened by the consequences of climate change) and/or moral commitment (for instance, because conservation is a national value). Generalised local support for far-reaching public measures may, for fundamental and/or practical reasons, go together with a call for protectionist measures to guard national or regional economies against competition from 'free-riding' states.

Global market

In this scenario, many nation states commit themselves to action, which is organised through the private sector. This scenario may materialise if and when an (amended) Kyoto regime enters into force. One variant is that Russia ratifies and the US also joins the Kyoto bandwagon (for example, because non-adhesion would be costly to US businesses with international operations). Another variant is that the Kyoto Protocol does not enter into force, but is renegotiated and accepted with amendments instead. In the latter case, the UN Climate Convention is likely to be the basis for constructing such a new regime. The main objective of amending the Kyoto regime would be to get major 'missing parties' on board. In addition to the cooperation of Annex I countries that have not ratified the Protocol (especially the US, Russia, and Australia), the participation of relatively large and industrialised developing countries, which are responsible for significant and increasing contributions to GHG emissions, would then be sought. This concerns in particular China and India, but also Brazil, Indonesia, Mexico, and South Korea. The realisation of such a broad climate coalition is most likely when the ambition level of (individualised) targets is relatively low – the more easily nation states can comply, the higher their propensity to join – or when participation is attractive (for example, because it is related to significant

[39] Restrictive international trade regimes may clash with the rules of the World Trade Organization (WTO). While the regulation of environmental protection with trade-distortive effects is not forbidden by principle, it is banned when there is a suspicion of protection of local industry. As countries with much international trade wish to avoid conflict with the WTO, they will be reluctant to adopt trade-distortive measures.

[40] Public goods and club goods share the characteristic of non-rival consumption, but differ as to the excludability dimension: producers who do not join the club are excluded (Kölliker, 2002). To the extent that certain costs can be shared among the different participators, club goods lead to higher utility levels than do public or private goods (Schelling, 1978). It should be noted that the nature of goods is not necessarily given by technical characteristics; but it is also shaped by human intervention (Kaul et al., 2003).

CDM projects). Thus, a soft global regime emerges, characterised by a multilateral, market-based framework with relaxed targets and/or timetables for the present 'dropouts', and largely symbolic commitments for the new entrants.

Markets are the main mechanisms to implement a global climate agreement. They enable economic agents to settle their commitments in the most cost-effective ways. The Kyoto mechanisms lead to the emergence of new international markets, where private financial considerations determine when, where, and how much is traded and invested.

Global collectivity

In this scenario, widespread international cooperation, as favoured by the EU, takes the form of networks of public arrangements which lead to a high level of concerted action at the global level. This regime may arise as a result of increased scientific evidence of the negative consequences of climate change or the recurrence of high-impact natural catastrophes whose causes are attributed to climate change. Nation states pursue significant emission reductions, which go far beyond those agreed in the Kyoto framework and which may involve considerable financial sacrifice. Yet, there are no competitive disparities owing to global participation. Furthermore, parties without commitments in the Protocol also assume emission caps. Thus, a global regime emerges, which is based on multilateralism. The targets are not necessarily phrased in terms of national emission caps; they may also concern relative energy efficiency or focus on emission-intensive sectors.

Actions are initiated and supervised at the central level by a global climate organism. Climate policy is regarded as a public responsibility owing to a high level of (perceived) awareness of the climate-change problem. Therefore, there is general acceptance of the adverse effects on certain economic sectors and of measures involving more austere consumption practices.

Policy Instruments and Parties

The overall human impact on climate is the product of global population size, economic activity per person, the energy intensity of the average economic activity, and the carbon intensity of energy supply.[41] Considering population size and the level of economic activity to be exogenous factors, public policy to mitigate climate change should thus focus on reducing the energy and carbon intensities of human activities.

---

[41] Holdren, 2003.

Climate policy should also be tailored to the characteristics of the prevailing scenario.[42] When intervening, national governments should take into account the specific possibilities and constraints of the scenario that materialises. The degree of international participation determines the extent to which national commitments can be realised in a global framework. The allocation mechanism determines the extent to which climate policy is to be implemented through private or public actions. It should be noted, though, that a minimal degree of political willingness to take action is required in *any* scenario; if societal actors (in particular, producers and consumers) are not sensitive to the climate-change issue, all government policies and instruments are powerless. Assuming that societal actors are – or can be rendered – sufficiently sensitive, the policy implications for the different scenarios are as follows:[43]

Local market

In the absence of international cooperation, government has recourse to domestic measures. These measures should not be costly to industry because of international competition effects. Nor should they interfere with the sovereignty of consumers and producers. The most suitable instruments in this scenario are national communication, national subsidies for applied research and development (R&D) of low-emission technologies, national carbon trade, and a national carbon tax. Communicating 'no regret' measures to producers and consumers involves awareness-raising, and is a way of realising both environmental and economic gains. Technology subsidies, such as those the Canadian government has granted to develop the hydrogen cell, stimulate business to realise low-emission innovations without adversely affecting existing competitive positions; they may not only absorb additional costs but also stimulate the exploration of new business opportunities. National carbon trade allows for cost-effective reductions among domestic producers without requiring changes of production modes; non-compliers can simply buy off their commitments. Finally, a carbon tax provides financial incentives for consumers to adopt lower-emission lifestyles without forcing them to change their consumption practices. This instrument does not particularly harm domestic producers, as imports are also affected.

The design and implementation of these policy instruments involve a small number of governing parties, in particular, the Ministry of the Environment,

---

[42] In this subsection, it is assumed that the climate policies of national governments aim at minimising the net national level of GHG emissions.

[43] Future public policy options are discussed in: Aldy et al., 2003; Barrett, 2003; Bodansky, 2003; Charnovitz, 2003; Den Elzen, 2002; Grubb et al., 2003; Heller and Shukla, 2003; ICCEPT, 2002; Kemp, 1997; Metz and Berk, 2001; Müller et al., 2003; Pershing and Tudela, 2003; Victor, 2001.

the Ministry of Finance, the Ministry of Economic Affairs, and the business community of nation states.

Local collectivity

In a polarised world with divergent climate regimes, proactive national policies are crafted unilaterally or in consultation with a small number of like-minded countries. Climate policy takes the form of collective action. Public policy instruments are national communication, national or regional energy-efficiency standards, national or regional covenants (i.e., negotiated agreements with industry in which targets are specified in exchange for exemption from legislation),[44] and national or regional partnerships (i.e., voluntary agreements involving business, civil society, and – often – government). Communication aims at making the climate-change issue a matter of (higher) common concern. Meeting stringent efficiency standards forces many producers to restructure the design of their products, during both the production phase and the consumption phase.[45] Covenants also require active industry commitment, as technical measures have to be agreed upon that go beyond current practices in order to reduce GHG emissions; in contrast to the carbon-trade system, this always requires technical and organisational measures. Finally, effective partnerships require the genuine willingness and active commitment of different societal actors to engage in open discussions and accommodate their behaviour to the outcomes of such processes.

The application of these instruments involves a considerable variety of parties: the Ministries of the Environment, Economic Affairs, and Finance of a country and/or like-minded nations, national and/or regional business, national and/or regional knowledge centres, national and/or regional NGOs, and the national and/or regional public at large.

Global market

The high degree of international participation allows for the international implementation of national commitments, which take the form of market-based actions such as envisaged in the Kyoto Protocol. Suitable policy instruments include national communication, participation in global carbon-trade schemes, adhesion to a global carbon-tax regime, and contribution to a global R&D fund for low-emission technologies. Market-like instruments

---

[44] Covenants may include 'bench-marking' agreements between government and local industry, indicating that local producers will be among the (world's) most energy-efficient producers in their sector in exchange for abstention from legislation.

[45] It should be noted that product standards are easier to control than process standards, as the former focus on characteristics of domestically available products, whereas the latter require physical controls of (foreign) production sites.

such as JI and CDM are applied to the extent that they are financially rewarding. Communication aims at enabling producers and consumers to make better decisions. Global carbon trade involves financial exchanges among buyers and sellers of hot air on global markets, with the possibility of enhancing cost efficiency. A global carbon tax may induce price-sensitive consumers to render their purchasing behaviour less carbon intensive, though the political feasibility of such a tax may be low. Constituting a global research fund enhances the likelihood of realising breakthrough technologies; the concomitant economies of scale are important to finance the costly R&D in this field. JI and CDM are widely used in this scenario because of their cost-effectiveness.

These instruments require the involvement of national Ministries of the Environment, Economic Affairs, Finance, and Foreign Affairs, national and international business, as well as supranational administrative and financial bodies (such as the UNFCCC Secretariat and the GEF).

Global collectivity

As this scenario involves the highest level of international cooperation, national actions are aligned with those agreed at the global level. They are implemented through public actions. Governments intervene through national communication and participation in global energy-efficiency standards, global covenants with multinational companies, and global partnerships. Communication is geared towards informing societal actors about our common present and future. Applying stringent global standards to energy efficiency in high-impact areas is likely to have a major impact: it forces producers to 'clean up' their practices and thus creates major incentives to engage in R&D and to meet the expected stricter standards of future generations. Global covenants with multinationals allow for regulation of 'footloose' companies which individual national governments cannot control because their activities exceed national borders. The same holds for global partnerships, in which actors have a greater propensity to take voluntary steps.

This scenario is the most comprehensive one, involving a large number and variety of parties: national Ministries of the Environment, Economic Affairs, Finance, and Foreign Affairs, national and international business, national and international knowledge centres, supranational administrative and financial bodies, national and international NGOs, and the public at large.

Table 23.1 summarises the different policy instruments and parties involved in the four scenarios.

Table 23.1 Instruments and parties in climate policy scenarios

Scenario	Policy instrument	Party
Local market	Communication	Ministry of the Environment
	R&D subsidies	Ministry of Economic Affairs
	National carbon trade	Ministry of Finance
	National carbon tax	National business
Local collectivity	Communication	Ministries of the Environment
	National/regional standards	Ministries of Economic Affairs
		Ministries of Finance
	National/regional covenants	National/regional business
		National/reg. knowledge centres
	National/regional partnerships	National/regional NGOs
		National/reg. public at large
Global market	Communication	Ministries of the Environment
	Global carbon trade	Ministries of Economic Affairs
	Global carbon tax	Ministries of Finance
	Global R&D subsidies	Ministries of Foreign Affairs
	JI	(Inter)national business
	CDM	Supranational bodies
Global collectivity	Communication	Ministries of the Environment
	Global standards	Ministries of Economic Affairs
	Global covenants	Ministries of Finance
	Global partnerships	Ministries of Foreign Affairs
		(Inter)national business
		(Inter)national knowledge centres
		Supranational bodies
		(Inter)national NGOs
		Global public at large

Creating Leverage

So far, it was assumed that a minimal willingness-to-act is present in all scenarios, though its extent was considered to be exogenously given. We next considered the motivation of societal actors to be endogenous, and explored ways of enhancing their commitment to climate-relevant measures. The literature provides two relevant ways of creating leverage. One way (issue linkage) originates in regime theory, while the other (bandwagon creation) is rooted in organisational theory (in particular, institutional theory).

Linking issues

When the climate-change problem is low on the agenda of societal actors, its position can be raised by linking it to issues that are perceived as important and urgent, provided the different issues are compatible and interdependent.[46] Issue linkage has the obvious drawback of rendering a complex issue even more comprehensive, but has important potential advantages. First, when pay-offs exist in several areas, the costs of realising them can be shared. Second, connecting issues which are attributed varying degrees of importance by different parties provides the opportunity to create leverage, because parties commit themselves to acting also on low-priority issues that are tied to issues which are given high priority. Third, scope enlargement involves more parties, and thus increases possibilities to enlarge coalitions. Fourth, linking issues allows solutions to deadlocked problems to be sought by widening the scope of topics covered.

Applied to the present case, Ministries of the Environment should link climate policy to issues high on the agendas of societal actors. National climate policy may enhance economic competitiveness. Links with economic policy can be sought in terms of higher performance at the micro level (corporate cost savings resulting from the implementation of energy-efficient techniques and new market opportunities emanating from the creation of markets for low-emission products)[47] and the macro level (the use of climate-relevant levies for cutting distortionary taxes).[48] A double-dividend policy can be pursued to realise both environmental and economic benefits.[49] This linkage strategy would be especially appealing in the local-market scenario, where economic factors are important drivers of decisions by private actors.

National climate policy can also be linked with social issues, such as health and safety. For instance, the recurrence of floods in residential areas and the death of large numbers of people from extreme temperatures enhance the domestic basis for action, not for environmental but for social reasons. In these cases, climate policy should not be put forward as a global environmental issue, but as a national political priority of an economic and/or a social nature. The global and distant nature of the climate issue can also be circumvented by tying climate policy to environmental problems which have predominantly local and short-term effects (such as air quality). While the two are not perfectly congruent,[50] benefits in both areas can be realised when

---

[46] Gray, 1999; Gupta and Tol, 2003; Kemfert, 2004; Keohane and Nye, 2001; Kroeze-Gil, 2003; Susskind, 1994.

[47] Porter and Van der Linde, 1995.

[48] Bovenberg and De Mooij (1994) discussed the macro-economic advantages resulting from using environmental levies to reduce the gap between labour costs and net wages.

[49] The principle of such 'no regret' or 'eco-efficient' measures is not new (Grubb et al., 1999; Hall and Roome, 1996), but has not been sufficiently pushed to overcome behavioural inertia.

[50] For example, electric vehicles may lead to better local air quality (as power is generated outside the focal area), but may adversely affect climate policy (because of a relatively low energy efficiency).

the consumption of fossil fuels decreases.[51] This type of linking may be particularly fruitful in the local-collectivity scenario, where social issues are the domain of collective action.

Links can also be established at the international level, whereby Ministries of the Environment align their actions with those of the Ministries of Foreign and Economic Affairs. Connections can be made with international financial assistance and the reduction of trade barriers in exchange for commitments in the climate field, especially in the global-market scenario. When bilateral and multilateral financial assistance to developing and transition countries are made contingent on the climate impact of the recipient projects, economic development with a relatively moderate climate impact may be spurred.[52] Likewise, nations may make their support of social issues that other countries perceive as important (for instance, security: joint abatement of terrorism or assistance in armed conflicts) contingent on collective action in the climate field; this is particularly relevant in the global-collectivity scenario. Therefore, leverage at the international level can be obtained by stressing other (than environmental) benefits of climate policy and by making national support in other areas contingent on the performance of counterparts in the climate field.

Creating bandwagons

Bandwagons are diffusion processes whereby innovations are adopted because of the sheer number of other participants.[53] Once the required critical mass has been reached, the number of adopters may rise quasi-automatically.[54] Reasons for adoption include the belief that economic opportunities are otherwise foregone and the fear that others perceive non-adoption negatively. Furthermore, adoption may be a recognition of the expertise of the initiators. It may also give rise to economies of scale: high adoption rates of standards decrease relative production costs and yield positive network externalities, thus reducing the (transaction) costs stemming from different standards (such as incompatibility, additional coordination, and limited learning effects).[55]

The national basis for action in a climate-sceptical country may be raised once a participation threshold has been met. Ministries (of the Environment) can initiate this process of creating momentum. In the local-market scenario, a (temporarily) favourable fiscal treatment of energy-efficient technology may

[51] European Environment Agency, 2004.

[52] Similarly, the transfer of best practices and techniques provides both environmental and economic benefits.

[53] Abrahamson and Rosenkopf, 1993.

[54] DiMaggio and Powell, 1983; Tolbert and Zucker, 1996. Examples include the expansion of supranational organisations such as the UN, the WTO, and the EU.

[55] Brunsson and Jacobsson, 2000.

lead to a sufficient number of early adopters to trigger a self-reinforcing process of diffusion. In the local-collectivity scenario, national champions (i.e., highly visible, knowledgeable, and proactive societal actors) may raise the collective willingness to engage in socially or environmentally inspired actions.

At the international level, a sufficiently large number of proactive countries can induce laggards to join the bandwagon. A high degree of international participation is most likely when adhesion is beneficial to potential participants. In the global-market scenario, participation in international R&D consortia is attractive for companies in order to share the high costs of breakthrough innovations, but also to make sure that economic opportunities are not foregone because of the lack of access to new technologies.[56] In the global-collectivity scenario, stringent collective technology standards in a sufficiently large group of proactive, wealthy countries would induce others to adhere, thus creating a self-enforcing international regime.[57] For example, if major consumer markets (say, the EU plus Japan) can be served only when products meet stringent emission requirements, foreign companies and multinationals are forced to adopt the more stringent standards in order to access these markets, and they may also apply them to other markets in order to reap the production and marketing economies of scale which stem from having uniform (environmentally stringent) product and process standards.[58]

CONCLUSIONS

We have examined the possibilities of conducting an effective future climate policy given the characteristics of the present Kyoto regime. When this regime was designed, the global-public-good nature of the climate issue had to be coped with, requiring the commitment of the global community. A diversity of – often conflicting – interests and viewpoints had to be reconciled. The ensuing Kyoto Protocol reflected the different desiderata and was, therefore, characterised by considerable complexity and ambiguity. It was signed by virtually all nation states but committed only a minority of them, thus leading to (perceived) moral and competitive disparities. Binding targets and timetables were formulated, but the Protocol's targets were

[56] This implies that corporate R&D is turned from a private into a club good: the costs and benefits no longer accrue to individual companies but to all who joined the research consortium, thus raising the utility levels of participating actors.

[57] Barrett, 2003. Free-riding behaviour is dissuaded, because actors have a personal interest in joining the relatively stringent regime; it is a precondition to entering major markets.

[58] Grubb et al., 1999; Vogel, 1995.

insufficient to halt the cumulative process of human-induced climate change that had started over a century earlier and the agreement's time horizon was too short; besides, compliance with commitments has faced enforcement problems. Innovative, flexible instruments were designed, allowing for cost effectiveness to be enhanced and technical measures to be avoided. Developments since the Protocol's conclusion have led to global political polarisation: while some parties (especially the EU) have taken political and institutional steps towards implementation, others (the US and Australia) have formally withdrawn from the agreement or withheld their support (as with Russia), thus rendering Kyoto's global prospects highly uncertain. It should be noted, though, that adherence does not guarantee progress and withdrawal does not preclude parties from taking climate-relevant (technical) actions. Should the Protocol not enter into force, the flexible instruments, together with the solid institutional infrastructure, are the elements that are most likely to survive the present regime.

Future climate policy is contingent on a variety of factors, which affect the willingness and ability to mitigate the human impact on climate change; these include political developments (especially those in large and wealthy nation states), evolutions in the field of low-emission technologies, and the occurrence and nature of climate-related catastrophes. We explored four scenarios of future climate regimes. The local-market scenario may arise if the Kyoto regime falls apart. From the perspective of abating human-induced climate change, it is the least favourable scenario: the scope of climate-relevant actions is limited because of national sovereignty and measures are mainly confined to those that yield private economic benefits because of the dominance of market-based decisions. The global-collectivity scenario, involving a high degree of international cooperation and collective allocation decisions, may materialise after major natural calamities. This regime is relatively favourable, because concerted, politically inspired actions lead to stringent standards and sectoral agreements at the global level. The local-collectivity scenario may occur if a failure of the Kyoto regime leads to global bifurcation. In terms of climate impact, it is an intermediate option: a small number of countries prescribes relatively stringent measures, but the (geographic) scope of these actions is limited; besides, to the extent that trade-distortive effects arise, these actions may need to be mitigated to avoid economic retaliation by non-participating countries. Finally, the global market scenario may be the outcome of a renegotiated or adapted Kyoto Protocol. It is also an intermediate regime: the leverage resulting from global participation and the realisation of scale economies may lead to important progress, though actions are predominantly confined to those that are economically rewarding.

The policy measures that national governments should take are contingent on the prevailing climate regime. A local-market scenario calls for local, mainly financial policy instruments that do not significantly distort the functioning of markets; local producers and consumers retain their sovereignty. In a local-collectivity regime, more commanding instruments can be applied in concert with like-minded nations, though the (geographic) scope of relatively stringent measures is limited and governments operate in hostile international environments. International instruments can be applied in a global-market regime, though they are, by and large, still compatible with market conditions; national climate policies become subordinate to a global, mainly financial, (Kyoto-like) regime. Finally, in the global-collectivity regime, both mandatory and voluntary actions are taken at the global level; national policy is an integral part of such measures. In the 'minimal' (local-market) scenario, relatively few actors are affected by government interventions. The variety of actors increases as decisions become more collective, while the number of parties rises with the degree of internationalisation.

Whichever future regime materialises, public climate policy will have to bear in mind what the Kyoto regime has failed to do: to capitalise on the *incentives* of the actors involved. The present stalemate in global public climate policy is largely the result of the perception that Kyoto commitments involve only burdens and that national compliance cannot be enforced in the absence of a powerful global regime. In order to overcome indifference and the temptation to indulge in free-rider behaviour, the incentives of the actors concerned should be aligned with the objectives of climate policy. Wherever possible, national governments should wield both the stick and the carrot to realise changes of production and consumption practices. When the climate problem has low political priority, it should be linked to 'hot' issues in other areas with direct and local pay-offs. Positive results may then stem not from climate-inspired actions, but from incentives to realise linked economic, social, and environmental objectives: reduced costs, new market opportunities, fewer victims of extreme weather events, improved air quality, etc. Linkage may help for developing and transition countries to raise investment and assistance funds. Connecting climate policy with air policy and atmospheric conditions is relevant in many industrialised nations.

Alternatively, actors may feel compelling incentives to participate. Once a critical threshold has been reached (for example, following commitment by the EU, Japan, and proactive US states), other private and public actors may join the climate bandwagon because they are forced to or do not wish to be excluded, thus turning the climate issue from a free-rider-struck public good into an incentive-rich club good. Powerful, self-enforcing instruments to do so are the prescription of stringent energy-efficiency standards (creating a high demand for low-emission technologies) and the establishment of R&D consortia (enhancing the likelihood of realising technological breakthroughs). When government interventions pay more heed to providing the most appropriate incentives, future public climate policies will be more effective and levered than the present ones.

REFERENCES

Abrahamson, E. and L. Rosenkopf (1993), 'Institutional and competitive bandwagons: Using mathematical modeling as a tool to explore innovation diffusion', *Academy of Management Review,* **18**(3), 487-517.

Aldy, J., R. Baron, and L. Tubiana (2003), *Addressing Cost: The Political Economy of Climate Change,* Arlington: Pew Center on Global Climate Change.

Barrett, S. (1998), 'Political economy of the Kyoto Protocol', *Oxford Review of Economic Policy,* **14**(4), 20-39.

Barrett, S. (2003), *Environment & Statecraft: The Strategy of Environmental Treaty-Making,* Oxford: Oxford University Press.

Bartsch, U., B. Müller, and A. Aaheim (2000), *Fossil Fuels in a Changing Climate: Impacts of the Kyoto Protocol and Developing Country Participation,* Oxford: Oxford University Press.

Bodansky, D. (2003), *Climate Commitments: Assessing the Options,* Arlington: Pew Center on Global Climate Change.

Bovenberg, L. and R. De Mooij (1994), 'Environmental levies and distortionary taxation', *American Economic Review,* **94**(4), 1085-1089.

Brunsson, N. and B. Jacobsson (2000), *A World of Standards,* Oxford: Oxford University Press.

Charnovitz, S. (2003), *Trade and Climate: Potential Conflicts and Synergies,* Arlington: Pew Center on Global Climate Change.

Den Elzen, M. (2002), *Exploring Post-Kyoto Climate Regimes for Differentiation of Commitments to Stabilise Greenhouse Gas Concentrations,* report 728001020/2002, Bilthoven: RIVM (National Institute for Public Health and the Environment).

De Mooij, R. and P. Tang (2003), *Four Futures of Europe,* The Hague: CPB (Netherlands Bureau for Economic Policy Analysis).

Depledge, J. (2000), *Tracing the Origins of the Kyoto Protocol: An Article-by-Article Textual History,* Bonn: UNFCCC.

DiMaggio, P. and W. Powell (1983), 'The iron cage revisited: Institutional isomorphism and collective rationality in organizational fields', *American Sociological Review,* **48**(April), 147-160.

European Environment Agency (2003), *Greenhouse Gas Emission Trends and Projections in Europe 2003: Tracking Progress by the EU and Acceding and Candidate Countries towards Achieving Their Kyoto Protocol Targets,* Environmental Issue report 36, Copenhagen: European Environment Agency.

European Environment Agency (2004), *Exploring the Ancillary Benefits of the Kyoto Protocol for Air Pollution in Europe,* technical report, Copenhagen: European Environment Agency.

Fahey, L. and R. Randall (1998), *Learning from the Future: Competitive Foresight Scenarios,* New York: John Wiley & Sons.

Gray, B. (1999), 'The development of global environmental regimes: Organizing in the absence of authority', in D. Cooperrider and J. Dutton (eds), *Organizational Dimensions of Global Change,* Thousand Oaks: Sage.

Grubb, M., C. Vrolijk, and D. Brack (1999), *The Kyoto Protocol: A Guide and Assessment,* London: Royal Institute of International Affairs/Earth Scan.

Grubb, M., T. Brewer, B. Müller, J. Drexhage, K. Hamilton, T. Sugiyama, and T. Aiba (2003), *A Strategic Assessment of the Kyoto-Marrakech System: Synthesis Report,* briefing paper 6, London: Royal Institute of International Affairs.

Gupta, J. (2001), *Our Simmering Planet: What to do about Global Warming?,* London: Zed Books.

Gupta, J. and R. Tol (2003), 'Why reduce greenhouse gas emissions? Reasons, issue-linkages and dilemmas', in E. van Ierland, J. Gupta, and M. Kok (eds), *Issues in International Climate Policy: Theory and Policy,* Cheltenham: Edward Elgar.

Haas, P., R. Keohane, and M. Levy (eds) (1993), *Institutions for the Earth: Sources of Effective International Environmental Protection,* Cambridge: MIT Press.

Hall, S. and N. Roome (1996), 'Strategic choices and sustainable strategies', in P. Groenewegen et al. (eds), *The Greening of Industry Resource Guide and Bibliography,* Washington: Island Press.

Heller, T. and P. Shukla (2003), *Development and Climate: Engaging Developing Countries,* Arlington: Pew Center on Global Climate Change.

Holdren, J. (2003), *Risks from Global Climate Change: What Do We Know? What Should We Do?,* presentation at the Institutional Investors' Summit on Climate Risk, 21 November, New York: United Nations.

ICCEPT (2002), *Assessment of Technological Options to Address Climate Change: A Report for the Prime Minister's Strategy Unit,* London: Imperial College Centre for Energy Policy and Technology.

IPCC (1995), 'Second assessment report of the Intergovernmental Panel on Climate Change', *http://www.ipcc.ch/pub/sa(E).pdf.*

IPCC (2001), 'Third assessment report of the Intergovernmental Panel on Climate Change', *http://www.ipcc.ch/pub/un/syreng/spm.pdf.*

Kaul, I., I. Grunberg, and M. Stern (eds) (1999), *Global Public Goods: International Cooperation in the 21st Century,* New York: Oxford University Press.

Kaul, I., P. Conceição, K. Le Goulven, and R. Mendoza (eds) (2003), *Providing Global Public Goods: Managing Globalization,* New York: Oxford University Press.

Kemfert, C. (2004), 'Climate coalitions and international trade: Assessment of cooperation incentives by issue linkage', *Energy Policy,* **32**, 455-465.

Kemp, R. (1997), *Environmental Policy and Technical Change: A Comparison of the Technological Impact of Policy Instruments,* Cheltenham: Edward Elgar.

Keohane, R. and J. Nye (2001), *Power and Interdependence,* 3rd ed., Boston: Little and Brown.

Kölliker, A. (2002), *The Impact of Flexibility on the Dynamics of European Unification,* PhD thesis, Florence: European University Institute.

Krasner, S. (ed.) (1983), *International Regimes,* Ithaca: Cornell University Press.

Kroeze-Gil, J. (2003), *International Environmental Problems, Issue Linkage and the European Union,* PhD thesis, Tilburg: CentER, Tilburg University.

Leal Arcas, R. (2001), 'Is the Kyoto Protocol an adequate environmental agreement to resolve the climate change problem?', *European Environmental Law Review,* **October**, 282-294.

Levy, D. and D. Egan (2003), 'A neo-Gramscian approach to corporate political strategy: Conflict and accommodation in the climate change negotiations', *Journal of Management Studies,* **40**(4), 803-829.

Löschel, A. and Z. Zhang (2002), 'The economic and environmental implications of the US repudiation of the Kyoto Protocol and the subsequent deals in Bonn and Marrakech', *Weltwirtschaftliches Archiv,* **138**(4), 711-746.

McGivern, B. (1998), 'Conference of the Parties to the Framework Convention on Climate Change: Kyoto Protocol: Introductory note', *International Legal Materials: Current Documents,* **37**(1), 22-28.

Meyer, J. and B. Rowan (1977), 'Institutionalized organizations: Formal structure and myth and ceremony', *American Journal of Sociology,* **83**(2), 340-363.

Metz, B. and M. Berk (2001), *Beyond Kyoto: Can we Stabilise at 450 ppmv CO_2?,* presentation at COP 6bis, Bonn, 25 July.

Müller, B., J. Drexhage, M. Grubb, A. Michaelowa, and A. Sharma (2003), *Framing Future Commitments: A Pilot Study on the Evolution of the UNFCCC Greenhouse Gas Mitigation Scheme,* Oxford: Oxford Institute for Energy Studies.

Oberthür, S. and H. Ott (1999), *The Kyoto Protocol: International Climate Policy for the 21st Century,* Berlin: Springer Verlag.

Pershing, J. and F. Tudela (2003), *A Long-Term Target: Framing the Climate Effort,* Arlington: Pew Center on Global Climate Change.

Phillips, N., T. Lawrence, and C. Hardy (2000), 'Inter-organizational collaboration and the dynamics of institutional fields', *Journal of Management Studies,* **37**(1), 23-43.

Phylipsen, G., J. Bode, K. Blok, H. Merkus, and B. Metz (1998), 'A Triptych sectoral approach to burden differentiation: GHG emissions in the European bubble', *Energy Policy,* **26**(12), 929-943.

Porter, M. and C. van der Linde (1995), 'Green and competitive: Ending the stalemate', *Harvard Business Review,* **September**, 120-134.

Prakash, A. (2000), *Greening the Firm: The Politics of Corporate Environmentalism,* Cambridge: Cambridge University Press.

Schelling, T. (1978), *Micromotives and Macrobehavior,* New York: Norton.

Susskind, L. (1994), *Environmental Diplomacy: Negotiating More Effective Global Agreements,* New York: Oxford University Press.

Tolbert, P. and L. Zucker (1996), 'The institutionalization of institutional theory', in S. Clegg, C. Hardy, and W. Nord (eds), *Handbook of Organization Studies,* London: Sage.

UNEP (2002), 'Global Environment Outlook 3', *http://www.unep.org/GEO*, Nairobi: United Nations Environment Programme.

UNFCCC (2003a), *National Communications from Parties Included in Annex I to the Convention: Compilation and Synthesis of Third National Communications,* FCCC/SBI/2003/7, Bonn: United Nations Framework Convention on Climate Change.

UNFCCC (2003b), 'Kyoto Protocol: Status of Ratification', *http://www.unfccc.int/resource/kpstats.pdf*, Bonn: United Nations Framework Convention on Climate Change.

Van der Heijden, K. (1996), *Scenarios: The Art of Strategic Conversation,* Chichester: John Wiley & Sons.

Victor, D. (2001), *The Collapse of the Kyoto Protocol and the Struggle to Slow Global Warming,* Princeton: Princeton University Press.

Vogel, D. (1995), *Trading Up: Consumer and Environmental Regulation in a Global Economy,* Cambridge: Harvard University Press.

Wallström, M. (2003), *COP 9 and Beyond,* speech at the Joint Informal Meeting of Environment and Energy Ministers, Montecatini, 20 July.

Weitzman, E. and M. Miles (1995), *Computer Programs for Qualitative Data Analysis: A Software Sourcebook*, Thousand Oaks: Sage.

Wijen, F. and K. Zoeteman (2004), 'Final report of the study "Past and Future of the Kyoto Protocol"', *http://www.uvt.nl/globus/publications/publications04/publ04.01.html*, Tilburg: Globus, Tilburg University.

Yamin, F. (1998), 'The Kyoto Protocol: Origins, assessment and future challenges', *Review of European Community and International Environmental Law,* 7(2), 113-127.

Young, O. (1994), *International Governance: Protecting the Environment in a Stateless Society,* Ithaca: Cornell University Press.

Young, O. (ed.) (1999), *The Effectiveness of International Environmental Regimes: Causal Connections and Behavioral Mechanisms,* Cambridge: MIT Press.

PART V

New Directions

24. Globalisation and Environmental Protection: A Global Governance Perspective

Daniel Esty and Maria Ivanova

SUMMARY

In this chapter, we disaggregate the impact of globalisation on the environment into economic, regulatory, information, and pluralisation effects. We complement this structure with an analysis of how national and global environmental policies affect globalisation. We then argue that there is a need for a revitalised governance regime to organise and sustain environmental cooperation at the global level. Such a global environmental mechanism (GEM) would provide a new model for collaboration, overcoming the shortcomings of existing bodies. The GEM's core elements would be a global information clearing-house that provides a data and analytic foundation, a global technology clearing-house to highlight tools and strategies, and a global environmental bargaining forum. We conclude that the GEM approach with a 'light' institutional architecture that relies on global public policy networks and modern information technologies offers the potential of improved results and greater institutional legitimacy because of its response speed, flexibility, cost-effectiveness, and potential for broader public participation.

INTRODUCTION

Globalisation has ushered in an era of contrasts, one of fast-paced change and persistent problems. It has spurred a growing degree of interdependence among economies and societies through transboundary flows of information, ideas, technologies, goods, services, capital, and people. In so doing, it has challenged the traditional capacity of national governments to regulate and control markets and activities. The rapid pace of economic integration has led to interlinked world markets and economies, requiring a degree of synchronisation of national policies across a number of issues. One dimension of this coordination concerns the environment, from shared natural resources such as fisheries and biological diversity to the potential for transboundary pollution spillovers across the land, over water, and through the air. We now understand that governance approaches that are bounded by the traditional notion of national territorial sovereignty cannot protect us from global-scale environmental threats. An effective response to these challenges will require fresh thinking, refined strategies, and new mechanisms for international cooperation.

In this chapter, we address the relationship between globalisation and the environment, seeking to answer four key questions: (1) How does globalisation affect the environment? (2) Conversely, how does national environmental regulation affect globalisation, particularly economic integration? (3) When is a degree of international cooperation useful or even necessary? (4) What institutional structure would best manage interdependence and foster the opportunities that globalisation has the potential to provide?

Globalisation can have both positive and negative environmental consequences. But the same forces can exacerbate existing environmental problems and create new ones, as well as run down stocks of non-renewable natural resources.[1] Economic integration and trade liberalisation can generate new resources that permit investments in environmental protection as well as faster and broader dissemination of pollution control technologies and new policy ideas.[2] Environmental choices can, likewise, shape the path of globalisation. National regulatory choices may act as barriers to liberalised trade, or they may trigger a convergence towards harmonised international standards.[3] The broad range of recent 'trade and environment' disputes at the World Trade Organization (WTO) – over beef hormones, asbestos regulation, genetically modified food, shrimp fishing, and endangered sea turtles, to name a few – highlights the dynamic complexity of these issues. For policy-makers, the core challenge lies in finding an appropriate mix of competition and cooperation, market forces and intervention, and economic growth and environmental protection.[4]

To maximise globalisation's upside potential, a fundamental reform of global governance structures in general, and of the international architecture for *environmental* cooperation in particular, will be required. Building greater

---

1 Dua and Esty, 1997.
2 Anderson et al., 1999; Jobes, 2003; Speth, 2003
3 Esty, 1994; Rodrik, 1997; Vogel, 1994.
4 Esty and Geradin, 2001.

environmental sensitivity into multilateral trade and financial institutions is necessary but insufficient. An equally broad-scale reform of the global environmental governance architecture is needed. We propose the creation of a Global Environmental Mechanism (GEM) to facilitate efforts to: manage global-scale environmental risks; support bargaining and negotiation; promote sound management of the global commons; and advance dissemination of information, 'best practices' in policy-making, and new technologies.

EFFECTS OF GLOBALISATION ON THE ENVIRONMENT

Globalisation presents a mixed blessing for the environment. It creates economic opportunities, but also gives rise to new problems and tensions. By increasing the volume and decreasing the cost of information, data, and communications, globalisation also offers expanded access to knowledge, new mechanisms for participation in policy-making, and the promise of more refined and effective modes of governance. Understanding this array of economic, regulatory, information, and pluralisation effects is essential if one is to make sense of globalisation's impact on the environment.

Economic Effects

Environmental impacts of expanded economic growth and trade can be understood in terms of scale, income, technique, and composition effects. Scale effects refer to increased pollution and natural-resource depletion due to increased economic activity and greater consumption. Income or wealth effects appear when greater financial capacity results in greater investment in environmental protection and new demands for attention to environmental quality. With higher income, we observe two other, related phenomena: technique and composition effects. Technique effects arise from tendencies towards cleaner production processes as wealth increases and, as trade intensifies, better access to new technologies and environmental best practices. Composition effects take place as the economic base evolves towards a less-pollution-intensive high-tech and services-based set of activities. The overall environmental impact of economic growth depends on the net impact of these four effects. If the income, technique, and composition effects overwhelm the negative scale effect of expanded economic activity, then the impact of growth will ultimately be positive. But in the early stages of industrialisation, it may well be that environmental conditions deteriorate.

The precise shape, duration, and applicability of the resulting inverted U-shaped environmental Kuznets curve has generated considerable debate.[5] The critical income level at which pollution begins to diminish is about 5,000 US dollars (USD)/year per capita.[6] In trying to separate out the various environmental effects of economic growth, it was found that a 1 per cent increase in the scale of economic activity raises pollution concentrations by 0.25-0.50 per cent, but the accompanying increase in income drives concentrations down by 1.25-1.50 per cent via a technique effect resulting in improved conditions overall.[7] It appears, however, that expanded trade and economic activity may worsen environmental conditions in other circumstances.[8] Carbon dioxide emissions do not, for instance, appear to fall at any known income level.[9]

Economic theory suggests that the free market can be expected to produce an efficient and welfare-enhancing level of resource use, production, consumption, and environmental protection if the prices of resources, goods, and services capture all of the social costs and benefits of their use.[10] However, when private costs – which are the basis for market decisions – fail to include social costs, market failures occur, resulting in allocative inefficiency in the form of suboptimal resource use and pollution levels. Market failures are a hallmark of the environmental domain. Many critical resources such as water, timber, oil, fish, and coal tend to be underpriced. Ecosystem services such as flood prevention, water retention, carbon sequestration, and oxygen provision often go entirely unpriced. Because underpriced and unpriced resources are overexploited, economic actors are often able to ignore part or all of the environmental costs they generate. Globalisation may magnify the problem of mispriced resources and the consequent environmental harms.

Regulatory Effects

A primary goal of trade liberalisation is the reduction of barriers to market access. Thus, trade agreements often include 'disciplines' on how the parties

---

5 Antweiler et al., 2001; Esty, 2001; Grossman and Krueger, 1995; Selden and Song, 1994; Shafik, 1994. In effect, the goal is to shorten the length and flatten the amplitude of the environmental Kuznets curve, which represents the path taken by countries undergoing economic development. It is an inverted U-shaped curve illustrating that pollution will increase during early stages of development, level off, and then decrease after a certain income threshold has been reached.

6 Grossman and Krueger, 1995.

7 Antweiler et al., 2001.

8 Esty, 2001.

9 Dua and Esty, 1997.

10 Anderson, 1992, 1998; Panayotou, 1993.

will regulate. Some environmental advocates decry this loss of regulatory sovereignty.

Perhaps more importantly, freer trade promotes competition. Increased competitive pressure may manifest itself in industrial or governmental efforts to reduce pollution-control compliance costs. This political dynamic could trigger a regulatory 'race to the bottom' in which jurisdictions with high environmental standards relax their regulatory regimes to avoid burdening their industries with pollution-control costs higher than those of competitors operating in low-standard jurisdictions.[11] While there is little evidence that environmental standards are actually declining,[12] the concern is not literally about a race *to* the bottom, but rather about a race *towards* the bottom that translates into suboptimal environmental standards, at least in some jurisdictions. Ample evidence exists to support the existence of a regulatory dynamic in which standards are set strategically with an eye on the pollution-control burdens in competing jurisdictions. The outcome may well be a 'political drag' which results in weaker environmental laws than might have otherwise been adopted and, perhaps more importantly, lax enforcement of existing rules or standards.[13]

Diverse national circumstances generally make uniform standards less attractive than standards tailored to local conditions and preferences.[14] But not always. Divergent standards across jurisdictions may impose market-access barriers on traded goods that exceed any benefits obtained by allowing each jurisdiction to maintain individualised requirements. In some cases, producers vying for access to high-standard jurisdiction will drive upward harmonisation (a 'race to the top').[15] But this logic applies only to product standards. Standards for production processes or methods (PPMs) are not subject to the same market pressures.

In an interdependent world, production-related externalities cannot be overlooked. Semiconductors produced using chlorofluorocarbons (CFCs), which contribute to the destruction of the ozone layer, should be treated as contraband. Where international environmental agreements are in place, as with the Montreal Protocol regulating the use of ozone-depleting substances, a recognised standard is available. In such cases, trade rules should be interpreted to reinforce the agreed-upon standards. Recrafted trade principles and WTO rules that accept the legitimacy of environmental controls aimed at transboundary externalities would make global-scale trade and environmental policies more mutually reinforcing and reduce the risk of the trade regime

[11] Esty, 2001; Klevorick, 1996.
[12] Wheeler, 2002.
[13] Esty, 2001; Esty and Geradin, 2001.
[14] Anderson, 1998; Mendelsohn, 1986.
[15] Vogel, 1994.

providing cover for those shirking their share of global environmental responsibilities.

Information Effects

One of the key features of globalisation is the expansion of communication networks across the globe. The increasing speed and decreasing cost of communication have virtually eliminated the traditional concept of distance. The Information Age has thus transformed space and time, drawing the world into networks of global communication, though some parts are more tightly linked than others.[16] This communication revolution has dramatically increased the intensity of national interdependence, fomenting a greater sense of international community and a foundation of shared values.[17] In turn, the incipient sense of a world community provides citizens with a basis for demanding that those with whom they trade meet certain baseline moral standards, including a commitment to environmental stewardship. As economic integration broadens and deepens, and information about one's partners becomes more readily available, what citizens feel should be encompassed within the set of baseline standards tends to grow.[18]

Increased access to data and information on economic and environmental performance allows for faster problem identification, better issue analysis, and quicker trend spotting. It can also aid the identification of leaders and laggards in the international arena relative to various environmental or social criteria, and spur competition (and thus improved performance) among nations, companies, or even communities. Information in and of itself is not, however, necessarily beneficial. Information overload could lead to a cacophony of voices in the policy realm and result in paralysis instead of action. Such risks need to be kept in mind as the volume of internationally shared information continues to increase and appropriate devices for sifting through and filtering relevant and accurate information become necessary.

Pluralisation Effects

Intensified interaction in the economic and political spheres coupled with rapidly diminishing costs of communication has increased the number and diversity of participants in global networks. This 'pluralisation' is evident in the exponential growth of non-governmental organisations (NGOs), their heightened levels of activity, and their increased access to the policy-making process at both the national and international levels. In 1990, there were

---

[16] Esty, 2004.
[17] Thompson, 2003.
[18] Dua and Esty, 1997; Rodrik, 1997.

about 6,000 international NGOs. By the year 2000, that number had reached 26,000.[19] An elaborate organisational or institutional infrastructure is no longer necessary for an entity to have a global reach.

With global interconnectedness on the rise, transparency, participation, and democratisation have also increased, providing a broader constituency of concerned groups and individuals with access to global decision-makers. While national governments remain central to global-scale policy-making, many new actors now play a role and the governance process has become much more complex.[20] For example, the Landmine Treaty resulted from an internet-based campaign started in 1991 by several NGOs and individuals. Today, the Treaty has been ratified or acceded to by 141 countries. An NGO network, representing over 1,100 groups in over 60 countries, is now working locally, nationally, regionally, and internationally to implement the ban on antipersonnel landmines. The significance of NGOs as actors on the global stage was recognised in 1997, when the International Campaign to Ban Landmines and its coordinator, Jody Williams, received the Nobel Peace Prize.[21]

The downside of pluralisation is that ability to participate in the policy process remains asymmetrical. Constituencies start out with unequal resources and the influence of special interests – which are often well financed and organised – may be magnified. Globalisation by no means implies the end of politics. Quite to the contrary, power relations remain important and mechanisms for levelling the playing field become increasingly necessary.

ENVIRONMENTAL EFFECTS ON GLOBALISATION

Just as globalisation will shape environmental protection efforts, so may environmental choices affect the course of globalisation, particularly efforts to liberalise trade and investment flows. At one extreme, a rigid harmonisation of policy approaches and regulatory standards could run roughshod over diverse environmental circumstances, resource endowments, and public preferences.[22] At the other extreme, uncoordinated national environmental policies might become non-tariff barriers to trade that obstruct efforts to open markets. In these ways, national-level environmental policies may influence international action. Similarly, ecological realities may require policy coordination and collective action on the global scale.

---

[19] Keohane and Nye, 2003.
[20] Slaughter, 1997.
[21] ICBL, 2004.
[22] Bhagwati, 1993, 2000.

National Activities with International Effects

National environmental performance may have international impacts. In an increasingly interconnected world, environmental harms (such as greenhouse gas emissions) left unattended at the local and national levels may result in global-scale problems (such as sea level rise, increased intensity of wind storms, and changed rainfall patterns that may come to pass as a result of climate change). The failure to address such spillovers of harm creates a risk for the international economic system of being weighed down by market failures. Transboundary pollution spillovers, which result in 'super externalities', are especially difficult to manage.[23] The need to bring multiple countries together in a common response represents a much more difficult problem to address than national-scale environmental protection. As with any global public good, where costs are borne locally and benefits spread across the world, no single jurisdiction has an incentive to regulate such harms optimally. In the case of *regular* externalities (i.e., harms within one nation), there are many reasons why governments may not optimally regulate emissions or other harmful practices, but at least they have an incentive to do so in the face of the welfare losses of their own citizens. When harms span multiple jurisdictions or even the entire world, there is an increasing likelihood that the government whose facility is causing the negative impact will choose not to act because its own cost-benefit calculus does not justify intervention.

Tensions are also likely to occur when national-scale regulatory policies differ widely among countries that are closely integrated economically. Deeper economic integration makes countries more sensitive to the regulatory choices and social policies of their trade partners. For instance, in the 1970s, when China's trade with the United States (US) totalled less than 1 billion USD a year, few US citizens had reason to care about China's labour or environmental policies. Today, as China emerges as a major trade partner and competitor and US-China trade has increased almost 100-fold to 92 billion USD in 2002, these policies are subject to much greater American interest and concern. A key focus of trade policy-making thus centres on non-tariff barriers to trade and the need for a 'level' playing field in the global marketplace.[24]

Because many domestic regulations could act as non-tariff trade barriers, trade agreements now routinely include market-access rules and disciplines that create a framework for national regulation. Public health standards, food safety requirements, emissions limits, labelling policies, and waste

[23] Dua and Esty, 1997.
[24] Esty, 1994, 2001.

management and disposal rules – all national measures – may shape the flow of international trade. For example, the import ban of the European Union (EU) on genetically modified food has led to a 55 per cent decrease in US corn exports to Europe over the past five years and strenuous US objections to the EU treatment of genetically modified food.[25] Similarly, Venezuela objected to the discriminatory approach of the reformulated gasoline provisions of the US Clean Air Act of 1990 and won a WTO dispute settlement case restoring its access to the US gasoline market.[26] From the 'Tuna/Dolphin' case[27] of the early 1990s to the recent 'Shrimp/Turtle' dispute,[28] the number of trade-environment flash points has continued to grow. As noted earlier, environmentalists fear that liberalised trade might make it harder for high-standard countries to keep their stringent environmental requirements in the face of market-access demands from trade partners.

The essential difficulty lies in separating legitimate environmental standards from protectionist regulations advanced under the guise of environmental protection. Few would argue, for example, that emission-control standards for cars are an unwarranted barrier to trade. However, the fear of protectionism in an environmental disguise is not unfounded and needs to be addressed, particularly if developing countries are to retain confidence in the fairness of the international trade system. The smooth

[25] US Trade Representative, 2003.

[26] In the 'Reformulated Gasoline' case, Venezuela and Brazil brought a complaint against the US alleging that the 'Gasoline Rule', promulgated by the US Environmental Protection Agency (EPA) under the Clean Air Act, which excluded importers from exercising two alternatives for determining the appropriate fuel content that were available to domestic refiners, violated the General Agreement on Tariffs and Trade (GATT) as an unjustifiable barrier to trade. In 1996, the Appellate Body of the WTO determined that the 'reformulated' gasoline rule did violate the GATT as it subjected Venezuelan and Brazilian refiners to potentially more stringent requirements for fuel emissions than domestic refiners and was, therefore, in violation of Article XX exceptions. Following the Decision, the countries agreed on a 15-month phase-out of the illegal regulation.

[27] In the 'Tuna/Dolphin' case, US import restrictions on tuna caught with unsafe nets and techniques were struck down under the GATT rules as an illegal barrier to trade. Under the Marine Mammals Protection Act of 1972, the US restricted the importation of tuna caught using methods that killed dolphins. The restrictions effectively imposed a barrier to trade on tuna caught in Mexico as a result of the ban on such importation. Mexico successfully argued that the ban served as an illegal barrier to trade under the GATT and that the US could not extraterritorially regulate in the name of the environment.

[28] In 1996, the US Court of International Trade ordered the prohibition of shrimp importation from all countries that had not adopted harvesting methods comparable to the US methods, which included Turtle Exclusion Devices to prevent further mortality of endangered sea turtles. India, Malaysia, Pakistan, and Thailand brought issue with these Guidelines at the WTO. In 2001, upon Appellate review, the WTO issued the ruling in the 'Shrimp/Turtle' case, upholding, for the first time in GATT history, unilateral trade restrictions to conserve extraterritorial natural resources. The restrictions were upheld under the General Exceptions in GATT Article XX. The outcome is contrary to that in the 'Tuna/Dolphin' dispute as sea turtles had been listed by the United Nations (UN) as threatened with extinction.

functioning and efficiency of the international economic system cannot be maintained unless there are clear rules of engagement for international commerce, including environmental provisions.

Global Environmental Policies

Globalisation is, in part, an ecological fact. There exist a series of environmental challenges that span multiple countries and even the globe. Polluted waters, collapsing fisheries, invasive species, and the threat of climate change are all realities that have been exacerbated by globalisation. But ecological realities also affect the pace and pattern of globalisation. Scarce environmental resources (such as water) shape countries' perceptions of their independence or interdependence and, consequently, influence their economic and political interactions within the global community. The value that citizens around the world place on nature and biodiversity within foreign jurisdictions may spur international political pressures that limit a country's economic and regulatory choices. Protection of the shared resources of the global commons (including the oceans and the atmosphere) provides a rallying point for NGOs aiming to promote worldwide collective action. Increased understanding of the interdependence of ecological systems contributes to establishing a more robust global environmental regime.

Clearly, the primary responsibility for environmental protection rests with national governments and local communities. But some problems are inescapably regional or global in scope and cannot be addressed without international cooperation. Yet, incentives to pursue behaviour that is individually rational but collectively suboptimal are especially strong with regard to the depletion of natural resources, which may be seen as belonging to everybody and nobody. It is economically rational for a fisherman, for example, to try to maximise his personal gain by catching as many fish as possible as quickly as possible. Collectively, however, such a strategy leads to overexploitation of the resource and a 'tragedy of the commons', leaving the entire fishing community worse off than if it had found a cooperative arrangement to manage the fishery on a sustainable basis. When extended to the global scale, the problem becomes even more acute and intractable in the absence of clear rules and institutions ensuring sustainable resource management. Such global-scale issues require responses aggregated beyond the level of national jurisdictions or, at the very least, coordinated national action.

While not strictly necessary, international cooperation is helpful in addressing a set of *common* problems encountered locally all across the globe and thus of concern to policy-makers the world over. These problems (control of air and water pollution, waste disposal, etc.) should be dealt with by local or national authorities. There is no inherent need for global-scale cooperation. But the fact that many countries face problems in common creates another logic for cooperation: the potential to gain from sharing data, information, and policy experiences. Comparative analysis often helps to illuminate issues and highlight best practices – policies and technologies – to be deployed in response. To the extent that problems require substantial scientific or technical analysis, cooperation may also generate economies of scale in data collection, analysis, and other research functions which both benefit from globalisation and contribute to a deepening of interconnectedness and interdependence.

GLOBALISATION AND GLOBAL GOVERNANCE

Without effective international-scale governance, globalisation may intensify environmental harms wherever national regulatory structures are inadequate.[29] In strengthening competitive pressures across national borders, economic integration may help consumers by lowering prices, improving service, and increasing choice.[30] But these same pressures at times threaten to overwhelm the regulatory capacities of national governments and thus necessitate intergovernmental coordination of domestic policies and cooperative management of the global commons. As shown above, some problems are local and can best be addressed on that scale. But even in these cases, there is a clear advantage of learning from other countries and localities that have managed to address similar issues. In other cases, the problems are so inextricably international that a coordinated multi-country response is required. This response, however, must always be backed up by effective action at the national and local levels.[31]

Theory suggests that the solution to this policy dilemma lies in a structured programme of *collective action*. But overcoming the collective action problem is especially difficult in the international realm. There is no Leviathan or overarching authority. And while the number of beneficiaries and potential contributors to a global public good may be much larger than on the national scale, so too is the number of potential contributors to a public 'bad'. The spatial and temporal distribution of causes and effects makes it hard to identify those who fail to cooperate. Moreover, in the absence of an international authority, even if defectors were detected, there are scant means of discipline and sanction. The problem, therefore, is one of *organising and maintaining* cooperation. Absent institutional support and efforts at collective

---

29 Nordstrom and Vaughan, 1999.
30 Bhagwati, 1993.
31 Kaul et al., 1999.

action tend to degrade towards what is called in game theory a 'lose-lose' or 'Nash equilibrium'. The situation must be converted from one in which decisions are made independently, based on narrow self-interest, to one in which actors adopt cooperative solutions serving a broader, common interest.[32]

The traditional policy prescriptions – a set of taxes or subsidies to internalise externalities – cannot be easily applied to a multi-jurisdictional context with a fragmented institutional structure. Successful intervention requires some mechanism for promoting collective action.[33] Fragmentation, gaps in issue coverage, and even contradictions among different treaties, organisations, and agencies with competing responsibilities have undermined effective, results-oriented action in the domain.[34] As pointed out by Charnovitz,[35] '[l]ike a city that does not have zoning ordinances, environmental governance spreads out in unplanned, incongruent, and inefficient ways'. A pervasive lack of data, information, and policy transparency adds to the challenge. An institutional structure is necessary that can provide: the data foundation needed for good environmental decision-making; the capacity to gauge risks, costs, benefits, and policy options comparatively; a mechanism to exert leverage on private and public resources deployed at the international level; and means to improve results from global-scale environmental spending and programmes.

Environmental and Economic Governance: Whose Reform?

While the United Nations Environment Programme (UNEP) lies at the centre of the environmental regime, international environmental governance falls within the mandate of multiple organisations in the UN system. Hampered by a difficult mandate, a modest budget, and limited political support, UNEP competes with more than a dozen other UN bodies, including the Commission on Sustainable Development, the UN Development Programme (UNDP), the World Meteorological Organisation (WMO), and the International Oceanographic Commission on the international environmental scene. Adding to this fragmentation are the independent secretariats to numerous conventions, including the Montreal Protocol (ozone-layer protection), the Basel Convention (hazardous-waste trade), the Convention on International Trade in Endangered Species (CITES), and the Climate Change Convention, all contending for limited governmental time, attention, and resources.

---

32 Ostrom, 1990.
33 Baumol and Oates, 1988.
34 Esty and Ivanova, 2002a: 182-188.
35 Charnovitz, 2002.

The existing international environmental system has failed to deal adequately with the priorities of both developed and developing countries. The proliferation of multilateral environmental agreements has placed an increasing burden of collective obligations and responsibilities on member states. The toll on developing countries has been especially heavy, as little assistance in the form of financing, technology, or policy guidance has been forthcoming. The inadequacy and dispersion of the existing financial mechanisms – scattered across the Global Environment Facility (GEF), UNDP, World Bank, and separate funds such as the Montreal Protocol Finance Mechanism – reinforce the perception of a lack of seriousness in the North about the plight of the South. Furthermore, fundamental principles of good governance such as participation, transparency, and accountability are still at issue in many of the institutions with environmental responsibilities. These procedural shortcomings undermine the legitimacy of the system as a whole.

In the absence of a functioning global environmental management system capable of addressing the growing number of international environmental issues, environmental groups have directed efforts towards the reform of international economic bodies, including the World Bank and the WTO. The WTO has been of particular interest, as it has assumed responsibility for integrating the policy realms of environment and trade. Although the WTO has a Committee on Trade and Environment that has been meeting for a number of years, the WTO dialogue has been dominated by trade experts, has demonstrated little understanding of the impact of trade on environmental policy, and has almost nothing in the way of results to show for its efforts.[36] The role of the WTO as the principal forum for the discussion and resolution of trade and environment concerns has been contested by both the environmental community and developing countries. Environmentalists perceive the WTO as an organisation charged narrowly with the promotion of trade liberalisation and argue that any attempt to mainstream environmental issues within the WTO inevitably privileges economic concerns over the environment. Free traders, on the other hand, regard the WTO as an inappropriate forum for environmental issues, which they see as burdening the trade regime. Developing countries, too, see the inclusion of environmental rules among the responsibilities of the WTO as a complication and a threat,[37] potentially creating an excuse for protectionism and the exclusion of Southern goods from Northern markets. Nevertheless, discussion is taking place within the WTO, and pressure to 'green' the organisation has resulted in a number of notable reforms.[38]

---

[36] Esty, 1999.
[37] Williams, 2001.
[38] Wofford, 2000.

Recognition of the WTO's lack of capacity for addressing environmental issues and the undermining of its efficacy and legitimacy whenever the organisation is forced to make decisions that go beyond the scope of its trade mandate and expertise have led a number of trade experts to call for the creation of a more robust environmental governance structure. The former WTO Director-General, Renato Ruggiero, and the current Director-General, Supachai Panitchpakdi, have both urged for the creation of a World Environment Organisation to help focus and coordinate worldwide environmental efforts. During the World Summit on Sustainable Development in 2002, French President Jacques Chirac called for the creation of a Global Environment Organisation that would bring greater balance to a multilateral system excessively focused on the economy. Similar calls have come from Mikhail Gorbachev, Lionel Jospin, The Economist magazine, and others.[39] It is becoming increasingly clear that successful reform of the trade and finance system needs to be coupled with an equally rigorous and fundamental reform of the global environmental regime.

GOVERNANCE ALTERNATIVES

Collective action in response to global environmental challenges continues to fall short of public needs and expectations as a result of the deep-seated weakness of the existing institutional architecture. The question, therefore, is not *whether* to revitalise the global environmental regime, but *how*. The integrated and interdependent nature of the current set of environmental challenges contrasts sharply with the nature of the institutions we rely upon for solutions. These institutions tend to be fragmented and poorly coordinated, with limited mandates and impenetrable decision-making processes.

Shifting from a 'prisoner's-dilemma' world of free-riding and lose-lose outcomes to one where reciprocity is recognised and collaboration understood will require careful institutional realignment. We need an approach that acknowledges the diversity and dynamism of pollution control and natural-resource-management problems and recognises the need for specialised responses.[40] The multi-faceted nature of the environmental

[39] For the text of the speeches, see: Jospin, 2002; Panitchpakdi, 2001; Ruggiero, 1998. For arguments in favour of a World/Global Environment Organisation, see: Biermann, 2000; Charnovitz, 2002; Esty, 1994, 2000a, 2000b; Runge, 2001; Whalley and Zissimos, 2001. For the opposing view, see: Juma, 2000; Von Moltke, 2001.

[40] Esty and Ivanova, 2002b.

challenge requires a multi-layered institutional structure that can address issues on various geographic scales[41] and with a variety of policy tools.[42]

Functions at Various Levels of Governance

We argue that there is a spectrum of global-governance responses ranging from very light to fairly robust. Amenable to a regime at the light end of the spectrum lie problems that are local in scope but can be found around the world (local water and air pollution, for example). As we move towards the more demanding side of the spectrum, regional issues such as international water-bodies pollution or regional fisheries management arise. At the most difficult end of the spectrum are issues that likely require a strong structure of global collaboration (such as climate change, ozone layer depletion, and ocean pollution). A number of functions need, therefore, to be performed at the various levels of governance by different institutions.

When dealing with global-scale problems, institutions need to possess several capacities, including the ability to identify and define problems, raise awareness about them in various forums, draft rules and create norms for behaviour leading to the solution of these problems, formulate policy options, facilitate cooperative actions among governments and other actors, finance and support activities, and develop management systems. As elaborated below, we see an information clearing-house, a technology clearing-house, and a policy forum as central elements to the effective functioning of a global regime for resolution of environmental problems. Global institutions also have an important role to play when the problems are primarily national in scale. They can serve as facilitators of information and knowledge exchange, promoting learning across contexts and among actors. The exchange of data, best practices, policies, and approaches could be an important tool in problem-solving at the national level.

National institutions also have roles to play, both at domestic and global levels of governance. National governments remain the primary actors charged with regulatory and enforcement powers to solve environmental problems. Functions such as standard-setting, policy formulation, compliance monitoring, and evaluation are among their responsibilities. When the problems are of a global character, national governments are again key actors. Implementation of multilateral agreements is ultimately their responsibility. They also engage in information-sharing and exchange in the process of arriving at agreement on the global problems to be addressed, the policies necessary for their resolution, and the actions to be undertaken domestically.

---

41 Esty, 1999; Karlsson, 2000; Ostrom, 1990; Vogler, 2000.
42 Esty, 2004.

An effective response to both the common elements of national problems and the special demands of transboundary issues requires a deft and agile structure able to hone in on the nature of problems and produce the right scale of activity while promoting worldwide cooperation. There is no silver bullet. Various institutional and organisational designs are possible. We believe that the best strategy centres on a new environmental mechanism at the global level. Conceptually, a GEM would fundamentally need to focus on promoting collective action on the international scale. Practically, it offers the chance to build a coherent and integrated environmental policy-making and management framework that addresses the challenges of a shared global ecosystem.[43]

We see three core capacities as essential: (1) provision of adequate data and information that can help to characterise the problems to be addressed, reveal preferences, and clarify reciprocity; (2) creation of a policy 'space' for environmental negotiation and bargaining; and (3) sustained support for national efforts to address issues of concern and significance. We identify data collection, monitoring, and scientific assessment as central in the information domain. A forum for issue linkage and bargaining, a mechanism for rule-making, and a dispute-settlement framework are essential to ensuring cooperative solutions. The continual development of technical, financial, human, and institutional capacities for addressing diverse challenges is another critical function requiring effective institutional mechanisms at the global level.

At present, various institutions and agencies ostensibly have many of the identified capacities. But the reality often falls short of the promise. And some are flagrantly absent. For example, a host of international organisations, scientific research centres, national governments, and environmental convention secretariats are carrying out data collection, scientific assessment, financing, and technology transfer with little coordination across jurisdictions. Compliance-monitoring and -reporting are unsystematic, scattered, and largely informal. The participation of non-state actors requires further structural elaboration and institutionalisation, along with procedures for rule-making. A forum for issue linkage, bargaining, and trade-offs, as well as a dispute-settlement mechanism, is lacking. A more robust policy space for the environment is necessary to sustain efforts at environmental advocacy within the broader system of global governance and to ensure that environmental concerns are integrated into sustainable development policies.

Building on the expertise and capacities of existing institutions and creating new mechanisms where functions are not currently performed, we see three institutional elements as central to a successful global environmental

[43] Esty and Ivanova, 2002b, 2003.

system. A Global Information Clearing-House might represent a first step towards improved global environmental governance, through provision of comparable data on environmental quality, trends, and risks. The coordination of existing institutional mechanisms for data collection, scientific assessment, and analysis might attract broad-based support. A Global Technology Clearing-House, focusing on information sharing, performance measurement and benchmarking, and dissemination of best practices, might also be launched as an early initiative with likely broad appeal. With competence established in these areas, a Global Bargaining Forum might be initiated with the capacity for rule-making and facilitation of burden sharing. Progressive development over time, as the new system proves its capacity and value, is likely to make any reform strategy more acceptable to nations reluctant to yield responsibility or control to any global entity.

Global Environmental Information Clearing-House

Better environmental data and information make it easier to identify problems and trends, evaluate risks, set priorities, establish policy options, test solutions, and encourage technology development.[44] A global information clearing-house providing timely, relevant, and reliable data on environmental issues and trends could transform the policy-making process on the global scale. Better data, science, and analysis could shift assumptions, highlight preferences, and sharpen policies. In the case of acid rain in Europe, for example, knowledge of domestic acidification damage allowed for refined policies that triggered emission reductions in several countries.[45] Simply put, data can make the invisible visible, the intangible tangible, and the complex manageable.

Information on how others are doing in reducing pollution and improving resource productivity tends to stimulate competition and innovation. Comparative performance analysis across countries – similar to the national PROPER scheme in Indonesia[46]– could provide much greater transparency, reward policy leaders, and expose laggards.[47] Just as knowledge that a competitor in the market place has higher profits drives executives to redouble their efforts, evidence that others are outperforming one's country on environmental criteria can sharpen the focus on opportunities for improved performance. The attention that the Yale-CIESIN-World Economic Forum

[44] Esty, 2002, 2004.
[45] Levy, 1993.
[46] PROPER (Program for Pollution Control, Evaluation, and Rating) is Indonesia's innovative programme for reducing pollution by rating and publicly disclosing the environmental performance of industrial facilities.
[47] Afsah et al., 2000.

Environmental Sustainability Index has generated demonstrates this potential.[48]

Data-gathering should primarily be the function of local or national organisations. But a central repository for such information and a mechanism for making the information publicly available could generate significant economies of scale, efficiently generate relevant comparisons, and expose slack performance.[49] An information clearing-house would not centralise science policy *functions,* but create a centralised source for coordinating information flows among the institutions responsible for performing scientific aspects of policy-making.[50]

Global Environmental Technology Clearing-House

Globalisation is fuelled by and plays a central role in the diffusion of technologies. Technological advances are often the key to environmental gains.[51] However, industrialised countries dominate the technology market and the generation of innovations. Some technologies and their environmental features may, therefore, be inappropriate for the economic and environmental circumstances of less developed countries.[52]

Most multilateral environmental agreements contain provisions related to technology transfer as part of the incentive packages for developing countries to meet their obligations under the conventions. The Basel Convention on the Control of Transboundary Movements of Hazardous Wastes and their Disposal, the Montreal Protocol on the Ozone Layer, the Convention on Biological Diversity, the Framework Convention on Climate Change and its related Kyoto Protocol all cite technology transfer as a critical method for achieving concrete environmental improvements. Agenda 21 also underscores the importance of technology transfer to sustainable development. The existing strategies for technology transfer have, however, been less than effective. A new mechanism to bring technologies to developing countries must be part of any strategy to improve international environmental policy results. Establishing such a mechanism, however, presents a significant challenge.

The empirical evidence shows that the gains from such cooperative arrangements have indeed been significant and beneficial for the

---

[48] Seelye, 2002; Yeager, 2002; see Yale Center for Environmental Law and Policy, 2004 for more information on the Environmental Sustainability Index.
[49] Chayes and Chayes, 1995.
[50] UN University, 2002.
[51] Chertow and Esty, 1997.
[52] Karlsson, 2002.

environment.[53] For example, the technology panel convened under the Montreal Protocol to report on the availability of CFC substitutes and the feasibility of larger production cuts generated new knowledge and new commercial opportunities for CFC reduction in a highly collaborative process.[54] Most technologies are, however, owned by private companies rather than governments. So some efforts need to be put into structuring incentives to motivate the private sector to disseminate technological advances optimally. An effective environmental technology clearing-house is thus not only necessary but also possible. It could guide nations towards the use of appropriate technologies, support North-South partnerships, and provide a forum for coordinating financial assistance to developing countries. It would contain information on best practices around the world and facilitate technology development and continuous learning.

Global Bargaining Forum

Successful responses to global-scale environmental problems depend on effective international agreements. To be workable, any such agreement must equitably distribute the burden of international collective action. Developing countries will often need support, subsidies, and other incentives to encourage their efforts to internalise externalities. In the past, issue linkage has been avoided in favour of lowest-common-denominator programmes in the absence of funding to support those least well positioned to act. Yet, there would be great value in a forum for the facilitation of international deals on the environment that improve quality and result in positive cash flows to custodians of environmental assets.[55]

A global bargaining forum could act as a catalyst for action, facilitating financial discussion among countries or private entities. A government in one country might, for example, negotiate a deal to preserve a particular natural resource in another country in return for a sum of money or other policy benefits. Brazil might, for instance, commit to certain limits on development in the Amazon in return for guaranteed access to EU and US markets for its orange juice. The forum might also provide mechanisms for verification, financial transfers, and dispute settlement.

Moving Forward

In designing a new global environmental architecture, form should follow

[53] For an analysis of technology transfer as a means of successful integration for developing countries into the global economy, see UNCTAD, 2003.
[54] Parson, 1993.
[55] Whalley and Zissimos, 2002.

function. We envision a 'light' institutional superstructure providing coordination through a staff comparable in size and quality to the WTO secretariat in Geneva. The secretariat's primary role would be to promote cooperation and achieve synergies across the disparate multilateral environmental agreements and other international institutions with environmental roles. A properly designed structure would provide a counterpart as well as a counterweight to the WTO and an alternative forum for addressing tensions over divergent environmental values and approaches. The GEM we envision would neither add a new layer of international bureaucracy nor create a world government. Quite to the contrary, movement towards a GEM should entail consolidation of the existing panoply of international environmental institutions and a shift towards a more modern 'virtual' environmental regime.

At the centre of our proposal lies a global public policy network drawing in expertise from around the world on an issue-by-issue basis. By utilising the resources of national governments, NGOs, private-sector enterprises, business and industry associations, think tanks, research centres, and academic institutions on an 'as needed' basis, the GEM would have far broader issue expertise and analytic capacity than has the existing environmental regime. Such a system for advancing international environmental agenda-setting, analysis, negotiation, policy formulation, implementation, and institutional learning would be more flexible, cost-effective, fleet-footed, and innovative. The benefits of such a structure are increasingly clear.[56] Global public-policy and issue networks respond to an ever more complex international policy environment, taking advantage of Information Age communication technologies to draw in relevant expertise, analyse problems from multiple perspectives, and build new opportunities for cooperation.

Streamlining the environmental system would be especially beneficial to the South. In particular, a single venue for negotiations and international coordination would make it much easier for the overstretched environment ministries of the developing world to monitor the spectrum of environmental issues at play and to contribute thoughtfully to the global-scale debate, even with a relatively small international policy-making team.[57] There would be no need to traipse around the world trying to keep up with an ever more extensive list of separate bodies and meetings. A network approach, drawing in diverse perspectives and expertise and using the internet, could facilitate greater developing-country participation in the international policy-making process.

[56] Reinicke, 1998; Reinicke and Deng, 2000; Rischard, 2002; Witte et al., 2003.
[57] Biermann, 2002.

Who will pay for global-scale environmental problem-solving stands out as a matter of particular importance to developing countries. Globalisation, as noted above, puts increasing pressure on national governments to become more competitive in the global marketplace. Expending scarce financial resources for environmental protection is, therefore, often regarded as counterproductive by developing countries, especially if there is no urgent demand from domestic constituencies. By placing the principle of common but differentiated responsibilities at the centre of the new mechanism, along with a real forum for bargaining and trade-offs, efforts to strike a fair balance of rights and responsibilities with regard to transboundary environmental issues might meet with increased success. A more carefully considered and coherent set of international environmental standards would also alleviate fears in the South that the industrialised world seeks to impose unreasonably high standards – and perhaps trade penalties for non-compliance – on developing countries, all of whom have many competing demands for limited public resources. Moreover, mechanisms to support technology transfers and to subsidise developing countries' environmental initiatives in pursuit of global environmental goals would help to alleviate North-South tensions.

A related question concerns the values to be promoted in a strengthened international environmental regime. It is essential that a GEM be seen as a transparent and inclusive forum that seeks to build consensus on a basis that respects the diversity of views across the world. Properly managed public policy networks create 'virtual public space' that is easier to enter than the established physical forums where decisions are currently made.[58] An Information Age set of outreach mechanisms could also decrease the distance between decentralised constituencies and global decision-makers, making it easier to insert into the policy process the broad array of values, perceptions, and perspectives that are now often overlooked or incompletely considered. At the same time, these mechanisms would facilitate public understanding of the issues addressed and decisions made on the global scale.

CONCLUSION

Both economic and ecological interdependence require rigorous national policies and effective international collective action. Our increasingly globalised world makes new thinking about international environmental cooperation essential, both in its own right and to undergird further economic integration. An extraordinary mix of political idealism and pragmatism will be required to coordinate pollution control and natural-resource-management

[58] Streck, 2002.

policies on a worldwide basis across diverse countries and peoples, political perspectives and traditions, levels of wealth and development, and beliefs and priorities. But the gains to be achieved go beyond the environmental domain. Indeed, coordinated pollution-control strategies and natural-resource-management standards provide an important set of ground rules for international commerce, serve as an essential bulwark against market failure in the international economic system, and make it more likely that globalisation will yield broad benefits.

It is time to re-craft the environmental regime, aiming for a new, forward-looking, sleeker, and more efficient architecture that will better serve environmental, governmental, public, and business needs. A new global environmental system need not compete with efforts to strengthen national pollution-control and natural-resource-management programmes. It should, in fact, reinforce such efforts. Success in the environmental domain depends on a multi-tier governance structure supporting vibrant efforts on the local, national, and global scales.

The logic of a GEM is straightforward: a globalising world requires thoughtful and modern ways to manage interdependence. The world community would benefit from a systematic mechanism to promote environmental cooperation in the international arena, a recognised forum for national officials and other stakeholders to debate and address global-scale issues, and an institutional mechanism designed to make economic progress and environmental protection mutually reinforcing.

REFERENCES

Afsah, S., A. Blackman, and D. Ratunanda (2000), *How Do Public Disclosure Pollution Control Programs Work? Evidence from Indonesia*, Washington, DC: Resources for the Future.

Anderson, K. (1992), 'The standard welfare economics of policies affecting trade and the environment', in K. Anderson and R. Blackhurst (eds), *The Greening of World Trade Issues*, Ann Arbor: University of Michigan Press.

Anderson, K. (1998), 'Environmental and labor standards: What role for the WTO?', in A. Krueger (ed.), *The WTO as an International Organization*, Chicago: University of Chicago Press.

Anderson, S., J. Cavanagh, and T. Lee (1999), 'Ten myths about globalization', *The Nation*, **269**(19), 26-27.

Antweiler, W., B. Copeland, and S. Taylor (2001), 'Is free trade good for the environment?', *American Economic Review*, **91**(4), 877-908.

Baumol, W. and W. Oates (1988), *The Theory of Environmental Policy*, Cambridge: Cambridge University Press.

Bhagwati, J. (1993), 'The case for free trade', *Scientific American*, **269**(5): 42-49.

Bhagwati, J. (2000), 'On thinking clearly about the linkage between trade and the environment', *Environment and Development Economics*, **5**(4), 485-496.

Biermann, F. (2000), 'The case for a World Environment Organization', *Environment*, **42**(9), 22-31.

Biermann, F. (2002), 'Strengthening green global governance in a disparate world society: Would a World Environmental Organization benefit the South?', *International Environmental Agreements*, **2**, 297-315.

Charnovitz, S. (2002), 'A World Environment Organization', *Columbia Journal of Environmental Law*, **27**(2), 323-362.

Chayes, A. and A. Handler Chayes (1995), *The New Sovereignty: Compliance with International Regulatory Agreements*, Cambridge: Harvard University Press.

Chertow, M. and D. Esty (eds) (1997), *Thinking Ecologically: The Next Generation of Environmental Policy*, New Haven: Yale University Press.

Dua, A. and D. Esty (1997), *Sustaining the Asia Pacific Miracle: Economic Integration and Environmental Protection*, Washington, DC: Institute for International Economics.

Esty, D. (1994), *Greening the GATT: Trade, Environment, and the Future*, Washington, DC: Institute for International Economics.

Esty, D. (1999), 'Economic integration and the environment', in N. Vig and R. Axelrod (eds), *The Global Environment: Institutions, Law, and Policy*, Washington, DC: CQ Press.

Esty, D. (1999), 'Towards optimal environmental governance', *New York University Law Review*, **74**(6), 1495-1574.

Esty, D. (2000a), 'Stepping up to the global environmental challenge', *Fordham Environmental Law Journal*, **8**(1), 103-113.

Esty, D. (2000b), 'Global environmental agency will take pressure off WTO', *Financial Times*, 13 July.

Esty, D. (2001), 'Bridging the trade-environment divide', *Journal of Economic Perspectives*, **15**(3), 113-130.

Esty, D. (2002), 'Why measurement matters', in D. Esty and P. Cornelius (eds), *Environmental Performance Measurement: The Global Report 2001-2002*, New York: Oxford University Press.

Esty, D. (2004), 'Environmental protection in the information age', *New York University Law Review,* **79**(1), 115-211.

Esty, D. and D. Geradin (2001), 'Regulatory co-opetition', in D. Esty and D. Geradin (eds), *Regulatory Competition and Economic Integration*, New York: Oxford University Press.

Esty, D. and M. Ivanova (2002a), 'Revitalizing global environmental governance: A function-driven approach', in D. Esty and M. Ivanova (eds), *Global Environmental Governance: Options & Opportunities*, New Haven: Yale School of Forestry & Environmental Studies.

Esty, D. and M. Ivanova (eds) (2002b), *Global Environmental Governance: Options & Opportunities*, New Haven: Yale School of Forestry & Environmental Studies.

Esty, D. and M. Ivanova (2003), 'Towards a global environmental mechanism', in J. Speth (ed.), *Worlds Apart: Globalisation and the Environment*, Washington, DC: Island Press.

Grossman, G. and A. Krueger (1995), 'Economic growth and the environment', *Quarterly Journal of Economics,* **110**(2), 353-377.

ICBL (2004), *http://www.icbl.org*, International Campaign to Ban Landmines.

Jobes, P. (2003), 'Globalization and regional renewal revisited', *Australian Journal of Social Issues,* **38**(1), 73-79.

Jospin, L. (2002), 'French Prime Minister calls for creation of new World Environment Organization', *International Environment Reporter,* **25**(5), 213.

Juma, C. (2000), 'The perils of centralizing global environmental governance', *Environment Matters, Annual Review*, Washington, DC: The World Bank.

Karlsson, S. (2000), *Multilayered Governance: Pesticides in the South – Environmental Concerns in a Globalised World*, Linkoping: Department of Water and Environmental Studies, Linkoping University.

Karlsson, S. (2002), 'The North-South knowledge divide: Consequences for global environmental governance', in D. Esty and M. Ivanova (eds), *Global Environmental Governance: Options & Opportunities*, New Haven: Yale School of Forestry & Environmental Studies.

Kaul, I., I. Grunberg, and M. Stern (eds) (1999), *Global Public Goods: International Cooperation in the 21st Century*, New York: Oxford University Press.

Keohane, R. and J. Nye (2003), 'Globalisation: What's new? What's not? (And so what?)', in D. Held and A. McGrew (eds), *The Global Transformations Reader: An Introduction to the Globalization Debate*, Cambridge: Polity Press.

Klevorick, A. (1996), 'The race to the bottom in a federal system: Lessons from the world of trade policy', *Yale Law and Policy Review and Yale Journal on Regulation*, **14**(Symposium Issue), 177-186.

Levy, M. (1993), 'European acid rain: The power of tote-board diplomacy', in P. Haas, R. Keohane, and M. Levy (eds), *Institutions for the Earth: Sources of Effective International Environmental Protection*, Cambridge: MIT Press.

Mendelsohn, R. (1986), 'Regulating heterogeneous emissions', *Journal of Environmental Economics and Management*, **13**(4), 301-313.

Nordstrom, H. and S. Vaughan (1999), *Trade and Environment*, Geneva: World Trade Organization.

Ostrom, E. (1990), *Governing the Commons: The Evolution of Institutions for Collective Action*, 10th ed., New York: Cambridge University Press.

Panayotou, T. (1993), *Green Markets: The Economics of Sustainable Development*, San Francisco: Ics Press.

Panitchpakdi, S. (2001), 'Keynote address: The evolving multilateral trade system in the new millennium', *George Washington University International Law Review*, **33**, 419-443.

Parson, E. (1993), 'Protecting the Ozone Layer', in P. Haas, R. Keohane, and M. Levy (eds), *Institutions for the Earth: Sources of Effective International Environmental Protection*, Cambridge: MIT Press.

Reinicke, W. (1998), *Global Public Policy: Governing without Government?*, Washington, DC: Brookings Institution Press.

Reinicke, W. and F. Deng (2000), *Critical Choices: The United Nations, Networks, and the Future of Global Governance*, Ottawa: International Development Research Council.

Rischard, J. (2002), *High Noon: 20 Global Problems, 20 Years to Solve Them*, New York: Basic Books.

Rodrik, D. (1997), *Has Globalization Gone too Far?*, Washington, DC: Institute for International Economics.

Ruggiero, R. (1998), *A Global System for the Next Fifty Years*, address to the Royal Institute of International Affairs, London.

Runge, F. (2001), 'A global environmental organization (GEO) and the World Trading System', *Journal of World Trade*, **35**(4), 399-426.

Seelye, K. (2002), 'Study puts Finland first, and US 51st, in environmental health', *The New York Times International*, 2 February.

Selden, T. and D. Song (1994), 'Environmental quality and development: Is there a Kuznets curve for air pollution emissions?', *Journal of Environmental Economics and Management*, **27**(2), 147-162.

Shafik, N. (1994), 'Economic development and environmental quality: An econometric analysis', *Oxford Economic Papers*, **46**(0), 757-73.

Slaughter, A. (1997), 'The real new world order', *Foreign Affairs*, **76**, September/October, 183-190.

Speth, J. (ed.) (2003), *Worlds Apart: Globalization and the Environment*, Washington, DC: Island Press.

Streck, C. (2002), 'Global public policy networks as coalitions for change', in D. Esty and M. Ivanova (eds), *Global Environmental Governance: Options & Opportunities*, New Haven: Yale School of Forestry & Environmental Studies.

Thompson, J. (2003), 'The globalization of communication', in D. Held and A. McGrew (eds), *The Global Transformations Reader: An Introduction to the Globalization Debate*, Cambridge: Polity Press.

UNCTAD (2003), *http://www.unctad.org/en/docs//iteipc20036_en.pdf*.

UN University (2002), *International Environmental Governance: The Question of Reform: Key Issues and Proposals: Preliminary Findings*, Tokyo: United Nations University Institute for Advanced Studies.

US Trade Representative (2003), *2003 National Trade Estimate Report on Foreign Trade Barriers: United States Trade Representative*, Washington, DC: Executive Office of the President of the United States.

Vogel, D. (1994), *Trading Up: Consumer and Environmental Regulation in a Global Economy*, Cambridge: Harvard University Press.

Vogler, J. (2000), *The Global Commons: Environmental and Technological Governance*, 2nd ed., Chichester: Wiley.

Von Moltke, K. (2001), 'The organization of the impossible', *Global Environmental Politics*, **1**(1), 23-28.

Whalley, J. and B. Zissimos (2001), 'What could a World Environmental Organization do?', *Global Environmental Politics*, **1**(1), 29-34.

Whalley, J. and B. Zissimos (2002), 'Making environmental deals: The economic case for a World Environment Organization', in D. Esty and M. Ivanova (eds), *Global Environmental Governance: Options & Opportunities*, New Haven: Yale School of Forestry & Environmental Studies.

Wheeler, D. (2002), 'Beyond pollution havens', *Global Environmental Politics*, **2**(2), 1-10.

Williams, M. (2001), 'Trade and Environment in the World Trading System: A decade of stalemate?', *Global Environmental Politics*, **1**(4), 1-9.

Witte, J., C. Streck, and T. Benner (eds) (2003), *Progress or Peril? Partnerships and Networks in Global Environmental Governance: The Post-Johannesburg Agenda*, Washington DC/Berlin: Global Public Policy Institute.

Wofford, C. (2000), 'A greener future at the WTO: The refinement of WTO jurisprudence on environmental exceptions to the GATT', *Harvard Environmental Law Review*, **24**(2), 563-592.

Yale Center for Environmental Law and Policy (2004), 'Environmental Sustainability Index', *http://www.yale.edu/envirocenter*.

Yeager, H. (2002), 'Scandinavia tops world league table on the environment', *Financial Times*, 3 February.

25. Governments and Policy Networks: Chances, Risks, and a Missing Strategy

Charlotte Streck[1]

SUMMARY

Government has engaged in a multitude of networks in order to respond in a flexible way to its rapidly changing political environment. A variety of international networks, fulfilling different roles, have been created between governments, governmental agencies, and private actors. This chapter describes and analyses different types of networks in which governments participate, and gives examples of governmental networks, transgovernmental networks, and public-private networks. A case is also made for a proactive governmental strategy with respect to these new tools of global environmental governance. Recognising both the opportunities (in terms of flexibility and efficiency) and threats (especially with respect to control and legitimacy) of international networks, government has to decide in which initiatives to participate, and how to manage, monitor, and evaluate them. Finally, global networks should be complements rather than substitutes of formal national authority.

INTRODUCTION

For the longest time, international environmental governance, firmly based on treaty law, has been the unchallenged arena of diplomats and negotiators representing the interests of sovereign states. However, times have changed. At the 2002 World Summit for Sustainable Development (WSSD), alliances

[1] This chapter represents the personal view of the author and should in no way be taken to represent the official view of any institution for which she works or with which she is associated.

between businesses and non-governmental groups for the first time took centre stage, when governments agreed to include partnerships between different sectors in the official outcomes of the Summit.

This official recognition of partnerships as part of the broader picture of environmental governance mirrors a change in the elements that constitute the system of global governance. Today, powerful transnational corporations dominate the global markets, and international non-governmental organisations (NGOs) claim increasing participation in international decision-making processes. These two phenomena equally force the state to redefine itself and its role in international relations and politics. In this context, the international debate on environmental governance and its regimes has moved from a focus on governments to a focus on a multitude of partners; from governance at the international level to governance at multiple levels; and from a largely formal, legalistic process to a less formal, more participatory and integrated approach.

Recent years have witnessed the mushrooming of a variety of more-or-less formal alliances between public and private entities. These initiatives, which build on different types of networks, spread around the globe as their creation turned out to be easier and their learning opportunities better than those of the traditional mechanisms of international cooperation. Globalisation and the revolution of information technology have made networks the preferred choice of cooperation for a rapidly changing environment. The term 'network' itself has been used in a variety of ways spanning a number of academic disciplines.[2] Networks are used in describing physical networks, in connection with lobbying and advocacy, with respect to partnerships and political alliances and in describing institutions. Indeed, the concept of networks appears in connection with such a wide range of initiatives that the flexibility and broadness of the 'network' phenomenon seems to be its main characteristic (see Box 25.1).

While some authors have embraced the emergence of this wide variety of networks and partnerships as the solution for almost all of the governance problems of the 21st century, others have been more sceptical and claim that the reliance on increasingly decentralised structures would lead to a privatisation of international relations. The government sceptics claim that governments lack the capability and flexibility to adapt to changes, and argue that sluggish bureaucracies are not able to keep up with the speed of the globalised world – in short, the state is slow, old-fashioned, and can no longer

---

2 Agranoff and McGuire, 2003; Dean, 1999; Jordan and Schubert, 1992; Marin and Mayntz, 1991; Milward and Provan, 2000; Reinicke, 1998; Reinicke et al., 2000; Reinicke et al., 2001; Rhodes, 1997.

Box 25.1 Types of networks

Advocacy networks are more or less loose alliances between NGOs, pressure groups, or businesses, all of which want to achieve a common set of objectives (for example, the Climate Action Network, Transparency International, International Campaign to Ban Landmines).

Civil-society networks are groups with loose ties, often between transnational civil society (NGOs, individuals) which can pursue different goals and play different roles. Civil-society networks can form part of advocacy networks in cases where they pursue a common policy goal (e.g., Africa's NGO Environment Network, the Asian NGO Coalition for Agrarian Reforms and Rural Development).

Business networks are alliances between businesses whereby resources, capabilities, and core competencies are combined to pursue mutual interests (for instance, business associations).

Global public policy networks are the collaborations between actors from different sectors (public as well as private), based on the more efficient allocation of complementary resources among the different partners (for example, World Commission on Dams (WCD), CIGAR, the Prototype Carbon Fund of the World Bank (WB)).

WSSD partnerships, as a sub-set of policy networks, are usually created around a specific issue, with a specified mandate, and a limited, clearly identified number of partners. In the context of the WSSD, Type II partnerships have been understood as policy networks or multi-sectoral alliances designed specifically to implement legal and political agreements in the area of sustainable development.[3]

Transgovernmental networks are informal networks of government officials or transnational public agencies (e.g., the Basel Committee, City Alliances).

Government networks provide platforms on which governments cooperate. International organisations are formalised forums of cooperation (such as the OECD and the Organization of American States). Government Networks also include consultation or negotiation networks (for example, the G8, the G77, or negotiation groups like the Umbrella Group in the context of climate-change negotiations), which are less formal and pursue common interests.

Institutional networks can refer to networks of institutions but also to institutions made of networks. Whereas networks consisting of institutions include transgovernmental networks, institutions of networks often provide increased stability, since they have been created in a process of defining win-win situations for different actors (e.g., the IUCN, the Global Environment Facility).

Scientist networks are networks of scientists to promote science and create a platform for debate and exchange of views. In some instances, the aim may be to find consensus (for instance, the Intergovernmental Panel on Climate Change).

be the dominant player in shaping international affairs.[4] The network sceptics, on the other hand, stress the danger that lies in uncontrolled and

[3] See United Nations, 2004 for the list of partnerships.

[4] Ohmae, 1995; Peters, 1997; Sassen, 1996; Strange, 1996.

unaccountable networks being increasingly substituted for binding commitments embodied in international law, which could thus threaten the legitimacy of international processes. In the same context, there is a concern that developments in the past decades would help to reduce state power as well as the power of national legislatures and international organisations, 'while private power (that of corporations rather than NGOs) is taking up even more of the slack left by the emergence of the minimalist state'.[5] And if indeed states are no longer in the position to administer and govern an increasingly interdependent and globalised world, the question arises which entity would bridge the gap and guarantee the representation of public and common interests.[6]

However, a closer look at the reality shows that neither the network glitter is all gold nor that the nation state is an outdated concept of the past. State sovereignty has, indeed, proven its resilience as an organising element of the world order. What is more, most governments and government agencies have not passively endured change but have actively responded to their evolving milieu by entering into an active dialogue with other players that have appeared on the stage of international politics. Yet, the fact that the public sector has adapted to the changing environment and has proved to be less sluggish and old-fashioned than is argued by many has not received a great deal of attention. Whereas the phenomenon of globalisation is well-described and we are aware of many of the consequences it entails (such as the porosity of national borders, the fading importance of states, and the shift of power and allegiance to non-state actors and international organisations),[7] comparatively little has been written on how governments reply to the challenge of globalisation and how they use informal and flexible systems to increase their responsiveness in addressing global problems.

On the international stage, governments have a role not only in setting the agenda on policy priorities and on creating binding law, but also in coordinating compliance through network approaches, promoting the sustainable development agenda through 'coalitions of the willing', and in facilitating action and responses through networks between experts and agencies. Today's picture of governance in general, and environmental governance in particular, is a highly complex mosaic of interactions, where governments have not only relations with other governments but also with a multitude of national and international actors representing diverse interests and stakeholders. Networks have been created between governments, between governmental agencies, and between governments and private actors. All of these networks fill different niches and fulfil different functions.

5 Alston, 1997: 442.
6 Cerny, 1995.
7 Barfield, 2001.

This chapter is aimed at describing the different functions governments fulfil when they participate in networks. It also analyses the opportunities as well as the risks that accompany a more active involvement of governments in different types of networks. Based on this analysis, I make a case for a proactive government strategy towards new tools of global governance, that would help governments to decide in which initiatives they have to participate, which ones they have to manage, monitor, and evaluate, and, finally, which public tasks should remain exclusively in the domain of the government's responsibilities.

Below, I first give a short introduction on the recent developments in international governance and on the emergence of networks. Three examples of networks and the roles governments have assumed in these networks are then shown. The first example centres around networks formed exclusively by governments. Government networks have resulted in the creation of more formalised platforms, mostly as international institutions, but they have also created flexible, open, and issue-driven arrangements. These arrangements are not always driven by diplomats; they can also emerge between technical staff and specialised agencies of different countries. The second example looks at such transgovernmental networks consisting of governmental agencies, departments, and/or single individuals at the subnational level. The final example discusses the emergence of partnerships between private and public actors, which have been described as global public policy networks. Thirdly, I highlight the opportunities and limits of governments participating in networks, especially in networks in which governments partner with non-state actors. Fourthly, the involvement of governments is described more specifically in environment-related networks. In an outlook at the end the challenges that governments in all three types of networks have to come to grips with are summarised. Governments have to develop a strategic approach towards their engagement in networks, and they have to take into account the different possibilities for holding networks accountable as well as develop mechanisms to ensure accountability.

BACKGROUND: WHY NETWORKS?

We live in an increasingly complex world, in which traditional forms of governance have reached a limit and are no longer sufficiently effective in governing our social and natural environments. Most importantly, the effects of globalisation have made obvious the shortfalls of traditional policy approaches: problems become increasingly international and transcend national borders; they require quick responses and effective decision-making; and they are complex and, therefore, can only be addressed through multi-actor cooperation. A single national government cannot solve these problems. Often, not even a group of governments is sufficient in proving the necessary policy responses.

Liberalisation of economies and a revolution in communication technologies have compressed distances and communication times, which, in turn, has helped to produce a global market place by dissolving borders for merchandise. The perforation of borders has primary effects on the notion of nation states, which are traditionally defined through a territory and borders. Consequently, globalisation, characterised by the everyday increasing flow of goods, communication, and monies, seems to threaten the role of sovereign states which depend on maintaining static territories as their defining attributes. Business and civil society have created powerful networks through which they can effectively lobby and put issues on the international agenda.

In recent decades, it has become obvious that an explosion of the world's population, the interdependence of economies and the rapid increase in international trade, an unprecedented movement of people, and an increasing strain on the Earth's natural resources has led to an internationalisation of the world's problems. Complex problems with international dimensions, like the spread of HIV or other epidemics, but also criminal activities such as international money laundry or illegal trade in arms, cannot be solved at the level of the nation state, thus requiring international cooperation. Due to their transboundary and global nature, environmental problems such as transboundary pollution, management of global commons and provision of global public goods figure prominently among the problems characterised by an increasing complexity and by a dense web of interconnections of international and national policy responses.[8]

Traditionally, international governance has been viewed primarily as a relationship between sovereign states. While government and governance both refer to systems of rule, the notion of government suggests activities that are backed by formal authority and police powers to ensure the implementation of policies. Governance refers to activities backed by shared goals that do not necessarily rely on the exercise of authority to attain compliance.[9] The growing importance of non-state actors, such as civil-society NGOs, interest groups, academia, and the private sector (in particular, multinational corporations and the global capital market), has transformed the system of global governance. The increasing influence of informal actors has come to typify the non-hierarchical structure of the current governance

---

8 Problems such as global warming and loss of biodiversity may highlight the complexity of the issues at stake. Both problems can only be addressed through international and intersectoral policy responses, including civil society and business in the formulation and implementation of policy solutions.

9 Hierlmeier, 2002.

regime.[10] Hence, governance is 'the sum of the many ways individuals and institutions, public and private, manage their common affairs. It is a continuing process through which conflicting or diverse interests may be accommodated and co-operative action may be taken. It includes formal institutions and regimes empowered to enforce compliance, as well as informal arrangements that people and institutions either have agreed to or perceive to be in their interests.'[11] Today, the exclusive understanding of governance as a formal, legalistic process dominated by the public sector has given way to a new conception that emphasises a less formal, more collaborative and integrated approach.

The idea of multi-level governance challenges the hierarchical, state-centric approach of international politics, as it reflects an increasingly complex, multi-layered policy. In this context, the instruments of international policy-making currently at our disposal (international treaties, cooperation through institutions, and agencies) have proven insufficient to meet the requirements of an increasingly interdependent world in a timely and efficient manner. The negotiation processes that lead to international agreements are cumbersome and usually span several years. Compliance is unsatisfactory and enforcement is weak. In sum, the traditional system has not always been able to react adequately to the challenges of our globalised world. In this context, concerns have been expressed that these developments lead to a decline in state power, as not only international and supranational organisations but also private actors and various types of networks compete with states for power and influence.

However, it has also become obvious that states themselves are not stagnant and actively participate in new forums of policy-making and international cooperation. State sovereignty is the attribute of states which legitimises their participation in intergovernmental forums. As globalisation literally turns the world inside-out by nationalising international law and internationalising national law, the opportunities for such participation expand exponentially.[12] Territorial boundaries have diminished in importance, and shared global or regional problems have further expanded the benefits as well as the necessity of cross-border cooperation. Together with private actors, states can capitalise on, rather than be circumvented by, the information age. The new speed of information also suggests new forms of organisation.[13]

---

[10] Hierlmeier, 2002; Reinicke et al., 2000; Strange, 1997.
[11] Commission on Global Governance, 1995: 1.
[12] Slaughter, 2000.
[13] Raustiala, 2002.

International cooperation has been based on networking between governments since its inception. Governments learned centuries ago to forge alliances and negotiate accords in order to achieve and maximise benefits that cannot be achieved single-handedly. Whereas these accords initially centred around issues of national security and interest politics, the emergence of global threats today requires more complex multilateral responses, usually orchestrated by international organisations and based on increasingly sophisticated regimes.

In the past few decades, a decreased financial capacity of states and the limits of technical expertise that they can hold, together with the acknowledgement that in some instances new platforms and new partners are better placed to fulfil certain functions, have led governments to explore more flexible channels of international and, to a lesser extent, national politics. In the search for more flexible forums of cooperation, they established consultation and negotiation networks, known as the G77, G8, G10, or G22. These forums are highly influential, though they are not founded on any legal structure. In parallel, government officials and technical experts have started to liaise with their colleagues around the world, creating transgovernmental networks which are based on technical cooperation and common interests. Additionally, multisectoral networks have emerged as another means of cooperation: coalitions of private and public sector actors that deal with specific issues or problems in an effort to crystallise scientific or political consensus, influence political negotiations, and generate momentum around the implementation of the agreed outcomes. This form of network responds to the increased need to find compromises between all segments of society in order to address the global challenges of the decades to come. Reflecting the changes that have taken place in international governance as a whole, these networks allow the different actors to bring their specific resources, advantages, and concerns to the table. Such initiatives, known as global public policy networks, complement the architecture of global governance.

Networks typically emerge in situations where traditional means of cooperation and problem solving are perceived as no longer being sufficient in addressing particular issues. Networks, for example, are created in deadlock situations, where there is a necessity for cooperation and policy coordination, or in cases of insufficient treaty implementation. In short, they emerge wherever formal agreements and relationships are not perceived to be necessary or advantageous, as relationships between the different actors are secured through trust and common interest, if not the same resources. As long as trust and common interest (and expected benefit) are in place, networks can be comprised of a large variety of partners.[14]

Today, we are surrounded by an increasingly complex fabric of networks, which appear to be the natural way to govern our complex, globalised world. Although the forms of government cooperation are very different in size, scope, and purpose, they are also characterised by a number of common features:

- Networks are based on informal arrangements instead of legally binding agreements.
- Cooperation in networks is based on trust and not on enforceable obligations.
- Cooperation in networks is voluntary in its nature.
- Networks are open to allow other partners/actors to join.
- The partners in a network bring different resources and assets to the table.
- Networks are loosely structured.
- Networks evolve over time.

The spontaneous creation of new governance mechanisms such as networks, alliances, and partnerships leads to situations which are increasingly difficult to manage. The difficulties in controlling policy approaches that are not based on formal intergovernmental cooperation relate more generally to the question of legitimacy and accountability in international governance. These concerns have been fuelled by the expanding influence of non-state actors and their participation in decision-making processes, and they are likely to grow if more authority continues to shift from the national to the international level and from states to non-state actors. If networks which have been created outside of treaty-based international diplomacy become a permanent part of the system of international environmental governance, rather than a passing fashion, the legitimacy of their place in the broader picture of governance needs to be addressed. Whereas the efficiency and the outcomes of networks may confer a certain legitimacy, mechanisms that ensure accountability should be put in place. Transparency and accountability enhance the legitimacy of policy bodies, whether these bodies represent electorates or special public interests. As a minimum mechanism of accountability, citizens around the world should, therefore, have the right and the opportunity to review the actors, the processes, and the results of such networks.

---

[14] Coleman (1988) argued, however, that social norms and shared values are also required in order to obtain network cohesion.

ROLES AND FUNCTIONS OF NETWORKS

The promise of networks lies in two central domains. First, through their ability to formulate quick responses to urgent problems, networks offer the opportunity to close the operational gap that characterises international, and especially international environmental, policy today. Second, through their non-hierarchical structure and their ability to involve non-state actors, networks promise to bridge the participation gap that is often the main reason behind international political deadlocks.[15]

Since networks are primarily characterised by their informality and flexibility, a consistent pattern of network-building under specific circumstances and conditions has yet to be observed. However, it is possible to highlight different functions that networks perform, even though no simple typology can do justice to the full range of network activities. Networks may perform one or several of the following functions:

- Strengthening of negotiation power. Networks bring together different actors with similar interests and thus increase the leverage of the arguments put forward.
- Coordinating policy approaches. Governments choose to cooperate in networks to coordinate policy responses with regard to specific issues, thereby increasing the effectiveness of responses.
- Bolstering institutional effectiveness. Networks can facilitate the building and effectiveness of institutions and broaden their constituency base.
- Implementing policies and agreements. Networks are also formed with the specific purpose of translating the results of intergovernmental negotiations into concrete activities and improving the willingness and capacity for compliance of different stakeholders.
- Generating and disseminating knowledge. Networks can serve as tools for gathering existing knowledge in a fast and efficient manner and can even generate new knowledge where gaps are identified.

The success of networks depends to a large degree on the common goal defined by the network and shared among its participants. It is only as long as the individual participants perceive that the benefits outweigh the costs of the cooperation that they will cooperate in achieving the common goal. In fact, as has been amply demonstrated by various empirical studies, success or failure of partnerships is, to a large degree, contingent on the existence of trust among partners, the level of transparency, and the way partnership initiatives deal with power asymmetries – all of which depend on the effective

[15] Streck, 2002a.

application of a minimum set of rules.[16] This does not mean, however, that networks must consist of partners which fulfil equal roles or have equal statuses. On the contrary, individual and institutional leadership as well as the different resources of participants are crucial for cooperation in networks. In sum, many networks only add value if they generate benefits – both for individual participants and for networks as a whole – that go beyond the sum of their parts.

NETWORKS BY, OF, AND WITH GOVERNMENTS

Networks of Governments

International diplomacy is the formalised version of governmental networking. Cooperation between states based on multilateral treaties and the establishment of international organisations has been the prevailing form of international policy in the 20th century. During the last century, states have been challenged to an increasing extent by global and transboundary problems. They have responded to this challenge by creating intergovernmental organisations. Whereas it is true that the growing influence of inter- or supranational organisations, such as the World Trade Organization (WTO) or the European Community, threaten national sovereignty and challenge state power, international organisations themselves have only limited power.

In order to consult and coordinate politics, governments have created groups outside of institutions and international organisations. The most influential and powerful group of countries is represented in the Group of Eight (G8), formerly the G7. The G7 goes back to an invitation of the then President of France Valéry Giscard d'Estaing to an informal gathering at the chateau of Rambouillet, near Paris in 1975. The idea was to discuss world issues of the day, at the time an agenda dominated by the oil crisis. Since then, a group of seven countries[17] has met for annual consultations. Today, the G8 has a assumed a wider spectrum of roles, reaching beyond the initial focus of economic problems. The G8 does not have a permanent secretariat or staff. Rather, it works like a club of leading industrialised countries, regularly meeting and consulting to enhance their friendship and synchronise their points of view as regards the major international economic and political issues. The organisation of the meeting is in the hands of 'sherpas', who are

[16] Nelson, 2001; Nelson and Zadek, 1999; Reinicke et al., 2000.

[17] Valéry Giscard d'Estaing invited the leaders of the United States (US), Japan, Germany, the United Kingdom (UK), and Italy. Canada joined the following year. Russia formally joined the group, which then became the G8, in 1998.

the G8 leaders' personal representatives. The sherpas also oversee the implementation of commitments made at summits. The umbrella of the G8 not only hosts summits, but also extends to working groups between different ministries. The Environment Ministers Working Group, created in 1992 and institutionalised in 1994, was the first of a generation of G8 ministerial bodies. Since then, ten other working groups have been created and deal with issues that were once enshrouded in national sovereignty.[18] Additionally, the G8 has launched a multitude of special initiatives, such as the 'heavily indebted poor countries' debt-reduction initiative, the Kananaskis G8 Africa Action Plan, and the Evian G8 Clean Water Initiative and Fund. Ministerial meetings help to ensure coordinated approaches to common and/or international problems. In recent years, the G8 has also opened its meetings to representatives from developing countries.

Also within the United Nations (UN) system, governments have established negotiations networks to enhance their position and make their voice heard. 'The outstanding fact about the way States associate in the General Assembly is the tendency of Member States to affiliate differently for different purposes.'[19] The oldest group of countries that decided to establish a permanent discussion forum is the Group of 77 (G77), which was created following the Joint Declaration of 77 developing countries issued at the end of the UN Conference on Trade and Development in 1964.[20] It represents the common interests of its members and enhances the negotiating capacity on all major international issues debated in the UN system. The G77 has also been used as a negotiations framework for environmental and other issues. It has, for example, played a crucial role in the negotiations of the UN Conference on Environment and Development (UNCED).

Today, we not only have the G77 and G8 but also different groups that consult on particular issues. The G77 and G8 have been complemented by the G10,[21] G24,[22] and G15.[23] An interesting recent addition to this list is the Group of 20 (G20), a forum of Finance Ministers and Central Bank

[18] These issues include employment, information technology, terrorism, crime, energy, labour, health, and development.

[19] Bailey, 1960: 28.

[20] Today, the G77 is made up of 133 emerging and developing countries, but its original title still stands due to its historical significance.

[21] The G10 consists of the Finance Ministers and Central Bank Governors of the G7, Belgium, Switzerland, the Netherlands, and Sweden, adding up to 11 countries, but the original name still stands.

[22] The G24 consists of a sub-group of the G77, which was established in 1971 to coordinate the position of the developing countries on issues related to the international monetary and finance system: Algeria, Argentina, Brazil, Colombia, the Democratic Republic of Congo, Egypt, Ethiopia, Gabon, Ghana, Guatemala, India, Iran, Ivory Coast, Lebanon, Mexico, Nigeria, Pakistan, Peru, the Philippines, South Africa, Sri Lanka, Syria, Trinidad and Tobago, and Venezuela.

[23] The G15 is the group that represents the G77 in the Bretton Woods organisations.

Governors which was created at the September 25 meeting of the G7 in 1999.[24] It was created as 'a new mechanism for informal dialogue on key economic and financial policy issues among systemically significant economies and to promote cooperation to achieve stable and sustainable world growth that benefits all'.[25] The newly formed group serves as an adequate forum for dialogue on the core financial issues, cooperates in the framework of the Bretton Woods institutions, and conducts consultations in integrating the work of its members. The G20 liaises closely with the G7 Finance Ministers' Meeting. Critics have, therefore, claimed that the G20 was created to legitimate G7 initiatives to the wider world by securing a broader consensus for G7 ideas.[26] However, the breakdown of the WTO trade talks in Cancún, Mexico, in September 2003, has made it clear that combining the voice of leading developing countries has made them far more influential than the industrialised powers realised when the group submitted its first joint paper.[27]

All of these groups are neither official institutions nor international organisations. They do not represent a legal entity. Due to its membership, the G8 is the most influential of these groups. Whereas critics oppose the G8 as a forum of special-interest politics of the rich, supporters of the G8 argue that the coordination of the most powerful nations in the world reflects the special responsibilities of these countries and contributes to the smooth running of the more formalised cooperation in international organisations. The G8 has set itself the goal of providing the essential coordination needed between countries whose economic and political weight makes them inevitable players in global governance. Decisions taken by this small group of countries bypass UN procedures. However, since the G8 has no implementation capacity, it puts its initiatives forward as a part of the activities conducted by the existing international organisations. Once these initiatives have been formally integrated in the work programme of these institutions, the G8 can no longer dictate the agenda. Furthermore, there is a realisation in international civil society that the significance of G8 Summits has increased. Therefore, NGOs have influenced the Summits with both concrete proposals (including the Jubilee 2000 campaigning, which was the

---

[24] The G20 consists of the G8 plus Argentina, Australia, Brazil, China, India, Indonesia, Mexico, Saudi Arabia, South Africa, South Korea, Turkey, as well as the country holding the presidency of the European Union (EU). The European Central Bank, the Managing Director of the International Monetary Fund (IMF), the Chairperson of International Monetary and Financial Committee of the IMF, the President of the WB, and the Chairperson of the Development Committee of the WB and the IMF also sit on the G20.

[25] G7, 1999.

[26] Kirton, 1999.

[27] For submissions of the G20, see ICTSD, 2003a, 2003b.

main driving force behind the HIPC initiative) and fierce protest (which culminated in the violence in Genoa, in 2001).

The G8 also pursues more ambitious goals. Since its inception, the G8 has moved from being a consultation club to becoming an ambitious group of countries which aim to regulate globalisation. This project would involve the design and establishment of a new generation of institutions which would complement, or even replace, the post-war international structure led by the UN. The system proposed would build on 'a new generation of inclusive, multistakeholder plurilateral and multilateral institutions to govern globalisation's critical areas, including the environment and energy'.[28] Critics of the UN system, which is based on the principle that every country's vote has the same weight, claim that a group of countries that represents about 80 per cent of the world's gross domestic product (GDP) would find itself in a better position to respond to the challenges of globalisation and changes in the global community in the post-cold-war era. Irrespective of whether this is desirable, or whether the G8 will succeed with this ambitious project, the project itself illustrates that states are far from giving up their regulative power and leaving it to private players to regulate globalisation.

The formation of interest groups can be observed in all negotiation processes. In conference diplomacy, in fact, groups and coalitions have become an essential feature. States sharing common interests form bargaining groups in order to organise themselves to maintain their negotiating positions. Examples of such groups in the context of the UN Framework Convention on Climate Change (UNFCCC) are the Alliance of Small Island States (AOSIS) and the Umbrella Group, a coalition of non-EU industrialised countries.[29] Without these consultation networks, international agreements would be impossible to forge. They serve to pool resources, focus issues and interests, reduce complexity, and make information and communication more manageable. Negotiation networks can be institutionalised, such as the EU and the Organization of Petroleum Exporting Countries (OPEC). Alternatively, they can be issue-specific, like the Umbrella Group. In general, more homogeneous and cohesive networks tend to perform more effectively than less homogeneous and cohesive groups. Also, negotiating networks tend to become more effective and cooperative over time, at least if they have a history of working together successfully.

An example of an institutionalised network is the Global Environment Facility (GEF), which has attempted to operationalise a unique and integrative governing structure combined with a structural flexibility that has

---

[28] Kirton, 2003: 1.

[29] The Umbrella Group consists of Japan, the US, Canada, Australia, Norway, New Zealand, Russia, and Ukraine.

a profound ability to adapt to changes.[30] The GEF answers new challenges of international public policy with a new type of international institution which bridges the traditions of the UN and the Bretton Woods agencies.

Networks between governments exist in all forms and varieties. Sometimes, they have resulted in the establishment of organisations and institutionalised forms of cooperation. In other instances, they do not rely on formalised processes but on a limited group of members pursuing common goals. Finally, they may exist only temporarily in special contexts. Governments can also seek less formal channels of communication and rely on experts or agency cooperation. An example of a network which consists of a specialised forum of cooperation is the G20, made up of Finance Ministers and Central Bank Governors. In the past few decades, we have witnessed the advent of an increasing number of these specialist, transgovernmental networks.

Transgovernmental Networks

A series of recent publications describes a shift in powers within governments, away from formal forums of cooperation towards cooperation between agencies, departments, and civil servants of different governments.[31] Slaughter analysed how states are disaggregating into their component institutions to form hydra-headed entities, represented and governed by multiple institutions in complex interaction with one another abroad as well as at home. These multi-faced states are represented through their components (agencies, departments, and individuals), all of which together form the mosaic of the governing sovereigns. While states have always interacted with their citizens through their different branches in the fulfilment of public functions, those same states have traditionally interacted with outsiders as single sovereign entities. However, economic liberalisation and the means of the information revolution invited not only private actors but also public officials to build interest groups and alliances to mutual benefit. The chosen vehicles of cooperation are often loosely structured, peer-to-peer networks, developed through frequent interaction rather than formal negotiation.[32] These alliances are commonly called transgovernmental networks, because they involve specialised domestic officials, with or without minimal involvement of the official lines of diplomacy.[33]

---

30 The Global Environment Facility can be interpreted as a formalised intergovernmental network (Streck, 2002b).

31 Raustiala, 2002; Slaughter, 2000, 2001.

32 Raustiala, 2002.

33 The concept of transgovernmentalism builds on, among other things, the pioneering work by Keohane and Nye (1974).

Compared with the often cumbersome and formal international negotiating procedure, transgovernmental networks pave the way to a more cost-efficient and flexible form of cooperation between public-sector representatives. In most cases, transgovernmental networks are built on soft law and soft power. Instead of treaties, they define the scope of their cooperation in non-binding Memoranda of Understanding.[34] These networks are based on voluntary forms of cooperation where all actors and partners decide to work together for mutual benefit. Guidelines or recommendations developed by the networks will only be implemented to the extent that they fit the specific circumstances of the countries. In cases where transgovernmental networks are formed by the same officials who make and implement regulations domestically, these officials simply extend their normal domestic functions to transgovernmental activities. The effectiveness in implementing consensus increases through a direct involvement of the executive powers. To the same extent, however, democratic accountability diminishes through the increasing distance from parliaments and ratification procedures.[35]

In parallel, the growth of legislative networks of parliamentarians suggests that public institutions with a more direct representative mandate are also participating in these new forms of governance. These networks establish links between those individuals that directly ensure democratic accountability. In this sense, promoting contact between communities and local or regional governments can help address the democratic deficit being observed at the supranational and international levels, as it allows democratic input through government institutions closest to the electorate.

Transgovernmental networks can be based on coalitions between thematic, regional or sectoral partners. They can also differ with respect to duration, membership, function, and scope.[36] Duration indicates whether networks are ad-hoc creations or built to pursue longer term interests. Rules on membership, if there are any, determine how open networks are. The function of networks refers to the outcomes that members expect from the operating networks. Networks can be based on common interests to lobby for or against certain issues, or they can be triggered by longer-term interests in consulting or exchanging ideas. The scope of networks indicates whether networks aim to pursue single issues or whether they are based on multiple issues.

---

[34] Slaughter, 2000.

[35] If the governments and legislators that bind sovereign states through the process of signing and ratifying legal instruments are put in place through democratic elections and legitimatised through transparent and democratic processes, treaty law can also provide democratic legitimacy on the international level.

[36] Ward and Williams, 1997.

Networks between agencies tend to emerge around issues that demand central regulation, such as banking or insurance supervision. In this context, the Basel Committee on Banking Supervision, the International Organization of Securities Commissions, and the International Association of Insurance Supervisors have been quoted as examples of transgovernmental networks with a different degree of formalisation. Additionally, there are increasing numbers of networks of legislature members in charge of key committees that oversee domestic regulatory agencies with the potential to improve accountability and legitimacy at the global level.[37]

To a lesser extent, transgovernmental networks have also emerged in fields with a more diffused regulatory power, such as the environment. The global environment tends to be regulated by a multitude of treaties rather than cooperation between oversight agencies; as treaties are still the core approach taken in environmental rule-making, the main focus of transgovernmental networks in the environmental field is to enhance the capacity of governments to implement and enforce environmental regulation.[38] Examples of transgovernmental environmental networks include the International Network for Environmental Compliance and Enforcement, and the Global Legislators Organization for a Balanced Environment.[39]

However, transgovernmental networks have also given rise to concerns and criticism, and are decidedly controversial.[40] Critics charge that networks reduce transparency and political accountability. They can provide states with a way of escaping or circumventing undesirable aspects of cooperation within the framework of treaty law and international organisations. They may fuel the fears that their members are engaging in politics of insulation from the international community. Because networks choose their participants, they often reinforce the dominance of the powerful. Some networks promote the export of a specific regulative system[41] and help to determine areas of influence of powerful partners. Through such networks, powerful states can exercise 'soft' power, bypassing the traditional safeguards and procedures built into the processes of international negotiations and laws. In this fashion, they may be used to penetrate the traditional defenses of sovereign states, imposing the will of the more powerful states on the weaker members of the international community. In general, national officials do not want to compromise their own national systems. If they are powerful enough, they may choose the soft and persuasive route in order to convince other, weaker, partners of the virtues of their systems. Through the export of their legal and

---

[37] Slaughter, 2000.
[38] Raustiala, 2002.
[39] GLOBE, 2003; INECE, 2003.
[40] Alston, 1997; Howse, 2000.
[41] Raustiala, 2002.

economic systems, industrialised countries continue to exercise transnational pressure on post-colonial societies. In cases where transnational networks are dominated by a few powerful nations, there is a danger that the variable cultural, economic, and political circumstances of countries and communities will be neglected.

Governmental networks can also be used to bypass central functions, such as national governments. Legal or regional authorities can, for example, bypass national decision-makers in their search for adequate solutions to certain problems. In some cases, national governments are seen as hostile or unsympathetic towards the interests of subnational entities, such as states or provinces. In these instances, international organisations may be better allies in promoting certain policies than national governments.[42]

Despite the potential to exclude national governments from the process of decision-making, the unbundling of states and the reconnection of their parts across national borders generally create a conceptual reconfiguration of state power that retains states as the pivotal actors in the international system.[43] Central to the success of networks are transparency and an examination of the plurality and complexities of the different partners that are brought together.[44] If such analysis takes place and networks are planned with care in order to allow for the most effective use of all resources, by giving voice to all partners, networks can provide 'the terrain for elaborating strategies of selfhood – singular or communal – that initiate new signs of identity and innovative sites of collaboration and contestation, in the act of defining the idea of society itself'.[45] This is valid not only for networks between government agencies, but also where the public sector opens its files for a broader form of cooperation with private actors.

Governments in Public-Private Networks

Sovereign states are entrusted with military and police power – they collect taxes, ensure that democracy and fundamental rights are protected, and build social-safety nets. However, ensuring welfare, security, health, or a clean environment has become increasingly difficult for state actors to accomplish alone. In order to address complex problems (such as the management of transboundary pollution,[46] the management of the global commons and the

[42] Ward and Williams, 1997.
[43] Raustiala, 2002.
[44] For an analysis of environmental law in postcolonial societies, see Richardson (2000).
[45] Bhabha, 1994: 1-2.
[46] Examples are the control of chemicals found in the Basel Convention on the Control of Transboundary Movements of Hazardous Wastes and their Disposal (1673 U.N.T.S. 57) and the Stockholm Convention on Persistent Organic Pollutants (U.N.Doc.UNEP/POPS/CONF/4, not yet in force).

provision of global public goods,[47] and ubiquitous environmental problems with worldwide implications),[48] international alliances need to be forged, involving not only intergovernmental organisations but also civil society and business representatives. Modern governance requires the participation of all four major players on the international scene: states, international organisations, business, and civil society. Whereas each of these sectors has an important role to play in international politics, none of them is a sole dominant power.[49] Sustainable solutions to complex international problems imply a broad consensus from both state and non-state actors.[50] Since governments have accepted an enhanced role of private actors in the formation of regimes, the result has been a growing set of hybrid regimes that have the active participation of both state and non-state actors.[51]

Over the past decade, networks that involve not only the public but also the private sector have grown in number, organisational variety, and scope. This development is particularly obvious in the field of international environmental politics, an area characterised by a multitude of decentralised functions and structures embedded in a complicated system of treaties, administrative structures, and implementation mechanisms. However, despite a proliferation of treaties and secretariats, agencies, and institutions around the globe, the architecture of international environmental governance has not lived up to its task, and the state of the global environment has not improved. It is in this context that the emergence of networks as new governance structures needs to be analysed.

At its last meeting before the WSSD, the UN General Assembly encouraged 'global commitment and partnerships, especially between Governments of the North and the South, on the one hand, and between Government and major groups on the other'.[52] These partnerships became known, in UN jargon, as 'Type II outcomes' and were described as 'specific commitments by various partners intended to contribute to and reinforce the implementation of the outcomes of the intergovernmental negotiations of the WSSD (Programme of Action[53] and Political Declaration)[54] and to help

[47] Global agreements include protection of the climate and the atmosphere, articulated in the UNFCCC, the Kyoto Protocol to the UNFCCC (Doc. FCCC/CP/1997/L.7/Add.1, not yet in force), the Montreal Protocol on Substances That Deplete the Ozone Layer (1513 U.N.T.S. 293), and the UN Convention on the Law of the Sea (1833 U.N.T.S. 41).

[48] A prominent example is biodiversity in general and tropical forests in particular, such as found in the Convention on Biological Diversity (1769 U.N.T.S. 79).

[49] Kondo, 2003.

[50] For the classification of the different environmental problems, see Esty and Ivanova, 2002; Haas, 1991.

[51] Clapp, 1998.

[52] UN General Assembly, 2001.

[53] The Programme of Action later became the Johannesburg Plan of Implementation (WSSD, 2000).

achieve the further implementation of Agenda 21 and the Millennium Development Goals'.[55] The recognition of partnerships as official Summit results reflects the transition from pure intergovernmental conference diplomacy to a more inclusive notion of international environmental governance.

Partnerships presented at the WSSD fall under the broader umbrella of global public policy networks or multi-sectoral partnerships. Both terms have often been used interchangeably, and, in general, both terms refer to the voluntary collaborations between actors from different sectors (public as well as private), based on the more efficient allocation of complementary resources among the different partners.[56] However, whereas the term network emphasises the open, informal, and flexible structure of these alliances,[57] the term partnership stresses a more proactive, problem-oriented approach. Such alliances have appeared on both the national and regional levels, but they are of special importance at the international level, where a constant need for policy solutions and the lack of a central government have left room for invention and innovation. In many cases, multi-sectoral initiatives have developed in response to the failure of traditional state-centred governance to provide solutions to complex problems with international dimensions.

Global public policy networks are aimed at minimising hierarchy through the involvement of multiple stakeholders across many sectors. The network participants bring complementary resources to the process, allowing for synergies and more effective responses.

> A typical network (if there is such a thing) combines the voluntary energy and legitimacy of the civil-society sector with the financial muscle and interest of businesses and the enforcement and rule-making power and coordination and capacity-building skills of states and international organisations. Networks create bridges that enable these various participants to exploit the synergies between these resources. They allow for the pooling of know-how and the exchange of experience. Spanning socioeconomic, political, and cultural gaps, networks manage relationships that might otherwise degenerate into counterproductive confrontation.[58]

The ideally trisectoral global public policy networks are characterised by collaboration between government, civil society, and the for-profit private sector. In the model case, they are inclusive towards the South and the North, and integrate international, regional, national, and local actors.

---

[54] The traditional intergovernmental negotiated results of the WSSD are Type I outcomes.
[55] Kara and Quarless, 2002: 1.
[56] Mitchell et al., 2001; Reinicke et al., 2000; Wolf, 2001.
[57] Reinicke et al., 2001.
[58] Reinicke et al., 2000: 24.

In many cases, existing policy networks have emerged in the shadow of traditional structures and began as social and organisational experiments. Networks are most likely to emerge in situations of political deadlock. They can help to put issues on the international agenda and then kick off a discourse in which to debate that agenda. They include actors of different sectors and are typically organised in informal or loosely structured frameworks that allow the networks to learn and adapt to changing circumstances. Flexibility is crucial for the success of policy networks. The network structure is prone to fulfil different functions, such as facilitating international processes, structuring politically contentious multi-stakeholder relationships, setting global standards, disseminating knowledge, and addressing participatory shortcomings. In doing so, they also bridge the operational and institutional gaps, two main weaknesses of the international environmental architecture.

In the past decade, governments have increasingly collaborated with private entities, in different forms of partnerships and networks. Creating networks is one means of involving non-state actors in addressing environmental problems. They demand commitment from all actors involved and give affected stakeholders an active role in promoting the success of treaties or political targets beyond the process of stakeholder consultations or lobbying for specific outcomes. In order to explore this role, governments need to develop a clearer picture as to which functions should be assumed by networks as compared to functions that need to be coordinated by intergovernmental processes. In the process of stock-taking, sovereignty, efficiency, accountability, and flexibility need to be balanced. Networks can only be seen as legitimate, more flexible, and efficient mechanisms if they do not prejudice principles or rights established under international or national law. Where a shared understanding between actors as to the different roles evolves, the traditional international process can be supplemented by an increased involvement of networks and partnerships in the process of international decision- and rule-making – as foreseen in the process that led to the WSSD – in translating these decisions into concrete action. In a process that gives appropriate room to both the public and the private sectors, the following complementarities should be explored:

1. Networks can help address an implementation deficit on the national, regional, and international levels. Alliances can be formed with the specific purpose of translating the results of intergovernmental negotiations into concrete action and improving the willingness and capacity for compliance of different stakeholders. On an international level, such implementation networks can be forged around mechanisms

foreseen in treaties,[59] emerge spontaneously to overcome deficits in the implementation as orchestrated by governments,[60] or they can take the lead in areas where governments have failed to reach agreement for coordinated action.[61]

2. Along the lines of the principle of subsidiarity, networks can help governments to address problems with international implications at the appropriate level. The idea that the responsibility for certain tasks should rest primarily with the levels of society and/or governance which are nearest to individuals and are best equipped to render specific results or certain services, applies here.[62] Networks respond effectively to the need for delegating policy processes to the governance levels that can most effectively formulate and implement policy solutions. Participating in networks can help focus concern on the legitimate roles and functions of the respective levels of governments.[63]
3. Networks can help governments in organising the exchange of information and in structuring consultation processes. Whether the exchange of views and opinions forms part of a formal process or whether it constitutes an informal process, networks are open to new actors and offer policy mechanisms adaptable to constantly changing circumstances. Different approaches of policy-making and varying cultural perspectives increasingly demand recognition and integration. Networks provide a vehicle for incorporating such diverse perspectives, including local knowledge, and involving affected communities in the problem-solving processes.
4. Another role for policy networks is the development of guidelines or standards which complement sustainable-development objectives as included in negotiated intergovernmental instruments. Where conferences of parties and other fora of international negotiations are not efficient enough to formulate quick policy responses on urgent issues, governments can convene 'networks of the willing' and formulate policy responses which may provide guidance on how to implement policy principles.

---

59 For instance, the Joint Implementation and the Clean Development Mechanism as defined under Articles 6 and 12 of the Kyoto Protocol.

60 Governments and private entities that want to advance implementation through voluntary action are likely to structure their cooperation around networks. The Prototype Carbon Fund was created by the International Bank for Reconstruction and Development with the aim of its acting as a catalyst for private- and public-sector investment in the Kyoto Protocol's flexible mechanisms. See IBRD, 2003.

61 The Forest Stewardship Council is one means of promoting the sustainable management of forests, created by private forces in response to the failure of the international community to set in place an effective system of forest protection.

62 See also Principle 10 of the Rio Principles, which states that environmental issues are best handled at the appropriate level and with the participation of all concerned citizens.

63 Ward and Williams, 1997.

However, despite all the benefits, network approaches to governance alone will never be a substitute for binding international commitments by governments. Nor would such a substitution be desirable. The success of policy networks depends to a significant degree on the willingness of governments to set ambitious binding targets. The legal and political frameworks create the nurturing context in which partnerships can develop. Networks can be one means through which such targets are effectuated. Within the framework of international politics and law, networks can complement conventions and protocols at different stages in the policy cycle. They can help to forge scientific consensus or start debates that eventually bring governments to the table to discuss specific treaties. They can also help to overcome difficulties in implementing treaties, and can aid evaluating and monitoring the success in their implementation.

Examples of mechanisms that foster the creation of implementation networks are the Clean Development Mechanism (CDM) and Joint Implementation (JI) – the project-based mechanisms of the Kyoto Protocol. With the establishment of these mechanisms through Articles 6 and 12 of the Kyoto Protocol, the parties to the UNFCCC established a platform that allows public-private networks to develop, execute, finance, and supervise projects. CDM and JI are designed to scale up cooperative climate-protection projects in fields such as renewable energy, waste management, and carbon sequestration. Both JI and CDM not only define a new method of cooperation between developed countries and developing countries or countries with economies in transition, but also offer new venues for the private sector and civil society to participate in such projects. The different stages of the project cycle involve a broad range of actors from developed and developing countries as well as from international development and finance institutions. The design of these new institutional mechanisms allows for the emergence of international implementation networks.[64]

Another example is the WCD, which served as a negotiation and consensus-building network. Large dams bring together many of the issues central to conflicts over sustainable development at the local, national, and international levels. The WCD demonstrates the potential of multi-sectoral networks to contribute to international consensus-building and standard-setting.[65]

However, different types of partnerships also have different implications for concerns about legitimacy, accountability, transparency, and power asymmetries, which need to be addressed in each case. Transnational networks operate, at least in part, beyond the reach of the specific

---

[64] Streck, 2002c.

[65] Dingwerth, 2003; Dubash et al., 2001.

governments and individuals whom they most affect. Control mechanisms have to be created to monitor and evaluate the implementation and execution of the network objectives in accordance with the different responsibilities that the actors participating in partnerships have assumed.

HOW TO MANAGE NETWORK PARTICIPATION

Internationally, structures that monitor networks and mechanisms that help to hold networks and network participants accountable have to be put in place. Transparency of networks, including their partners, financing, and goals, is a basic condition of such accountability. Recognising that it is necessary to maintain flexibility and openness with regard to the types of rules that have to be developed, three sets of issues figure prominently on the management agenda: accountability (as an instrument for addressing concerns about legitimacy), monitoring and evaluation (as instruments for addressing concerns about legitimacy and fostering compliance), and capacity building (as a mechanism for overcoming power asymmetries).[66]

Accountability has been defined as 'the obligation to present an account of and answer for the execution of responsibilities to those who entrusted those responsibilities',[67] which in essence requires the possibility of holding 'individuals and organisations responsible for performance'.[68] And, while accountability is no substitute for truly representative democracy, it still can contribute to the democratisation of the policy-making process. In the international context, accountability cannot rely on the command-and-control concept applicable in the national context. Instead, there is a need for 'more imagination in conceptualising, and more emphasis on operationalising, different types of accountability'.[69] In the absence of a global political structure that could facilitate controls and institute checks and balances for global environmental governance, national governments and international organisations should advocate a pluralistic system of accountability. The basis of this system would be the natural checks and balances provided by the participation of diverse actors and incentive mechanisms designed to generate compliance with a broad set of rules. Mechanisms of control should include professional/peer accountability, public reputational accountability, and market accountability – where participants in global governance are also market participants. Prerequisite for all these forms of accountability is the transparency of networks and their objectives.

[66] Witte et al., 2002.
[67] Gray, 1998: 22.
[68] Paul, 1992: 1047.
[69] Keohane and Nye, 2001: 8.

In a next step, monitoring and evaluation mechanisms need to be put into place to endow the legally non-binding rules of partnerships with sufficient strength to accomplish their mandates. It is in this area that governments and international organisations will have to make their greatest effort to ensure the viability of the partnership approach to sustainable development.

Monitoring and evaluation are critical for a number of reasons. Both, if properly managed, facilitate learning from experience, which is a crucial precondition for future improvements of partnership processes and outcomes. Evaluation, in particular, is a crucial device for analysing the costs and benefits of networks and determining whether they accomplish their objectives. Many observers have questioned whether crucial resources such as time, money, and personnel should be directed towards governance mechanisms that do not promise hard and fast results. Proper evaluation is needed to assess whether networks are a fitting and necessary governance mechanism, or whether they simply waste resources. Monitoring and evaluation also help to improve the transparency of network proceedings, and are, therefore, the most important – if not the only – instrument for outsiders to arrive at informed judgments on the legitimacy, effectiveness, and efficiency of specific partnerships. Finally, monitoring and evaluation help to identify 'free-rider' and 'rent-seeking' behaviour within partnerships.[70]

A fair and transparent process of network coordination recognises the obstacles posed by varying degrees of institutional and financial capacity.[71] As one observer notes, 'partnerships are 'nested' within local, national and international policy frameworks that either enable equitable conditions for partnerships, or exacerbate power asymmetries'.[72] As trust is the glue that holds networks together, they need to ensure that the power asymmetries are bridged. Such asymmetries exist between the different sectors as well as between different representatives of one sector. Civil-society representatives often do not have the resources to engage in more than a few partnerships and, therefore, cannot afford failure as easily as their private-sector counterparts can.

NGOs participating in policy networks generally have more to lose than business representatives. It is usually their organised opposition towards certain practices which give advocacy groups their strength, and NGOs risk losing some of their credibility (towards their own membership) as well as their edge (towards the groups whose practices they oppose). They risk compromising themselves, their members, and their objectives.

---

70 Witte et al., 2002.
71 Steiner, 2002; Witte et al., 2002.
72 Weitzner, 2002.

However, power asymmetries also exist between representatives of the same sector. The NGO system is far from representative of a 'global civil society', despite such claims by NGOs.[73] Instead, Northern and Western NGOs dominate the international NGO fora, and some Southern observers have characterised relations between Northern and Southern NGOs as 'emerging colonialism' in which 'Third World NGOs have had to suit their agendas to the agendas of Northern NGOs'.[74]

There are three basic strategies for addressing power asymmetries in partnerships and networks. First, actors can be empowered to participate effectively and make their voices heard. This can be accomplished through capacity building and resource endowment. Second, rules can be set to ensure that those who do not have access to financial or other resources are not disadvantaged in the partnership process. At both ends, governments and international organisations can make important first steps. Third, governments need to act as arbiters in conflicts that may arise between private-sector and NGO representatives.[75]

THE ROLE OF GOVERNMENTS IN GLOBAL NETWORKS

The track record of the system of international environmental governance is poor. Numerous international agreements have been concluded over the past few decades, and institutions and secretariats have been established. Yet, environmental quality on a global scale has deteriorated. Legal and institutional arrangements for environmental protection have not lived up to their task. Despite the plethora of treaties, agreements, and an expanding array of international agencies, the evidence suggests a continuing decline of biodiversity, increasing global warming, depletion of the world's forests, and widespread chemical pollution. The time is ripe to complement the traditional governance system with innovative elements, bringing together governments, public agencies, private businesses, and the not-for-profit sector.

Networks provide governments with opportunities. They promise to provide governments with a tool to react flexibly, efficiently, and swiftly to the challenges of a globalised world. Networks equip public actors with the ability to interact meaningfully with the different levels of the international and national constituencies. The old and new partners of national agencies, departments, and ministries include supranational, national, and regional levels as well as private for-profit and not-for-profit entities. Modern state actors exercise their power by different means and through different channels.

---

73 Martens, 1993.
74 Brown and Fox, 1998: 339.
75 SDIN, 2003.

The state and state agencies that compete with, complement, and even bridge the gaps to networks of (supranational and subnational) public and private actors open the door to a host of new ways in which state actors can address global problems.[76]

But networks not only bring different sectors together, they can also provide a vehicle for incorporating diverse perspectives, such as sharing local knowledge and involving affected communities in the problem-solving processes. Local communities have the closest physical contact with environmental issues and they are most likely to be affected by public environmental policies. Whereas the global cooperation between governments in international institutions and through the context of less formal consultation platforms can provide for a framework of action and cooperation, transgovernmental and multisectoral policies can provide room for flexible and alternative policy solutions and implementation activities. Networks, alliances, and partnerships potentially represent the most positive developments for institutionally combining local communities, organisations, and authorities at the global level. Their value lies in the exchange of information, management precedents, and advice.[77]

Yet, networks also pose risks. Networks may help to sideline elected governments and replace binding commitments by informal and vague expressions of intention. They may undermine legitimate formal processes and traditional forms of accountability. Networks cannot replace formal governmental and legislative action, which is crucial in determining the framework in which networks operate. In order to manage network risks, governments need to define general objectives, set binding targets, and define the broad lines of politics. Governments also have to ensure the accountability and transparency of networks in which they participate. They have to assess their involvement in partnerships and draw a line where direct and exclusive public action is required.

It is now time for governments to react to this new situation and develop a strategy on how to participate in different types of networks. Whereas they have been cooperating in traditional government networks for a long time, transgovernmental and public policy networks are relatively recent additions to the system of global governance. The change in governance requires a change in thinking: there must be a readiness to renounce governance pretensions and exchange these for a readiness to cooperate with other actors. By engaging with these other actors, it is subsequently possible to achieve something positive. For governments, this means abandoning the idea that

[76] Slaughter, 2000.
[77] Richardson, 2000.

they are the only safeguards of the environment.[78] These shifts in international governance will ultimately have to involve efforts to determine which organisational forms are best suited to which governance tasks. The public sector needs to delegate some aspects of public policy-making to non-state actors. Besides, national, regional, and local levels have to be part of the process wherever necessary and possible.[79] Additionally, the vertical application of the principle of subsidiarity entails that governments will have to delegate policy processes to the governance level that can most effectively formulate and implement policy solutions. Such solutions can only be found if implementation mechanisms draw on the skills and resources of a diversity of people and institutions at many levels.

Governments need to develop a clearer picture of which functions should be assumed by partnerships and which functions need to be coordinated by intergovernmental processes. Instead of creating networks on an ad hoc basis and participating in partnership structures in an opportunistic manner, governments have to start analysing their involvement in different initiatives. In this process, sovereignty, efficiency, accountability, and flexibility need to be balanced. Partnerships can only be seen as legitimate, flexible, and efficient mechanisms if they do not prejudice principles or rights established under international or national law.

In some cases, governments will choose to actively participate in networks. In other cases, they may just elect to monitor and evaluate. They may also endorse and accept regimes created by non-state actors in their own regulatory structures.[80] Networks provide a means of involving non-state actors in addressing environmental problems. Networks demand commitment from all actors involved and give affected stakeholders an active role in promoting the success of treaties or political targets beyond the process of stakeholder consultations or lobbying for specific outcomes. In order to explore this role, governments have to assess the resources that they need to put into networks (in terms of money, time, and expertise) and evaluate the network results. Based on that type of examination, they need to decide whether cooperation was successful or whether another means of governance, or another set of actors, would have accomplished more.

Networks need to be managed. They provide, on the one hand, opportunities to react flexibly to changing circumstances, but are, on the other hand, also disorganised, hard to control, and even chaotic.[81] Networks create

---

[78] Ward and Williams, 1997.

[79] Reinicke and Witte, 2000.

[80] The ISO 14000 is an example of a series of environmental-management standards developed by a public-private hybrid organisation and has been adopted as an official set of international standards (Clapp, 1998).

[81] Susan Strange described this situation as chaotic because there is no hegemony, and termed these developments 'the retreat of the state' (Strange, 1996).

the image of a menagerie of diverse and contesting policy discourses. It is the role of governments to manage this menagerie and restore confidence in governance for those who feel threatened by the symptoms of globalisation. Governments have to put in place mechanisms through which they can manage their own participation and the involvement of their different branches in networks. In order to ensure the legitimacy of networks, they need to define who is taking which decisions, what the processes of decision-making are, and how different stakeholders can participate in these processes. They have to find a proper balance between what needs to be established as legally binding and enforceable processes and obligations on the one hand, and what is open to action through partnerships in given societies on the other hand.[82]

Where networks fulfil governance roles, they should function within basic agreed-upon rules.[83] The success or failure of networks and partnerships is contingent, to a large degree, on the existence of trust among partners, the level of transparency, and the way partnership initiatives deal with power asymmetries – all of which depend on the effective application of a minimum set of rules.[84] Governments and international organisations are not the only players that have to respond to this pressing agenda. Business and civil-society organisations are equally challenged to work with the public sector to apply basic rules to their activities and to monitor and enforce good behaviour. Yet, governments and international organisations have a particular responsibility vis-à-vis their citizens, who rightly demand effective as well as transparent, accountable, and legitimate instruments of global environmental governance.

CONCLUSION AND OUTLOOK

A modern system of international governance integrates network structures in the traditional system of formal relationships between governments, and thus complements the system by integrating public and private entities in the architecture of international politics. Today, corporate and financial interests, as well as consumer and environmental groups, have not only gained a stronger voice in the negotiation of international decisions; they have also become fundamental in implementing these decisions through advocacy and specific activities. This fundamental change in the perception of governance has had a profound effect on how governments behave internationally.

---

[82] Giscard d'Estaing, 2003.
[83] Giscard d'Estaing, 2003.
[84] Nelson, 2001; Nelson and Zadek, 1999; Reinicke et al., 2000.

We have seen that states are already active partners in different sorts of networks, and that they have complemented diplomacy and intergovernmental negotiations with an array of less formal and more flexible mechanisms of governance. Thus far, this involvement has been opportunistic, driven by single actors on an ad hoc basis. In an attempt to restore confidence in governments and international processes, governments should develop a strategic approach towards their involvement in networks, which allows them to make their policies more effective.

Environmental governance needs champions, it does not need more institutions. Internationally, the UN and the WB have declared a desire to make networks and coalitions for change a central part of their strategic orientation for the future.[85] In a next step, governments and international organisations need to define a clear strategy with regard to networks and global programmes, and develop a framework for their own roles in fostering the establishment of and coordination between different networks. It is time to elaborate and implement an overall strategy and coordinate the efforts of international organisations in participating in and developing networks with regard to their comparative advantage. It would also be useful to identify a clearing-house and a centre for knowledge management which could coordinate this work and disseminate the lessons learned into networks around the world. The new UN Partnership Office that will be created as part of the UN Secretary-General's reform agenda could play an important role in this respect.[86]

The future does not lie in 'governance without governments' but in a networked governance in which governments take an active part and which is open to initiatives by and partnerships between international and national actors. Governments are crucial in ensuring that networks promote the ultimate goal of sustainable development. They have to take the necessary steps that help to ensure that networks and partnerships do not result in a disconcerted system of governance which dilutes the efforts to prioritise and synchronise international action.

Let me conclude with the words of Kofi Annan and the vision he formulated in his Millennium Report:

> If we are to get the best out of globalisation and avoid the worst, we must learn to govern better, and how to govern better together. That does not mean world government or the eclipse of nation states. On the contrary, states need to be strengthened. And they can draw strength from each other, by acting together within common institutions based on shared rules and values. These institutions must reflect the realities of the time, including the distribution of power. And they must serve as an arena for states to co-operate with non-state actors, including

[85] Annan, 1999; Wolfensohn, 1999.
[86] Annan, 2002.

global companies. In many cases they need to be complemented by less formal policy networks, which can respond more quickly to the changing global agenda.[87]

REFERENCES

Agranoff, R. and M. McGuire (2003), 'Governance networks', in J. Rabin (ed.), *Encyclopedia of Public Administration and Public Policy*, New Greenham Park: Marcel Dekker.

Alston, P. (1997), 'Myopia of the handmaidens: International lawyers and globalization', *European Journal of International Law*, **8**, 435-442.

Annan, K. (1999), 'A compact for the new century', address of United Nations Secretary-General Kofi Annan to the 1999 World Economic Forum Meetings in Davos, Switzerland, *http://www.un.org/partners/business/davos.htm#speech*.

Annan, K. (2002), *Strengthening of the United Nations: An Agenda for Further Change; Report of the Secretary-General*, United Nations General Assembly, New York, A/57/387, 9 September.

Annan, K. (2003), 'We the people, the role of the UN in the 21st century', Millennium Report of the Secretary-General of the United Nations, *http://www.un.org/millennium/sg/report/*.

Bailey, S. (1960), *The General Assembly of the United Nations*, London: Stevens.

Barfield, C. (2001), *Free Trade, Sovereignty, and Democracy: The Future of the World Trade Organization*, Washington, DC: AEI Press.

Bhabha, H. (1994), *The Location of Culture*, London: Routledge.

Brown, L. and J. Fox (1998), 'Accountability within transnational coalitions', in J. Fox and L. Brown (eds), *The Struggle for Accountability*, Cambridge: MIT Press.

Cerny, P. (1995), 'Globalization and the changing logic of collective action', *International Organization*, **49**, 595-624.

Clapp, J. (1998), 'The privatization of global environmental governance: ISO 14000 and the developing world', *Global Governance*, **4**(3), 295-316.

Coleman, J. (1988), 'Social capital in the creation of human capital', *American Journal of Sociology*, **94**(Supplement), 95-120.

Commission on Global Governance (1995), *Our Global Neighborhood: The Report of the Commission on Global Governance*, Oxford: Oxford University Press.

Dean, M. (1999), *Governmentality: Power and Rule in Modern Society*, Thousand Oaks: Sage Publications.

Dingwerth, K. (2003), *Globale Politiknetzwerke und ihre Demokratische Legitimation: Analyse der World Commission on Dams (Global Public Policy Networks and their Democratic Legitimacy: Analysis of the World Commission on Dams)*, Global Governance working paper no. 6, Potsdam, Berlin, Oldenburg: The Global Governance Project.

Dubash, N., M. Dupar, S. Kothari, and T. Lissu (2001), *A Watershed in Global Governance? An Independent Assessment of the World Commission on Dams*, Washington, DC: World Resources Institute.

Esty, D. and M. Ivanova (eds) (2002), *Global Environmental Governance: Options and Opportunities*, New Haven: Yale Center for Environmental Law and Policy.

Giscard d'Estaing, O. (2003), 'The needs and challenges of world governance', Transcript of a speech delivered at the 2003 G8 Pre-Summit Conference

[87] Annan, 2003.

'Governing Globalization: G8, Public and Corporate Governance', *http://www.library.utoronto.ca/g7/conferences/2003/insead/insead_papers/giscard_transcript.html*.

GLOBE (2003), *http://www.globeinternational.org*.

Gray, A. (1998), 'Management, accountability and public sector reform', in R. Boyle and T. McNamara (eds), *Governance and Accountability – Power and Responsibility in the Public Service*, Dublin: Institute of Public Administration.

G7 (1999), 'Statement of G7 Finance Ministers and Central Bank Governors', *http://www.library.utoronto.ca/g7/finance/fm992509state.htm*, 25 September, Washington, DC.

Haas, P. (1991), *International Environmental Issues: An ACUNS Teaching Text*, Waterloo, Ontario: Academic Council on the United Nations Systems.

Hierlmeier, J. (2002), 'UNEP: Retrospect and prospect – options for reforming the global environmental governance regime', *Georgetown International Environmental Law Review*, **14**, 767-768.

Howse, R. (2000), 'Transatlantic regulatory cooperation and the problem of democracy', in G. Bermann et al. (eds), *Introduction to Transatlantic Regulatory Cooperation: Legal Problems and Political Prospects*, Oxford: Oxford University Press.

IBRD (2003), *http://www.prototypecarbonfund.org*.

ICTSD (2003a), 'Annex A: Framework for establishing modalities in agriculture', *http://www.ictsd.com/ministerial/cancun/docs/G21_ag_text.pdf*.

ICTSD (2003b), 'Agriculture – Framework proposals', *http://www.ictsd.com/issarea/ag/resources/G-20%20official%20 text.pdf*.

INECE (2003), *http://www.inece.org*.

Jordan, G. and K. Schubert (1992), 'A preliminary ordering of policy network labels', *European Journal of Political Research*, **21**(1,2), 7-27.

Kara, J. and D. Quarless (2002), 'Explanatory note by the vice-chairs: Guiding principles for partnerships for sustainable development', PrepCom IV, Bali, *http://www.johannesburgsummit.org/html/documents/prepcom4docs/bali_documents/annex_partnership.pdf*.

Keohane, R. and J. Nye (1974), 'Transgovernmental relations and international organizations', *World Politics*, **27**(1), 39-62.

Keohane, R. and J. Nye (2001), 'Democracy, accountability and global governance', Harvard working papers, *http://www.ksg.harvard.edu/prg/nye/ggajune.pdf*.

Kirton, J. (1999), 'The G7 and China in the management of the international financial system', G7 Info Center, Toronto: University of Toronto, *http://www.library.utoronto.ca/g7/scholar/kirton199903/china5.htm*.

Kirton, J. (2003), 'Governing globalization: The G8's contribution to the twenty-first century', G8 Info Center, Toronto: University of Toronto, *http://www.g8.utoronto.ca/scholar/kirton2003/kirtonmoscow2003.pdf*.

Kondo, S. (2003), 'Governing globalization: G8, public and corporate governance', speech held at the 2003 G8 Pre-Summit Conference, *http://www.g8.utoronto.ca/conferences/2003/insead/insead_papers/kondo.html*.

Marin, B. and R. Mayntz (eds) (1991), *Policy Networks*, Frankfurt: Campus Verlag.

Martens, J. (1993), 'Dabeisein ist noch nicht alles: Die NGOs in den Vereinten Nationen: Akteure, Kritiker, Nutzniesser' ('Being present is not everything: The NGOs in the United Nations: Actors, critics, and layabouts'), *Vereinte Nationen*, **41**(5), 168-170.

Milward, H. and K. Provan (2000), 'How networks are governed', in C. Heinrich and L. Lynn (eds), *Governance and Performance*, Washington, DC: Georgetown University Press.

Mitchell, J., J. Shakleman, and M. Warner (2001), 'Measuring the "added value" of tri-sector partnerships', working paper no. 14, Business Partners for Development (BPD), Natural Resources Cluster, *http://www.bpd-naturalresources.org/media/pdf/working/wp1ba.pdf.*

Nelson, J. (2001), *Cooperation Between the United Nations and all Relevant Partners, in Particular the Private Sector*, Report of the Secretary-General A56/323.

Nelson, J. and S. Zadek (1999), *Partnership Alchemy: New Social Partnerships in Europe*, Copenhagen: The Copenhagen Centre.

Ohmae, K. (1995), *The End of the Nation State: The Use of Regional Economics*, London: Free Press.

Paul, S. (1992), 'Accountability in public services: Exit, voice and control', *World Development*, **20**(7), 1047-1060.

Peters, B. (1997), 'Can't row, shouldn't steer: What's a government to do?', *Public Policy & Administration*, **12**(2).

Raustiala, K. (2002), 'The architecture of international cooperation: Transgovernmental networks and the future of international law', *Virginia Journal of International Law*, **43**(1), 1-92.

Reinicke, W. (1998), *Global Public Policy, Governing without Government?*, Washington: Brookings Institution.

Reinicke, W. and J. Witte (2000), 'Interdependence, globalization and sovereignty: The role of non-binding international legal accords', in D. Sheldon (ed.), *Commitment and Compliance – The Role of Binding Norms in the International Legal System*, Oxford: Oxford University Press.

Reinicke, W. and F. Deng (2000), *Critical Choices: The United Nations, Networks and the Future of Global Governance*, Ottawa: IDRC.

Reinicke, W., T. Benner, and J. Witte (2001), 'Global public policy: Globalisierung gestalten durch Politiknetzwerke' ('Global public policy: Shaping globalisation through public policy networks'), in H. Oberreuter and M. Piazolo (eds), *Global Denken: Die Rolle des Staates in der internationalen Politik zwischen Kontinuität und Wandel*, München: Olzog.

Rhodes, R. (1997), *Understanding Governance: Policy Networks, Governance, Reflexivity and Accountability*, Buckingham: Open University Press.

Richardson, B. (2000), 'Environmental law in (post) colonial societies: Straddling the local-global institutional spectrum', *Colorado Journal of International Environmental Law & Policy*, **11**(1).

Sassen, S. (1996), *Losing Control? Sovereignty in an Age of Globalization*, New York: Columbia University Press.

SDIN (2003), paper no. 1, Sustainable Development Issues Network, *http://www.sdissues.net/sdin/docs/takingissue-no1.pdf.*

Slaughter, A. (2000), 'Governing the global economy through government networks', in M. Byers (ed.), *The Role of Law in International Politics: Essays in International Relations and International Law*, Oxford: Oxford University Press.

Slaughter, A. (2001), 'Globalization, accountability, and the future of administrative law: The accountability of government networks', *Indiana Journal of Global Legal Studies*, **8**, 347-356.

Steiner, A. (2002), 'Creating new platforms for dialogue', in J. Witte, C. Streck, and T. Benner (eds), *Progress or Peril – Partnerships and Networks in Global*

Environmental Governance; The Post-Johannesburg Agenda, Berlin: Global Public Policy Institute.

Strange, S. (1996), *The Retreat of the State: The Diffusion of Power in the World Economy*, Cambridge: Cambridge University Press.

Strange, S. (1997), 'Territory, state, authority and economy', in R. Cox (ed.), *The New Realism: Perspectives on Multilateralism and World Order*, Houndmills: Palgrave MacMillan.

Streck, C. (2002a), 'Global public policy networks as coalitions for change', in D. Esty and M. Ivanova (eds), *Global Environmental Governance, Options & Opportunities*, New Haven: Yale Center for Environmental Law and Policy.

Streck, C. (2002b), 'The Global Environment Facility – A role model for international governance?', *Global Environmental Politics*, **1**(2), 71-94.

Streck, C. (2002c), 'The Clean Development Mechanism: A playing field for new partnerships', in F. Biermann et al. (eds), *Proceedings of the 2001 Berlin Conference on the Human Dimension of Global Environmental Change 'Global Environmental Change and the Nation State'*.

UN General Assembly (2001), Resolution 56/226.

United Nations (2004), 'List of partnerships for sustainable development', *http://www.un.org/esa/sustdev/partnerships/list_partnerships.htm*.

Ward, S. and R. Williams (1997), 'From hierarchy to networks? Sub-central government and EU urban policy', *Journal of Common Market Studies*, **35**(3), 439-444.

Weitzner, V. (2002), 'Partnerships at the World Summit – reflections on a new buzzword', The North South Online Opinion, *http://www.nsi-ins.ca/ensi/news_views/oped45.html*.

Witte, J. et al. (2002), 'The road from Johannesburg: What future for partnerships in global environmental governance?', in J. Witte, C. Streck, and T. Benner (eds), *Progress or Peril – Partnerships and Networks in Global Environmental Governance: The Post-Johannesburg Agenda*, Berlin: Global Public Policy Institute.

Wolf, K. (2001), *Private Actors and Legitimacy of Governance Beyond the State*, paper prepared for the workshop 'Governance and Democratic Legitimacy', ECPR Joint Sessions Grenoble, TU Darmstadt.

Wolfensohn, J. (1999), *Coalitions for Change*, World Bank Group/IMF Annual Meetings Speech, 28 September, *http://www.imf.org/external/am/1999/speeches/pro2e.pdf*.

WSSD (2000), *Report of the World Summit on Sustainable Development; Plan of Implementation of the World Summit on Sustainable Development*, Articles 7, 8, 25, 31, and 40, 4 September, Doc A/CONF.199/20.

26. Globalisation and Environmental Policy Design

Konrad von Moltke

SUMMARY

Numerous forces contribute to the process of globalisation, including the dynamic of economic liberalisation and the need for environmental management. The environment is an agent of change in economic globalisation. Over the past twenty years, sustainable development has emerged as an alternative to the Washington Consensus as a guide for globalisation. Environmental policy design must be situated in the field of force defined by globalisation, as it is itself an active force in this process. The challenge is to develop international institutions that are capable of balancing the overlapping needs of economic liberalisation and environmental management. This is hindered by the structural differences between (the institutions associated with) the hierarchical system of economic policy and the subsidiarity-ruled environmental system. When designing more effective governance systems, the environmental dimension should be more fully integrated into the process of economic globalisation, especially in the critical area of investment.

INTRODUCTION

The relationship between globalisation and environmental policy design depends on how one understands the process of 'globalisation' and in particular the role of environmental issues within it. Globalisation is widely understood to be synonymous with economic liberalisation, with the World Trade Organization (WTO), the International Monetary Fund (IMF), and the World Bank (WB) at its centre. Besides, there are numerous other forces that create significant pressure for globalisation, not least among them the need to protect the environment.

The significant role that the environment plays in promoting globalisation is not widely understood, because the needs of environmental management are frequently seen as being in conflict with economic processes of globalisation. In practice, however, conflicts that exist between international economic policy and environmental management derive principally from differences in their approaches to their respective processes of globalisation. In a nutshell, economic policy is a hierarchical system based on universal principles, whereas environmental management is a system of subsidiarity based on an entirely different set of universal principles.[1]

These differences have required a painful process of mutual adjustment in both international economic policy and in environmental management systems. This process is still in progress, although certain important conclusions can by now be drawn from the experiences of the past twenty years.

ECONOMIC AND ENVIRONMENTAL GLOBALISATION

Economic Globalisation

The 'Washington Consensus', involving ten interrelated policies, is widely accepted as forming the core agenda of economic globalisation:[2]

- Fiscal discipline.
- A redirection of public expenditure priorities towards fields offering both high economic returns and the potential to improve income distribution, such as primary health care, primary education, and infrastructure.
- Tax reform (to lower marginal rates and broaden the tax base).
- Interest rate liberalisation.
- A competitive exchange rate.
- Trade liberalisation.
- Liberalisation of inflows of direct investment.
- Privatisation.
- Deregulation.
- Secure property rights.

Since this set of policies was formulated, the WTO, an almost inadvertent creation of the Uruguay Round,[3] has become the institution where the legal

[1] Subsidiarity is the principle of seeking to deal with issues at the lowest level of governance consistent with effectiveness. It is an explicit principle of the EU Treaties.

[2] Drawn from Williamson, 2000.

[3] Jackson, 2000.

framework for economic globalisation is elaborated. There appears to be a concerted effort to gather all the strands of international economic policy in the WTO and to negotiate rules that will promote their respect. This is not the place to consider whether the WTO is institutionally capable of taking on this task, but there are reasons to doubt it.[4]

The Uruguay Round lasted from 1986 until 1995, and resulted in the most important set of economic agreements since the end of World War II. Nine years of intensive negotiations brought major changes to the multilateral trade regime.[5] A dramatically strengthened dispute settlement system is arguably the most important.[6] Agriculture was for the first time brought under the umbrella of the General Agreement on Tariffs and Trade (GATT).[7] The Uruguay Round launched a major new agreement on trade in services.[8] It included a wide-ranging agreement on intellectual property rights.[9] The Uruguay Round expanded the provisions on subsidies[10] and on non-tariff barriers to trade,[11] and opened the door to negotiations on foreign direct investment.[12] Towards the end of the Uruguay Round, a decision was taken to create the WTO, drawing together the agreements that had been negotiated under the GATT and solving some institutional problems that dated back to the strange origins of the GATT itself.[13]

Environmental Globalisation

In parallel to the Uruguay Round, but largely independent of it, countries negotiated an extraordinary range of international environmental agreements, beginning with the Montreal Protocol. By the end of the Uruguay Round, new agreements existed on the transboundary transport of hazardous wastes, biological diversity, climate change, and desertification. In Europe, the Convention on Long Range Transboundary Air Pollution (LRTAP) had slowly mutated into a remarkably effective tool of environmental policy. Following the dramatic transformation of Central and Eastern Europe, a process was under way to develop common environmental rules for the continent. The United Nations Conference on Environment and Development (UNCED) had given high visibility to the idea of sustainable development,

---

4 Von Moltke, 2001.
5 WTO, 2004.
6 Understanding on Rules and Procedures Governing the Settlement of Disputes (DSU).
7 Agreement on Agriculture (AoA).
8 General Agreement on Trade in Services (GATS).
9 Agreement on Trade Related Intellectual Property Rights (TRIPS).
10 Agreement on Subsidies and Countervailing Measures (SCM).
11 Agreement on Sanitary and Phytosanitary Measures (SPS) and Agreement on Technical Barriers to Trade (TBT).
12 Agreement on Trade Related Investment Measures (TRIMS).
13 Jackson (1991) spoke of 'the flawed constitutional beginnings of GATT'.

the only widely accepted vision of the global economy that provided an alternative to the Washington Consensus. Indeed, the preamble to the WTO Agreement makes explicit mention of sustainable development as a substantive goal of the organisation. Taken together, these developments in international environmental policy represent no less important a set of economic agreements than the Uruguay Round itself. The question is how the two bodies of international law relate to each other, and what that means for environmental management.

The WTO and the structure of international environmental agreements create the framework within which national environmental policy must evolve. By now, there is virtually no field of environmental policy that is not affected by international agreements. In the European context, there is the added dimension of the European Union (EU), which has exclusive competence for external commercial policy and shared competence for environmental policy.

In the light of the extensive economic consequences of environmental regimes and of the potential environmental impacts of economic regimes, it is reasonable to consider whether these could not have been developed in a much more integrated manner. This is an objective that has eluded most countries. At the international level, it is rendered additionally difficult by the lack of an integrated governance structure. International agreements, and the regimes that arise from them, stand by themselves, and the management of overlaps – synergies as well as conflicts – is one of the most intractable problems of contemporary international governance. In addition, differences in the attribution of competences reflect the structural differences of international economic and environmental policies. Exclusive competence for commercial policy flows from the hierarchical nature of economic policy. It is largely a matter of power and efficiency. In the EU, this has led to the assignment of exclusive competence for commercial policy to the supranational level, since shared competence would reduce effectiveness. On the other hand, exclusive competence for environmental policy is unthinkable at any level of international society. It would remove critical decisions too far from the specific environmental realities that need to be taken into account. Shared competence is the institutional response to subsidiarity.

The EU needs significant institutional resources to achieve a proper balance between its economic and environmental dimensions. In particular, the balancing of policy areas with exclusive competence against those with shared competences is a demanding task, not impossible but necessitating significant institutional resources. Comparable issues exist at the broader multilateral level,[14] yet this is an environment that does not command comparable institutional resources. As a result, the relationship between economic and environmental policies goes through a cycle of conflicts.

The Environment in Multilateral Economic Policy

The environment has emerged as a major agent of change in international economic negotiations. The first area to be affected was development policy, in large measure because the WB in particular proved to be a convenient medium for environmental issues. Moreover, development assistance does not engage domestic economic interests in the same way as most other areas of international economic policy-making do, leaving more space for the accommodation of environmental concerns. This has led to two contradictory outcomes: the elaboration of the concept of sustainable development and enduring scepticism on the part of developing countries towards the environmental agenda.

The first practical encounter of many developing countries with modern environmental policy came through development assistance, where it appeared more like a conditionality imposed by donors than a matter of concern for the countries themselves. This attitude also affected the position of developing countries in most international environmental negotiations, where the view prevailed that this was a problem of developed societies. It is hard to be critical of the environmentally oriented use that was made of development policy, since this has permitted a level of interaction on environmental issues with policy-makers in developing countries that could hardly have been achieved by other means. Yet, the result continues to be deep-seated resistance to environmental issues on the part of developing countries in most economic negotiations, a resistance that will require much effort to overcome.

The formulation of the concept of sustainable development was in many ways a response to the hesitation of developing countries' governments who were confronted with environmental requirements. By 1990, most developed countries had recognised that, contrary to initial expectations, environmental protection could be achieved without sacrificing economic growth or employment. Sustainable development represents an attempt to develop a broader base of theory and practical policy prescriptions that permits a proper balancing of the priorities of environmental protection, economic growth, and more equity in international policy-making.[15] It remains the most important

---

[14] 'Multilateral' is used here in the meaning of the trade regime to denote the broadest current consensus on economic policy.

[15] First formulated systematically in the 'Brundtland Report' (WCED, 1987), the concept of sustainable development has been the subject of numerous studies. Yet, the articulation of

conceptual alternative to policies based on the limited economic concepts of the Washington Consensus. Yet, its realisation continues to challenge policy-makers at all levels.

Sustainable development is fundamentally about structural economic change. Not the size of an economy – in terms of goods and services traded – is critical, but its composition. Clearly, no current economy can be viewed as sustainable; consequently, policy (in particular, environmental policy) needs to promote structural economic change. In this respect, environmental policy is surprisingly congruent with trade policy and economic liberalisation, since these also result in structural economic change. The difference is that sustainable development applies qualitative criteria to such change, while traditional economic policy does not. This difference represents the core conflict between environmental policy and economic globalisation.

ENVIRONMENTAL POLICY DESIGN

Responses to the Environmental Challenge

The need to confront the environmental challenge has resulted in significant institutional change in the WTO. The fundamental reason why the environment has proven to be an agent of change in the trade regime is that both include a major international dimension that is an inescapable part of their agenda. Both seek to promote structural economic change. Both require international agreements to succeed. Yet, the structure of environmental agreements is dramatically different from that of economic agreements, and the search for common ground has proven elusive.

Governments are constantly confronted with policy objectives that are entirely legitimate, yet in conflict with each other. Finding acceptable solutions and building consensus around them is a central task of the public institutions of governance. A number of factors render that task particularly difficult when it comes to reconciling the goals of environment and economy. The most intractable is the fact that the environment itself does not respond directly to policy measures as the economy does. Environmental measures are determined by scientific assessment and are often presented as non-negotiable. For policy-makers seeking balance and compromise, that is a highly undesirable situation. The solution has been to seek those areas where environmental imperatives and economic priorities are congruent, in practice a larger domain than is often realised, provided economic policies are

---

practical policy prescriptions that have the clarity of those underlying the Washington Consensus remains a distant challenge.

carefully crafted. It is, however, this very congruence that makes conflicts between environment and economy particularly unpleasant: they operate in overlapping domains and have the ability to create real problems for each other.

The WTO Agreements permit countries to choose the level of environmental protection that they desire. This is, however, limited by the fact that most environmental decisions are constrained by international environmental agreements, and are required to be in conformity with international economic agreements, those of the WTO in particular. In practice, few countries can exercise much choice, and the relationship between trade policy and international environmental agreements is untouched by this perspective.

The first major conflict between trade agreements and the environment concerned the interpretation of the former and the implications of these interpretations for the environment and sustainable development. A GATT panel in the tuna/dolphin dispute between Mexico and the United States (US) posited that distinctions between otherwise 'like products' based on process and production methods (PPMs) were inadmissible under the GATT. This interpretation focused on the definition of like products, the central concept that underlies the principle of non-discrimination embodied in the GATT: most-favoured-nation treatment and national treatment apply to like products.[16] The use of the ambiguous word 'like' in so critical a context was presumably intentional, leaving the interpretation of what products are to be considered 'like' to subsequent practice.

The ability to distinguish between otherwise like products on the basis of PPMs is of critical importance to the attainment of sustainable development. It embodies many of the qualitative criteria that distinguish sustainable development. Environmental interests mounted a challenge against an interpretation of the GATT that would largely have avoided the need to take into account the environment and sustainable development. This challenge was the first of its kind: until the tuna/dolphin panel report, no interpretation of the GATT by a dispute panel had been challenged, neither from the inside nor from the outside. In the event, the critics of the tuna/dolphin panel report

---

[16] GATT Article I.1: 'any advantage, favor, privilege or immunity granted by any contracting party to any product originating or destined for any other country shall be accorded immediately and unconditionally to the like product originating in or destined for the territories of all other contracting parties.' GATTArticle III.4: 'The products of the territory of any contracting party imported into the territory of any other contracting party shall be accorded treatment no less favorable than that accorded to like products of national origin'. The French and Spanish texts of the GATT translate 'like' as 'similar' since both languages lack a word that is comparably situated between similar and identical.

appear to have prevailed.[17] They also contributed to the rapid opening up of the trade regime to broader scrutiny.

For a long time, it was assumed that trade law would take priority over international environmental law, presumably because the trade regime was widely seen as powerful and the environmental regimes as weak. The trade regime has turned out to be less potent and many international environmental regimes more robust than many assumed. It also appears that an approach to interpreting trade law that posits its priority over other international law is inconsistent with well-established traditions of interpretation of international law.[18] In the absence of any clear statement of priority in either trade or international environmental law, interpreters have an obligation to seek readings that give due respect to both bodies of law. The mention of sustainable development in the Preamble to the WTO Agreement lends particular force to this argument. The Appellate Body created by the Dispute Settlement Understanding has brought the kind of discipline and precision to the interpretation of WTO law that was lacking in the tuna/dolphin panel report.[19] Indeed, one of the most striking aspects of the new dispute settlement procedure introduced by the Uruguay Round has been the degree of fallibility of legal interpretation by trade dispute panels, leading to frequent corrections by the Appellate Body.

A vivid example of the difficulties that must be managed in the relationship between the trade regime and international environmental regimes concerns the issue of genetically modified organisms (GMOs). This has for some time been a contentious issue within the WTO, although not until recently did it rise to the level of a formal dispute.[20] At the Seattle WTO Ministerial Conference in 1999, one of the first compromises agreed between negotiators from the US and the EU was a decision to establish a WTO Working Group on Biotechnology. The likely effect of this decision would have been to abort, or at least severely hinder and delay, the negotiations for a Biosafety Protocol within the Convention on Biodiversity, which were at a particularly delicate juncture at the time. The decision to establish the Working Group was taken by trade negotiators without consultation with those responsible for the biosafety negotiations, even though the Executive Director of the United Nations Environment Programme (UNEP) and the

---

[17] Howse, 2002; Howse and Regan, 2000.

[18] The Vienna Convention on International Treaties is often cited, with its rules governing the precedence of one treaty over another. But the Vienna Convention should only be applied if an interpretation that accommodates apparently conflicting bodies of law cannot be found – and that has not thus far happened with respect to trade and the environment.

[19] Mann and Porter, 2003.

[20] Inside US Trade, 2003.

chair of the Cartagena Protocol negotiations were both in Seattle at the time.[21] In response to this action, the EU Environment Ministers present in Seattle convened a meeting of the EU (Environment) Council and voted unanimously to reject the decision; a move that was ignored by the EU trade negotiator, who was at that moment more concerned about his credibility as a negotiator than about a looming interinstitutional conflict within the EU. The collapse of the Seattle Ministerial avoided a serious institutional confrontation within the EU between the Council and the Article 133 Committee, which oversees trade negotiations. Not long after this, the Seattle negotiations for the Cartagena Protocol were concluded successfully, including intensive discussions about its relationship to the trade regime. At least some trade negotiators had come to the realisation that they needed the Cartagena Protocol to provide structure to an issue that otherwise would have threatened the stability of the trade regime. The Protocol will now form the basis of decisions in trade disputes that relate to matters that it covers.

The WTO and Multilateral Environmental Agreements

An emerging line of analysis views the WTO system as reflecting 'distributed governance': a structure where the WTO articulates economic rules but relies on other international institutions such as the World Intellectual Property Rights Organization (WIPO), the International Organization for Standardization (ISO), and multilateral environmental agreements (MEAs) to address matters of substance in their areas of competence.[22]

Certain WTO Agreements mention some of these external institutions. The relationship between the WTO and MEAs, however, poses a range of particular challenges. These arise from the structural differences between the trade regime and international environmental regimes. The challenge of capturing the environment in a limited number of international agreements has resulted in the formation of a large number of regimes. While the current number of regimes is not optimal, it is also clear that more than one regime is needed. The institutions required to manage the transboundary movement of hazardous waste are different from those for the protection of endangered species. The problem structures of two closely related issues such as climate change and stratospheric ozone depletion are so manifestly different that it is hard to argue that they should be part of a single regime.[23]

---

[21] The Cartagena Protocol on Biosafety to the Convention on Biodiversity creates rules concerning the international movement of living GMOs.

[22] Abbott, 2000.

[23] Stratospheric ozone depletion involves the control of a limited number of industrial chemicals, most of which have substitutes. Climate change involves the allocation of emissions of greenhouse gases, many of which have no substitutes and most of which also occur naturally in large but unpredictable quantities.

The relationship between the WTO and MEAs is the subject of two items in the Doha Ministerial Declaration that are part of the 'single undertaking'.[24] The Doha Round as a whole cannot conclude unless solutions to these problems have been found that satisfy all parties. Paragraph 31(i) concerns 'the relationship between existing WTO rules and specific trade obligations set out in multilateral environmental agreements (MEAs). The negotiations shall be limited in scope to the applicability of such existing WTO rules as among parties to the MEA in question. The negotiations shall not prejudice the WTO rights of any Member that is not a party to the MEA in question.' This paragraph presents numerous problems and no resolution is in sight. In the light of recent Appellate Body decisions, there is significant scope for a negotiating outcome that would be less desirable from an environmental perspective than the current situation. Paragraph 31(ii) concerns 'procedures for regular information exchange between MEA Secretariats and the relevant WTO committees, and the criteria for the granting of observer status'. Despite some difficulties, there does not appear to be any obstacle to resolving this issue – except that no negotiator is willing to commit to any compromise at this stage of the Doha process.[25]

It is possible to view the trade and the environmental regimes as part of a system of distributed governance. Nevertheless, the structural differences between them render the practical management of the relationship difficult. It is made additionally contentious by the widespread suspicion on the part of developing countries that environmental measures (including international environmental measures) are yet another pretext for the protectionism of developed countries.

Environmental Policy Design in an Era of Globalisation

Environmental policy design must of necessity be situated in the field of force defined by globalisation, with the understanding that it is itself an active force in this process. This has required a range of responses, the most important being the growing preference for market-based policy instruments. The difficulties that have been – and will continue to be – encountered in introducing such instruments are rooted in the nature of environmental policy itself, which requires multi-dimensional policy solutions. The dominant realities of environmental policy are the extent of our ignorance concerning the environment and the impossibility of controlling the environment and, consequently, effecting environmental change in a direct manner.

---

[24] Doha Ministerial Declaration paragraphs 31(i) and 31(ii) (WTO, 2004).
[25] Von Moltke, 2003.

Ignorance

Thirty years have now passed since the beginnings of environmental policy, that is, the attempt to view the natural environment itself as the object of policy concern rather than the surrogates that had determined policy previously, namely, the protection of human health from the effects of pollution. An extraordinary research effort, distributed over many countries (including most major developing countries) has accompanied the development of environmental policy. The reasons are simple: there can be no environmental policy without research, because we need research to make the environment speak. The results of this research have been impressive, and yet the dominant reality remains our lack of knowledge about the environment, or, more accurately, our constant need to act on the basis of partial knowledge – first indicators of impending trouble or even no more than a robust hypothesis. Action to protect the stratospheric ozone layer was initiated on the basis of a research hypothesis that had not yet been verified by the time of the adoption of the Montreal Protocol, even though there was evidence that something unusual was occurring above the Antarctic. Similarly, measures to control persistent organic pollutants received new impetus from a hypothesis concerning endocrine disruption that has not been fully proven.[26] More recently, the Cartagena Protocol was adopted prior to the existence of evidence concerning the environmental consequences of GMOs. When it comes to ecosystems not inhabited by people capable of articulating their concerns (such as tropical forests or marine ecosystems), the balance of research and ignorance is particularly unsatisfactory.

The need to act without clear evidence has been the source of much confusion at the international level.[27] To some extent, this confusion is likely to be temporary. All governments face the problem that they must act in the absence of sufficient scientific evidence. Yet, the ways in which they deal with this problem differ widely, reflecting political and administrative traditions and the structure of the relevant research communities as much as the nature of the problem that is being considered. The use of formalised risk assessment happens to be particularly suited to the US structure of governance; European countries, with a much more unified system of governance, have no need of such procedures but apply the precautionary principle, which may or may not include risk assessments as the situation dictates. It is likely that the actions of responsible governments will ultimately

[26] The endocrine disruption hypothesis suggests that certain persistent organic substances that are by now ubiquitous in the environment and accumulate in living organisms can disrupt the endocrine system by mimicking certain hormones.

[27] Von Moltke, 2000a.

converge, but it will also take time to recognise any convergence.[28] In the interim, interested parties on all sides of an issue can exploit the variability in responses for their own ends.

Impossibility of controlling the environment

Just as important as lack of knowledge is the fact that environmental policy measures never change the environment directly. Environmental policy is like shooting around corners. Measures may change human behaviour – although that is also uncertain – and these changes may result in desirable environmental responses. It is, consequently, extremely difficult to assess the effectiveness of (international) environmental measures. Environmental policy depends on a chain of causality, beginning with recognition of the problem, identification of surrogate standards,[29] definition of possible responses, enactment of appropriate measures, and enforcement of these measures with consequent changes in human behaviours – which are assumed to feed all the way back along the chain to the environment. Environmental measures are not necessarily effective when the original problems are solved, because the issues of causality may remain uncertain: improvements may originate in the measures that caused the problems or some other, possibly natural, causes. But measures that change human behaviours are also effective if they do not solve the underlying problems. For example, dramatic reductions of atmospheric emissions from automobiles per kilometre driven are a significant success, even though an increase in the total number of kilometres driven may negate the benefits – as is currently happening in Beijing. Environmental measures are typically effective in one or more ways. They may change behaviours and reduce emissions, but still not solve the underlying environmental problems.[30]

When human impacts on the environment are gross, this complexity poses few problems: knowledge is hardly an issue and the impacts must be controlled. Rules can be developed that will have the desired results. But to achieve a minimum of certainty, such rules will almost always be command-and-control measures. Yet, it is not acceptable to wait until environmental impacts are gross beyond question – and there will always be residual resistance to the taking of measures, even when these are inescapable. That is the nature of democratic and market institutions. Moreover, the absence of

---

[28] A similar problem existed in chemicals policy twenty years ago, until it was realised that different institutional procedures in the US and the European Community produced largely comparable results (Gusman et al., 1980).

[29] Thus, air pollution is defined as a specified concentration of certain pollutants, since clean air cannot be measured.

[30] There is an extensive literature on the effectiveness of international environmental agreements. The approach outlined here is based on Young, 1989; see also Miles et al., 2002.

gross environmental damage is not the same as acceptable environmental quality.

Most recent environmental policy is situated in a grey area where impacts are uncertain and measures become increasingly controversial. Under those circumstances, it has become necessary to develop a range of new policy tools that will contribute incrementally to the attainment of environmental goals. It has become inappropriate to use a single policy instrument to achieve desired results, because such instruments are almost always blunt, have uni-dimensional distributive effects, and attract a corresponding degree of resistance.

Ignorance and the indirect effect of policy measures on the environment combine to produce environmental policy that is increasingly complex and that employs a growing range of policy instruments, none of which can be expected to achieve the desired environmental results by itself. Perhaps most important among these new instruments are a range of economic policy measures that create incentives rather than mandate certain outcomes. While it is widely recognised that these economic incentives will not normally achieve the necessary environmental results by themselves, their introduction can create additional room for the adjustment of existing measures, most of which are command-and-control measures, reflecting the urgency of action in the first phases of environmental management.

The complexity and indirectness of environmental policy are a major source of frustration from the perspective of economic globalisation. There is a noticeable desire to deal with the environment in a simple, coherent fashion. At the WTO, this has resulted in the creation of a Committee on Trade and Environment (CTE). Nevertheless, environmental issues keep cropping up in unexpected places. Thus, the Doha Ministerial Declaration contains two paragraphs that are devoted to environmental matters. But environmental issues arise in ten other places, including the most sensitive parts of the agenda: agriculture and investment.[31]

Procedural Environmental Policy

A further characteristic of environmental policy is its procedural nature. Rather than prescribing specific measures or outcomes, it has become customary to require a number of procedures to be followed, including monitoring, environmental assessment, public information, and public participation. These procedures are designed to produce a broadly based understanding of the goals that are being pursued. In addition, widely participatory procedures provide a degree of protection against the problem

[31] See Von Moltke (2003) for a brief overview.

of ignorance, as affected persons are more likely to be aware of environmental conditions and will be in a position to articulate them, or at least to articulate questions and hypotheses that require answers. Moreover, the procedural approach permits the continuous adjustment of measures to the emergence of knowledge about environmental conditions and the environmental effects of measures that have been adopted. In many ways, transparency and participation are core elements in ensuring the effectiveness of environmental policy. Procedural environmental policy has the added advantage of creating options for affected persons to articulate the social dimension of environmental management, moving the discourse closer to the ultimate goal of sustainable development.

The procedural aspect of environmental policy is presumably the dimension that is most readily assimilated to the requirements of economic policy. The trade regime has responded to the need to ensure that measures (technical standards and SPS in particular) do not constitute hidden protectionist measures by emphasising the procedural dimension of the development of such standards. By and large, the requirements of the TBT and SPS agreements are not that certain standards be respected nor that they be harmonised between countries, but that the process for developing the standards meet certain requirements in terms of transparency and the use of scientific evidence.

It has often been said that there is no conflict in principle between economic globalisation and the environment. Like many such statements, this observation fails to provide any guidance as to the measures that must be adopted to ensure that there is no conflict. What is certain is that the application of the Washington Consensus without the inclusion of environmental safeguards is insufficient to avoid wrenching conflicts. Similarly, the pursuit of environmental goals by individual countries without proper coordination and cooperation with other countries that are or may be affected by that problem leads to a disruption of economic relations with an attendant penalty in economic efficiency. It is also unlikely to solve the underlying environmental problems.

We must assume that environmental policy will occur in an internationally defined framework. The integration of this framework with the structures of international economic policy-making remains one of the more challenging political tasks of our time. It will not be accomplished without the continuous and active participation of those responsible for the environment at all levels. The most important difficulty resides in the need to accommodate structures of subsidiarity in a system that is designed to operate on universal principles. In this context, it is important to recognise that some markets are by now so large that segmenting them does not impose significant economic penalties. Thus, global soya markets have continued to clear despite the existence of the

EU moratorium on the importation of GMO soya. While there may be a modest price penalty for those who want (segregated) non-GMO soya, the overall system remains remarkably efficient, in particular since one-way substitution of non-GMO products for GMO products is freely possible.

Investment: The Touchstone of Sustainable Development

The touchstone of the relationship between economic globalisation and environmental protection is investment. It is marked by conflicts between economic globalisation and environmental globalisation. For example, Chapter 11 of the North American Free Trade Agreement (NAFTA) deals with investment.[32] From its inception, it has generated (environment-related) problems. The parties have already tried to fix one aspect of Chapter 11, but in truth there is not a single provision of that chapter that is not currently subject to challenge.

Negotiations for the Multilateral Agreement on Investment (MAI), modelled on NAFTA Chapter 11, collapsed after environmentalists dragged it into the public eye and numerous other interests discovered how much they disliked it. In the end, it was killed by a long-running dispute between the French film industry and the US Motion Picture Association. This result, itself hardly an endorsement of NAFTA, may yet be seen as the beginning of the end for the Washington Consensus as the sole criterion for globalisation. The EU member states, the principal proponents of the MAI and largely responsible for running the negotiations, have lost control of the investment agenda to the European Commission.

It is hard to overstate the importance of investment for sustainable development.[33] Investment provides a long-term economic indicator. While investment alone does not guarantee growth, it is a good indicator of expectations of growth. Perhaps more importantly, investment shows the direction of growth. It is customary to emphasise the short-term nature of business, with a focus on the next quarterly report. Yet, investments are routinely undertaken in a long-term perspective, frequently ranging up to thirty years. Even investments in the dynamic technology sector are taken within a five-year time horizon. Investments in agriculture, forestry, or mining may take as long as ten years to show a positive return. Some investments establish patterns of production that can persist for decades, even centuries. Most power plants will be rebuilt and refurbished over and over again. Some of the sites in the US Midwest that are the source of acid rain in the Northeast were established more than a hundred years ago. Investments to transform

[32] IISD, 2001.

[33] For an introduction to international investment negotiations and sustainable development, see Von Moltke, 2000b.

virgin land for agricultural production or forestry change that land for centuries: crops may vary but the original ecosystems will not return. Mining operations can create impacts that are irreversible within a human time scale. Thus, the time frame of investments can range far, and investors routinely plan for the long term. Any policy that looks to the longer term – as does sustainable development – must look to investment as one of the central factors determining the chances of attaining the goals that have been set.

The most basic rule of investment is that it must be profitable. It must at least generate returns that cover the cost of amortisation plus the (variable) cost of capital. There can be no guarantee that this will be the case, so investment is linked to risk, which in turn affects the expected returns. The assessment and control (or distribution) of risk are central to the process of investment. International investment is no different in this regard than other investment, except that additional risk factors relating to currency fluctuation and other risks associated with specific locations may enter the calculation.

The goal of sustainable development requires structural economic change. It has been established that the current pattern of economic activity is unsustainable, in the double sense that it is environmentally destructive and not sufficiently supportive of the promotion of equity. Consequently, the pursuit of sustainable development is the pursuit of a changed economy. In a market economy, this structural change can only be brought about by investment, indeed by profitable investment. Thus, the promotion of sustainable development is ultimately the promotion of profitable investment that improves sustainability. In this sense, there is no economic activity that is more important to the attainment of sustainable development than investment.

Investment is distributed unevenly. For productive investments, access to inputs and (skilled) labour is of critical importance, as is access to markets for the products and services generated by investment. The existence of reliable infrastructure and the quality of services (transport, banking, insurance, and government services in particular) are important. To a significant degree, investment attracts investment; that is, investing where others have already invested is generally a lower-risk business strategy. In other words, investment decisions are influenced by a large number of factors whose weighting varies depending on the circumstances of each individual investment. The existence of a market for the goods or services that are produced is fundamental to all investment.

Investment decisions are fraught with conflict, arising from the possibility of competition between jurisdictions to attract investments, from the need to balance private economic rights and public goods that may be affected, and from the existence of risk associated with investment. Risk persists once investment decisions have been made, and there can be further conflict over the fulfilment of obligations undertaken in connection with investments or the assignment of responsibility for the failure of investments to achieve anticipated profits. In some instances, this responsibility is purely commercial. In others, it can involve public authorities and the discharge of their obligations. Most countries have developed complex institutional provisions to avoid such conflicts in the first place and to manage them when they occur. They continue to work to ensure the legitimacy of these institutions. These include regulations governing public goods, administrative procedures, requirements for transparency and accountability, provisions for hearings, notification requirements and public information, and judicial institutions that frequently involve several levels of review. These regulations are the bedrock of environmental management.

International investment agreements are part of this landscape. They are affected by the manner in which public authorities in different countries respond to the challenge of balancing conflicting priorities – and the agreements in turn affect these activities. Negotiators must take due care to ensure that the agreements do not have unintended consequences; that is, they must be aware of all the ramifications as they negotiate. It does not appear that this has always been the case.

Despite the existence of large numbers of such agreements, it remains difficult to pinpoint what their purpose is. The interests of the parties are contradictory: the primary interest of home countries is to protect their investors by avoiding losses and enabling investors to repatriate profits. Host countries seek to attract investments that they would not otherwise obtain. The existence of contradictory interests in international agreements is not unusual. However, these interests are neither explicitly articulated in the agreements nor reflected in their specific provisions. The agreements assume that desirable outcomes will occur more or less automatically, a characteristic of many international economic agreements that tend to reflect unarticulated assumptions about economic theory. These assumptions may not be tenable in the light of experience with relations between states: it is important to ensure through explicit legal provisions that desired outcomes are actually achieved. In the absence of provisions in the agreements themselves that are designed to ensure that the differences in expectations and in power between the parties that exist at the outset are counterbalanced, the likely result of such agreements is to reinforce power relations, including the distribution of benefits that may be available. Indeed, this appears to have been the overall result of the existing investment agreements.

It has become necessary to move beyond general declarations concerning the desirability of economic liberalisation, which is not in dispute, and to review in detail the provisions of international investment agreements such as the NAFTA as well as the proposals for new agreements in the WTO, in the framework of the Cotonou Agreement,[34] in EU-Mediterranean and EU-Mercosur negotiations,[35] and in the negotiations for a Free Trade Area of the Americas (FTAA). Several provisions in these agreements appear not to be contributing to the attainment of the goals that were articulated by negotiators. The ultimate goal is to identify measures that can be included in international investment agreements and that promise to actively promote the sustainable development outcomes that are sought.

CONCLUSION

Environmental policy is a crucial factor in the process of globalisation, impacting on the rules for economic liberalisation that are being adopted and in turn influenced by them. At the present time, the framework of rules for globalisation is incomplete. Negotiations continue in a large number of forums, including the WTO, the EU, bilateral negotiations between the EU and the Cotonou countries, Mercosur, and the Mediterranean, as well as the FTAA in the Americas. All of these negotiations will further flesh out the framework for environmental policy.

As it is developed, this framework will further constrain the scope for action by individual countries. It is important to ensure that the sustainable development dimension is fully integrated into this process. Of all the current economic and environmental negotiations at the international level, those aimed at foreign direct investment are clearly the most significant from the perspective of sustainable development and ensuring that local communities and countries have the necessary discretion to properly balance private rights and public goods.

Environmental policy-makers must be active participants in the numerous discussions and negotiations concerning globalisation. Without such participation, the new rules that are being elaborated will remain incomplete. More significantly, it is likely that they will constrain domestic environmental decision-making in ways that are unacceptable.

---

34 The Cotonou Partnership Agreement defines the relationship between the EU and a large group of African, Caribbean, and Pacific (ACP) countries, most of which were former colonies of one of the EU member states. It represents an ambitious attempt to integrate development assistance policies with economic liberalisation measures in a framework that is consistent with the WTO rules.

35 Mercosur is the common market of four countries of South America (Argentina, Brazil, Paraguay, and Uruguay). Chile and Bolivia are associate members, and discussions are held concerning the inclusion of other South American countries. The EU is negotiating an interregional agreement with Mercosur.

REFERENCES

Abbott, F. (2000), 'Distributed governance at the WTO-WIPO: An evolving model for open-architecture integrated governance', *Journal of International Economic Law*, **3**(1), 63-81.

Gusman, S., K. Von Moltke, F. Irwin, and C. Whitehead (1980), *Public Policy for Chemicals: National and International Issues*, Washington, DC: The Conservation Foundation.

Howse, R. (2002), 'The appellate body rulings in the shrimp/turtle case: A new legal basis for the trade and environment debate', *Columbia Journal of Environmental Law*, **27**, 491-521.

Howse, R. and D. Regan (2000), 'The product/process distinction: An illusory basis for disciplining unilateralism in trade policy', *European Journal of International Law*, **11**(2), 249-289.

IISD (2001), 'Private rights, public problems: A guide to NAFTA's controversial chapter on investor rights', *http://www.iisd.org/trad/pubs*, Winnipeg: International Institute for Sustainable Development.

Inside US Trade (2003), *U.S. Announces Panel on EU GMO Moratorium as Grassley Warns Egypt*, 20 June.

Jackson, J. (1991), *The World Trading System: Law and Policy of International Economic Relations*, Cambridge: MIT Press.

Jackson, J. (2000), *The Jurisprudence of GATT & the WTO: Insights on Treaty Law and Economic Relations*, Cambridge: Cambridge University Press.

Mann, H. and S. Porter (2003), 'The state of trade and environment law', *http://www.iisd.org*, Winnipeg: International Institute for Sustainable Development.

Miles, E., A. Underdal, S. Andresen, J. Wettestad, J. Skjærseth, and E. Carlin (2002), *Environmental Regime Effectiveness: Confronting Theory with Evidence*, Cambridge: MIT Press.

Von Moltke, K. (2000a), 'The precautionary principle, risk assessment and the World Trade Organization', *http://www.iids.org/trade/pubs.htm*, Winnipeg: International Institute for Sustainable Development.

Von Moltke, K. (2000b), 'An international investment agreement? Implications for sustainable development', *http://www.iids.org/trade/pubs.htm*, Winnipeg: International Institute for Sustainable Development.

Von Moltke, K. (2001), 'Trade and ...: The agenda of trade linkages', in A. Liberatore and N. Christoforides (eds), 'Global trade and globalising society: Challenges for governance and sustainability: The role of the EU', *http://www.iisd.org/trade*, Brussels: European Commission.

Von Moltke, K. (2003), 'Information exchange and observer status: The World Trade Organization and multilateral environmental agreements, paragraph 31 (ii) of the Doha Ministerial Declaration', paper for EU China Programme for China's Accession to the WTO, *http://www.ecologic.de* and *http://www.iisd.org/trade.*

Williamson, J. (2000), 'What the World Bank should think about the Washington consensus?', *The World Bank Research Observer*, **15**(2), 251-264.

World Commission on Environment and Development (WCED) (1987), *Only One Earth*, London: Oxford University Press.

WTO (2004), *http://www.wto.org*, Geneva: World Trade Organization.

Young, O. (1989), *International Cooperation: Building Regimes for Natural Resources and the Environment*, Ithaca: Cornell University Press.

27. Effective Environmental Strategies for Small Countries in an Interconnected Global Setting

Pieter Winsemius

SUMMARY

A small country like the Netherlands has traditionally played a prominent role in (international) environmental policy. However, globalisation and the higher assertiveness and demands of citizens have eroded this leading role. National initiatives lose meaning in a world where citizens are more individualised, increasingly vote with their feet, and support good causes through non-governmental organisations (NGOs). Three schools in public administration shed a political light on environmental policy: the conservative school highlights the reinforcement of inherent and institutional trust, communitarians stress community trust, and liberals focus on personalised trust. The practice of environmental policy shows evidence of all three: legislation (as advocated by the conservatives), voluntary sectoral agreements (as espoused by communitarians), and corporate social responsibility (as promoted by liberals). In this context, the Netherlands has excellent assets to fulfil a future trigger role in the international arena. Using its strong home base of multinational companies and NGOs, the Netherlands can create a stimulating environment in which partnerships and civil-society initiatives prosper and spill over to other countries.

PROSPECTS FOR SMALL COUNTRIES

Ever since the early days of environmental policy, some thirty years ago, countries like the Netherlands and Sweden have played a prominent role in the development of international environmental policy. Countless times, for example, Dutch politicians and officials have chaired international conferences or intermediated in stubborn negotiations between 'large countries'. This is why the power or influence of small countries like Sweden, New Zealand, and the Netherlands has been markedly greater than what would be considered appropriate in terms of geopolitical indicators.

However, the international policy arena is changing.[1] Large countries, the United States (US) in front, are showing an increasing tendency to go alone. Decision-making in the matter of the 2003 war in Iraq is not an isolated occurrence: in environmental matters, the US has refused to join the ranks of the signatories of the Kyoto and Cartagena Protocols, and demurred in the preparations and decision-making procedures of the 2002 World Summit for Sustainable Development in Johannesburg. Within the European Union (EU), a new balance of power is also becoming manifest, with leading roles for countries like Germany, France and – occasionally acting as a curb – the United Kingdom (UK). With over twenty EU member states, the relative influence of small countries is diminishing, which is becoming evident in matters like the weight of national votes and the impact on choice of chairmanship. These soloist tendencies are not unwarranted. Decision-making in multilateral organisations like the United Nations (UN) turns out to be increasingly susceptible to excessive delay. In addition, the importance of the environment as a political priority decreased in the 1990s. All these factors are cause for concern about the effectiveness of national governments in an international setting.

A more positive assessment can be offered of the effect of globalisation on the environmental awareness of large multinationals. This is partly due to the greater potential influence of global consumers, who, by boycotting brands, can effect global consequences for multinationals. Many of these enterprises, however, had already made considerable progress in committing to tasks in the field of sustainable development that used to be considered beyond their responsibility. 'Responsible care' and subsequently 'responsible corporate citizenship' are in the process of becoming established notions.

The downward effect of foreign developments on the role of national governments as innovative forces is reinforced by the increasingly restrained position of governments in formerly proactive countries such as the Netherlands and Denmark, which now firmly pull on the public brakes. For example, the coalition agreement of the Dutch government that took office in 2003 contained a prohibition – which had never been expressed before – for Dutch policy-makers to take initiatives that are more stringent than required by EU standards. Despite these worrisome developments, there are also less gloomy sides to the new situation. The next generation of international environmental policy is facing new challenges. Especially small countries

---

[1] Zoeteman, 2003.

such as the Netherlands can play a major role in these. However, such a role requires thoroughly revised, proactive strategies.

Below, I discuss the changing public support for policy, including environmental management and nature conservation policies, to build on. Citizens are individualising, and their needs are changing accordingly; they have internalised environmental values and demand that institutions do likewise in their strategies and actions. Some private institutions, ranging from multinational companies to non-governmental organisations (NGOs), have appeared to be unable to adapt to these new requirements and have not survived. Other organisations, both established and new, have responded to these more pointed requirements from demanding citizens. Thus, the approach to international environmental issues can bank on much wider public support. A colourful range of environmental NGOs – which, with the aid of the internet, may form coalitions with other groups for particular purposes – offers new possibilities here.

I also look at the position of governments, which, against this backdrop, must formulate policy that will enable small countries like the Netherlands to mould their own sustainable future. Three administrative philosophies are clamouring for attention, each of them with its own proponents. In practice, however, they are complementary and only represent a shift of emphasis. In the final section, I sketch the outlines of environmental policy in small countries that can be effective in a globalising society.

CHANGING PUBLIC SUPPORT

More Assertive Citizens

The Social and Cultural Planning Office of the Netherlands has pointed to individualisation as the common denominator in the changes in Dutch society over the last 25 years of the 20th century.[2] Citizens are more affluent, better educated, and – owing to technological progress – better informed than ever before. They know more, have pushed back frontiers, and are better able to express what they want. They are also more articulate and more critical of 'regents' who, in a more hierarchical and compartmentalised society less than half a century ago, told them what to do. And, not unimportantly, they increasingly tend to vote with their feet: if providers cannot deliver what they

---

[2] SCP, 1998.

demand, they go over to others. They have become more assertive and more demanding citizens.[3]

The Widening Gap between People and Institutions

In the individualising society of the past decades, this had some painful consequences. While citizens were crying out for faster and better tailored responses from providers, these often opted to scale up for reasons of effectiveness and/or efficiency. The results were spectacular, especially in government. Policy, for example, was shifted to multilateral institutions such as the EU, which takes away market impediments and allows for coordinated action in the field of the environment. As a cost-cutting measure, the administrative scales of municipalities and schools were increased. However, this also increased mental and physical distances, made organisations too complex and hard to motivate, and made public spaces too anonymous. People withdrew from all fields of social action, including the environment. In his study of the Dutch citizens' trust in Dutch institutions, Dekker looked into sustainable development, which is pre-eminently built on trust: without trust in the contributions of others, self-restraint is pointless.[4] Without trust in the government, restrictive measures in favour of future generations are hard to accept, for government policy is the mainstay of joint action. The extremely low assessment of trust in governments and political parties, therefore, is cause for concern.

The development of citizens towards greater assertiveness is of great consequence for the shape of public administration, for such administration exists by virtue of the mutual bond of trust between citizens and governments: each party needs to know where it stands.

Restructuring Civil Society

As citizens have become more assertive and more demanding, this has thoroughly altered the structure and tasks of civil society.[5] Citizens have found new ways of dealing with their need for social contact or trust. A clear grasp of these changes escapes many people, which is partly due to confusion, as the term 'civil society' is defined in different ways. In the specific Dutch context, civil society is an umbrella term covering a wide variety of professionalised organisations engaged in policy areas like care, education, housing, welfare, and sports. In an empirical sense, therefore, civil society refers to the social mid-zone in which – again, especially in the Netherlands –

---

[3] Van den Brink, 2002.
[4] Dekker, 2001.
[5] WRR, 2002.

changes have been strongly impacted by processes of secularisation and the ensuing breakdown of religious and socio-political barriers.

At the same time, however, we have also witnessed a strong rise in the membership and variety of organisations pursuing good causes, often with social single-issue objectives. Some of the more established of these also modified their approach. Conservation groups, which used to be rather detached, stable, and respected institutions with considerable membership and funds, have moved more frequently into the political arena, where they have joined forces with environmental pressure groups, which tend to be more offensive and publicity-oriented. Their new members expect them to display a combination of more active opposition and cooperation vis-à-vis government and business. Organisations such as WWF, for example, have developed partnerships with leading multinational companies like Unilever in the fields of sustainable fishing and/or agriculture. In other sectors, such as human rights and development cooperation, business was taken to task in more direct ways. In areas where the international community was remiss in its duties, multinational NGOs like Greenpeace and Amnesty International carried on campaigns against negligent governments. Even multilateral institutions like the World Bank have had to justify unsound investments in developing countries.

The significance of corporate social responsibility is increasingly being underlined. In a new balance of power between governments, companies, and good causes, it is assumed, sustainable development can be promoted. The idea is that citizens, well-informed by the NGOs trailing the critical media in their wake, vote with their feet and can thus impact governments and enterprises. On the sidelines and scarcely identifiable as yet, private business is limbering up. Debates on the pros and cons of this development are heated, with the anti-globalisation movement in the lead. If governments do not manifest themselves effectively in the global arena, international business itself, supported by NGOs and the media, must offer the bulk of the requisite resistance against 'dodgers'. The question is to what degree these parties are capable of doing so.

Civil society appears to be reshaping itself once more. Old forms of organisations are being replaced by new ones; closed networks are being replaced by more open ones. Successful organisations in the current day and age are often highly professionalised and have a knack for publicity. Their members have virtually no face-to-face contact, and most of them are nominal members or contributors rather than activists. These organisations often lack a solid democratic foundation and operate on the fringes of the political arena. Old and new good causes fill the gaps in the overall domain of effective campaigning. For example, the Red Cross was joined by Médecins

Sans Frontières (MSF) to provide humanitarian aid in contingency situations. Even the greater scope obtained by multinationals has led to a counter-campaign. The anti-globalisation movement turned out to be surprisingly effective with large-scale demonstrations in Seattle and Genua.

All this may herald a new stage in the development of civil society, in which it is no longer the major, stable NGOs that call the tune, but rather the small professional campaigners. Statistics often fail to register such changes or register them too late. When new NGOs arrive on the scene now, their entrepreneurs focus on establishing a small but high-quality desk and on fundraising, and then proceed to target the media and decision-makers in government, business, and other NGOs. They have little time nor inclination to bother about their constituencies: 'Members are a not very lucrative department', as a North-American leader put it.[6]

Action-oriented organisations like Greenpeace used to be the only ones that were elusive because they refused to be drawn to the negotiating table, but the new NGOs are also elusive because they are virtually unidentifiable. Many new social movements are developing while remaining largely invisible.[7] New civil society manifests itself like a moorland fire: invisible on the surface for the 'regents' of democracy and civil society, but suddenly flaring up in many shapes and guises when people of manifold persuasions unite to defend causes they feel strongly about. Every now and again, the old NGOs will take the lead, with their stable, vertical-style composition. In the future, however, they will be supplemented or even succeeded by horizontal, virtual networks of small players with great clout.[8] This clout is determined by the support they manage to enlist for their single-event manifestation.

Once again, it is a case of citizens voting with their feet and making themselves heard. This time, however, it is all about personalised trust, and citizens appear to be saying: if you have nothing to offer us as individuals, we will not join you. They are individuals choosing from a menu that is presented to them by NGOs or obscure coalitions of these. Who staged the anti-globalisation demonstrations in Seattle and Genua or those against the war in Iraq? Who will be in charge of future protests against sound pollution caused by Amsterdam Airport Schiphol? In the past, such things used to be clear: it was Greenpeace, for example, which headed the campaigns against the Brent Spar or against the import of genetically modified organisms (GMOs). These days, however, this is no longer the case. The leadership of politics, business, and traditional NGOs feel disordered. They used to insist on behavioural codes for NGOs and tried to get NGOs to the negotiating table, but now they no longer know who their opponents are.

---

6 Skocpol, 2002: 134.
7 Poldervaart, 2002.
8 Rischard, 2002.

Many things are changing in civil society as it is reshaping itself. However, are citizens satisfied with the supply they are being offered by the private institutions? As far as I am aware, they have never been specifically asked. In the Netherlands, NGO membership in various fields went up between 1980 and 2000: abortion and euthanasia + 669 per cent, nature and the environment +524 per cent, international solidarity +137 per cent, health care +118 per cent, and consumer interests +51 per cent.[9] Dutch NGO fundraising figures more than quadrupled between 1975 and 1996.[10] If we also take into account the comfortable position of NGOs on various trust barometers, it seems we may conclude that the wide array of NGOs has sufficiently managed to cater to citizens' needs.

THREE MAIN SCHOOLS IN PUBLIC ADMINISTRATION

In the wake of this evolving public support in society, governments must formulate environmental policies that will be effective in a globalising society; the difficulty of this task was underlined above. Moreover, there is no agreement on what is the optimum public administration philosophy; the proponents of the three main schools in this field – the conservatives, the communitarians, and the liberals – contest each other with great zeal.[11] I now present an analysis of the characteristics – including objectives, lines of reasoning, and outlines of solutions – of these three main schools, and then consider the extent to which they are conflicting or complementary.[12] In the latter case, an international environmental policy could accommodate the strengths of each of the three.

Conservatives: Reinforcing Inherent and Institutional Trust

The conservative school traditionally focuses on the reinstatement of inherent and institutional trust. It endorses the importance of family units, breeding inherent trust. Government bodies such as the police and social institutions like education play a major role in actualising desirable social change, thus providing institutional trust. Without a doubt, politics or government has the opportunity and the legitimacy to impact such change. In the Dutch context, the institutions evolved from their denominational base to executive branches of government policy and then on to independent or even privatised but highly regulated units.

---

9 SCP, 2002.
10 Van den Brink, 2002.
11 I apply the North-American definition of the notion of liberals.
12 See also WRR, 2003.

Institutional trust must be reinstated, with an emphasis on safety issues, implemented through zero-tolerance policies. Adherents of this administrative philosophy are also law and order proponents, and cry out for greater police presence in the streets. Phenomena like waiting lists in health care are considered emblems of institutional failure. A strong state is considered necessary to put a stop to further erosion. If bridging is wobbly and parts of society feel distant from others, bridges need to be built and secured with society at large through formalised (government) institutions.

Communitarians: Reinforcing Community Trust

This school underlines that reinforcing community trust is the way to go.[13] Community trust, its adherents say, is largely built up in voluntary organisations and voluntary individual engagement. Citizens unite in societies and in institutions to fulfil social needs and to subordinate themselves to joint objectives, because together they stand stronger than alone. This reinforces their commitment to one another, establishing bonding ties and, to society at large, constituting bridging ties. Eventually, this leads to greater social capital. Social-capital theory appears to be aiming primarily at those major groups of people who do not (feel the need to) rely on governments, because they are part and parcel of bonding and bridging networks that have traditionally been strong. This school focuses on preventing people from withdrawing from their society.

Communitarians also focus on the reinstatement of a civil society that is more fully engaged as counterpart of and complement to governments. Citizens' involvement, in their view, prepares them for dealing with the institutions of democracy and turns them into more capable co-administrators. People voluntarily engage in basic types of democracy, and all this entails: give and take, decision-making procedures, an ability to empathise with others, the balancing of multiple options, and reconciling and continuing to work together when they do not get their way. Developing civic skills at basic levels may help to foster some empathy with politicians and other policy-makers who have to operate with processes like these at higher levels. It also provides active citizens with a way – through regional, national, or even international organisations – of impacting processes of thinking and decision-making in the higher echelons.

[13] De Tocqueville, 1835; Putnam, 2000.

Liberals: Creating Space for Personalised Trust

Whereas the glass is half empty for communitarians, it is half full for liberals, who believe that value changes in society have led citizens to change their position vis-à-vis politics and government.[14] They have become more assertive and more demanding,[15] or more critical.[16] Such a cultural shift, however, need not be the herald of bad news. More individualistic citizens have a need of personalised trust, which they attempt to satisfy through new kinds of bridging. If the new roads lead to better solutions and if a new generation of inspired pioneers rises to the top, society will regenerate itself.

A growing group of young postmodern citizens has different needs. As a result, traditional institutions have come under pressure; young people have less hierarchical attitudes, want to exert an immediate influence, and have their own networks to do so, which replaces membership of political parties or traditional NGOs. Their fine-meshed networks of ties, which are in themselves weak, offer them a greater challenge than the unwieldy structures of the past, which used to be ruled by elites. It is not lack of interest but the institutions that are the problem. As such values are internalised at an early age, their effect is only gradually becoming clear as the generations succeed one another.

These individualists are prepared to dedicate themselves to social goals, which may seem surprising at first sight.[17] They are looking for personal ways of shaping their lives. In housing, for instance, they are looking for neighbourhoods with some character: apartments on the Amsterdam canals, for example, with a wide range of small-scale entertainment facilities like restaurants, bars, and small shops in their immediate vicinity. They are interested in the physical layout and the accessibility of their local community, including parking facilities and day-care services. They are prepared to think about such issues, as long as they are taken seriously.

Shifting Paradigms

These three public administration schools would appear to be having rather divergent principles, but they have for a long time managed to coexist without serious conflict. The schools even cut through the boundaries of party politics, albeit as shifting paradigms: the different schools have successively gained the upper hand. The co-occurrence of greater prosperity, better education, and improved information supplies has led to rapid

---

[14] Florida, 2002; Inglehart, 1977, 1997, 1999.
[15] Van den Brink, 2002.
[16] Norris, 1999.
[17] Florida, 2002; Müller, 2002.

individualisation in the Netherlands: more assertive and more demanding citizens were increasingly able to help themselves, which allowed the government to take their hands off. Civil society and the strong state of the conservatives had more of a supporting role to play when the Dutch economy was burgeoning and the market of the liberals was reigning supreme. Conversely, when the economy recently began to decline and things were swinging the other way, safety issues and waiting lists in health care, for example, reintroduced the call for a strong state. The institutional trust of the conservatives and the community trust of the communitarians had been eroded to such an extent that restoration was called for. The three schools, even if the loud protests of their fervent proponents might suggest otherwise, have always maintained their complementary character.[18]

As policies mature, they almost unvaryingly make their round of these three schools in succession.[19] The first generation of environmental policies in the 1970s was based to a high degree on legislation and regulations, the policy instruments of the conservatives par excellence. In the 1980s, this shifted, to some extent, to the favourite instruments of the communitarians: covenants and financial incentives. In the last few years, responsible corporate citizenship has made its appearance, with an emphasis on individual responsibility that fits the liberals. The three schools have coexisted all this time; the best way of dealing with some environmental issues, for example, remains stringent legislation and regulations.

It would seem advantageous to build a sustainable environmental policy on a shockproof foundation that incorporates the strong points of each of these three schools. An excessive inclination towards one philosophy will lead to vulnerability in case of economic or social discontinuities. Such a balanced policy may also be successful in an international perspective, as argued below.

OUTLINES OF A NEW ENVIRONMENTAL POLICY

How can a small country like the Netherlands make a contribution to an effective international environmental policy? It should be stressed that we do not have to start from scratch. Citizens have internalised nature conservation and environmental protection issues, and regard objectives in these fields as elements of common decency: they separate household waste, for example, without calling its necessity into question. These selfsame citizens also expect others, including government and business, to take action. NGOs and

[18] WRR, 2003.
[19] Winsemius, 1986.

multinational companies build on these foundations. The environmental movement in the Netherlands is exceptionally vigorous: nowhere else can NGOs boast such large numbers of members and supporters. Leading multinational companies like Unilever and Shell are global trendsetters for thought and action in the sustainability framework.

Over the last thirty years, moreover, the Dutch government has developed a series of initiatives that have impressed the international community. It is on the basis of this unique volume of strength and experience that the Netherlands can take the lead in the future. This mainly requires us to highlight approaches based on the three public administration schools.

A Trigger for Governments

We may expect with good reason that many international policy measures – within the EU, for example – are founded on the institutional trust of the conservatives. The requirements of unanimity, which are often predicated on the desire to prevent distortion of competition, often cause one to resort to a classic policy repertory. Legislation and regulations, moreover, are less controversial than more experimental alternatives, such as built-in market incentives, which inevitably cause some countries to fall out of line. Many policy-makers, in addition, have a preference for the certitude of clear-cut rules and clearly agreed penalties, and do not consider it their problem that enforcement mechanisms are often lacking. NGOs, too, tend to prefer the verifiableness of formal rules to the ambiguity of the market instruments of the communitarians or the social responsibility of the liberals: governments can be held accountable more easily, and politicians are more sensitive to their arguments than defiant hardliners in the world of business. The anti-globalist alarm at the idea of multinational companies playing the principal part on the global stage is not without historical foundation. If we add to this the increasing popularity of the conservative body of ideas among the electorate, it will be clear that we have to keep paying close attention to legislation and regulations. This is the foundation of trust, on which more innovative kinds of policies can be built.[20]

The process of legislation and regulations, however, would also benefit from a nation's trigger role. People, including politicians and civil servants, often only stir themselves if there is a spark to kindle their flame. Some pioneer country needs to ignite this spark through a combination of inspiration and mutual trust. This trigger is an inspirational force with bold ideas and good examples, which initiates reflection, lubricates consultation processes, and mediates in solution scenarios. In all of these ways, the

[20] Zoeteman, 2003.

Netherlands has a track record of making an above-average contribution to international policy-making: it generated ideas on carbon dioxide emission trading and, in the 1980s, on subsidising unleaded petrol by imposing a levy on leaded petrol. In consultation processes, we are thinking not only of Dutch minister Jan Pronk, who launched many initiatives in the field of climate change, but also of Dutch officials who held top positions in international organisations such as the UNEP, the OECD, and the secretariat of the UN Climate Convention, or who served as key figures in preparations for major conferences. In addition, the Netherlands hosted a great many of such meetings on climate, water, and biodiversity, for instance. In the EU Environmental Council, the Netherlands was in a position to act as initiator and mediator in implementing the Basel Convention on hazardous waste and in accomplishing the Kyoto Protocol. A more dated example, 20 years ago, would be the many attempts at mediation undertaken by the Netherlands to try and unite Germany, France, and the UK – in what was then the European Community – on the matter of a joint policy on acid rain or to try and separate the US and Canada, who were fierce opponents, at the major global conferences on this issue. The Netherlands did not always have its way, but it was consistently an interlocutor at the negotiating table when the cards were dealt for the very last time.

Such initiatives will continue to be significant in the near future, on the condition that the catalyst has earned his position of trust and is not a nine days' wonder. High-quality preparations, readiness to share and take the interests of other parties into consideration, and creativeness in breaking through stalemates: these are vital ingredients for success. The main key, however, is a carefully maintained network with political and official players in foreign capitals. And this requires, both in its literal and its figurative sense, boundless dedication.

Designer of a New Balance of Power

As of the 1980s, the Netherlands has gained an international reputation for developing and actually implementing communitarian kinds of policies. Ensuing from the unique community trust that tied together government, business, and the environmental movement, an entirely new set of legislative instruments saw the light of day. At first, legislation and regulations focused on the licensing applicants' own responsibility. For example, environmental impact assessments and toxic substances legislation required applicants to assess possible consequences before making major investments or marketing new substances. When experiences in several delicate practical cases proved to be positive, multinational companies in particular took the lead in devising the most efficient ways of meeting agreed environmental targets within their own subsidiaries. They opened themselves up to accountability by the government and, often by going public, by NGOs. Concepts such as environmental action plans, covenants, and voluntary long-range agreements appeared on the scene in an attempt to find solutions in which both sides might win. This also helped to contain additional expenditures on environmental protection measures, because companies themselves could draw on their own knowledge of technical processes and markets to select the best ways of meeting targets. In some cases, the government was prepared to adjust the time schedule for implementing proposed measures to suit the investment rhythm of enterprises, and NGOs were willing to suspend deployment of their legal or publicity-based channels of protest.

In the course of time, community trust was buttressed by positive experiences. As thousands of people concluded that they had ownership of problems in their working environment, integrated solutions were developed rather than the usual end-of-pipe filtering techniques. This is how environmental aspects were internalised in operational management. Nevertheless, the limitations of this approach also became clear. Trust, for example, is fragile: it is hard to earn and easy to lose. Some covenants were half-baked and caused frictions. Occasionally, the government proved to be rigid when unforeseen problems arose, or, conversely, was too eager to accommodate the business community, which exasperated the NGOs. Political and personal preferences of members of the government, and their comings and goings, also played a role. Yet, the government has pursued a consistent course in the covenants approach, and some cabinet members overcame their initial scepticism and decided to let the advantages outweigh the disadvantages. As all parties were aware of the price of an error, they were extra careful investing in the other parties concerned. Moreover, it was a first-rate asset that senior executives of the major enterprises themselves were directly involved and rapidly retrained in environmental policy issues. Indeed, one of the most hopeful aspects of this public administration philosophy was the personal dedication of senior executives when consultations between parties had ended in deadlock.[21]

A major limitation of this approach, however, is that it has a relatively small geographical range of action. In many countries, community trust is simply less well developed than it is in the Netherlands, which makes it a less suitable entity for international application. In a prominent country like Germany, for example, negotiations between government and business are razor-edge affairs, but once a decision has been taken, there is hardly any scope for flexibility in its implementation. In the US, a considerable number of multinational companies is still antagonistic to public environmental

[21] Winsemius and Guntram, 2002.

policy, and links with NGOs are feeble. New legislation and regulations are often elaborated in detail through case law; battles between opponent lawyers and concurrent sleaze campaigns by NGOs rarely help to foster mutual ties based on trust. Though the Dutch approach has drawn great international attention of foreign policy-makers, therefore, its emulation has remained restricted.

As strong community trust ties its key players together, this has secured for the Netherlands a lead role in the design of global sustainability policies. Progressive communitarians in the Netherlands advocate a new balance of power, in which governments, multinationals, and NGOs balance each other in growing towards a sustainable future. National governments and multilateral organisations such as the UN prove to be increasingly inept at developing hands-on policies in areas like climate change, biodiversity, and the approach to large-scale health threats. The international community, inspired by NGOs and the media, therefore tends to focus more often on the multinational companies and to pressurise them to change their ways, using negative publicity as a means of coercion. Though these kinds of result-oriented tactics do not always shrink from deploying dubious means, great pressure is getting more and more results. Greenpeace, itself only a small organisation, was quite effective in its campaigns on the import of GMOs into Europe and the intended sinking down of the oil platform Brent Spar. Nike, faced with charges of child labour being used by its Asian suppliers, revised its internal control procedures. Heineken decided not to invest in Burma. And the major pharmaceutical companies decided, without being legally forced to do so, to make their anti-HIV therapies available to developing countries at much lower prices.

These are, however, kill-or-cure remedies that exact change and are a sign of less sustainability-friendly attitudes than used to be common in Western Europe. The Dutch experience from the early stages of environmental action plans and covenants is relevant to the global arena. If the millions of people in multinationals feel they have ownership of the problem of sustainability approaches in their own enterprises, they will find better solutions and be proud of them into the bargain. If NGOs are prepared to follow multinationals constructively rather than cynically and cooperate with them in forming grassroots alliances in distant areas, win-win partnerships will flourish. If governments are prepared to stimulate such developments, they would help to implement a new set of policy instruments that could be effective at the global level. This last aspect especially is not self-evident, because it requires governments to be prepared to surrender the helm and their ability to wield the stick of legislation and regulations to force those who are unwilling to cooperate. Agreements should rather be arranged in direct interaction between multinational companies and NGOs, which, if necessary, can use the instrument of public campaigning.

Nevertheless, it is precisely the Dutch government that is in a unique starting-position to achieve global environmental goals in this way. There is no other nation in the world that serves as a home base for both strong multinational companies and NGOs in quite the same way. Five partly Dutch multinationals rank high in the top of Fortune's survey of the world's largest corporations. Their total worldwide turnover is much larger than the Dutch gross national product. The head offices of NGOs such as Greenpeace International, Friends of the Earth, and the Global Reporting Initiative are located in Amsterdam, while the Dutch branches of WWF and Greenpeace are among the largest within their international umbrellas. Dutch environmental organisations have approximately 2.5 million members and contributors, representing one in three or four households, which is again unparalleled in virtually any other country.

It is perhaps even more to the point to observe that there is no other country in the world that can claim an equal measure of experience in successful tripartite cooperation. Multinationals like Unilever and Shell are global leaders in taking business-initiated sustainability measures. Their top people do not hesitate to speak out on responsible corporate citizenship and function as pioneers in global associations like the World Business Council for Sustainable Development (WBCSD). Unilever has taken initiatives in the fields of sustainable fisheries and sustainable agriculture. Shell's reports in the field of sustainability are held up as examples for competitors; the charity objectives of the Shell Foundation are exceptional, certainly within Europe. Though they may be open to further reinforcement, the ties of trust between the three types of key players in a new balance of power are stronger than anywhere else. The like-minded attitude of these key players offers opportunities for achieving cross-border kinds of cooperation that will support the development of sustainability.

Is the Dutch government able and willing to utilise this unique basis as a springboard for an effective international environmental policy? This would mean stimulating the thoughts and actions of multinational companies and NGOs that are firmly rooted in the Netherlands in directions that fit the national policy agenda. The government would also have to have something to offer: joint think tanks, for instance, which would explore issues in depth and make representatives of the partners involved better prepared and give them a headstart at the international negotiating tables. What could the government do, for instance, to stimulate Unilever's promising initiatives in the field of sustainable agriculture and to reinforce international support for these initiatives? What could it do to institutionalise Shell's lead in the field of social reports? How could WWF be supported in dealing with large-scale deforestation in tropical rainforests? Are any efforts being made, for example, to harmonise agendas with the Ministry of Foreign Affairs?

Without pretending to pursue completeness, I would like to present an idea in this field. Innovation is branded as a policy spearhead in many places. Climate change, for example, cannot be solved if we do not make a massive transition away from fossil fuels towards more sustainable sources of energy. A multinational like Shell underlines this importance in words and in action, and is considered one of the world leaders in exploring technological avenues. As far as I know, however, this has hardly led to any serious cooperation initiatives with the Dutch government, which, in theory, is pursuing the same objectives. Would it be conceivable, for instance, for Shell to take a 50 per cent share in the Energy Research Centre of the Netherlands (ECN) or in a joint research programme dealing with this topic, with the ambition of producing the best research worldwide? What would be needed to make this come true? As a rough estimate, a research group of about 200 top researchers would probably be a sufficient guarantee for success. The annual budget of such an institute will not exceed a generous estimate of 50 million euros. Is this not an inconsiderable investment to make in a policy area in which the structural stakes are so high? And would a partnership between government and business not contribute to the realisation of governmental objectives in other policy areas, such as power supplies, technology and science, industrial policy, and the management of nature reserves?

Facilitator of Wide-Ranging Involvement

In the communitarian approach, governments need to be prepared to share their lead in policy-making, but the challenge of letting go is extraordinary if one opts for the set of instruments of the liberals, in whose conception governmental tasks are limited to fostering a climate that allows citizens to vote with their feet in all segments of social action. It is far from self-evident, however, how we are to flesh out such a task, which requires new and targeted policies.

More assertive and more demanding citizens increasingly want to be able to make their own choices. Citizens can now choose from a wide range of options, much more than half a century ago, and can easily switch suppliers in case of dissatisfaction. Individualistic citizens are less ready to commit themselves to organisations, be they political parties, enterprises, or NGOs. In each case, they make a selection out of a range of topical bargains. They are also prepared to punish providers who misbehave in one particular product or particular market by repudiating them in other markets.

Providers, therefore, have had a rough time of it. In order to survive in rapidly changing circumstances, they have to act swiftly. For them, citizens have become more elusive: their loyalty has to be sought over and over again. Their need of more person-related kinds of trust has also made new demands. Citizens want to be able to address the leadership of their providers directly, and sometimes they go for charismatic leadership. Citizens also wish to be able to identify with the providing organisations. If companies do not perform well, young people refuse to buy their products and, perhaps even more significantly, do not want to work there. If they are not given the opportunity to fulfil their personal ambitions in their job, they look around for other positions, where culture does allow them to do so. If firms ignore their shareholders, investments go elsewhere. The growing success of sustainable investment funds is showing a parallel trend. If companies turn out not to be good neighbours, citizens in the neighbourhood will be uncooperative in their expansion plans. Citizens also want to live in a setting that suits their lifestyle and expect their employers to make it come true. People have options and like to be proud of the providers they do business with and identify with them; if not, they turn to others.

The fundamental change providers need to address in their own organisations lies in citizens' unpredictable behaviour. Above, I described the changes that are taking place in civil society. Individualistic citizens enter into a smaller number of close ties with people in their immediate environment and compensate for this by maintaining a larger number of weak ties with many people, often at some distance. These networks are loosely structured and notoriously difficult to map. However, news travels faster than ever and reaches partners in the network, often complete strangers, with unfailing accuracy. The fashion business learned its lessons the hard way. Levi's (jeans) and Nike (trainers), for instance, missed a fashion swing ten years ago, because their traditional market research channels were no longer tuned into the news gathering ways of their young target groups. The organizers of the World Trade Organization (WTO) conference in Seattle were completely surprised by the coordinated action of their anti-globalist opponents: again, antennas were not properly tuned in.

How can governments of small countries like the Netherlands take advantage of such movements in designing an effective international environmental policy? The global networks need to be fed: possible partners in single-event actions get together, because some parties – old NGOs or coalitions of new networks – make them offers they decide to accept. Such parties then act as triggers for joint action. After heated spot actions, the network continues to exist, and the fire goes underground like a moorland fire. In quite a different composition, it may flare up again in entirely different locations and possibly with quite different means and goals. Somewhere in the network, however, there are the architects: the people who forge the coalitions, plan the actions, and who, better than others, can trace how these networks develop and sense what is needed. They build onto their ideas and form schools of thought like artists, scholars, and other creative professionals do. These schools interact with their environments. Some environments are inhospitable and do not welcome freethinkers. Others have more to offer and pull creative people towards them like a magnet. What if a large number of these were accommodated in the Netherlands and joined up in a kind of Silicon Valley of NGOs? Would they not be inspired by the Dutch ideas on environmental issues, develop them, and disseminate them through their networks?

What could a small country like the Netherlands do as a host country to attract such welcome guests? Lessons learned in stimulating new enterprises might hold out an option. Loans for start-ups with tax advantages attached and a venture capital fund for starting NGOs may lower the threshold for establishing NGOs. Time and again, inspired initiators get bogged down in their having to keep the pot boiling: environmentalists cannot survive on the beauty of their ideas. Established organisations can turn to existing funds for project funding. New arrivals, however, need to spend a lot of time and energy on acquiring basic funding to raise a minimum infrastructure and feed a starting network. Their earnings, if any, are usually limited and offer virtually no basis for normal funding. The possibility of tax deductions for financiers if starters fail, smoothes the way for starting entrepreneurs. Subsequent funding would then have to be obtained from venture capital organisations. The government can offer a helping hand in this matter by way of taxation measures and revise the current 'green' arrangements, for example. More direct inducements are also conceivable, for example, by endowing NGOs with a considerable share of national lotteries.

Experience with starting enterprises has shown that venture capital, though indispensable, is not sufficient. Inspired beginners are commonly overstretched in the areas of housing and administration, and lack access to an infrastructure of legal, tax, and other advisors who can show them the way. They also require support in drafting a business plan. This is no different for NGOs. The government does not have to deliver all this itself, but it could function as a trigger for others to take the lead in certain areas. In Amsterdam, for example, the city council is experimenting with so-called 'breeding sites', offering spaces and basic infrastructures to starting entrepreneurs and artists on moderate conditions. In the framework of the New Ventures initiative, large companies have stimulated their young employees to coach starters in drafting their business plans and to function as sounding boards. Would not creative people in multinational companies be inspired by the idea of being involved in social initiatives in the field of sustainability? For they, too, are often part of the same underground networks as the starting NGO initiators. Would not government and business profit from an arrangement that stimulates, say, six-month sabbaticals in these organisations, if only because this would provide them with a firmer foothold in those networks?

Love must always be reciprocated, and this also goes for the relationship between NGOs and consumers in their networks. A caring environment will act like a magnet for the NGOs. A final idea, therefore, might be to promote the establishment of a think-tank in a small country like the Netherlands: university-based centres, for example, aiming to develop theories and generate ideas on sustainable development and the role that NGOs and multinational companies could play. Perhaps we could establish an annual prize for the best NGO or the most worthwhile initiative. Or we might stimulate consumers with ideas, for instance, by inviting three superpowers each year to write a report on one of the twenty biggest threats to humanity that Rischard identified.[22] Topics like climate change, the decline of biodiversity, water shortage, poverty alleviation, education for all, intellectual property rights, and rules for biotechnology are relevant here.

Are these tasks appropriate to governments or should they be left to citizens themselves and to private initiatives, just like the liberals would want it? Should governments take the risk of such direct involvement in an NGO 'underworld', which many believe to be aiming at disrupting the status quo? These are big questions that do need to be addressed. In any case, it is certainly conceivable for small countries like the Netherlands to take the lead in helping to bring about a better, more sustainable world by taking the indirect way of stimulating and challenging conceptualisations and public opinions; this represents ambitious, challenging policy-making.

REFERENCES

Dekker, P. (2001), *Vertrouwen in de Overheid: Een Verkenning van Actuele Literatuur en Enquêtegegevens (Trust in the Government: An Exploration of Current Literature and Polls)*, Tilburg: Globus, Tilburg University.

De Tocqueville, A. (1835), *De la Démocratie en Amérique*, English edition (2000), *Democracy in America*, Chicago: University of Chicago Press.

Florida, R. (2002), *The Rise of the Creative Class*, New York: Basic Books.

Inglehart, R. (1977), *The Silent Revolution: Changing Values and Political Styles among Western Publics*, Princeton: Princeton University Press.

Inglehart, R. (1997), *Modernization and Postmodernization: Cultural, Economic and Political Change in 43 Societies*, Princeton: Princeton University Press.

[22] Rischard, 2002.

Inglehart, R. (1999), 'Postmodernization erodes respect for authority, but increases support for democracy', in P. Norris (ed.), *Critical Citizens: Global Support for Democratic Governance*, Oxford: Oxford University Press.

Müller, T. (2002), *De Warme Stad: Betrokkenheid bij het Publieke Domein (The Warm City: Involvement in the Public Domain)*, PhD thesis, Amsterdam: Uitgeverij Arkel.

Norris, P. (ed.) (1999), *Critical Citizens: Global Support for Democratic Governance*, Oxford: Oxford University Press.

Poldervaart, S. (ed.) (2002), *Leven volgens je Idealen: De Andere Politieken van Huidige Sociale Bewegingen in Nederland (Living According to your Ideals: The Other Politics of Contemporary Social Movements in the Netherlands)*, Amsterdam: Aksant.

Putnam, R. (2000), *Bowling Alone: The Collapse and Revival of American Community*, New York: Simon & Schuster.

Rischard, J. (2002), *High Noon: Twenty Global Problems, Twenty Years to Solve Them*, New York: Basic Books.

SCP (1998), *Sociaal en Cultureel Rapport 1998: 25 Jaar Sociale Verandering (Social and Cultural Report 1998: 25 Years of Social Change)*, The Hague: Social and Cultural Planning Office.

SCP (2002), *Zekere Banden: Sociale Cohesie, Leefbaarheid en Veiligheid (Certain Ties: Social Cohesion, Livability, and Security)*, The Hague: Social and Cultural Planning Office.

Skocpol, T. (2002), 'United States: From membership to advocacy', in R. Putnam (ed.), *Democracies in Flux: The Evolution of Social Capital in Contemporary Society*, New York: Oxford University Press.

Van den Brink, G. (2002), *Mondiger of Moeilijker? Een Studie naar de Politieke Habitus van Hedendaagse Burgers (More Assertive or More Difficult? A Study into the Political Habits of Contemporary Citizens)*, WRR Voorstudies en Achtergronden V115, The Hague: Sdu Uitgevers.

Winsemius, P. (1986), *Gast in Eigen Huis*, Alphen aan den Rijn: Samsom Tjeenk Willink; English version (1990): *Guest in our Own Home*, Amsterdam: McKinsey & Company.

Winsemius, P. and U. Guntram (2002), *A Thousand Shades of Green: Sustainable Strategies for Competitive Advantage*, London: Earthscan.

Wetenschappelijke Raad voor het Regeringsbeleid (WRR) (Netherlands Scientific Council for Governmental Policy) (2002), *De Toekomst van de Nationale Rechtsstaat (The Future of the National Constitutional State)*, The Hague: Sdu Uitgevers.

Wetenschappelijke Raad voor het Regeringsbeleid (WRR) (Netherlands Scientific Council for Governmental Policy) (2003), *Waarden, Normen en de Last van het Gedrag (Values, Norms, and the Burden of Behaviour)*, report no. 68, Amsterdam: Amsterdam University Press.

Zoeteman, K. (2003), 'State governance of sustainable development and globalization', in P. van Seters, B. de Gaay Fortman, and A. de Ruijter (eds), *Globalization and its New Divides: Malcontents, Recipes, and Reform*, Amsterdam: Dutch University Press.

Index